1950
700

MICHIGAN STATE UNIVERSITY

OVERDUE FINES:
25¢ per day per item

RETURNING LIBRARY MATERIALS:
Place in book return to remove
charge from circulation records

GULL LAKE LIBRARY

JAN 3 1 2002

W. K. Kell...   for

D1249211

HANDBOOK OF FRESHWATER FISHERY BIOLOGY

A publication of the Iowa Cooperative Fishery Research Unit, sponsored by the Iowa State Conservation Commission, the Fish and Wildlife Service (U.S. Department of Interior), and the Iowa Agricultural and Home Economics Experiment Station, Iowa State University.

HANDBOOK OF FRESHWATER FISHERY BIOLOGY VOLUME TWO

Life History Data on Centrarchid Fishes of the
United States and Canada

KENNETH D. CARLANDER

THE IOWA STATE UNIVERSITY PRESS, AMES, IOWA

ABOUT THE AUTHOR

KENNETH D. CARLANDER is Charles F. Curtiss Distinguished Professor
in Agriculture, Professor of Fisheries, Department of Animal Ecology,
Iowa State University, where he teaches courses in fishery management
and population ecology. He has served as President of the Iowa Academy of
Science and the American Fisheries Society and has been a United States
Representative, International Association of Theoretical and Applied Limnology, attending Congresses in Madison, Wisconsin; in Poland; in Israel;
and in Winnipeg, Canada. He served as a consultant for the Ford Foundation in Alexandria University, Egypt, in 1965-1966. Besides this book his
publications include chapters in books concerned with conservation, limnology, and fishing, as well as articles for professional journals.

© 1977 The Iowa State University Press
Ames, Iowa 50010. All rights reserved
Composed and printed by Cushing-Malloy, Inc.
First edition, 1977

Library of Congress Cataloging in Publication Data

Carlander, Kenneth Dixon, 1915-
    Life history data on centrarchid fishes.

    (His Handbook of freshwater fishery biology; v. 2)
    Includes bibliographical references and indexes.
    1. Centrarchidae.   2. Fishes—North America.
I. Title.
QL625.C373   vol. 2   [QL638.C3]   597'.0929'7s
ISBN 0-8138-0670-4   [597'.58]   77-4176

662346

Contents

Preface

WHEN VOLUME 1 was published in 1969 it was planned that a second volume would complete the revision. The amount of data available on the sunfish-black bass family, the Centrarchidae, and the time it would take to complete the revision of the data on the other Perciform fishes prompted the decision to publish Volume 2 at this time. The third volume will probably take from three to four years to prepare. It is anticipated that manuscript copies of sections will be made available as completed, as was done with this volume.

In addition to bringing the coverage of the available literature up to mid-1975, this revision includes data on several phases of life history not covered in the first two editions (1953 and 1956) of Volume 1. Some of these phases are not as completely covered because it was not practical to review all the literature already covered to abstract the data on the additional phases.

It is not practical to list the many persons and organizations who have helped in the preparation of this handbook. Special mention should be made, however, of the support of the Iowa Agricultural and Home Economics Experiment Station which has provided support through Projects 1767 and 2002. Special thanks are due also to the many individuals associated with typing portions of the manuscript.

HANDBOOK OF FRESHWATER FISHERY BIOLOGY

VOLUME TWO

Introduction

VOLUMES 1 AND 2 of this handbook are first of all an index to the literature on certain aspects of the life histories of freshwater fishes of the United States and Canada. Much of the literature of fishery biology is scattered and difficult to locate. Much of it is in separate leaflets, bulletins, or periodicals published by state conservation commissions, universities, research laboratories, and other agencies. *Biological Abstracts, Zoological Record,* and, in particular, *Sport Fishing Abstracts* summarize and provide help in locating some of this literature, but these do not provide as complete an indexing as is attempted in this handbook. Many of the life history items are in papers dealing with several species or with broader ecological studies, and thus the life history items for the various species often are not indexed in the abstracting journals.

Some of the information in the handbook comes from papers which cannot accurately be described as published: mimeographed or otherwise duplicated bulletins, theses, and typed manuscripts. Criteria for inclusion or exclusion of such data have not always been consistent. In general, availability of the manuscripts to the compiler has been the first criterion. Much other equally or more significant data are in the files of research laboratories around the country. The progress and final reports of federal aid projects have a great deal of life history data, only a small part of which is included in the handbook. These data are particularly difficult to summarize because many are provisional or "in progress" and because it is often difficult to determine which have also been included in other reports. I will attempt to provide Xerox or other duplicator type copies of papers not otherwise available, at the cost to me of about 10¢ per sheet. A sheet may often take two pages of a publication.

References to the sources of the data in the tables and text are indicated by code numbers consisting of the first letter of the author's (or first author's) last name and a number. These code numbers lead to the full citation in the Citations section. The code numbers are not always consecutive. The missing numbers refer to papers which are not cited in this volume but which appear or will appear in other volumes.

In addition to serving as an index to the literature, the handbook summarizes, for quick reference, the available data on several aspects of the life histories of each species. The student is advised to go to the original paper for further information and for verification. It is hoped that these summaries also will indicate the gaps in our knowledge and lead to further research and publication of the needed information. On some species, the general aspects of life history and growth are so well documented that there is little need for further compilation.

I do not wish to imply that the lack of information in this handbook necessarily indicates a lack of available information. While I have tried to give as complete coverage as time and facilities permitted, I can make no claim that the handbook is complete. Many significant papers have undoubtedly been overlooked, and I will appreciate having them called to my attention. Coverage is more complete on the length conversions, weight-length relationships, and age and growth sections than on other aspects of life history. In the early stages of preparation of the handbook these were the only data recorded, and when it was decided to expand the areas of life history to be summarized it was not feasible to go back over the previously reviewed papers. In general, the data included on these other aspects were only the data uncovered in looking for the age and growth summaries, and little effort was made to adequately search the literature in these other areas.

Age, growth, weight-length, and condition factor data are valuable in describing the general life history of a fish. They are more valuable from a management viewpoint, however, when they can be compared with similar data from other populations. The tabulations in this handbook provide such quick comparison and guide the biologist to the original papers for more detailed analyses.

The needs of fishery biologists and others working in fish conservation and management were of prime interest in selecting the life history areas for inclusion in the summaries. Numerical data which could be more readily included in tabular form were given precedence over material which would require extensive descriptive analysis.

As a measure of the variation in the relationships, ranges are given wherever possible. In some cases, the ranges as given are merely the ranges of means from various lakes, ponds, year classes, etc., and do not accurately depict the entire range shown in the samples. Range is not an entirely satisfactory measure of variation. Standard deviation and analysis of variance give better measures of variation and also permit statistical treatment to determine the significance of differences between sets of data.

Fish show great variations in growth, weight-length relationships, condition, and other measurements. Even within a given population or year class in a lake the variation may be large. For this reason, the application of data from one body of water to a population in another body of water, or to the same population at another time, should be made with great care and with full recognition of possible variation and probable errors.

Small samples may give erroneous pictures of growth or weight-length relationships. The size of sample needed depends upon the degree of accuracy desired and the variation within the sample.

The metric system of measurement is used throughout the tabulations unless otherwise indicated. Lengths are in millimeters (mm), weight in grams (g), standing crops or production in kilograms per hectare (kg/ha), and temperatures in degrees Celsius (C). Because much of the data were first published in the English system of measurement, the class ranges are a bit unnatural for the metric system. Tables for conversion from English to metric are given for reference.

The species are listed alphabetically according to their scientific names. Under each species the data are listed in the following order:

Range and habitat (These are usually given in general terms and are pre-
    sented only for quick reference. They do not represent extensive liter-
    ature review.)
Conversion factors for various length measurements
Weight-length relationships
Ponderal indexes or condition factors
Observed lengths and weights at various ages
Calculated lengths and weights at various annuli
Discussion of growth data
Age at maturity and reproduction data
Food habits
Population estimates
Sizes taken in various mesh sizes of gill nets

Occasionally, other data on mortality, boood counts, temperature tolerance, and management are briefly recorded. In some species where the total
description is short, the above order may not be strictly adhered to. The following discussions will help in interpreting the various sections.

CONVERSION FACTORS FOR LENGTH MEASUREMENTS

Unfortunately, no generally accepted method of measuring fish is available despite several papers on the subject (R74, R31, R41, C37, H68, S508,
K132). At least eight standard lengths, one fork length, and two total lengths
have been used in fishery work (H68). Total lengths are used throughout this
text except as otherwise indicated.

Standard length (SL) has been variously defined, but is essentially the
length of a fish from the tip of the snout to the end of the vertebral column.
Standard lengths are extensively used in taxonomic studies, but now are rarely
used in other aspects of fishery biology.

Fork length (FL) is measured from the tip of the snout to the end of the
rays in the center of the caudal fin. Fork lengths are the usual lengths used in
marine fisheries and have been used extensively in fresh water fisheries in
Canada, Europe, and some states in United States. Often they are reported as
total lengths, "measured to the center of the fork of the tail." The fork (Smitt
or median) length is reported to be the most frequently used length measurement in fishery work (R315), but in United States freshwater fisheries total
length is more widely reported.

Total length (TL) as defined by Hile (H68) is "the distance from the tip of
the head (jaws closed) to the tip of the tail with the lobes compressed so as to
give the maximum possible measurement." Total length to the tip of the tail
in its "natural position" should be dropped from usage.

In many cases, no indication of the length measurement used is given in
the original papers, and it has been necessary to arbitrarily assign the data to
one of the lengths. When an author or group of workers in a given research
unit has used a given method of measurement in one paper, the same method
is assumed to have been used in other papers which do not indicate the type of
length used.

Factors for converting one type of measurement to another are given for
each species when such data are available. In most cases, the conversion factor has been determined by dividing the TL (or FL) by the SL. Where the
ratio varies as the size of the fish increases, separate conversion factors

often have been determined for fish in various size ranges. Better statistical methods of describing the relationships between SL and TL or FL are available, but present evidence indicates that differences in the ways in which various workers measure fish are more significant in causing errors of conversion than are the methods of calculating the conversion factors (C32).

WEIGHT-LENGTH RELATIONSHIPS

Since it is frequently necessary to estimate the weights of fish from their lengths, data have been compiled on the weight-length relationships of many of the species.

For several species the data from several populations are combined rather than listing data from each paper. In these compilations the mean values from each population or each paper are weighted as follows:

Weight	Data based on:
1	1 fish
2	2-5 fish, or number of fish not indicated
3	6-10 fish
4	11-99 fish
5	over 100 fish

While this weighting system is somewhat arbitrary, it is believed to be superior to an unweighted mean since small samples are not given equal weight with large samples, and at the same time, this weighting prevents the average from being unduly affected by a large sample from a single population. Furthermore, this system permits the inclusion of data for which the size of the sample is not indicated. The weighting system is used both in deriving the mean of the means and in determining the central 50 percent. The central 50 percent should not be considered as including 50 percent of the fish of the size range, but 50 percent of the means. In some cases, the mean of the means may not be included in the central 50 percent, because the mean of the means is being affected by a widely divergent mean. In these cases, the median might be a more representative average, and this value can be approximated by taking the center of the central 50 percent range.

Several authors have determined the mathematical relationships between length and weight for various populations using the formula:

$$W = cL^n$$

where

$$W = \text{weight}$$
$$L = \text{length}$$
$$c \text{ and } n = \text{constants}$$

The value of the constant n will usually be near 3.0 since the weight of an object will vary as the cube of its length if shape and specific gravity remain the same. Since this formula must be calculated in the logarithmic form and is most usable in this form, the weight-length relationships in these compilations are given in the logarithmic form:

$$\log W = \log c + n \log L$$

With the conversion of all measurements to the metric system, it has also been necessary to convert the regressions based on English measurements. A method for conversion is given in Volume 1, p. 7.

Most weight-length regressions have been calculated as predictive regressions of log W on log L. Ricker (R314) recommends the geometric mean (GM) estimate of the functional regression of log W on log L as being superior because it does not assume that there is no error in log L and because it does not have a systematic bias related to the range of lengths represented in the sample. I believe that all values given in Volumes 1 and 2 of the handbook are predictive regressions, but care will be needed in the future to be sure which regression has been used. The slope of the GM functional regression, v, will always be higher than the slope of the predictive regression, n (R314). In one example (R315 pp. 214-5) n was 3.572 and v was 3.636, with the difference being less than the standard deviation. When the regressions were computed for each of the 5 age groups individually, the mean n was 2.976 and mean v was 3.303, a greater difference because the length ranges were shorter. In another example with largemouth bass (V95) n was 3.078 and v was 3.122, a difference of 0.044; whereas for 31 samples of individual age groups in different months the mean n was 0.220 less than the mean v with v being 0.005 to 0.549 higher than the corresponding n. In a study of walleye (S598) with 72 samples by age groups and months the mean n was 3.1345 and the mean v was 3.1853 with v being 0.0012 to 0.2849 higher than the corresponding n. The difference between the slopes increased as the range of lengths in the samples decreased (0.01224 per 100 mm increase in range of lengths), but the variance was large.

It has been stated that the slope will usually be above 3.0 because most fish become plumper as they grow. If the entire life span were considered this generalization probably would be true because fry and young fingerlings are usually quite slender; but most weight-length relationships do not include these early stages of growth, and changes in body form are less characteristic during most of the life span. The mean slope of 398 populations in Volume 1 was 2.993, but more of the slopes were above 3.0 than below. The mean slope of 415 centrarchid populations reported in this volume is 3.070 with the mode in the 3.076 to 3.225 class (Fig. 1).

The mean slope of 65 bluegill populations was 3.11 and that of 46 populations of other *Lepomis* was 3.08 (Fig. 2).

The mean slope of 116 largemouth bass was 3.08 and that of 35 smallmouth bass was 3.04 (Fig. 3).

The mean slope of 41 populations of white crappies was 3.19, a slope somewhat higher than that of 75 populations of black crappies, which was 3.11 (Fig. 4).

I have thought of white crappie as usually thinner than black crappie, and the average unweighted mean of the mean K (TL) values of white crappies was 1.30 compared to 1.44 of the black crappies. The same values for K (SL) were 2.70 compared to 3.18. In a few cases the intercept values for similar slopes were higher for black crappies than for white crappies, but most populations of black crappies did not show higher intercepts (and thus plumper fish) than did white crappies (Fig. 5).

The regressions with slopes of less than 2.5 and more than 3.6 merit discussion. The slope of 1.34 referred to largemouth bass fry from 3 to 6 mm, and that of 3.90 to fry 6 to 12 mm, (K78). Samples of a single age group accounted for slopes of 1.76, 2.11, 3.61, 3.78, and 4.06 (V95) and of 2.35 and 2.45 (Z7). Within age-groups, weight-length relationships may not indicate growth

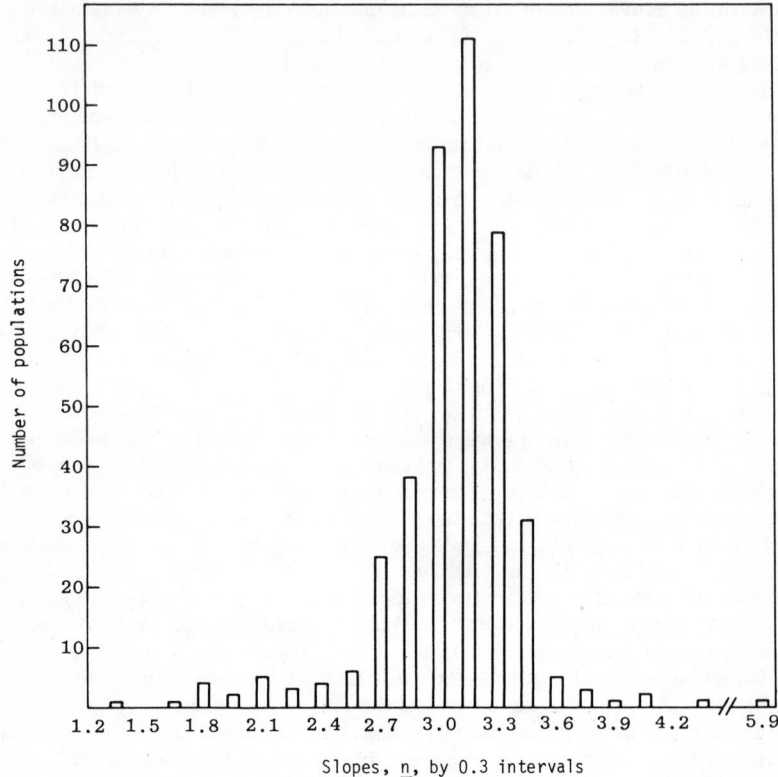

Fig. 1. Frequency distribution of slopes, *n*, of weight-length regressions of centrarchid fishes from 415 populations.

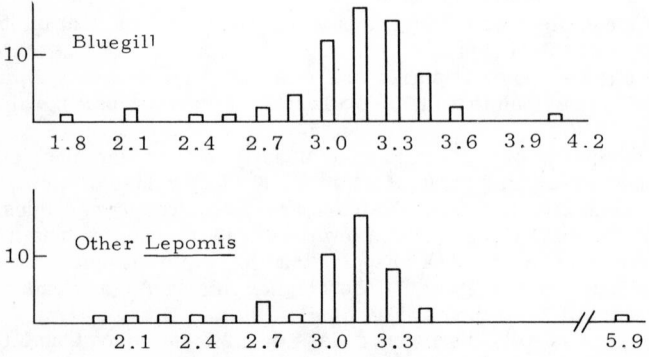

Fig. 2. Frequency distribution of slopes, *n*, of weight-length regressions of bluegills and other *Lepomis*.

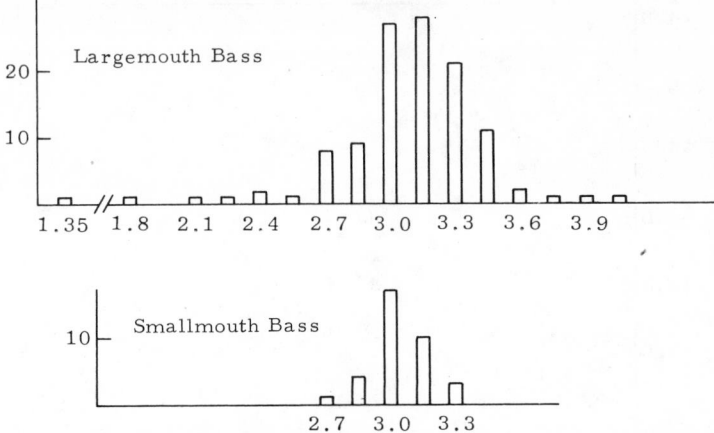

Fig. 3. Frequency distribution of slopes, *n*, of weight-length regressions of largemouth and smallmouth bass.

(R315) but rather differences between the slower growing and faster growing individuals.

Slopes of 3.62, 3.69 and 4.31 were based upon samples respectively of 11, 6, and 3 black crappies among 49 monthly samples (V91). Slopes of 1.76, 1.80, 1.86, 1.80, 2.02, 2.06, 2.07, 2.17, 2.26, 2.33, 2.34, and 3.85 all came from P208. The reasons for their variation from normal are not evident, but errors of measurement may be involved. The other extreme values not accounted for are slopes of 4.11 (G163) and 5.91 (P122).

PONDERAL INDEXES OR CONDITION FACTORS

Since the weight of a fish varies with the cube of its length, provided the shape and specific gravity remain the same, any change in the shape or relative plumpness of a fish will cause a change in the value of c in the formula:

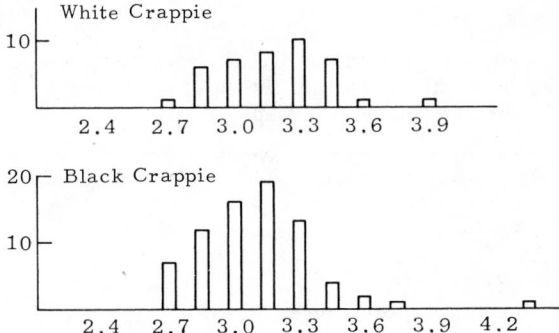

Fig. 4. Frequency distribution of slopes, *n*, of weight-length regressions of white and black crappie.

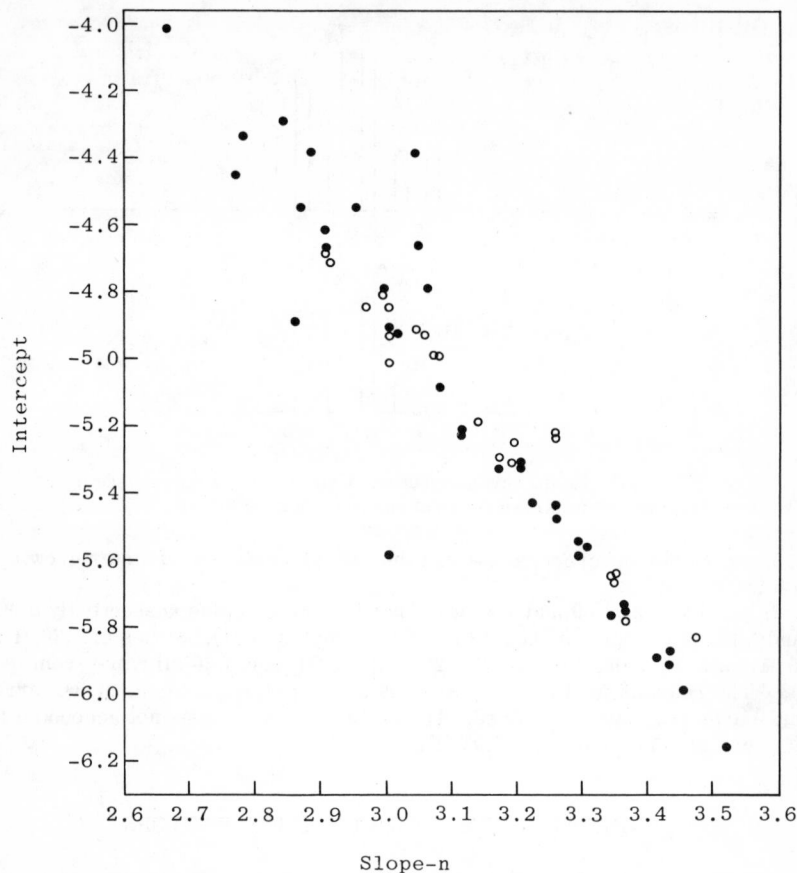

Fig. 5. Intercept values plotted on slope values for weight-length regressions of white crappie ● and black crappie ○.

$$W = cL^3$$

Fishery biologists have used this fact in describing the condition, plumpness, or well-being of a fish. The coefficient of condition, K, or Fulton's condition factor (R315), has been widely used:

$$K = \frac{W\ 10^5}{L^3}$$

where

W = weight in grams
L = length in millimeters
and 10^5 is a factor to bring the value of K near unity.

The method of length measurement is indicated as K(SL), K(FL), or K(TL). The same formula, using the English rather than the metric system, can be used to calculate the condition index, C. The value of C is usually calculated from lengths in inches and tenths and weights in hundredths of pounds and is usually adjusted so that there are two digits in front of the decimal point, instead of one as in K. In this handbook all ponderal indexes are given as K, but conversions can be made as follows:

$$K = 0.277C$$

The K values are reported according to the type of length measurement used, SL, FL, or TL. If the ratio, r, of SL to FL or TL is known, conversion can be made as follows (K16, B30):

$$C(TL) = 3.61 \, r^3 \, K(SL)$$

To eliminate the trends in condition associated with increase in length, LeCren (L213) proposed a relative condition factor K_n where

$$K_n = W/aL^n$$

or where the observed weight is divided by the expected weight as computed from the weight-length regression for the population. While the relative condition factor, also referred to as the allometric condition factor (R315), is useful in certain studies, it is not suitable for comparisons among populations; and it assumes that the weight-length relationship remains constant over the period of study.

Weight-length relationships and condition factors vary with season, sex, sexual maturity, age, and various other factors. These relationships are reported for each species when they have been discussed in the reviewed papers or were evident from the data. In the compilation of the data, however, these trends are often not discernible, and one should refer to the original papers for more detailed analysis.

OBSERVED LENGTHS AND WEIGHTS AT VARIOUS AGES

Throughout the compilations, the ages are indicated by Roman numerals. A numeral III indicates that the fish has passed through three winters. In some of the reviewed literature, such a fish would be referred to as being in its fourth summer. In accordance with a suggestion by Hile (H68), the first of January is taken as the birthday of all fish, and fish taken in the early part of the year are therefore credited with an annulus at the edge of the scale even though that annulus was not yet completed. For fall spawning fish it is assumed that the eggs do not hatch until after the first of the year—even though they may do so, particularly in hatcheries. These fish are listed as age 0 until January of the next year, when the fish are just over one year old.

The ages of the fish were determined by one of the following methods (L135):

Direct observation: This method applies when the age of the fish is known because the fish have been under observation, as in hatcheries and rearing ponds; or when fish of known age have been tagged or otherwise marked, so that they can be recognized upon recapture; or when fish are introduced at a known age into waters where the species was not previously present and as long as the introduced fish can be distinguished from their progeny.

Length-frequency or Petersen method: Age groups can often be separated and identified as peaks in frequencies when the numbers of fish are plotted against their lengths. In later years of life, the sizes may overlap so much that the peaks cannot be detected, but the method is usually useful during the first year or two. Where growth is fairly uniform, peaks may be followed several years.

Growth rings: Scales, otoliths, vertebrae, spines, fin rays, opercles, and other bones frequently show annual growth rings by which the age of the fish can be determined. Examination of such annual rings or annuli is the source of most of the growth data past the second year of life.

It must be recognized that the estimation of age from length frequencies and from growth rings may involve some errors, particularly among older fish. In the compilations, where the mean of means and central 50 percent ranges are given, the data are weighted as described under weight-length relationships. Also, comparisons of lengths and weights at a given age are somewhat approximate since some of the fish may have been measured during the early part of the age range and others toward the end. For example, some fish listed as age I may have been measured in March before the season's growth started; and others, in November after an additional season's growth.

CALCULATED GROWTH

Measurements made on scales, bone, or spines are used to estimate the size of a fish at certain times in its early life. Measurements of the growth rings are usually assumed to be directly proportional to the length of the fish at the time that the growth ring was formed. The lengths at each annulus are then computed by the Dahl-Lea method:

$$\frac{Sn}{Sc} = \frac{Ln}{Lc}$$

where

Sn = scale (or bone) measurement to a given annulus, n
Sc = scale measurement to edge
Ln = length of fish at the time of annulus, n, formation
Lc = length of fish at capture

It has been found that the growth of body length and of scales or bones is not directly proportional, and that more accurate calculations can be made using other formulas (L135, H465). The most common correction is the Fraser (V5, p303; R282, p324) or Lee Method (L135, p152), which assumes a straight line body-scale regression with an intercept at some place other than zero on the ordinate.

$$L = a + bS$$

$$\text{or} \quad Ln = a + \frac{Sn}{Sc}(Lc - a)$$

The value of a has frequently been interpreted as the length of the fish at the time of scale formation (R282, p324), but this interpretation is not necessarily correct. The value of a is the intercept that will give the best straight line relationship. Early growth of the scale usually is not proportional to that of the body even though scale growth may be proportional through much of life.

More accurate estimation of past growth may come from more precise description of the body-scale relationship (the Sherriff, Monastyrsky, etc., methods in L135).

The past growth histories estimated by these various methods are usually comparable except for the first year or two of life, and the data are thus tabulated together. The method of computation, if not the Dahl-Lea method, is usually mentioned in the discussion which follows the tabulation.

The increments given in the tables are often not the increments as reported in the original paper but were computed from data given in the paper. The increments are computed to include data only from the fish completing the year's growth. Frequently, increments are recorded by subtracting the mean length at the end of one year of life from the mean length at the end of the next year of life even though the first mean may include several fish not included in the second mean. These growth-curve increments can be readily computed from the average growth data given in the tables, and there is no reason to repeat them. More importantly, it is recognized that growth potential is determined to some extent by size at the beginning of the period (L91, P123), and therefore the average increment should be computed using at the beginning of the year only the fish that also complete the year. The size of the fish to which the increments apply can be determined by subtracting the average increments from the average length of the fish at the end of the year of growth. This length may differ somewhat from the mean length reported for the end of the previous year because the latter mean includes fish which did not complete the increment under consideration. Summation of annual increments (as suggested by L1, page 156, and some other papers) is not recommended since it applies the increment to fish at a different length than the fish which actually made the increment.

For some species, increment data are given as percentages of length at the beginning of the year (relative growth as defined in R282, p314 or R315, p207). Since these data were derived by dividing the mean increment by the mean length at the beginning of the year, they are only approximate measures of relative growth since the sizes may have varied considerably around the mean.

Calculated growth in weight is given for some populations. In these cases the calculated lengths are converted to weights through use of a weight-length regression for the given population. If necessary, growth in weight of other populations may be estimated from the calculated lengths and an acceptable weight-length regression.

DISCUSSION OF GROWTH DATA

The discussions which follow the growth tabulations cover a variety of topics. First, they may describe the methods of age and growth determination, the characteristics of annuli, and evidence that the scale method is valid. In general, scale measurements are along the anterior radius unless otherwise indicated in this discussion.

Lee's phenomenon of apparent change in growth rate is characteristic of much growth data. The calculated lengths of the older fish in the earlier years of life are systematically lower than those of younger fish at the same age. Several causes of this phenomenon have been proposed (see particularly V5,

pages 302-6, 328-44), but usually it is the result of (1) failure to use a corrected body-scale relationship in computing the lengths, (2) selective sampling,
or (3) differential mortalities of the fish in relation to their growth rates.
Many sampling techniques tend to select only the larger fish in younger age
groups. Faster growing fish are frequently shorter lived than the slow growing fish which, thus, constitute most of the fish in the older age groups.

Growth compensation is the tendency for slower growing fish to grow
more rapidly in later years than the fish which showed faster early growth.

Wherever possible, the regional, genetic, and environmental factors affecting growth rates are discussed. There is, however, a basic problem in
evaluating the effects of environmental factors upon the average growth rates
of fish in a body of water (C348). Most environmental factors control the
growth of the biomass of the fish population rather than the average growth
rate. Where the population numbers can be kept constant or are known, the effects of environmental factors might be readily detected if only one factor at a
time is modified. But in most natural conditions, several factors fluctuate in-
dependently or in various interdependent relationships, and the changes in
population numbers are not known.

The potential change in biomass growth brought about by any factor can be
met by change in population numbers or by change in the average growth rates
of the individual fish or by combination of the two. In many situations the population changes, which are more difficult to document and measure, are much
greater than the changes in average growth rates. Population density is so
often the most significant factor affecting growth that fishery biologists consider slow growth and low condition factors as indications of overpopulation.
The problem is further complicated by the fact that it is not absolute population density of the fish which has such a dominant effect on growth, but it is the
population density in relation to the carrying capacity of the body of water.
Growth is a function of the degree to which biomass is below carrying capacity.
Carrying capacity is in turn difficult to measure.

AGE AT MATURITY AND REPRODUCTION DATA

Centrarchids mostly mature at ages I to III unless growth rates are slow.
Males usually prepare a cavity or cleaned area for the deposition of eggs; and
the males may attend the eggs and young, but details vary with species and
general habitat. Information on courtship, spawning, and nesting behavior are
only briefly described in these summaries, and the reader is referred to other
references for further information. No attempt has been made to include
many of the references given in B358.

Since the spawning season is fairly extended for most centrarchids all of
the ova for a given year may not ripen at the same time, and counts of ova
may not be representative of the annual production of eggs. Nevertheless, the
available data on fecundity are summarized.

FOOD HABITS

Data on food of fishes from various studies are difficult to tabulate or to
summarize because of differences in the method of reporting (e.g., frequency
of occurrence, percentage of volume, number, or fullness) and in the cate-
gories of food reported. The summaries provided here merely suggest some

of the findings and lead to some of the original papers. Many other references are available and can often be located in the bibliographies of the papers listed.

STANDING CROP AND HARVEST DATA

The population estimates and harvest data summarized in this handbook were mostly added in some of the final stages of compilation and editing. Therefore, additional data in earlier papers have been left out. A more comprehensive study of population density and harvest is planned.

ABBREVIATIONS

In the tables, P., L., R., and C. refer to pond, lake, river, and creek. Reservoirs are usually referred to as lakes but sometimes Res. is used as the abbreviation. The two-letter abbreviations for states and U.S. possessions are those recommended by the American Fisheries Society in their computorized membership list, and abbreviations for Canada have been made up on the same model.

United States and Possessions

Alabama	AL	Kentucky	KY	Ohio	OH
Alaska	AK	Louisiana	LA	Oklahoma	OK
Arizona	AZ	Maine	ME	Oregon	OR
Arkansas	AR	Maryland	MD	Pennsylvania	PA
California	CA	Massachusetts	MA	Puerto Rico	PR
Colorado	CO	Michigan	MI	Rhode Island	RI
Connecticut	CT	Minnesota	MN	South Carolina	SC
Delaware	DE	Mississippi	MS	South Dakota	SD
District of Columbia	DC	Missouri	MO	Tennessee	TN
Florida	FL	Montana	MT	Texas	TX
Georgia	GA	Nebraska	NB	Utah	UT
Guam	GU	Nevada	NV	Vermont	VT
Hawaii	HI	New Hampshire	NH	Virginia	VA
Idaho	ID	New Jersey	NJ	Virgin Islands	VI
Illinois	IL	New Mexico	NM	Washington	WA
Indiana	IN	New York	NY	West Virginia	WV
Iowa	IA	North Carolina	NC	Wisconsin	WI
Kansas	KS	North Dakota	ND	Wyoming	WY

Canadian Provinces and Territories

Alberta	AT	Northwest Territories	NT	Ontario	ON
British Columbia	BC	Dist. of Franklin Bay	FB	Prince Edward I.	PE
Manitoba	MB	Dist. of Keewatin	KE	Quebec	QU
New Brunswick	NK	Dist. of Mackenzie	MC	Saskatchewan	SA
Newfoundland	NF	Nova Scotia	NS	Yukon Territory	YU

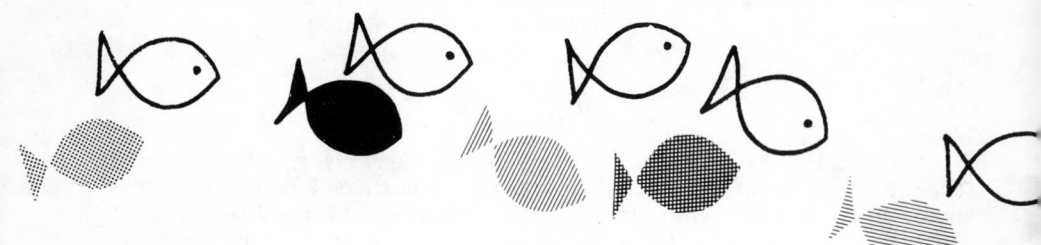

LIFE HISTORIES BY SPECIES

MUD SUNFISH, *Acantharchus pomotis* (Baird)

The mud sunfish is a little known secretive centrarchid found in lowland streams of the coastal plain from New York to Florida (M172). Adults frequently rest head down, almost perpendicular, in weeds and are most active at night (R285). They have been reported to make a grunting sound, probably by grating the pharyngeal teeth (R285).

M172 MD 14 fish 111-170 mm.   TL = 1.281SL

The caudal fin is not forked.

SL.	111	114	118	119	119	123	123	124	126	149	151	158	162	170
Wt.	48	68	71	65	74	78	78	78	91	142	142	139	156	190

E105 MD   206 mm TL mud sunfish weighed 190 g.

	Mean calculated SL at each annulus							
	No.	1	2	3	4	5	6	7
B72 NY Post Brook	10	39	52	69	82	94	106	115
M172 MD Chambers L.	13	22	50	87	108	130	145	157

Few nests have been observed. One in early June in New Jersey was about 30 cm in diameter in sand surrounded by mud and aquatic plants (R285).

Chromosome numbers are reported as 48 (R286).

ROANOKE BASS, *Ambloplites cavifrons* Cope

The Roanoke bass, described by Cope in 1867, was not recognized in North Carolina until 1963 (S526), although it was reported from the coastal plain from New York to Florida (M420). The largest collected was 330 mm and 770 g, but larger bass are reported by anglers (S526).

Age 1-April		No.	TL	No.	Weight
S526 NC	Fayetteville Hatchery	25	86±4	25	8.8±1.6
	Table Rock Hatchery	25	107±3	25	20.9±2.2
	Selected group	72	149		

Calculated TL at each annulus

	No.	1	2	3	4	5	6	7
S579 NC streams	72	83	144	191	227	261	287	288

The growth data given in S526 was questioned by the same author in S579.

The growth of young Roanoke bass in ponds was less than that of redbreast sunfish and rock bass under the same conditions (S526). No sex difference in growth was reported (S579).

Females mature at age II or at about 100 g (S579). The number of eggs in gonads ranged from 21.7 to 29.4 per g of body weight; and there were 3,500 and 5,400 eggs per age IV female, 8,200 eggs per age V female, and 9,600 per age VII female (S526).

Aquatic insects were the major food of fingerlings, but crayfish and fish were the only foods found in stomachs of 20 adults (S526, S579).

ROCK BASS, *Ambloplites rupestris* (Rafinesque)

The rock bass is usually found in cool weedy lakes and rocky streams from Manitoba to Quebec south to the Gulf of Mexico (H382). One population has been established in England (W245, V92). The southern variety has been at times considered a separate species, *A. ariommus*. Rock bass are often associated with smallmouth bass. In the southern part of the range they are primarily stream fish. They handle well as test animals (W214). Rock bass have been suggested for introduction into slightly alkaline waters, not over 900 ppm total alkalinity, nor over 250 ppm total carbonates, nor over 200 ppm K + N (M552). Rock bass were listed as strictly freshwater fish (B399).

	Lengths	No.	FL/SL	TL/SL
C37 MN	100-149	80	1.157	1.207
	150-249	337	1.157	1.193
C22 MN L. of Woods	140-229	101	1.164	-
C35 MN L. Vermilion	-	-	1.174	1.200
C30 IA	-	1	1.16	1.22
H65 WI Nebish L.	under 100	29	-	1.222
	males over 100	136	-	1.217
	females 100-159	109	-	1.217
	females over 160	32	-	1.206
B30 MI	under 118	1,066	-	1.268
	over 118	925	-	1.247

When data have been recorded as SL or FL, they have been converted to TL on the basis of SL = 0.820 TL and FL = 0.95 TL. The lengths indicated are the lower limits of the size class, not the class centers.

SL	FL	TL	No.	Mean of means	Weight Central 50%	Range	Citations
21	24	25	2	1	-	0.4	B30, S472
42	48	51	33	4	-	2.5-6	B30, H66, S472
62	72	76	548	17	14-17	6-50	B30, C20, H65, H66, M20, S472
83	97	102	1,184+	30	26-31	15-51	B30, B192, C20, H65, H66, M20, M136, R24, S130, S472

				Weight			
				Mean of	Central		
SL	FL	TL	No.	means	50%	Range	Citations
104	121	127	1,454+	56	45-60	37-77	B30, B192, C20, H65, H66, L113, M20, M136, M499, P113, R24, S130, S472, V21, W132
125	144	152	1,607+	87	77-100	45-122	B30, B192, C20, C182, H65, H66, M20, M136, M499, P113, R24, S130, S472, V21, W132
167	193	203	1,067+	196	170-227	136-263	B30, B192, C182, H65, H66, E32, M20, M136, M499, P113, S130, S472, W132
187	218	229	343+	267	255-301	204-322	B30, B192, C20, E32, H65, H66, M20, M136, M499, P113, R24, S130, S472
208	241	254	162+	364	357-386	284-397	B30, B192, C20, E32, E105, H66, M20, M136, M499, R24
229	265	279	23+	419	391-471	269-567	B30, C20, E32, M20, M136, R24, S130
250	290	305	1+	531	-	499-595	C20, M136
271	314	330	1	595	-	-	M136

	Length	No.	
H66 WI 3 lakes	-	-	Log W = -4.57398 + 3.057 Log SL
H66 WI Muskellunge L.	-	-	Log W = -4.48057 + 2.986 Log SL
H65 WI Nebish L.	-	-	Log W = -4.54000 + 3.003 Log SL
B29 MI Booth L.	-	-	Log W = -4.44406 + 3.025 Log SL
B30 MI	-	-	Log W = -4.319450 + 2.96914 Log SL
S130 IN Tippecanoe R.	127-211	1,134	Log W = -4.4255 + 2.923 Log FL
P113 Mo Meramec R.	-	-	Log W = -5.089 + 3.182 Log TL
B192 OH Little Miami R.	105-265	422	Log W = -5.200 + 3.2117 Log TL
P122 WI Flora L. 1952,	-	52	Log W = -4.921 + 3.055 Log TL
1956,	-	52	Log W = -3.688 + 2.498 Log TL
S472 AL	25-230	38	Log W = -3.818 + 2.58 Log TL
B376 IN	-	36	Log W = a + 3.192 Log TL

The weight-length relationship quoted in B376 was obviously erroneous and the slope should be 3.192. The length weight relationship in S472 was somewhat biased by the high recorded weight of the smallest fish. Thinning of the population of rough fish in Flora Lake (P122) did not affect the adjusted mean weights of rock bass, but the slope of the weight-length relationship flattened significantly.

	No.	SL	K(SL)
H65 WI Nebish L.	1,035	72-200	2.92
H62 IN	30	-	3.28
B29 MI Booth L. before thinning	13	-	3.07
after thinning	-	-	4.10
C34 MN Leech L.	81	100-249	3.72
C23 MN 1941-43	457	-	3.83
C35 MN L. Vermilion	22	-	4.03
E3 MN 1936-41	289	51-278	4.18
V20 MN Red Lake	5	170-176	4.19
C36 IA	1	162	4.23
B30 MI	-	30-231	4.07-4.32
C23 MN Standards poor			<3.1
average			3.5-4.1
excellent			>4.3

	No.	K(FL)
S120 IN	1,134	2.55

	No.		Mean	K(TL) Range
B30 MI means	-	-	-	2.05-2.13
M499 IL	50	127-254	2.40	1.86-2.49
S472 AL	108	50-230	2.02	1.64-2.28
C23 MN standards-poor				<1.80
average				2.02-2.38
excellent				>2.49

The condition factor increased irregularly with age in Nebish Lake (H65), and females tended to have slightly higher values than males at the same age. In Tippecanoe River K did not change with age (S130).

		TL				Weight	
	No.	Mean of means	Central 50%	Range	No.	Mean of means	Range
Age 0-June							
D43 TN streams	3	13	-	10-18	-	-	-
R71 OH	+	-	-	18-23	-	-	-
Age 0-July							
D43 TN streams	4	20	-	10-30	-	-	-
E32, R71 IN, OH	+	33	-	30-41	-	-	-
Age 0-Aug.							
R71 OH	+	-	-	41-56	-	-	-
Age 0-Sept.							
A1, M243 NY lakes	24	31	-	30-36	-	-	-
D43 TN streams	6	38	-	33-46	-	-	-
C13 MI lake	7	41	-	33-42	73	3	-
R71 OH	+	65	-	-	-	-	-
Age 0-Oct.							
G94 MI Loch Alpine	59	51	-	33-64	-	-	-
Age 0							
P30 MD Blade R.	11	30	-	20-38	-	-	-
B31, B73, E9, K137 lakes	17+	45	-	38-70	17	1	-

		TL				Weight	
	No.	Mean of means	Central 50%	Range	No.	Mean of means	Range
Age I							
E79 MD Deep Creek L.	11	38	-	-	-	-	-
A1, E9 NY lakes	26+	54	-	43-64	-	-	-
H62 IN	24	67	-	-	-	-	-
B192, H218, L129, P74 IL, OH, MO, NY streams	201+	73	48-89	28-119	-	-	-
R200 L. Erie	46	76	-	54-99	46	10	4-22
B25, C13, G94 MI lakes	139	84	76-86	61-106	396	8	6-20
B31, C117, E3, M20 MI, MN, WI	191	92	81-94	63-155	-	-	-
C176, D43, J25 OK KY TN streams	45	97	-	64-127	-	-	-
K41 IN ponds	+	-	-	-241	+	-	-227
combined	683+	78	54-93	28-241	442+	-	6-227
Age II							
E9 NY	+	79	-	-	-	-	-
E79 MD Deep Creek L.	11	79	-	-	-	-	-
B25, B31, C13, C117, E3, G94, H65, M20, N37 MI MN WI lakes	761	114	108-122	63-217	608	33	14-62
B192, C182, H218, L38, L129, M318, P74, IL IN MO OH WV streams	713+	121	90-132	43-170	4	11	8-14
E22 TN Norris L.	1	127	-	-	-	-	-
G36, H62, L14 IN OH	42	135	-	120-158	16	40	-
C176, J25 KY OK streams	51	145	-	122-157	18	65	-
R200 L. Erie	22	152	-	124-200	22	71	40-128
T44 OK Grand L.	3	174	-	-	+	77	-
K41 IN ponds	+	-	-	-213	+	-	-142
Combined	1,604+	122	108-137	43-217	668+	43	8-142
Age III							
E79 MD Deep Creek L.	10	114	-	-	-	-	-
C182, H218, L38, M317, P74 IL IN MO OH WV streams	261+	159	145-160	84-233	22	42	11-85
B192, G36, H62, L14, L129 IL IN OH lakes	449+	161	133-168	79-195	11	102	-
E27, J25, T44 OK TN reservoirs	6+	176	-	170-181	-	-	-
C176, D43, J25 KY OK TN streams	71	183	-	199-234	19	94	-
R200 L. Erie	26	190	-	160-219	26	153	94-207
Combined	1,939+	152	132-168	84-279	996	71	11-207
Age IV							
E79 MD Deep Creek L.	7	137	-	-	-	-	-
B25, B31, C13, C35, C117, C213, C215, E3, G94, H65, M20, N37, MI MN WI mostly lakes	1,467	161	148-173	107-279	1,015	87	51-170
V25 MS	4	162	-	157-168	4	87	76-99
B192, P74, C182, H218, L129, M317, IL IN MO OH WV mostly streams	283+	174	170-190	105-219	42	60	26-108
J25 OK Grand L.	+	178	-	-	-	-	-
G36, H62, L14, IN OH lakes	79	204	-	173-234	8	198	-

		TL				Weight	
	No.	Mean of means	Central 50%	Range	No.	Mean of means	Range
Age IV (cont.)							
C176, D43, J25 KY OK							
TN reservoirs	30	207	-	170-267	-	-	-
R200 L. Erie	9	213	-	165-244	9	193	170-224
Combined	1,879+	172	155-193	105-279	1,078	102	26-224
Age V							
E79 MD Deep Creek L.	3	152	-	-	-	-	-
S372 NY pond	1	155	-	-	-	-	-
B25, B31, C13, C35,							
C117, C213, C215, E3,							
G94, H65, M20, N37,							
W45 MI MN WI lakes	1,033	176	152-193	111-279	677	132	77-255
B192, H218, L108, M317,							
P74, IL IN MO OH WV							
streams	103	186	170-206	143-229	14	85	54-119
V25, V42 LA MS	3	187	-	181-190	3	122	113-127
G36, H62, L14, L129							
IN OH lakes	14+	230	-	198-249	8	244	-
D43, C176, J25 KY OK							
TN streams	6	233	-	204-262	1	207	-
R200 L. Erie	3	237	-	224-240	3	227	215-249
Combined	1,166+	186	168-206	111-279	706	143	54-255
Age VI							
E79 MD Deep Creek L.	1	196	-	-	-	-	-
B25, B31, C13, C35,							
C117, C213, C215, E3,							
G94, H65, M20, N37,							
W45, MI MN WI lakes	601	199	160-228	135-315	55	143	102-340
B192, C176, H218, M317,							
P74, IN KY MO OH WV							
streams	17	205	-	168-241	1	85	-
R200 L. Erie	1	238	-	-	1	261	-
H62, L129 IL IN lakes	1+	266	-	254-289	1	365	-
Combined	621+	204	183-228	135-315	58	162	85-365
Age VII							
B31, C13, C35, C117,							
C213, C215, E3, G94,							
H65, M20, W45 MI MN							
WI lakes	491	199	186-232	168-310	292	204	108-318
E79 MD Deep Creek L.	1	226	-	-	-	-	-
B192, C176 KY OH							
streams	10	239	-	219-260	-	-	-
R200 L. Erie	1	244	-	-	1	281	-
W13 CN	1	270	-	-	1	482	-
H62 IN lake	3	279	-	-	4	348	-
Combined	434	209	186-241	168-310	298	230	108-482
Age VIII							
B31, C35, C117, C215,							
E3, H65, M20, N37 MI							
MN WI lakes	371	228	216-238	174-340	355	212	164-326
E79 MD Deep Creek L.	1	238	-	-	-	-	-
R200 L. Erie	2	259	-	249-270	2	307	284-329
Combined	274	230	216-238	174-340	357	221	164-329

	No.	TL Mean of means	Central 50%	Range	No.	Weight Mean of means	Range
Age IX							
B31, C35, C117, C215, E3, G94, H65, M20 MI MN WI lakes	235	232	203-251	186-274	228	235	144-397
C176 KY stream	1	254	-	-	-	-	-
Age X							
B31, C35, C215, H65, M20 MI MN WI lakes	82	240	211-267	186-267	69	239	159-397
C176 KY stream	1	246	-	-	-	-	-
Age XI							
H65 WI Nebish L. females	16	219	-	203-229	16	179	-
males	3	237	-	235-244	3	224	-
C35 MN L. Vermilion	1	259	-	-	1	454	-
Age XII							
H65 WI Nebish L. females	4	219	-	210-236	4	179	-
M20 WI south	1	279	-	-	1	417	-
Age XIII							
H65 WI Nebish L. females	1	234	-	-	1	218	-

Rock bass lived 12 years (F7) and 18 years (B69) in captivity. The maximum age in nature was listed (L189) at 13 years and the maximum weight at 590 g.

Northern lakes

Mean calculated TL at each annulus

	No.	1	2	3	4	5	6	7	8	9	10	11	12
W45 WI Trout L.	108	24	44	68	83	123	149	173					
W45 WI Muskellunge L.	194	24	44	68	83	119	146	168					
B29 MI Booth L. 1943	266	39	61	81	94	105							
S372 NY pond	1	25	53	84	109	122							
B25 MI Standard L.	113	39	64	85	93	102							
B25 MI Standard L. after thinning	159	38	65	89	115	146	178						
P122 WI Flora L.	160	56	76	94	119	140	165	183	193	208	221		
after thinning	147	56	79	109	114	147	163	185	206	218	211		
K33 MN statewide (also M136, M258, B365)	907	41	76	114	149	181	211	232	244	257	267	274	274
C35 MN Vermilion L.	31	39	75	115	156	187	208	224	236	249	251	254	
B29 MI Booth L.	289	40	81	123	156	186	213						
E2 MN statewide	324	51	97	143	181	213	246	251					
D89 WI Murphy Flowage	+	69	109	160	175	203	216	234	231	234	241		

Mean calculated TL and increments at each annulus

		1	2	3	4	5	6	7	8	9	10	11	12	13
H66 WI Silver L.	females	38	59	83	106	136	162	187	211	221				
	Increments	38	21	24	23	32	25	25	23	10				
	Number	95	95	95	95	69	31	9	4	3				
	males	38	59	85	109	144	173	202	219	228				
	Increments	38	21	26	23	36	29	29	16	9				
	Number	59	59	59	41	37	8	3	2	2				
H66 WI Muskellunge L.	females	40	63	84	106	126	145	165	179	195	207	219		
	Increments	40	23	21	23	21	19	19	14	16	13	12		
	Number	313	313	313	300	272	244	199	162	69	25	3		
	males	40	63	85	107	131	151	174	191	210	220	228		
	Increments	40	23	22	23	23	20	22	18	18	10	8		
	Number	220	220	220	208	190	166	138	103	38	9	1		

Mean calculated TL and increments at each annulus (cont.)

		1	2	3	4	5	6	7	8	9	10	11	12	13
H66 WI Allequash L.	females	43	64	85	108	135	156	173	183	194	200			
	Increments	43	21	21	23	26	22	17	10	11	6			
	Number	9	9	9	9	8	6	4	1	1	1			
	males	44	72	92	114	143	170	185	199					
	Increments	44	28	20	23	28	26	16	14					
	Number	11	11	11	11	11	8	6	2					
H66 WI Trout L.	females	39	64	88	113	145	173	187	204	216	224			
	Increments	39	25	24	25	31	26	16	17	11	8			
	Number	52	52	52	50	45	23	19	12	5	1			
	males	39	61	85	113	149	187	208	228	237				
	Increments	39	22	24	27	36	36	21	18	11				
	Number	25	25	25	24	23	7	7	4	2				
G94 MI Lower Loch Alpine	females	48	76	91	104	122	145	157	180	188	196			
	Increments	48	28	20	23	23	25	23	20	18	8			
	Number	270	211	147	108	86	68	26	4	1	1			
H65 WI Nebish L.	females	41	76	109	137	159	173	183	190	198	204	212	219	228
	Increments	41	34	34	28	21	14	10	8	8	7	8	7	9
	Number	586	586	586	519	488	430	396	320	195	72	21	5	1
	males	42	76	111	145	170	186	198	208	216	225	234		
	Increments	42	34	35	32	21	16	13	9	9	8	9		
	Number	395	395	395	346	322	269	254	207	79	18	3		
E3 MN statewide		54	101	143	181	206	234	253	281	238				
	Increments	54	48	44	39	33	26	22	19	17				
	Number	469	461	442	380	211	102	30	4	1				
Unweighted mean–Northern lakes		42	70	99	124	150	180	196	211	219	222	237	246	228

OH, IN, MO, IL, KY streams	No.	Mean calculated TL at each annulus											
		1	2	3	4	5	6	7	8	9	10	11	12
R49 OH statewise	-	36	71	99	133	160	187	200	208	249			
B192 OH N. Fork, Massie	121	43	76	117	147	170	190	201					
B192 OH N. Fork, Miami	412	43	79	122	160	185	198	211					
B376 IN Big Walnut C.	38	52	88	134	155	201	221	247	257				
S130 IN Sugar C.	52	48	86	137	175	216							
P113 MO statewise	6689	41	86	140	178	203	216						
headwaters	4311	41	81	132	175	203	218						
middle	1682	43	91	142	178	198	213						
lower	696	46	94	145	188	208	211						
lowest average	-	33	69	104	145	173	190	190	168	211	221	231	254
best average	-	53	114	175	221	239	249	262	277	279	274	259	
B192 OH S. Fork, Miami	132	46	94	147	180	203	213						
L129 IL statewide	52	102	114	168	196	229	254						
C182 OH Sandusky R.	14	51	117	175	185								
T54 KY streams	50	84	140	178	203								

OH, IN, KY streams	No.	Mean calculated TL and increments at each annulus									
		1	2	3	4	5	6	7	8	9	10
P74 MO Black R.		40	76	119	152	185	185				
Increments		40	38	45	45	40	12				
Number		577	484	314	93	28	1				
S130 IN Tippecanoe R.	1291	46	82	130	173	203	211				
Increments		46	36	48	48	30	21				
Number		1288	1288	1216	596	68	6				
B192 OH Little Miami R.		42	79	132	173	198	211	221			
Increments		42	37	53	38	22	13	11			
Number		425	425	259	103	33	15	6			

Mean calculated TL and increments at each annulus (cont.)

	No.	1	2	3	4	5	6	7	8	9	10			
B192 OH Massie C.		43	84	135	170	190	206	213						
Increments		43	41	50	39	18	15	6						
Number		745	745	358	153	50	7	2						
C175, C176 KY 2 streams		36	89	145	168	180	185	216	211	229	231			
Increments		36	53	58	43	38	28	28	15	18	10			
Number		134	114	81	44	13	6	3	2	2	1			
Unweighted mean OH, IL, IN, KY, MO		49	91	139	170	194	206	216	225	229	231			

NC, VA, OK, TN Streams   No. Mean calculated TL at each annulus

	No.	1	2	3	4	5	6	7	8	9	10			
J42 OK Ft Gibson L.	17	43	91	140	168	196								
S192 VA Potomac R.	–	48	104	142	163	193	213							
S579 NC streams	40	85	128	161	178	196	216	228	258					
B380 TN Elk R.	18	76	155	211	244									

Mean calculated TL and increments at mean annulus

	No.	1	2	3	4	5	6	7	8	9	10			
OK streams														
J25 OK Illinois R.		43	107	147	188	206								
Increments		43	64	36	28	18								
Number		65	41	23	4	1								
Unweighted mean NC, VA, OK, TN		59	117	160	188	198	214	228	258					
Unweighted mean of regions		50	93	133	161	181	200	213	231	224	227	237	246	228

Average calculated weight in grams at each annulus

			No.	1	2	3	4	5	6	7	8	9	10	11
B29	MI	Booth L.	266	1	4	9	15	19						
H66	WI	Muskellunge L.	533	1	4	10	20	36	56	82	108	141	165	187
H66	WI	Silver L.	154	1	5	11	22	53	92	146	195	222		
H66	WI	Allequash L.	20	2	6	12	26	51	84	113	137	143	160	
H66	WI	Trout L.	77	1	4	12	28	62	113	154	200	232		
H66	WI	Nebish L.	981	1	7	22	46	72	95	113	128	143	160	179
B29	MI	Booth L.	289	2	12	43	87	148	221					
S130	IN	Tippecanoe R.	389	2	18	63	136	181						
		Increments		2	16	44	73	45						
		Number		389	389	389	191	29						

Although rock bass are fairly long lived in lakes, only 1 of 585 rock bass from Black River (P74) was beyond Age V; and only 16 of 3,237 from several streams (P74, C175, B192, J25, S130) were beyond Age VI.

Many of the calculated lengths were computed by direct proportion; but Fraser-type corrections were used in B29, B25, B192, G94 (23 mm), P122 (36 mm), P74 (18 mm), and a logarithmic correction was used in H65, H66, and S130.

One paper on rock bass, by Hile (H65), is a classic in the treatment of growth data and in the comparison of growth increments from different years. Another (H67) gives further mathematical interpretation. The validity of annulus interpretation was demonstrated in H65 and B192.

Average annual increments in length were greater in the 4th and 5th years than in the 2nd and 3rd in Allequash, Silver, and Trout lakes (H66), and Lower Loch Alpine (G94) and greater in the 3rd and 4th years than in the 2nd in Black River (P74) and in Ohio streams (B192); but in other waters the increments generally decreased with increase in age. H66 suggested that the differences in the form of the growth curve might be related to water hardness. Growth compensation was not evident at least in the 2nd year of life of rock bass in Nebish Lake but may be present later in life (H65). It did not appear until the 4th year of life in the Tippecanoe River (S130). Lee's phenomenon was evident in the growth data from Black River (P74) and Flora Lake (P122) and appeared in the Tippecanoe River sample only after the 3rd year (S130).

Age groups I-III had completed annulus formation by May 5 in Black River, MO (P74), but annuli did not develop in older rock bass until later in the season. About 75% of the annual growth of the younger fish was completed by the end of July, and even the older bass had much of their annual increment by that time: In Wisconsin lakes most of the season's growth was completed by late July (H65). In Ohio, growth was also more rapid in May-June than in July-August (B192).

Growth was reported to be slower in the north than the south in Michigan (L222) and Minnesota (E3), and the unweighted mean of the northern lakes was lower than those of the middle and southern part of the range. Growth of rock bass in the Illinois River, OK, was above that of the northern states, but the life span was shorter (J25). Growth in Lake Erie (R200) was faster than in most other waters. R285 suggested that the size of the body of water influenced growth of rock bass.

There were pronounced differences in the average growth in different tributaries of the Little Miami River, OH. Streams with lowest annual flow usually had the slowest growth (B192). Growth was fastest in upstream areas, where population numbers were less in Red Cedar River, MI (L222). In Missouri, headwaters showed the lowest average growth and downstream areas the best, with middle stretches showing intermediate growth (P113). In the Black River, MO, however, no trend in growth upstream or downstream was evident (P74). Growth varies so much in various parts of a stream system that average growth data mean little (L222, P113). In Ohio, rock bass confined to small streams are often dwarfed (T113). Growth in newly impounded Tenkiller Reservoir, OK, was slightly faster than in tributary streams (H150), but numbers of rock bass declined with impoundment. Decrease in numbers after impoundment was also noted by H66 and M517.

Growth was more rapid where carbonates and pH were higher (E3), where water was harder (H66), in areas richest in allochthonous materials (L222), and in areas with the most available food (R285). Growth was reduced when dissolved oxygen levels were reduced to 3 ppm at $21°-24°C$ for 8 hours per day (B415).

In Nebish Lake, WI, growth during the first year of life was not correlated with growth of older fish, but growth in the second and later years was positively correlated with summer temperatures, particularly June and September (H65). Similar correlation between annual growth and June temperatures was noted in Muskellunge and Silver Lakes but not in some other lakes, which indicates that local conditions may be more important than the temperature fluctuations (H65). Longer growing seasons seemed to result in greater annual growth in Minnesota (E3). Growth of rock bass in Ohio streams was positively correlated with April-October temperatures (B192). Growth was also better in years with heavier rainfall in June (H65).

One of the first demonstrations in which thinning a population resulted in increased growth was with rock bass (B25, B29). The population reduction had to be over 50% to increase growth and condition, however. Elimination of the population and restocking in Whippoorwill Creek, KY, resulted in a spurt of growth, but the growth rate decreased as the standing crop increased (C175, C176). The increased growth following thinning was more precisely demonstrated by analysis of the annual increments before and after thinning (P122). Rock bass seemed to respond more quickly to population reduction than did other species (P122).

Males grew more rapidly than females in Michigan (B30), in several Wisconsin lakes after the 3rd year of life (H66), and in the 2 Minnesota lakes for which sex data were available (E3). In Lower Loch Alpine, MI, there was little evidence of sex difference in growth, but females outnumbered males at least in the older age groups (G94). Superior late season growth of the males accounts for their more rapid growth (H65).

Both male and female rock bass were mature at age II (i.e., ready to spawn the next spring at age III) in Nebish L. but some did not mature until age V (H65). Both sexes first spawn in their fourth year in New York (R285). They mature at about 100 g (L189).

Spawning is initiated when water warms to $20.5°-21°C$ and may continue to $26°C$ (R285). Spawning is in May and early June in OK (J25), mid-May to mid-June in L. Maxinkuckee, IN (E32), May and June in IL (L38), May to July in MN (B365), April to June in NY (A1, R285) and to mid-July in northern New York (R285), and June and early July in NC (T129). At the Neosho, MO, hatchery rock bass spawn once per year, but they have been known to produce two separate broods in one season (B73). Pituitary injections induce spawning (R302).

Immediately preceeding and during the spawning season, adult females congregate in pools (T113). Breeding males are blackish (T113). Spawning is described by A1. The nest is on gravel near vegetation, and the young fish are usually found only in areas protected from waves (H262).

	No. females	Weight	Number of eggs per female mean	Range
V21 MN	10	110-230	3,000	3,000-5,000
	24	230-370	5,000	2,000-7,000
	17	370-450	5,500	3,000-8,000
	3	450-570	8,500	8,000-11,000

Strong year classes in Nebish L. were correlated with high temperatures and heavy rainfall early in the season, particularly in June (H65).

Rock bass feed mostly on the bottom but may feed on the surface (K129). They are reported to be equally nocternal and diurnal (K135) or to be inactive

at night (H262). During the daytime, rock bass in an Ontario lake schooled in
areas of rocky cover at depths of 1-7 m. The schools hung almost motionless
over deep water near rock cliffs. Foraging started toward dusk and reached a
maximum about 2 hours before darkness and ceased about 30 minutes before
darkness. Rock bass then settled onto rocks, logs, or plants and were relatively inactive at night (E153). Aggregation in shallow areas of rocky cover
was noted in November (E153).

The eyes of rock bass have considerable adaptive movement, mostly along
the rostral-caudal axis, which is associated with feeding on organisms picked
off the bottom (S580). Small rock bass, 15-20 mm, feed on copepods (H262);
23-35 mm, on entomostraca and algae (M117); 45-70 mm, on insects and crustaceans (K129); and 80-135 mm, on chironomids (N117). Few rock bass over
75 mm ate entomostraca (K153). Small forage fish are taken as rock bass
reach 75 mm (K129), and fish become a more important part of the diet as the
rock bass grow (J25, L38, L249, R285, E32). Adult rock bass show a marked
preference for crayfish (H262, M404, D183, E32) or fish and crayfish (P205),
but insects may be a major food in some waters (K125). A few molluscs were
also eaten by age V and VI rock bass (R200). Among the forage fish utilized
were alewives (G97), emerald shiners (M404), and darters (E32). In Lake
Oneida in early summer adult rock bass fed on worms, crustaceans, and insect larvae but later fed mostly on minnows and crayfish (A1). Amphipods
made up 60-75% of the stomach volume in rock bass (150-260 mm) in the winter in an Ontario stream, and they started feeding at about 8.5°C (K137). At
18°-22°C, rock bass consumed 0.04 g of food (dragonfly nymphs, crayfish, and
fish) per gram of body weight per day (K135).

The LD_{50} temperature for rock bass acclimated to 30°C was 35°C (T102).
At the Neosho hatchery, rock bass survived to 30.6°C (B73). Body temperatures of rock bass were 27° to 28°C, even though they were collected in thermal plume water at 34°C, which indicates that the rock bass were near the
bottom (N141). Preferred temperatures of rock bass were 25.5°-29°C in daytime and 27°-30.5°C at night. These temperatures were lower than bluegill
and largemouth bass and about the same as black crappie (N141). The average
temperature at which rock bass were found in Ontario streams was 20.8°C as
compared to 21.5°C for smallmouth bass (H461). The rock bass was also
found in areas with less flow than the smallmouth. In laboratory experiments,
pH values of 7 to 4 did not affect cover utilization by rock bass (C385). Rock
bass were collected in streams at pH 6.0 to 6.4 but not in acid mine drainage
areas of Pennsylvania at pH 4.6 to 5.9 (C386).

Dissolved oxygen levels of 3 ppm for 8 hours per day at 21°-24°C changed
plasma protein levels, increased hematocrits, increased breathing rates from
60-65/minute to 113/minute, and almost stopped feeding and growth (B415).

Most rock bass maintain themselves in a small range of movement (B252,
F158, G174, L222). In Flora L., WI, (P122) there were 10-54 rock bass (0.4-
3.5 kg) per hectare from 1952-56 and in Whitmore L., MI, (C215), 2.6 (0.34 kg)
per hectare. Fertilization of the Obed R., TN, increased the standing crop of
rock bass (P196). In some Ohio streams general fishermen caught 0.1 rock
bass per hour by angling; but smallmouth bass specialists reported 0.8 rock
bass per hour, and "cooperators" reported 0.4 (B192). Hellgramites, worms,
and spinners were the most successful baits.

Hoop nets were the most effective gear for collecting rock bass (L38,
F152), but electric seines were also effective (F152). Difficulty in recovering
all young after poisoning was reported (C13).

The sizes taken in gillnets of various mesh sizes were:

Stretch measure	No.	Mean TL	Central 50%	Citation
32	17	106	94-105	H65
38	172	112	107-113	H65
	1	94	-	C22
44	79	141	125-144	H65
51	92	164	144-189	H65
57	70	190	172-209	H65
64	224	198	190-209	H65
76	350	216	210-222	H65
	6	190	-	C222
95	1	211	-	C222
102	6	257	-	C222

Annual mortality rates were estimated at 55% from age II to III and at 72% from age IV to V in the Ohio streams (B192) and at 66%-79% for rock bass over age X in Nebish Lake with no fishing mortality (R196). Chromosome numbers were reported as 48 (R286).

Artificial hybridization with several other species in declining order of success were black crappie, bluegill, banded sunfish, warmouth, and largemouth bass (T147).

Additional weight-length and growth data in L223 were referred to in L222 but not quoted.

SACRAMENTO PERCH, *Archoplites interruptus* (Girard)

The only centrarchid native west of the Rockies has decreased in numbers in the Sacramento-San Joaquin River system and a few nearby river systems, where it was found when the area was first settled (M520, T130). Egg predation by exotic species may be a major cause of the decline (M85). Some were introduced into western Nevada in 1877 and into Arizona, Colorado, Nebraska, New Mexico, South Dakota, Utah, and Texas in the 1960's (M520). Sacramento perch are euryhaline and adaptable to inland waters too high in minerals for other game fishes. The perch have survived and reproduced in chloride sulfate waters with salinities near 17,000 ppm and have survived in 800 ppm of sodium potassium carbonate (M520). They were recommended for stocking in moderately alkaline waters, 1,200-1,800 ppm total alkalinity, 550-850 ppm carbonates, and 350-400 ppm sodium potassium carbonate (M552).

M85. Weight-length relationship, Clear L., CA

FL	150-74	175-99	200-24	225-49	250-55
Weight, range	50-100	95-157	150-250	150-300	275-300
Mean	75	130	180	235	287
No.	3	11	14	7	2

In the early days specimens up to 610 mm were reported (M85). In Pyramid Lake, NV, anglers have caught perch up to 405 mm and 1,360 g (T130).

		FL		TL		Weight	
	No.	Mean	Range	Mean	Range	Mean	Range
Age 0-June							
M85 CA Clear L.	+	20	-	-	-	-	-

	No.	FL Mean	FL Range	TL Mean	TL Range	Weight Mean	Weight Range
Age 0-July							
M520 NB	+	-	-	-	8-70	-	-
I18 CO	34	-	-	37	32-43	1.2	0.8-1.9
Age 0-Aug.							
M521 NB Smithy's Pond	+	-	-	-	18-28	-	-
M520 NB	+	-	-	-	10-28	-	-
I18 CO	77	-	-	53	20-79	6	0.9-15.5
Age 0-Sept.							
M520 NB	+	-	-	-	15-18	-	-
M521 NB Hatchery	127	-	-	68	40-86	-	-
I18 CO	59	-	-	106	88-108	30	13-31
Age 0-Oct.							
M85 CA Clear L.	+	-	x-89	-	-	-	-
T130 CA Clear L.	+	-	-	-	-	14	-
Age 0-Nov.							
M521 NB Smithy's Pond	61	-	-	49	34-62	-	-
Age I							
M520 SD White L.	+	-	-	71	-	-	-
M520 ND Round L.	+	-	-	81	-	-	-
M520 NV 5 lakes	+	-	-	81	69-102	-	-
G158 CO Newell L., April	1	-	-	97	-	91	-
M522 CA L. Anza	+	-	-	-	95-125	-	-
M520 NB 5 lakes	+	-	-	127	109-147	-	-
I18 CO	17	-	-	138	-	59	-
M521 NB	3	-	-	157	140-180	-	-
M85 CA Clear L.	10	173	152-193	-	-	-	-
T130 CA Clear L.	+	-	-	-	-	100	-
Age II							
M520 SD White L.	+	-	-	117	-	-	-
M520 NV 5 lakes	+	-	-	138	102-173	-	-
M520 NB 5 lakes	+	-	-	188	180-193	-	-
I18 CO	37	-	-	190	151-213	190	96-247
G158 CO Newell L.	1	-	-	193	-	165	-
M521 NB	6	-	-	206	195-218	-	-
M85 CA Clear L.	22	203	170-246	-	-	-	-
T130 CA Clear L.	+	-	-	-	-	170	-
Age III							
M520 NV 5 lakes	+	-	-	180	130-216	-	-
M520 NB 4 lakes	+	-	-	229	218-244	-	-
G158 CO Newell L.	2	-	-	237	231-244	237	220-255
M521 NB	6	-	-	250	240-265	-	-
M85 CA Clear L.	16	226	201-262	-	-	-	-
T130 CA Clear L.	+	-	-	-	-	227	-
Age IV							
M520 NV 5 lakes	+	-	-	224	157-267	-	-
M520 NB 4 lakes	+	-	-	264	241-284	-	-
M85 CA Clear L.	7	254	226-279	-	-	-	-
Age V							
M520 NV 5 lakes	+	-	-	269	216-305	-	-
M520 NB 2 lakes	+	-	-	291	257-325	-	-

| | | FL | | TL | | Weight | |
	No.	Mean	Range	Mean	Range	Mean	Range
Age VI							
M520 NV 5 lakes	+	-	-	302	262-325	-	-
M520 NB Clear L.	+	-	-	338	-	-	-
Age VII							
M520 NV 3 lakes	+	-	-	320	284-351	-	-
Age VIII							
M521 NV 3 lakes	+	-	-	340	312-368	-	-
Age IX							
M521 NV 2 lakes	+	-	-	378	363-394	-	-

Some difficulty was experienced in reading scales of Clear L., CA, perch because of "false checks" perhaps associated with spawning (M85). The first annulus was easily defined (M85, M521). Females are of a larger mean size than males due to faster growth and possibly lower mortality rate (M522).

Sacramento perch spawn in their second year of life (M85, M520, M522). Spawning behavior is much like other centrarchids except that they do not disturb the substrate in building a nest (M522, M85). Prior to spawning the perch school and move to shallow water (M85). The males defind territories and guard the eggs for 2 or 3 days after they are spawned (M522). In Clear Lake, however, males left the eggs shortly after spawning (M85), and it is usually stated that they do not guard their eggs (M140). Males guarded eggs in aquaria (C327, M522). The territories are 300-450 mm in diameter (M522). In Lake Anza, egg guarding males did not drive off plankton-feeding *Lavinia* but did attack possible egg predators (M522). Carp and catfish were reported to be the major predators on spawn (N142). In courtship the male continues to drive off the female much longer than in other centrarchids, and the female must persist to spawn (M522).

The adhesive eggs are deposited over aquatic plants or algae attached to rocks (M85, M520, M522) at depths of 30-90 cm. A 265 mm female snagged off a spawning bed had 84,000 unspawned mature eggs (M85). Spawning occurs at water temperatures of $22°-28°C$ (M520) or $22°-24°C$ (M522). The presence of small fry in September suggested that Sacramento perch may spawn twice, or possibly over much of the summer in the Nebraska waters (M520). In California they spawn principally in May and June; in Nevada, from mid-June into August (T130).

Young perch disperse to deeper water when 25-30 mm (M521) or about 50 mm (M85).

Sacramento perch up to about 125 mm feed primarily on invertebrates, including zooplankton, *Hyallela* (M520), *Chaoborus* (L183), odonates, hemipterans, and mayflies (M521, M571). Larger perch feed to a considerable extent on fish (M85, L183, M520, M571, T130, B305). A 318 mm perch had eaten a 100 mm hitch, *Lavinia exilicauda* (B305).

Failure to coexist with other species has been frequently mentioned (M85, M520), although several exceptions have been noted. Sacramento perch fingerlings were dominated by bluegill fingerlings in aquaria and in small ponds. This behavior may drive the perch fingerlings out of shallow weed beds to areas with less food and more danger from predators (M571).

FLIER, *Centrarchus macropterus* (Lacepede)
    The flier is found from Virginia to Florida and north in the Mississippi Valley to southern Illinois (E119), but not in brackish water (B399).

D105   Weight-length relationship in a good fertilized pond, AL

SL	45	54	64	74	84	92	105	113	123	132	143	160
W	2.5	4.5	8.8	14	22	40	47	65	91	99	113	176
K(SL)	2.77	2.97	3.37	3.51	3.75	5.21	4.31	1.76	4.85	4.23	3.91	4.27
No.	8	15	15	13	15	21	15	10	13	5	8	1

S472 AL

TL	51	76	102	127	152
W	5	12	25	50	68
K(TL)	3.46	2.73	2.43	2.33	1.92
No.	1	3	8	10	1

The fact that the K(TL) values dropped so markedly suggests that the recorded weights at the lower lengths are too high and that the following calculated formula is probably not very accurate (S472):

$$\text{Log W} = -5.63 + 2.51 \text{ Log TL}$$

The maximum weight is about 450 g (L189).

	No.	Mean TL	Mean W
D105 AL pond			
Age I-Sept.	+	-	71
Age II-Aug.	+	208	207
D68 NC Little R.			
Age V	13	147	-
Age VI	27	183	-
Age VII	30	193	-
Age VIII	18	198	-

		Calculated TL at each annulus						
	No.	1	2	3	4	5	6	7
D67, D68 NC Little R.	88	41	71	99	127	155	170	185
L157 NC Waccamaw L.	1	28	58	99	127			
L157 NC Salters L.	11	36	64	86	107	117	122	
L157 NC Black L.	10	41	74	89				
L157 NC Jones L	11	33	69	109	132	147	152	173
L157 NC Singletary L.	15	38	66	107	132	152		
L157 NC White L.	2	41	97	137	160			
R119 VA Drummond L.	+	71	114	142	168	185	193	

		Calculated weight at each annulus					
R119 VA Drummond L.	+	14	23	50	77	109	145

The annulus was formed in May in Alabama (D105). Scales formed when the young fish were 15-16 mm with the first scales forming on the midlateral region and spreading over the body in a different pattern than for other centrarchids (C351). No sexual difference in growth rate was noted (D105), but differences in coloration occur (L189).

Nesting was observed as early as February 13 in Alabama at 14°C (D105), but it usually occurs from March to May at 17°C. Nests are usually close together in colonies, and the male guards the nest and fry (L189). The first hatch usually occurs when the water at 15 cm is 15.5°-18°C (S334). Eggs hatched in 51 hours at 22°C (M522). The right ovary averaged smaller than the left, and the number of eggs per female was (D105):

	No. females	Number of eggs	
		Mean	Range
70-75	2	2,150	1,900-2,400
76-101	12	4,680	2,400-10,300
102-126	20	11,925	5,500-23,000
127-151	2	17,000	16,500-17,500
152-177	2	31,250	30,000-32,500
178-190	2	36,250	35,000-37,500

Survival to 112 days from the potential egg deposit was estimated at 0.9% (D105). The small fish were easily frightened. Adults school in winter (L189).

Cladocera, chironomids, and other aquatic invertebrates were the principle items of food (D105). They ate more hemipterans (mostly Corixidae) than other species in the study (F148).

Production was low, 36 kg/ha, even when fed in a pond (S307), but standing crops of 67-90 kg/ha and yields of 9-11 kg/ha were also reported (L189).

The number of chromosomes is 48 (R286), which is a common number in centrarchids.

EVERGLADES PYGMY SUNFISH, *Elassoma evergladei* Jordan

This sunfish is found in the swamps of southern Georgia and Florida (E119) with a maximum length of 38 mm. They do not enter brackish water (K151). In captivity these sunfish would eat only live food, brine shrimp, etc. (M528). The males are territorial during breeding, and the breeding behavior is similar to other sunfishes. With warm temperatures and a constant 15-hour photoperiod, they were kept in breeding condition throughout the year (M528). Larvae developed in 21°-25°C in aquaria (M528).

BANDED PYGMY SUNFISH, *Elassoma zonatum* Jordan

The banded pygmy sunfish is found in swamps and weedy ponds from southern Illinois to Texas and Georgia (E119). It does not enter brackish water (B399).

B18 LA	No.	Standard length
at hatching	+	2.7
Age 0-April	200	6
0-May	158	11
0-June	93	12
0-July	138	13
0-Aug.	254	15
0-Sept.-Nov.	273	17
I-March	404	25
II	+	34

Males were observed to protect territories over *Myriophyllum* but made no nest (T131). They spawned at 23°C and eggs hatched in 48-60 hours. At 18°C, incubation time was 7 days (T131). Spawning was intermittent (B18).

	SL	No. of females	Eggs per female Mean	Eggs per female Range
B18 LA	22-27	5	232	96-325
	29-35	3	682	500-970

They eat crustaceans and insects (B18).
The number of chromosomes is 48 (R286).

BLACKBANDED SUNFISH, *Enneacanthus chaetodon* (Baird)
This small sunfish is found in lowland streams from New Jersey to Florida (E119). A101 reported that it was not found above tidewater.

S381 MD   TL = 1.32SL

The weight-length relationship was described as

$$W = 0.19SL - 3.97$$

but this straight line makes a poor fit of the data on the graph (S381, p87).

The following data were approximated from the graph:

SL	No.	Weight	SL	No.	Weight
16-20	5	0.1-0.25	36-40	12	2.3-3.4
21-25	10	0.2-0.6	41-45	22	3.2-5.5
26-30	4	0.6-1.2	46-50	16	4.2-6.3
34	1	1.8	51-54	15	6.1-7.6

Using the midpoints of these classes gives:

$$\text{Log } W = -5.50 + 3.73 \text{ Log SL}.$$

Although a maximum length of 100 mm has been reported, the largest MD specimen was 69 mm TL (S381).

	No.	Calculated SL at each annulus 1	2	3	4
S381 MD Smithville Ponds					
1955	30	10	27	45	54
1958	56	25	35	41	44

These computations were not made on a direct proportion basis but on the basis of regression formulas, which differed for the 2 years and which gave higher calculated lengths at the 4th annulus in the 1955 sample than the lengths at capture. No sex difference was noted in lengths and weights (S381).

Aquatic insects, gammarids, filamentous algae, and plant leaves were the food of 90 sunfish collected in July and November (S381). The frequency of empty stomachs suggested that they are nocturnal feeders. Because chironomid larvae were the principal foods, A101 classed banded sunfish as surface feeders, but this does not suggest feeding at the water surface.

Blackbanded sunfish spawn in March in North Carolina (S381).

The number of chromosomes is 48 (R286).

BLUESPOTTED SUNFISH, *Enneacanthus gloriosus* (Holbrook)

This brightly colored sunfish is common on the coastal plain from southeast New York to Florida (E119, R285). A recently established population in the Lake Ontario drainage in New York has been discussed (W248). Bluespotted sunfish have not been found in brackish water (B399).

B72 NY	2.3 mm at hatching	4.5 mm at 120 hours
	3 fish age V averaged 62 mm	range 55-71

Calculated AL at each annulus

1	2	3
15	30	40-50

A maximum age of 5 years was reported (L189). A sample of 25 bluespotted sunfish 31-58 mm SL from Jamesville Reservoir, N.Y., were found to include age II-V (W248).

They spawn at approximately 50 mm and rarely exceed 100 mm in length (R285). The male guards an area and makes a circular next about 100 mm in diameter or spawns over a similar diameter of filamentous algae. The spawning season is protracted (B72). Eggs hatched in 57 hours at 22.8°C (B72).

Small crustacea were the main food items, but aquatic insects (F148, R285), plants, worms, and molluscs (B72) were also eaten.

The number of chromosomes is 48 (R286).

BANDED SUNFISH, *Enneacanthus obesus* (Girard)

The banded sunfish is found on the coastal plain from Massachusetts to Florida (E119). The maximum length is about 80 mm (R285), and its food habits and reproduction are similar to the bluespotted sunfish. Prolonged daylight and temperatures at 21.7°C cause ova to enlarge, testes to develop, and sexual colors to develop in December as they normally do in early summer (H432). The number of chromosomes is 48 (R286).

REDBREAST SUNFISH, *Lepomis auritus* (Linnaeus)

The redbreast is found from Maine to Florida in waters tributary to the Atlantic and westward along the Gulf Coast to Texas (E119, R285). It flourishes in a wide range of ecological conditions from headwater streams to coastal plain rivers and lakes (R285), from elevations of 3,500 feet to waters with 8% seawater equivalency (S528) and pH ranges of 4.8 to 8.4. They are found in trout waters, but had an LD_{50} of over 37°C (T102). They have been established in mountain lakes of northern Italy (V92).

I found no data for converting SL to TL, except 1 fish, TL = 1.22 SL (W13).

The data which have been converted from SL to TL using this factor are marked *.

TL	Mean	Weight Range	No.	Citation
25	0.9	-	31	S472 AL
45*	1.8	-	21	B222 NY
51	2.5	0.9-4.3	19	S472 AL
57*	4.0	-	25	B222 NY
69*	5.5	-	63	B222 NY

TL	Mean	Weight Range	No.	Citation
76	14	5-16	34	S472 AL
80*	8.4	-	97	B222 NY
91*	11.3	-	57	B222 NY
102	18	17-30	48	S472 AL
102*	16	-	2	B222 NY
126*	50	-	1	D179 TX
127	44	27-86	45	S472 AL
132*	43	-	1	B222 NY
140*	60	-	3	B222 NY
151*	72	-	2	B222 NY
152	77	59-104	34	S472 AL
169*	92	85-99	2	D179 TX
178	132	91-159	46	S472 AL
203	163	154-227	10	S472 AL
239	312	-	1	W13 CT
241	301	-	1	E105 MD

S472 AL   3,937 fish; Log W = -469 + 3.01 Log TL

The 25 and 76 mm fish seem to have too high weights, otherwise, the slope would have been slightly above 3.01.

The K(SL) decreased with age in Wolf Lake but not in Catlin and Rich Lakes, NY (B222). The means ranged from 2.7-4.1.

The K(SL) of 3 fish, 103-140 mm SL, were 3.30-4.48 in TX (D179).

The mean K(TL) of 65 fish, 25 and 76 mm classes, was 4.21 in AL (S472),
67 fish, 51 and 102 mm classes, was 1.90,
135 fish, 127-203 mm, was 2.20.

	No.	TL Mean	TL Range	Weight No.	Weight Mean
Age I					
B222 NY Wolf L.*	10	43	38-48	10	1
B222 NY Rich L.*	24	52	43-63	24	3
B222 NY Catlin L.*	10	60	49-72	10	5
O50 GA Sinclair L.*	211	73	65-78	-	-
R217 NC Hickory L.	5	89	-	-	-
R217 NC L. Lure	5	91	-	-	-
R217 NC James L.	8	104	-	-	-
V48 WV Shenandoah R.	+	104	-	-	-
R217 NC Lookout Shoals L.	4	112	-	-	-
R217 NC Rhodhiss L.	3	135	-	-	-
Age II					
B222 NY Wolf L.*	6	63	60-70	6	5
B227 NY Rich L.*	5	70	62-77	5	7
B222 NY Catlin L.*	4	77	72-89	4	8
O50 GA Sinclair L.*	12	83	74-106	-	-
R217 NC Lookout Shoals L.	14	109	-	-	-
R217 NC Hickory L.	12	124	-	-	-
R217 NC L. Lure	8	132	-	-	-
R217 NC James L.	9	132	-	-	-
R217 NC Rhodhiss L.	17	137	-	-	-
V48 WV Shenandoah R.	+	163	-	-	-

		TL		Weight	
	No.	Mean	Range	No.	Mean
Age III					
B222 NY Wolf L.*	34	68	56-82	33	5
B222 NY Rich L.	33	78	65-87	33	8
B222 NY Catlin L.*	20	88	78-99	17	11
R217 NC Lookout Shoals L.	11	140	-	-	-
R217 NC Hickory L.	22	145	-	-	-
R217 NC L. Lure	4	147	-	-	-
R217 NC James L.	18	157	-	-	-
R217 NC Rhodhiss L.	17	157	-	-	-
V48 WV Shenandoah R.	+	196	-	-	-
Age IV					
B222 NY Wolf L.*	30	77	63-96	30	7
B222 NY Rich L.*	15	88	78-112	12	11
B222 NY Catlin L.*	6	123	109-132	2	53
R217 NC Lookout Shoals L.	9	157	-	-	-
R217 NC Rhodhiss L.	3	165	-	-	-
R217 NC James L.	7	170	-	-	-
R217 NC Hickory L.	8	173	-	-	-
Age V					
B222 NY Wolf L.*	71	87	63-109	69	9
B222 NY Catlin L.*	1	109	-	-	-
B222 NY Rich L.*	3	110	105-112	-	-
R217 NC Hickory L.	2	180	-	-	-
R217 NC Lookout Shoals L.	1	183	-	-	-
Age VI					
B222 NY Wolf L.*	12	94	81-124	11	11
B222 NY Rich L.*	2	133	132-134	2	50
B222 NY Catlin L.*	6	137	116-155	3	68
R217 NC Lookout Shoals L.	2	188	-	-	-
Age VIII					
W13 CT	1	239	-	1	312

	Calculated TL at each annulus					
	No.	1	2	3	4	5	6
H475 GA West Point Res.	-	44	79	120	148	166	
S526 NC	123	47	84	122	156	168	211
L157 NC White L.	8	48	102	140	163		
S579 NC	119	94	140	161	188		

	Calculated TL and increment at each annulus					
O50 GA Sinclair L.*		32	58			
	Incr.	33	25			
	No.	223	12			
R217 NC Hickory L.		56	94	124	152	173
	Incr.	56	38	33	20	18
	No.	49	44	32	10	2
R217 NC L. Lure		74	107	135		
	Incr.	74	33	28		
	No.	17	12	4		
R217 NC Lookout Shoals L.		74	99	130	150	170	180
	Incr.	74	28	25	18	10	10
	No.	41	37	23	12	3	2

		Calculated TL and increment at each annulus					
		1	2	3	4	5	6
R217 NC Rhodhiss L.		81	114	137	150		
	Incr.	81	38	23	20		
	No.	40	37	20	3		
R217 NC Jones L.		102	127	145	160		
	Incr.	102	20	18	13		
	No.	42	34	25	7		

A group averaging 38 mm, 0.4 g, grew to 89 mm, 12 g, from May 1 to August 31 in a NC pond where fed on an automatic feeder (C352). Another group 76 mm, 0.8 g, grew to 109 mm, 30 g, in the same time.

Annuli formed from March to May in Georgia (H470). Lengths were calculated by direct proportion, except that an intercept of 5 mm was used for Hickory Res., 46 mm for Lookout Shoals Res., and 61 mm for James Res., (R217). No sexual difference in growth was noted (B222), but H470 reported a sex difference in growth of age II fish. The slow growth in Wolf L. NY was thought to be the result of overcrowding with few predators, and the faster growth in Catlin L. was associated with more vegetation and associated insects (B222). The growth in the New York lakes studied was slower than that farther south, but the one Connecticut fish showed good growth. Effects of thermal outflow were not evident on the growth of redbreast sunfish in Lake Sinclair (O50).

They reach maturity at about 23 g (L189) or in the 2nd year of life (H470). Nests have been found from early June to early August in New York (R285) and Connecticut (D210) and in mid-May and June in North Carolina (D210). In Florida most spawn from April to June, but spawning may occur in any month (H460). The first hatch is when waters reach $25°-27°C$ (L189), but spawning is reported at $20°-29°C$ (D210). Nests were 30-40 cm in diameter and only 30-60 cm apart (R285) in water 15-40 cm deep. Nests were usually in the shelter of a log or stump in sand or fine gravels (D210). The number of eggs per female increased with age and size (D210).

Age	L	Eggs/female, mean, and standard deviation
II	140-142	963±88
III	150-155	1000±436
IV	170-185	3563±763
V	183-203	5620±852
VI	229-236	8250±278

The redbreast was the only one of 7 species of *Lepomis* which did not produce sound during courtship (G191).

Redbreast sperm failed to fertilize walleye (*Stizostedion vitreum*) eggs (H458).

At temperatures below $5°C$, redbreast form dense hibernating schools in deeper water (R285) but otherwise show little evidence of schooling (L189).

Insects constituted most of the food (F148, D210). Redbreasts are good test animals, easily fed and cared for, although some difficulty has been reported in transporting them (B349).

The number of chromosomes is 48 (R286).

GREEN SUNFISH, *Lepomis cyanellus* Rafinesque

The green sunfish is found from the Great Lakes region south to Mexico (E119). It is strictly freshwater in distribution, not entering brackish water (B399), but green sunfish will tolerate alkalinities up to 2,000 mg/l. The green sunfish and Sacramento perch were the only centrarchids to survive more than a month in alkaline waters greater than 950 mg/l total alkalinity (M552). It has been introduced a few places east of the Appalachians (R285) and west of the Rockies (K61, W131) and in the Frankfort area of Germany (V92). They were accidentally introduced into California in 1891 (M526). The green sunfish was the most widely distributed introduced species in the Sierra Nevada foothills in Central California in 1970 (M566), and its abundance was negatively correlated with the abundance of most native fishes. Multivariate analysis indicated that in the California area they were abundant in small intermittent streams at lower elevations, especially in warm, turbid, muddy-bottomed pools with large amounts of vegetation and with largemouth bass and mosquito fish. Usually green sunfish do not become large enough to interest anglers.

	No.	Length	FL/SL	TL/SL	TL/FL
C30 IA	73	34-152 SL	1.18	1.23	-
C38 IA Little Wall L.	304	45-100 SL	-	1.239	-
	52	101-135 SL	-	1.209	
	24	136-180 SL	-	1.192	-
S278 IA Little Wall L.	426	11-180	-	1.218	-
R125 IA Ike L.	290	53-190 TL	-	1.241	1.035
L110 IL	81	51-199 SL	-	1.194	-
H264	1170	30-170 TL	-	1.253	-

No trend in conversion factors was evident in H264, but a trend was noted in S278 and C38. For the purposes of converting SL to TL, a factor of 1.24 was used.

TL	Weight Mean of means	Central 50%	Range	No.	Citation
25-50	0.5	-	0.3-1.0	3	IA: S278, C30
	0.65	0.5-0.7	0.5-0.9	136+	AL, OK: A40, L97, L189, J44
	1.4	-	0.009-3.9	5,097	AL: S472 (too high)
	1.4	-	-	33	UT: W131
	0.8	0.5-1.0	0.3-1.4	172+	Combined
51-75	2.5	2.3-2.6	0.02-17	4,610+	AL: A40, L189, S472, L97, A39
	4.1	-	3.6-5.0	80+	OK: J44, H392
	4.9	-	-	84	UT: W131
	5.2	-	2-7	93	IA: C30, C38, S278
	3.8	2.6-5.0	0.02-17	4,867+	Combined
75-101	7.1	7.0-7.7	3-29	2,880+	AL: A40, L97, A39, L189, S472
	11.7	-	6-23	717	IN, IL, IA: M499, B150, S278, C30, C38, K43, C285
	11.6	-	11-13	143+	OK: J44, H392

Weight

TL	Mean of means	Central 50%	Range	No.	Citation
	13.4	-	13-14	95	CO, UT: W131, L126
	10.5	7.3-13.0	3-19	3,835+	Combined
102-126	20.0	17.7-21.0	5-23	1,470	AL: S472, A39, L97, L189, A40
	24.6	22-28	14-41	479+	OH, IN, IL, IA: M499, S278, C38, C30, K43, R54, C285, I10
	25.5	-	24.5-28	229+	OK: J44, H392
	28.4	-	23-31	132	CO, UT: C208, W131, L126
	24.1	23-28	14-41	2,310+	Combined
127-151	39.9	37-45	10-68	486+	AL: A40, S472, L97, L189
	49.1	-	38-65	435+	OK: J44, H392
	50.0	-	48-54	76	CO, UT: L126, W131, C208
	52.4	50-55	27-85	223+	OH, IN, IL, IA: C285, M499, K43, S278, C30, C38, R54
	48.1	47-54	10-85	1,220+	Combined
152-177	70.3	68-73	36-113	123+	AL: A40, S472, L189, L97
	80.1	-	60-127	391+	OK: J44, H392
	92.1	-	91-93	27	CO, UT: L126, W131
	96.4	87-109	62-113	150+	OH, IA, IL: C30, C285, S278, M499, C38, R54
	86.1	74-93	36-127	691+	Combined
178-202	101.3	-	45-136	25+	AL: L97, S472, L189
	122.3	-	99-145	19	CO, UT: C208, W131, L126
	127.6	-	109-172	243+	OK: J44, H392
	155.7	136-164	111-212	30+	OH, IA, IL: C285, S278, C38, C30, R54
	131.6	100-145	45-212	308+	Combined
203-228	160.9	-	136-172	92	AL, OK: S472, J44
	211.0	-	207-215	11	IA; S278, C38
	227.0	-	-	3	CO: C208, L126
	408.0	-	-	1	MD: E105
	204.8	169-215	136-408	107	Combined
229-253	260.0	-	-	19	OK: J44
267	422.0	-	-	1	OK, J44

No sex difference in weight-length relationship was noted (C285).

H264 MO:
L. of the Ozarks   1078 fish   Log W = -4.3347 + 2.9483 Log SL
S278 IA:
Little Wall L.   426 fish   = -4.768 + 3.169 Log SL
L110 IL:
Hutchins and Clear R.   83 fish, 55-199 mm   = -4.89 + 3.19 Log SL
W131 UT: 402 fish   = -4.821 + 3.200 Log SL

J34 OK:

Ardemore City L.	64 fish, 112-206 mm	Log W = -4.978 + 3.093 Log TL
J44, H419 OK:	1631 fish, 46-267 mm	= -5.27 + 3.2345 Log TL
R125 IA:		
Ike L.	290 fish, 53-190 mm	= -5.280 + 3.290 Lot TL

The intercept in J44 and H419 did not seem correct when adjusted to the metric system, so a new intercept was estimated from the table of calculated weights.

		Length	No.	K(SL) Mean	Range
C23	MN	80-170	20	2.46	-
W105 IL	Crab Orchard L.		2	3.13	2.65-3.61
L110 IL	Hutchins and Clear R.	55-199	83	3.19	
E60 IL	2 ponds		12	3.25	-
S278 IA	Little Wall L.	21-100	305	3.26	3.16-3.40
		101-180	111	3.96	3.70-4.33
C38 IA	Little Wall L.	45-180	380	3.46	
W142 IL	Little Grassy L.		8	3.52	-
H264 MO	L. of the Ozarks	-	1,078	3.6	2.7-4.1
C36 IA		42-152	30	3.66	2.55-5.12
W131 UT		26-145	402	3.73	-
B368 AZ	Pena Blanca L.	93-178	195	3.87 ± 0.52	
D179 TX		54-149	17	4.01	2.92-5.06
W61 IL	Thompson's L.	47-111	43	4.18	-

		Length	No.	K(TL) Mean	Range
S581 WI	ponds	70-90	+	1.64	1.48-1.84
J96 OK	Rod and Gun Club L.	-	+	1.68	-
S472 AL		51-220	9,187	1.87	1.68-2.02
M499 IL		76-165	21	1.94	1.30-2.57
H392 AL	before treatment	-	+	2.32	2.22-2.41
	after Silvex treatment	-	+	1.88	1.63-2.19
	before treatment	-	+	1.93	1.80-2.00
	after Silvex treatment	-	+	1.68	1.55-1.77
R125 IA	Ike L.	76-138	175	1.96	
		140-190	20	2.18	
F105, S225, S226, S229, S276,					
N115 SD		100-180	59	1.99	1.92-2.44
B150 IA	Little Wall L.	76-145	3	2.02	1.80-2.13

Condition factors increased with increase in length in M499, H264, W131, S278, and H392, and the slope of the weight-length regressions also indicate increases in condition factors with length in J44, H419, R125, W131, L110 and J34. In Lake of the Ozarks (H264), such an increase was not evident when mean K's were highest in spring and early summer and in September-October, but it was evident when mean K's were low, as in July-August. Reduction of the population density had little effect on the K of green sunfish in Rod and Gun Club Lake, OK, but the green sunfish standing crop had increased with the general reduction of population (J96). The average K was 1.84 in an unheated pond with low population density compared to 1.68 in a heated pond with a higher density (S581). When population densities were equalized there were no differences between the K in the heated and unheated ponds. The condition factors were lower after treatment of two lakes with Silvex (H392).

The mean maximum depth at various lengths were determined (L189, L226) to estimate the sizes which could be taken by predators with various mouth sizes.

Mean TL

| 15 | 25 | 35 | 45 | 52 | 65 | 76 | 85 | 95 | 103 | 115 |

Mean maximum depth

| 3.3 | 8.1 | 10.1 | 12.8 | 15.7 | 19.7 | 22.7 | 27.4 | 30.5 | 32.7 | 37.6 |

No. measured

| 4 | 80 | 43 | 36 | 166 | 57 | 145 | 33 | 22 | 75 | 6 |

Mean TK

| 126 | 140 | 153 | 165 | 181 |

Mean maximum depth

| 43.1 | 49.2 | 52.1 | 57.0 | 66.0 |

No. measured

| 28 | 14 | 14 | 1 | 2 |

L226   TL = 11.7 mm + 2.6859D

The mean volumes of green sunfish in relation to the length of the last 5 vertebrae were determined, so that sizes of fish eaten could be determined from the remains including the 5 vertebrae (B368).

Length of 5 terminal vertebrae	2-3	3-4	4-5	5-6	6-7
Fish volume, in ml	1	1.5-3.5	3.5-7.0	7.4-11	9.5-13
No. examined	2	4	4	3	3

	No.	Mean	Range	No.	Weight Mean	Weight Range
Age 0-June						
S581 WI ponds	+	-	5-9	-	-	-
F64 OK Little R.	44	48	25-61	-	-	-
J44 OK	7	69	-	-	-	-
Age 0-July						
S581 WI pond	+	-	13-29	-	-	-
H264 MO L. of the Ozarks	3	51	46-58	3	1	-
J44 OK	11	76	-	-	-	-
Age 0-Aug.						
S581 WI pond, unheated	+	28	19-35	+	0.3	-
S581 WI pond, heated	+	40	25-47	+	1	-
W105 IL Crab Orchard L.	3	30	20-41	-	-	-
H264 MO L. of the Ozarks	105	48	30-69	105	2	-
J44 OK	10	86	-	-	-	-
Age 0-Sept.						
S581 WI pond, unheated	+	32	21-41	+	0.5	-
S581 WI pond, heated	+	43	30-56	+	1.3	-
C13 MI Deep L.	10	38	30-43	-	-	-
S225 SD Ft. Fandall L.	95	41	-	-	-	-
L108 IL Big C.	3	57	-	-	-	-
H264 MO L. of the Ozarks	44	61	41-79	44	6	-

	No.	Mean	Range	No.	Weight Mean	Range
Age 0-Oct.						
S581 WI pond, unheated	+	33	-	+	0.6	-
S581 WI pond, heated	+	43	-	+	1.3	-
C353 OK pond	63	25	±1.8	-	-	-
R32 IN	+	26	-	-	-	-
R38 IN ponds	600	45	31-58	-	-	-
R38 IN pond	52	63	39-79	52	3	-
H264 MO L. of the Ozarks	7	64	51-76	64	6	-
Age 0						
H90 MI	86	24	10-41	-	-	-
H265 IL	100	36	26-51	-	-	-
T44 OK	1	56	-	1	2	-
H264 MO L. of the Ozarks, Nov.-Dec.	19	61	36-86	19	4	-
B416 IN White R.	8	72	-	8	25	—
L126 CO 2 Butte L.	10	104	89-130	-	-	-

	No.	TL Mean of means	Central 50%	Range	No.	Weight Mean	Range
Age I							
NY: B7	87	50	-	38-95	-	-	-
CA Pardee L.: K61	+	55	-	-	-	-	-
MI: C13, H90, H93, H94	347	57	51-61	25-86	-	-	-
KS, MO, IL, IN rivers: M215, L108, P74, L110, B416	92	69	67-79	28-84	69	23	11-35
UT: W340, W131	116	73	-	64-75	-	-	-
IA Ike L.: R125	226	79	-	-	-	-	-
MO, IL, KS, IN lakes & ponds: R71, H332, L129, H264, H265, C285, R32, B86, B97	951+	87	69-95	36-165	490+	15	3-57
OK rivers: F64, J25	549	94	-	51-140	57	14	-
OK statewide: H419	+	97	-	46-165	+	14	-81
OK lakes: O37, J25, T44, J26, J42, H303, H150, P207	36+	100	86-109	56-147	23+	33	5-71
CO: P202, L126	19	118	-	86-190	1	20	-
Combined	2,423+	80	61-95	25-190	640+	19	3-81
Age II							
NY: B7	87	65	-	53-75	-	-	-
MI: C13, H90, H94	182	86	81-92	56-140	-	-	-
MO L. of the Ozarks: H264	348	97	-	64-127	348	17	11-37
CA Pardee L.: K61	+	98	-	79-116	-	-	-
KS, IL, IN, MO rivers: M215, P74, L108, L110, B416	65	103	99-117	53-117	31	36	28-44
UT: W131, M340	105	105	-	102-117	-	-	-
IA: I10, R125	498	109	-	55-200	4	32	18-41
KS, IL, OH, IN lakes & ponds: W105, C226, H332, L129, H265, C285, B86, R32, R38, T145	227+	111	99-131	66-183	-	-	-
OK lakes: B183, J42, O37, J25, T44	86+	140	127-160	76-173	68	62	32-102

		TL				Weight	
	No.	Mean of means	Central 50%	Range	No.	Mean	Range
Age II (cont.)							
OK rivers: F64, J25	432	141	-	102-193	148	43	-
OK statewide: H419	+	152	-	91-229	+	62	12-230
CO: L126, P207	11	154	-	114-216	1	40	-
SD Gavins Pt. L.: S276	1	168	-	-	-	-	-
Combined	2,042+	113	97-135	53-229	600+	47	11-230
Age III							
NY: B7	55	76	-	67-87	-	-	-
MI: C13, H90	92	111	-	72-147	-	-	-
CA Pardee L.: K61	+	118	-	-	-	-	-
MO L. of the Ozarks: H264	179	122	-	94-157	179	37	23-85
UT: M340, W131	76	122	-	114-145	-	-	-
IL, IN, KS, MO rivers: L108, L110, M215, P74, B416	86	126	115-137	79-155	13	67	62-68
SD: S276	1	130	-	-	-	-	-
IN, OH, IL, KS lakes: C336, H332, L129, H265, B86, G39, T145	63+	128	110-137	191-216	-	-	-
IA Little Wall L.: S278	40	144	-	127-172	-	-	-
OK lakes: B183, J42, O37, W7, J25, T44	49+	159	137-178	114-180	44	78	59-111
CO lakes: C208, L126	22	167	-	147-213	-	-	-
OK rivers: F64, J25	168	185	-	152-224	34	99	-
OK statewide: H419	+	185	-	124-239	+	117	32-266
Combined	831+	137	118-165	67-239	270+	74	23-266
Age IV							
NY: B7	38	87	-	75-102	-	-	-
MI: C13, H90	30	131	-	86-163	-	-	-
UT: W131	36	137	-	-	-	-	-
MO L. of the Ozarks: H264	48	140	-	124-157	48	54	48-65
KS, OH, IL lakes and ponds: C226, W142, H332, L129, H265, B86, T145	33+	143	127-157	118-160	-	-	-
MO, IL, KS rivers: L108, L110, M215, P74	21	167	152-183	137-202	7	111	-
OK lakes: J42, O37	21	176	-	152-211	20	118	-
CO lakes: L126, C208	4	197	-	193-218	-	-	-
OK statewide: H419	+	201	-	155-249	+	151	66-304
OK rivers: J25, F64	55	204	-	157-234	14	125	-
TX: C112	+	229	-	203-254	-	-	-
Combined	149+	156	137-180	75-254	34+	127	66-304
Age V							
NY: B7	9	104	-	89-117	-	-	-
CA Pardee L.: K61	+	131	-	-	-	-	-
UT: W131	19	148	-	-	-	-	-
MI: C13, H90	17	152	-	117-185	-	-	-
IL, KS, MO lakes: C226, H264, H332, W105, W142	14+	174	157-188	145-221	5	65	48-96
IL, KS, MO rivers: L110, M219, P74	8	180	-	152-194	4	136	-
OK: F64, O37, H419, J25	8+	221	-	175-267	6+	196	112-380
Combined	75+	171	148-188	89-276	15+	149	48-380
Age VI							
NY: B7	1	138	-	-	-	-	-
MI: H90	11	162	-	132-185	-	-	-

| | | TL | | | | Weight | |
	No.	Mean of means	Central 50%	Range	No.	Mean	Range
Age VI (cont.)							
UT: W131	4	176	-	-	-	-	-
IL, KS: L110, L129, C226	3+	224	-	188-236	2	224	-
OK: F64, O37	2	241	-	224-259	1	200	-
Combined	21+	191	158-224	132-259	3	216	200-224
Age VII							
MI: H90	2	188	-	180-196	-	-	-
Age IX							
NY: B7	1	255	-	-	-	-	-
Age V + ?							
UT: M340	1	279	-	-	-	-	-

	No.	1	2	3	4	5	6	7	8	9	10
C65 OH Armstrong L.	+	61	69	79	86	119	163				
H93 MI Wiards P.	20	33	65								
H265 IL	40	33	66	90	125						
A87 AR Bull Shoals L.	98	46	71	91	109	122					
K33 MN	184	43	74	99	127	150	168	183			
P113 MO statewide	1,390	41	81	114	140	157	170				
Headwater stream	615	41	76	107	132	147	145				
Middle stretches	515	38	79	114	142	163	168				
Downstream	260	46	94	127	147	160	190				
Poorest station	+	33	61	89	114	132	145	157	137	178	190
Best station	+	51	99	142	160	193	203	188	173	183	196
P74 MO Black R., Upper	52	43	84	117	142	183					
Middle & lower	30	41	69	97	124	165					
P112 MO Salt R.	232	43	81	119	150	193					
H264 MO Salt R.	+	43	81	119	150	193					
White R. Upper	+	38	71	107	130	142					
Middle	+	43	84	124	160						
Lower	+	48	97	142	165	178	203				
Merimac R. Upper	+	38	69	91	117	132	145	157	137		
Station 4	+	51	86	114	135	142					
Middle	+	41	81	122	160	180	203				
Lower	+	46	99	130	132	157					
I10 IA Snyder L.	4	56	94								
J27 OK Spavinaw L.	2	51	107	124	145						
J31 OK Grand L.	6	56	89	127	168						
J42 OK Ft. Gibson L.	23	53	94	130	157	183					
J37 OK Rod & Gun Club L.	12	66	107	135	155						
L129 IL statewide	1,312	86	109	137	160	185	224				
R54 OH	+	58	109	142	160	173					
R125 IA Ike L.	289	33	84	145							
J25 OK Illinois R.	258	53	102	145	175	208					
J45 OK Verdigris R.	15	69	112	152	193						
tributaries	76	56	104	142	175	183					
S391 OK Rock C.	35	76	117	152	173						
B86 IL Onized L.	95	51	107	155	150						
F62 OK Sub Prison L.	7	61	112	152	185	213	229				
restocked	32	89	168	226							
H150 OK Illinois R.	388	58	117	157	175	208					
E126 OK Rocket Plant L.	27	91	137	165	190						
E61 OK Salt C.	21	81	132	170							
F87 OK Little R.	119	69	132	185	229	249					
tributaries	743	56	114	163	198	218					

Calculated TL at each annulus

	No.	1	2	3	4	5	6	7	8	9	10
					Calculated TL at each annulus						
F87 OK Little R. (cont.)											
cutoff lakes	49	51	117	168	196						
J44 OK statewide	1,656	97	152	185	201	226	229				
slowest mean	+	46	91	124	155	183					
fastest mean	+	165	229	239	249	267					
20 reservoirs	476	99	163	178	203	239					
9 large lakes	128	91	142	170	201	239					
33 small lakes	245	97	150	183	190	213	229				
38 ponds	362	102	160	201	229	239					
10 streams	445	81	132	168	185	213					

		1	2	3	4	5	6	7
				Calculated TL and increment at each year of life				
L108 IL Big C.		33	66	89	183			
	Incr.	33	36	22	20			
	No.	41	36	26	1			
W131 UT		51	81	102	122	137	163	
	Incr.	51	31	26	23	21	21	
	No.	351	236	133	59	23	4	
H264 MO L. of the Ozarks		51	84	109	135	152		
	Incr.	51	30	28	28	25		
	No.	901	579	231	32	4		
W61 IL Thompson's L.		50	94	112	145	169		
	Incr.	50	56	33	30	20		
	No.	59	30	13	2	1		
S278 IA Little Wall L.		36	81	114	196	231		
	Incr.	36	47	30	58	20		
P71 Mo Clear water L.	Incr.	51	38					
E60 IL 2 ponds	Incr.	44	48	39	42			
	No.	21	19	5	2			
S255 SD Ft. Randall L.		43	102	119				
before impoundment	Incr.	30	51					
	No.	34	8					
after impoundment	Incr.	71	76	36				
	No.	17	26	8				
S229 SD Gavins Point L.		38	99	124	150	175		
	Incr.	38	61	33	28	28		
	No.	17	17	9	2	1		
S226 SD Ft. Randall L.		46	109	127				
	Incr.	46	63	41				
	No.	22	22	7				
O37 OK Heyburn L.		46	84	127	155	165	193	
	Incr.	46	41	48	30	15	20	
	No.	142	122	62	22	2	1	
L110 IL creeks		46	89	130	160	185	218	
	Incr.	46	43	39	29	25	19	
	No.	78	42	25	13	6	2	
B183 OK Canton L.		81	140	163				
	Incr.	81	64	41				
F64 OK Little R.		53	114	165	201	229	249	
	Incr.	53	61	46	33	25	30	
	No.	954	462	178	44	3	1	

		Calculated TL and increment at each year of life						
		1	2	3	4	5	6	7
C271 OK Canton L.		81	140	170	180			
	Incr.	81	64	30	18			
	No.	72	33	16	2			

		Mean calculated weight at each annulus						
		1	2	3	4	5	6	7
J44	OK: statewide	14	62	117	151	222	230	
	slowest	-	12	32	66	112		
	fastest	81	230	265	304	379		
R125 IA	Ike L.	0.5	11	67				

	Mean increments in weight in each year						
	1	2	3	4	5	6	7
L110 IL creeks	1	10	26	34	44	42	

F6 reports that an individual lived 7.5 years in captivity.

Evidence that the scale method was validly applied was presented in H90 and S278. Both papers reported some false annuli and indicated that some of these false annuli may be spawning marks. Most calculated growth was on a direct proportion basis, but Fraser-type corrections were used by C271, H264, J25, P74, P112, R125, S278, W61, and W131. The body-scale relationship of 20 green sunfish 17-46 mm gave an intercept of 11 mm (C353). The body was fully scaled at 17 mm and cteni formed at 22 mm (C353). Annuli were formed from early April to early July in L. of the Ozarks, with younger fish forming annuli before the older (H264). Growth compensation and Lee's phenomenon (believed to be the result of higher mortality of faster growing fish) were reported (H264).

The data tabulated above indicate slower average growth in NY, MI, MN, CA, and UT but rapid growth in OK and CO. Growth was more rapid in 2 rivers than in lakes (H264), but J44 reported stream growth slower than in lakes. Growth was more rapid in the main river than in tributaries in the Verdigris R. (J45) and Little R. (F87). It was also faster downstream than upstream in the Salt and Merimac R. (H264), but the reverse was reported for the Black R. (P74). Growth was possibly a little faster in clear than in turbid ponds (H332). Growth was faster in southern than in northern Michigan, where it was associated with the longer growing season (H90). The growing season was reported as April to October in Missouri (H264). Four yearling green sunfish in March were twice the length of the average size the previous October in an OK pond (C353), indicating winter growth. Growth was faster at 28°C than at 20.9°C and faster at both temperatures than at 13.2°C (J114). Growth was faster in an unheated pool than a heated pool, but the population density was low in the first because of winterkill (S581). When population densities were equalized, growth of adults did not differ in the unheated pool from the pool maintained at about 5°C higher temperature. Young of the year grew faster in the heated pool. In experiments young grew faster at 25°C than at 20°C, and this was even more evident when the tanks were well aerated (S581).

Males grow more rapidly than females (C285, H90, H264, S278), but the differences may be slight. Most of the older fish were males (H264).

Growth was more rapid after poisoning and restocking (F62) and after an almost complete winterkill (S278). There are circumstantial evidence that the slower growing individuals more successfully survived winterkill conditions (S278). Many populations showed evidence of stunting with overcrowding.

Green sunfish grew faster at 2 ppm of ammonia than did controls where ammonia was minimal (J114). When exposed to over 2 ppm, there was an initial decline in feeding and growth; and growth rates declined as ammonia concentration were increased to 24 ppm. Cadmium levels of 3 ppm or more slowed the growth of green sunfish, and growth decreased with increase in cadmium levels (J114). An LC_{50} value of 20.5 ppm Cd was obtained with continuous flow.

Growth of hyperthyroid fish was faster than that of hypothyroid fish. Food consumption and efficiency of conversion increased with thyroid activity (G122). Thyroid activity was increased by shortening the photoperiod or by injecting thyroxine.

Maturity occurs at age I in MO (H264), IL (D178), and IA (S278) but not until age III in MI (H90). Mature males are reported as small as 76 mm (C271) and 45 mm (S278) and females at 66 mm (C271) and 6 to 10 g (S529). A color mutant, referred to as the Texas golden green sunfish, reaches sexual maturity in less than 6 months, and being golden in color, is particularly susceptible to predation and may be a good forage fish (W250).

Long photoperiods (15-16 hrs.) and raised temperatures (20°-24°C) caused gonadal development of both male and female green sunfish in the winter, but neither photoperiod nor temperature alone were effective (K157). Injections of extracts of carp pituitary stimulated gonad growth in males and females in the winter at temperatures of 21°C but not at 10°C (K156). The gonads regress rapidly at temperatures of 24°C or higher (K158). Green sunfish reproduced where alkalinity fluctuated from 400 to 960 mg/l (M552). Green sunfish spawn from June to August in IL (H8), MD (H264), MI (H90), and IA (S278); from May to August in IL (C354); in late June through July in KS (M215) and in CA (S366) at temperatures of 22°-26°C. Nesting begins in May (S529, R285). The first hatch is when water temperatures at 15 cm are 20°-21°C (L189, S334). The male attends the nest, a circular cleaned area with eggs attached to gravel or peasized clay lumps (H8). Males may be readily caught off nests (D178). Males make a grunting sound while courting the female (G191).

Nests are frequently in large colonies in water less than 40 cm on gravel with maximum sunshine (M526, H436) or in water 15-25 cm deep (C388). Males may remove larger pebbles by mouth in clearing the next area, about 31 cm in diameter (C388). Courtship and parental care are described (H436). Ripe individuals from stunted populations were collected 2-4 weeks later than from unstunted populations (C354). Females are reported to produce 15,000 to 50,000 eggs, but the basis for this estimate was not stated (C388). Sperm survived, on the average, less than 1.0 minute, and a 50 percent reduction in fry viability resulted when eggs were held 47 minutes after release before fertilization (C354). These are shorter times than for warmouth and bluegill eggs in the same study. Eggs hatched in 50 hours at 23.8°C, 31.5 hours at 27.1°C, and 29.1 hours at 27.6 C; and the fry at hatching averaged 4.79, 4.72, and 4.64 mm respectively (C354). Eggs are 1.0-1.4 mm in diameter, and larvae were free swimming at 4.2-4.7 mm, two days after hatching (M525). At 9 days they were 5.4 mm; at 30 days, 10.2 mm; at 37 days, 15.1 mm, and at 57 days, 23.7 mm (M525).

Young green sunfish eat zooplankton (S581). The green sunfish has a larger mouth than most other sunfish of the same size and eats larger food items. A 97 mm sunfish had eaten a 51 mm carp, and a 168 mm sunfish had eaten two 50 mm crappies, two 50 mm gizzard shad, and one other fish (C271). The mouth gape (Y) was described in (W249) as:

$$Y = -2.635 \text{ mm} + 0.178 \text{ SL, with } r = 0.99.$$

Green sunfish eliminated *Gambusia affinis* and *Gasterosteus aculeatus* from a pool in a stream as water levels decreased, and 97.6% of the food of the green sunfish consisted of these two species in this situation (G192). Green sunfish ate more largemouth bass eggs and fry than other sunfish, but the numbers taken were not significant after mid-May (M524). Green sunfish were competitive with largemouth bass in feeding on invertebrates in the winter, but the greens, in turn, were the major food of the bass (B368). The food items were more diverse than those of bluegill or longear sunfish in a KY stream (M404). Crayfish often were important food items (M404, C271, M336, A87). One green sunfish had eaten a bat, *Tadarida mexicano* (J77). Green sunfish consistently released blister beetles when fed them (T134). Insects were the principal food in many waters (M523, M340, M336, E135). Terrestrial arthropods were significant food for stream populations (M404) in a new reservoir (M523, M524) and in lakes (E135).

Food conversion, with *Gambusia* as forage, was more efficient at 28°C than at 20.9°C and more efficient at both temperatures than at 13.2°C (J114). There was also evidence that the green sunfish became more efficient in utilization of *Gambusia* with time. Ammonia concentrations of over 2 ppm caused reduced food consumption at first, but as the green sunfish acclimated to higher concentrations, food consumption became normal. The LC_{50} value was 33 ppm of ammonia as N (J114). Green sunfish would tolerate a pH change from 7.2 to 9.6 or 8.1 to 6.0, at 17°-19.5°C, with 4-8 ppm O_2 (W237). A 10 g green sunfish utilized about 33% of the absorbed protein for growth as compared to 20% for a 55 g sunfish (G170). At 77 mm, 0.088 g of protein is needed to add 1 mm; at 87-99 mm, 0.148-0.159 g of protein; and at 138 mm, 0.266 g protein (G170).

Green sunfish usually remain hidden under available cover until some food item appears (D178, S529). Silt turbidity up to 14-16 Jackson Turbidity Units did not seem to affect feeding or attack behavior but affected social hierarchy and increased scraping behavior (H434). Green sunfish appeared to home more when displaced than did bluegills and pumpkinseeds (K139) and largemouth bass (H433).

Water dropping into a pond, as from a discharge pipe, stimulated green sunfish to jump. Green sunfish, 71-97 mm TL, could jump up to 10 times their body length, and jumping occurred more in the day than at night and increased with water flow (E154).

W247 OK ponds	Standing crops kg/ha	Net production kg/ha	Production × 100 / Standing crop
	0.43	0.2	51
	37.5	33.2	89
	73.5	45.4	62
	99.5	70.6	71

Preferred temperatures of green sunfish acclimated at various temperatures were (C396):

Acclimated	6	9	12	15	18	21	24	27	30
Preferred	15.9	18.2	21.1	22.7	25.2	28.1	30.4	30.7	30.6
95% CI	14.7-18.8	17.0-20.5	19.3-22.1	21.5-23.9	23.5-25.8	25.4-27.8	27.2-31.0	28.8-32.3	30.5-34.6

The final preferred temperature was 28.2°C but the width of the preferred range of individual green sunfish varied from 2.8°-5.1°C (B424). Preferred temperature did not differ with time of day. The greatest activity was at dawn and dusk (B424).

Green sunfish are easily handled and make good bioassay animals (W214). The hematocrit readings of 143 green sunfish ranged from 21%-53% and averaged 37%; the hemoglobin of 38 ranged from 6.0-10.0 gram% and averaged 8.0; and the blood plasma protein of 59 was from 4.0 to over 12.0 gram% and averaged 8.0 (H435). They survive at temperatures of 33°-34°C (S463) and 36°C (P208).

The chromosome numbers were 46 in cell cultures and 48 in testes (R286).

PUMPKINSEED, *Lepomis gibbosus* (Linnaeus)

The pumpkinseed is widely distributed in southern Canada, the upper Mississippi River system, the Great Lakes region, and south along the Atlantic Coast to Georgia (R285). In general it lives in cooler waters than other members of the genus. It has been widely introduced elsewhere in the United States and Europe (V92, W245). They have not become established in southern California, although stocked there (H437). Pumpkinseed prefer quiet clear water where the bottom has some organic debris and aquatic vegetation (R285), and they tend to inhabit larger and denser vegetation than do bluegills (B5, T113).

	Length, range	No.	FL/SL	TL/SL
C37 MN	70-129	155	1.146	1.196
	130-189	218	1.129	1.181
C30 IA	-	31	1.16	1.22
B30 MI	x-94	843	-	1.259
	95-x	1,460	-	1.239
S289 MA	40-199	1,169	-	1.278
D72 IA	38-165	150	1.193	1.247

Where data were converted from SL to TL a factor of 1.25 was used; and from FL to TL, a factor of 1.04.

TL	No.	Weight in grams Mean of means	Central 50%	Range	Citations
35-50	18	1.95	-	1.9-2.0	B222, B30
51-75	289	5.3	5-7	1-10	B30, B222, C30, D72, K63
75-101	326	14.8	14-16	8-27	B30, B222, C30, D72, H61, K63, M20
102-126	544+	34	28-41	14-54	B30, C30, H61, M20, S62, R101, K63, D72, M499
127-151	774+	63	50-68	36-117	as above + C20, D68, B222
152-177	648+	99	90-108	60-144	as above + R24, H62, N52
178-202	337+	157	138-173	113-193	R24, S62, B30, M20, M499
203-228	187+	216	196-233	181-241	S62, C20, B30, D68, M20
229-253	27+	256	235-293	42-326	R24, C20, S62, B30, M20, E105
254-278	6	227	-	-	C20
279-304	2	397	-	-	C20
305-329	1	454	-	-	C20
330-355	1	340	-	-	C20
381-405	1	454	-	-	C20

			No.	Length	
D72	IA	Clear L.	150	38-165	Log W = -4.740 + 3.1856 Log SL
B30	MI		+	-	Log W = -4.789 + 3.1986 Log SL
K63	CA	Lower Susan R.	+	-	Log W = -6.232 + 3.332 Log FL
S188	RI		+	-	Log W = -0.8569 + 3.070 Log TL
C355	PA	Alanconnie L.	629	30-221	Log W = -5.213 + 3.262 Log TL
P122	WI	Flora L., 1952	17	-	Log W = -4.990 + 3.123 Log TL
		1956	17	-	Log W = -10.878 + 5.912 Log TL
E136	MT	Horseshoe L.	331	55-200	Log W = -5.2231 + 3.238 Log TL

				K(SL)	
	Length	No.	Mean	Range
E20 MI Howe L.	yearlings	16	3.09	-
	adults	11	4.68	-
B30 MI	-	+	-	3.29-4.66
C34 MN Leech L.	70-189	69	3.98	-
C36 IA	56-135	11	4.02	3.41-4.70
C23 MN 1941-43	70-209	502	4.06	-
D72 IA Clear L.	54-135	134	4.32	3.78-4.76
E3 MN 1936-41	76-329	123	4.98	-
W25 MI	-	48	5.14	-
F97 MI Wintergreen L.	angling	32	5.23	-
May-Aug.	netting	172	5.01	-
C23 MN standards	poor			<3.1
	average			3.6-4.5
	excellent			>5.0

				K(TL)	
	Length	No.	Mean	Range
E136 MT Horseshoe L.	51-203	+	1.82	1.59-2.60
B30 MI	-	+	-	1.64-2.44
M499 IL	100-180	154	2.18	2.10-2.47
D72 IA Clear L.	67-134	+	2.23	1.94-2.47
C23 MN standards	poor			<1.88
	average			2.13-2.77
	excellent			>3.02

The condition factors increased with length (B30, B222, E20, D72, M499), and the weight-length relationship was above 3.00 in each of the populations studied. Fish caught by angling had a slightly higher average K than fish caught in nets (F97).

	No.	Mean TL	Range	No.	Mean W	Range
Age 0-July						
R22 NK Chamcook L.	103	15	-	-	-	-
P191 NY Hudson R.	57	17	10-24	-	-	-
G32 OH Buckeye L.	+	26	23-30	-	-	-
Age 0-Aug.						
P191 NY Hudson R.	167	24	10-44	-	-	-
H137 MN Deming L.	382	35	24-46	382	0.6	-
C284, G32 OH	+	38	36-41	-	-	-

	No.	Mean TL	Range	No.	Mean W	Range
Age 0-Sept.						
R22 NK Chamcook L.	+	25	-	-	-	-
C13 MI Deep L.	36	41	20-46	-	-	-
L132 MN Linwood L.	138	43	-	-	-	-
G32 OH Buckeye L.	+	43	-	-	-	-
Age 0-Oct.						
B93 MI N. Twin L.	183	38	-	-	-	-
L132 MN Linwood L.	23	43	-	-	-	-
K82 MI Loch Alpine P.	57	51	33-71	-	-	-
B4 IA Clear L.	3	58	54-60	-	-	-

		TL				Weight	
	No.	Mean of means	Central 50%	Range	No.	Mean	Range
Age 0							
MI lakes: B31, C72, E20, H93, H94	2,480	36	28-46	23-61	20	2	-
MD lakes: B218, M306	70	41	30-61	-	-	-	-
E9 NY	+	41	-	-	-	-	-
S289 MA ponds	1	56	-	-	-	-	-
Age I							
K63 CA Lower Susan R.	+	25	-	18-28	-	-	-
P192 Yugoslavia, Danube R.	51	43	-	35-54	-	-	-
R22 NK Chamcook L.	+	56	-	36-76	-	-	-
MA, NY: B7, B222, S289, E9	176+	62	54-74	41-88	69	4	3-6
MI, MN: B75, B77, C13, C72, E20, H93, H94, H137, B31, K82	1,957	67	56-80	28-131	794	8	1-12
MD: B218, M306	3+	69	-	64-79	-	-	-
E136 MT Horseshoe L.	40	71	-	-	-	-	-
R200 L. Erie	46	79	-	61-109	46	12	5-25
IA, IN, OH: D72, G39, O8, R34, W220	223+	103	100-119	56-139	-	-	-
S120 AL	-	-	-	-	+	31	-
Combined	2,496+	70	56-81	18-139	929+	8	1-31
Age II							
CA Lower Susan R.: K63	+	69	-	43-119	-	-	-
NK Chamcook L.: R22	+	76	-	58-94	-	-	-
MD: B218, M306	8+	89	-	71-109	-	-	-
MA, NY: B7, B222, G25, E9, S289	214+	91	75-102	60-152	16	6	3-7
MT Horseshoe L.: E136	78	99	-	-	-	-	-
MI, MN, WI: B93, B75, C72, C13, H94, B77, B31, E20, M20, E3, W25, C117, C213, H137, K82	1,149+	103	94-112	54-190	623	36	8-91
OR: O34	+	104	-	-	-	-	-
Yugoslavia: P192	54+	104	-	85-132	54+	38	13-55
IA, IN, OH: D72, G39, O8, R34, T145, U1, W220	229	121	119-137	76-185	8	14	-
L. Erie: R200	17	117	-	99-142	17	34	23-57

		TL				Weight	
	No.	Mean of means	Central 50%	Range	No.	Mean	Range
Age II (cont)							
AL: S120	-	-	-	-	+	45	-
Combined	1,749+	103	91-112	43-100	718+	30	3-91
Age III							
NK Chamcook L.: R22	+	95	-	81-109	-	-	-
MT Horseshoe L.: E136	13	110	-	-	-	-	-
OH ponds: T145	54	111	-	107-119	-	-	-
MA, NY: B7, B222, G25, S289, S372	279	115	92-114	65-176	20	21	6-86
CA Lower Susan R.: K63	+	117	-	81-145	-	-	-
MD, NC: D68, D167, M306	11	121	-	104-130	-	-	-
Yugoslavia: P192	17+	127	-	111-150	17+	61	43-85
MI, MN, WI: B31, B75, C13, C72, C117, C213, H94, H137, K82, M20, W25	1,869	131	117-155	71-223	1,356	52	17-125
IN, IA: D72, G39, R34, U1, W220	77	133	97-152	97-185	11	20	-
L. Erie: R200	4	140	-	130-140	4	65	57-68
AL: S120	-	-	-	-	+	105	-
Combined	2,324+	124	109-136	65-223	1,408+	47	6-125
Age IV							
NK Chamcook L.: R22	+	114	-	112-117	-	-	-
OH ponds: T145	20	119	-	114-128	-	-	-
MA, NY: B7, B222, G25, S289, S372	261	124	99-140	66-175	13	37	6-77
CA, OR, MT: E136, K63, O34	25+	135	-	132-142	-	-	-
MI, MN, WI: B75, B31, C13, C72, C117, C213, C215, E3, H94, K82, M20, W25	1,154	141	120-160	83-286	815	77	23-150
Yugoslavia: P192	2+	143	-	130-160	2+	106	91-112
MD, NC, D68, D167, B218, M306	45+	144	-	117-152	-	-	-
IN, OH: O8, U1, W222	9+	149	-	115-189	3	34	-
L. Erie: R200	5	155	-	145-170	5	91	77-111
AL: S120	-	-	-	-	+	112	37-187
Combined	1,521+	137	116-155	66-286	838+	74	6-187
Age V							
NK Chamcook L.: R22	+	114	-	107-122	-	-	-
MA, NY: B7, B222, S289, S372	159	141	130-150	109-171	10	46	40-48
AL: S120	-	-	-	-	1	34	-
IN: U1, W220	4	142	-	115-169	2	34	-
MT Horseshoe L.: E136	52	153	-	-	-	-	-
MD, NC: D68, D167, M306	24	161	-	145-185	-	-	-
L. Erie: R200	2	168	-	165-170	2	116	11-122
MI, MN, WI: B31, B75, C13, C72, C117, C213, C215, E3, K82, M20, W25	656	171	157-180	112-286	485	95	37-181
Combined	897+	160	145-175	107-286	500	82	34-181

	No.	TL Mean of means	TL Central 50%	TL Range	Weight No.	Weight Mean	Weight Range
Age VI							
NK Chamcook L.: R22	+	123	-	119-127	-	-	-
AL: S120	-	-	-	-	1	40	-
MA ponds: S289	70	165	-	-	-	-	-
MT Horseshoe L.: E136	56	172	-	-	-	-	-
MI, MN, WI: B31, B75, C13, K82, C72, C213, C215, E3, M20	298	184	168-205	119-318	248	114	54-318
MD, NC: D68, D167, M306	8	192	-	165-221	-	-	-
Combined	429+	179	165-203	119-318	249	108	40-318
Age VII							
NK Chamcook L.: R22	+	132	-	130-135	-	-	-
MT Horseshoe L.: E136	18	186	-	-	-	-	-
MA ponds: S289	27	175	-	-	-	-	-
MI: B31, C13, C72, C213, C215, K82	88	194	170-203	147-241	79	133	91-147
NC: D167	1	254	-	-	-	-	-
CT: W13	1	274	-	-	1	482	-
Combined	135+	191	170-203	130-274	80	183	91-482
Age VIII							
NK Chamcook L.: R22	+	137	-	-	-	-	-
MA ponds: S289	7	196	-	-	-	-	-
MI: B31, C72	28	199	-	189-214	26	179	-

A pumpkinseed lived 12 years in captivity (F7).

	No.	1	2	3	4	5	6	7	8	9	10
					Calculated TL at each annulus						
P159 MT 5 lakes	73	28	53	79	94	112	114	112	185	196	
H93 MI Wiards P.	133	29	55								
G160, G161 OR 6 waters	55	38	94	114							
H94 MI Crystal L.	206	31	54								
M342 PA 14 waters	623	25	61	94	122	135	140	145	163	185	180
lowest mean	+	15	36	58	76	99	107	117			
highest mean	+	41	104	135	155	168	183	165			
P122 WI Flora L.	84	53	79	104	127	150	170				
D68 NC	54	34	76	107	130	152	198				
B218 MD	+	69	71	97							
R101 OH	+	38	76	112	140	157					
G161 OR 6 waters	55	39	94	118							
S188 RI 26 ponds	258	41	79	112	140	157	170	180			
K33, M258 MN	1,582	43	79	109	140	163	183	196	206	216	
C72 MI Houghton L.	573	38	75	119	166	198	215	225			
B134 CO Loveland L.	1	58	107	127							
S62 MN	+	46	104	130	165	196	244				
L129 IL	515	89	112	135	142	163					
G77 OR 5 waters	+	36	107	137							
L157 NC L. Waccamow	15	36	102	142	163						
H158 DE	+	69	130	155	173	185					
B25 MI	4,184	84	124	150	173	190	203	206	213	218	
D89 WI Murphy Flowage	+	99	142	168	183	193	211	196	216		

		Calculated TL and increments each year of life						
		1	2	3	4	5	6	7
S372 NY pond		32	64	101	127	146		
	Incr.	32	32	38	31	41		
	No.	13	13	13	11	3		
D72 IA Clear L.		56	107	137				
	Incr.	56	51	36				
	No.	328	132	19				
E136 MT Horseshoe L.		53	83	106	125	143	162	178
	Incr.	53	30	24	18	19	18	15
	No.	260	220	142	129	106	54	18
D167 NC Kitty Hawk		36	66	102	132	170	193	234
	Incr.	36	30	36	33	36	23	18
	No.	11	11	11	10	4	2	1
K82 MI Lower Loch Alpine		41	71	86	97	109	122	135
	Incr.	41	33	18	15	15	18	20
	No.	237	183	138	90	23	8	3

	Mean weight in grams at each annulus						
	No.	1	2	3	4	5	6	7
E136 MT Horseshoe L.	260	2	10	20	37	58	86	115

One of the first applications of the scale method to centrarchid fishes was with pumpkinseed in 1926 (C72). The body-scale relationship was then described as curvilinear, but most growth studies have assumed a direct proportion in body-scale growth. Correction factors of the Fraser-type were used in D72, S188, P122, E136, and K82. Lee's phenomenon was evident in the Lower Loch Alpine data (K82).

Regional differences in growth were not very consistent in the data tabulated above. Better growth in Catlin than in Rich Lake, NY (B222) was associated with the greater amount of vegetation and associated insects. In deeper waters of Oneida Lake, NY, some are reported to reach almost 300 mm, and in some populations in the same lake few exceed 100 mm at IV to VII years (R285). Females averaged larger than males in each age group I-IV (P192), but males averaged slightly larger than females at all annuli 1-7 (E136). Growth in calendar years showed significant correlation at 0.7 level with mean May-June temperatures, mean July temperatures, mean May-September temperatures, and total April-July rainfall, but not with length of growing season or rainfall in May or June (E136). Growth rates increased with temperature (tested at 5°C intervals from 5°-30°C) with the greatest relative growth rate during the first 2 weeks of each experiment (P206). Unfortunately, the fish tested at different temperatures were not at the same initial weights, which affects interpretation of relative growth rates. In 15 weeks the pumpkinseeds increased an average of only 0.6 g at 5°C, 9.0 g at 10°C, 22.9 g at 15°C, 36.2 g at 20°C, 53.7 g at 25°C, and 51.9 g at 30°C (P206). Average food intake also increased with temperature. Slow growth in Lower Loch Alpine was associated with it being an old reservoir (K82). No correlation with growing season length was apparent (K82).

Increased harvest resulted in faster growth in Alanconnie Lake, PA; and pumpkinseeds showed the best response to thinning, although they never accounted for more than 13% of the total weight of the population (C355). Growth of pumpkinseeds declined as the carp population increased, however. Relative

condition and growth indexes for the various years did not show much correlation for pumpkinseeds, although they did when all species were combined.
Heavy infection with *Neascus* sp. reduced pumpkinseed growth rate (K143).

Annual mortality rates were estimated at 88% in Spear L. (R152) and at
80% for ages II-III and 95% for ages III-IV in Muskellunge L. (R34).

Sex of adults can be recognized on the basis of color, fins, and opercular
flaps in spring, fall, and winter (C365).

Pumpkinseeds mature at age I in Ohio (M248) and Iowa (H342), and faster
growing males mature at age I in New York with others maturing at ages II
and III (R285). In Yugoslavia they mature at age II (P192). Adults averaging
about 90 mm and age III-VI were observed spawning in one New York pond,
and in a lake the nesting males were 70-130 mm (R285).

	Age	Number of females	Eggs per female Mean	Range
U1 IN	II	8	1,034	600-1,684
	III	11	1,491	1,108-2,366
	IV	3	2,422	
	V	2	2,436	1,095-2,923
P192 Yugoslavia	II	23	8,590	3,220-14,667
	III	11	13,404	8,432-17,366
	IV	2	25,765	22,358-29,172
	TL			
P192	73-90	10	6,432	3,220-9,948
	91-110	19	11,614	6,926-17,366
	111-120	4	13,504	10,621-16,121
	121-131	3	22,066	14,668-29,172
	Weight			
P192 Yugoslavia	16-20	6	6,175	3,220-9,646
	21-40	6	8,434	3,583-12,918
	41-60	17	11,603	6,926-17,366
	61-80	4	13,505	10,621-16,128
	81-100	2	18,513	14,668-22,358
	112	1	29,172	-
L189	-	-	3,000	-

Pumpkinseeds spawn sometime between early May and August, with nest
building starting when water temperatures reach 13 -17°C (C365) or 15°-18°C
(M528). In the Delaware R. nesting was from May 12 to August 11 at temperatures of 20°-29°C (S546); in Wisconsin, from late May to July at 19°C (J105).
Nest attendance began May 28, 1970, and June 5, 1969; first eggs were June 4,
1970, and June 9, 1969; and first larvae June 11, 1970, and June 15, 1969, in
Lake Opinicon, ON, about 2 weeks before bluegills (A102). Water warmed
earlier in 1970 than 1969. They spawn about 3 weeks earlier than bluegills
(C365), but duration of care of the nest by the parent is dependent upon water
temperature. Pumpkinseeds nest nearer shore and in shallower water than
bluegills and are not colonial nesters (C365). Nests were in water 18-69 cm
deep in the emergent plant zone, avoiding full shade (I17). Males drove 300-
460 mm chain pickerel off the nest area (I17). Males are dominant, even when
immature, in confinement. Interrenal tissue was found to be negatively correlated with the number of aggressive encounters initiated, indicating that the

less aggressive individuals are under more stress (E137). Nesting and non-nesting male pumpkinseeds responded more to females of the same species than to female bluegills or longear sunfish (K140). In other experiments nesting male pumpkinseeds showed no preference between female pumpkinseeds and longears, but the females showed a preference for their conspecifics (S583). Gonadotropin stimulates courtship and androgen also shows some effect (K160).

The male builds and guards a nest, eggs, and newly hatched fry. Nests are usually close together in sand and/or fine gravel. Spacing of 2 to 3 m is preferred if there is adequate spawning space (K140, S530). Areas clear of woody debris were preferred as spawning sites, but when breeding pressures were greater at midseason, areas with woody debris were also used (C387). Woody debris which might be used to designate territories did not increase nesting density. Nest diameters were 3.5 times the length of the male (C387) and were smaller in cluttered areas than clear areas. A review of several papers indicated that pumpkinseeds spawn in shallower water than bluegills and that they avoid dense shade but may nest in partial shade from overhanging trees. They also avoid pure mud substrates and prefer thin silt substrates on rock or gravel (C387). Gonadectomized male pumpkinseeds maintained high levels of aggression but did not dig nests (S530). Castrates dug nests when treated with testosterone. Photoperiod seems to control gonadal androgens, and sunfish kept on short photoperiods failed to build nests (S531). Injection with human chorionic gonadotrophin caused nest building in short photoperiod fish at 25°C but not at 11°-13°C. The sunfish school in relatively deep water in winter. Rising spring temperatures increase aggressive behavior. Increased photoperiod induces androgen activity, movement to shallow water, and nest building (S531).

Several thousand eggs may be deposited in a nest by one or more females (R285). Incubation is about 3 days at 28 C (B370). In Michigan (C6), 1,509 and 14,639 fry were recovered from two nests. Fry at hatching were 3 mm long (B72) and grew to 5 mm in 5 days. Constant bath treatment of eggs with 0.01 ppm malachite green gave a hatch of 59.9%, and untreated eggs only hatched 17.8% (M527). Hatching success reduced rapidly at higher concentrations, but survival of fry was better at 0.005 ppm. Embryonic development was described (B286).

Pumpkinseeds seem to occupy home ranges in the summer to which they return if released elsewhere (S532, K139, R290).

Young pumpkinseeds schooled loosely near the surface among emergent aquatic plants in water 10-50 cm deep. Adults were found in deeper water over rocky or plant covered substrates, rarely in aggregations of more than 3 or 4 (E153). Almost 5 times as many young pumpkinseeds were taken in trawls at night than in the day in L. Oneida, NY (F164). In laboratories they selected water temperatures of 31°-31.7°C (R289) with a final preferred temperature of 31.5°C (F161). Pumpkinseeds are not as well adjusted to temperatures at 30°C or up than bluegills (O49). Metabolic rates of smaller pumpkinseeds were affected by temperature change than those of larger fish (O49). Pumpkinseeds have survived with body temperatures as high as 35.6°C (T102). When acclimated at 18°C the mean lethal temperature was 28°C, when acclimated at 24 C, 30.2°C (B200), and when at 25°-26°C it was 34.5°C (B201).

When Mill L., MI, was opened after 5 years of no fishing, there were 58 pumpkinseed/ha; and in 3 days 29% were caught with 96 hours of fishing/ha, about twice the percentage of bluegills caught (S595).

Pumpkinseeds feed mostly on insects, crustaceans, and snails which they apparently crush with their flat teeth in the throat (R285, R200, P205, K62, F148, E135, K137, S464, G133, E121, S357). Even at 18-19 mm they were feeding on chironomids (N117). Pumpkinseeds consume more snails than other sunfishes (B15, P165, R288, E135, S464). Pumpkinseeds were cited as the most diverse and versatile feeders in a lake, with a wide range of foods and of depths of feeding (K125). Larger pumpkinseeds occasionally take fish (R285, T113). Fish had been eaten by 10 of 1,319 pumpkinseeds 75-220 mm in MN. (S438). Pumpkinseeds 150-230 mm were eating alewife in NJ (G97). When feeding they move forward towards the substrate in short quick darts from a head down, tail-up position. At dusk they move to the bottom and rest on the anal, pectoral, and pelvic fins (E153). Lower body temperatures of pumpkin-seeds captured at night than in daytime indicated that the pumpkinseeds were in the deeper, cooler waters at night (N141). Pumpkinseeds were less active in the dark than in lighted experiments (K159), and the pattern of movement differed with light intensity. Bluegills on nesting sites were quiescent at night (I17). Pumpkinseeds are largely diurnal in feeding (K135, R288). A major di-urnal shift in kinds of food was noted, as bottom-dwelling organisms are taken mostly when light intensities are low and they leave the bottom. Pumpkinseeds apparently do not feed during the winter, as the stomach is shrunken and mu-cous filled. Feeding did not start until water temperatures reached 8.5°C (K137). Greatest feeding was in early summer (S464). Laboratory feeding tests showed definite dominance and subordination (M374). Pumpkinseed readily feed in the laboratory (M528).

Digestion of food and evacuation of the stomach took 15-20 hours at 22.7°C (K141). Digestion of food was 50% complete in 4-6 hours (S464). Food con-sumption was estimated (H288):

Weight of fish	% of body weight/day 10°C	20°C
13-15 g	1.0-2.6	4.8-6.6
44	1.2-1.5	2.3-3.0
68-87	0.6-1.3	1.3-3.0

The number of chromosomes is 48 (R286) or 46 (F165).

In Flora L., WI, (P122) the numbers of pumpkinseed and bluegills over 70 mm were estimated to decrease from 823 to 54 per hectare or 3.6 to 0.2 kg/ha from 1952 to 1956. In Cranberry P., MS, (R290) there were 402-1,396 blue-gill/hectare.

WARMOUTH, *Lepomis gulosus* (Cuvier)

The 1970 list of common and scientific names of fishes (B371) placed this species in the genus *Lepomis*, downgrading *Chaenobryttus* to subgeneric rank. The warmouth is found in ponds, lakes, and occasionally streams from Kansas and Iowa to southern Wisconsin, lower Michigan, Lake Erie, and west-ern Pennsylvania south to Florida and the Rio Grande (L86). It is strictly a freshwater species (B399) in Florida but has been reported at 4.1 ppt salinity in Louisiana (C368) and at 1/8 ppt in Florida (K151). It has also been intro-duced west of the Rockies and to New York (R285). Maximum length is 284 mm (T113).

	No.	TL	
L32 IA Red Haw L.	72	50-221	FL = 1.179 SL TL = 1.248 SL
H200 IA Ahquabi L.	203	50-206	SL = -4.3mm + 0.8536 TL
L86 IL	102	under 100	TL = 1.259 SL
	93	102-175	TL = 1.240 SL
	41	over 176	TL = 1.211 SL
	264	combined	TL - 1.240 SL

Where it was necessary to convert SL to TL, a conversion halfway between those of H200 and L86 was used. Where FL to TL conversion was needed, the bluegill conversion factors were used.

TL	No.	Weight Mean of means	Weight Central 50%	Range	Citations
25-50	4,707	1.2	-	0.8-1.5	A39, J44, S472
51-75	2,101	3.5	2.1-5	0.3-20.0	A40, J44, L32, S472
76-101	663	11	9-13	3-32	A39, J44, L32, M499, S472
102-126	887	27	24-31	8-36	J44, L32, M499, S472
127-151	657+	46	36-49	14-113	A40, D68, J44, L32, M499, S472
152-177	505+	85	77-88	23-172	D68, J44, L32, M499, S472
178-202	248+	163	140-180	59-272	D68, D179, E105, J44, L32, M499, S472
203-228	58+	214	200-214	136-690	D68, J44, L32, M499
229-253	7	316	-	-	J44
274	1	458	-	-	J44

	TL	No.	
L32 IA Red Haw L.	50-210	76	Log W = -4.683 + 3.138 Log SL
L86 IL	81-208	866	Log W = -4.499 + 3.049 Log SL

	TL	No.	
J44 OK	41-274	1,113	Log W = -4.487 + 3.2089 Log TL
J95 OK Ardmore City L.	89-234	36	Log W = -4.605 + 3.319 Log TL
J38 OR Rod and Gun Club L.	127-215	16	Log W = -5.054 + 3.491 Log TL
H200 IA Ahquabi L.	53-206	203	Log W = -5.094 + 3.204 Log TL
S472 AL	51-215	3,860	Log W = -4.841 + 3.08 Log TL

	SL	No.	K(SL) Mean	K(SL) Range
D179 TX	101-167	14	3.61	2.45-5.04
W105 IL	-	9	3.43	3.04-4.31
E60 IL 2 ponds	-	5	3.54	-
L32 IA Red Haw L.	40-177	76	3.95	-
W61 IL Thompson's L.	24-108	26	3.96	-
W142 IL Little Grassy L.	104-176	11	3.77	-

	TL	No.	K(TL) Mean	K(TL) Range
H200 IA Ahquabi L.	53-206	161	2.30	-
J38 OK Rod and Gun Club L.	127-215	16	-	1.66-2.11
Chickassaw L.	127-240	78	-	2.05-2.27
statewide	51-215	3,860	-	1.94-2.30
L86 IL	81-208	866	-	2.01-2.29
S472 AL	76-215	1,791	2.11	1.74-2.36
M499 IL	76-210	64	2.05	1.80-2.24
J96 OK	-	+	2.00	1.72-2.11

The average K(TL) increased with increase in length in J38 and L86, and the slope of the log W-log L relationship was above 3.0 in all cases. Only the data on fish over 75 mm were included in the average K given for S472, since the K given at 25-51 mm was 7.80 and at 52-75 mm, was 1.62. The calculated W-L relationship given for S472 is probably also in some error. The average K(TL) increased to 2.11 (compared to 1.72 before) in Rod and Gun Club L. 2 years after 90% of the population was removed, but it showed little or no change in Franklin P. with the same treatment (J96). However, the population of warmouth was estimated at 134 kg/ha at 2 years compared to 3.4 before in Franklin P., but it was only 3.4 compared to 4.5 kg/ha in Rod and Gun Club L.

Warmouth from 84-107 mm showed wide seasonal variation in average condition probably because of their dependence on cladocerans and certain insects which varied in abundance (L86). The larger warmouth generally had a decline in condition in early fall and again in winter (L86). No consistent differences in condition were evident between males and females (L86, L32).

Age	No.	TL Mean	TL Range	Citation
Hatching	+	3.3	-	L86
	50	4.73	-	C354
48 hours	+	4.6	-	L86
4 days	+	5.3	-	L86
10 days	+	-	5.5-6.0	C39
0-June	20	61	51-74 (3g)	J33, J44 OK
0-July	4	71	-	J44 OK
0-Aug.	4	-	29-37	W105 IL Crab Orchard L.
0-Sept.	3	81	-	J44 OK
0-Oct.	1	165	- (94g)	V44 LA Springhill L.
0-Dec.	+	140	-	J95 OK Ardmore City L.
0	78	-	32-106	B218 MD Redington P.

	No.	TL Mean of means	TL Central 50%	TL Range	Weight No.	Weight Mean	Weight Range
Age I							
B233 OR L. of Woods	3	61	-	-	-	-	-
IA, IL, IN: B86, G80, H200, L32, L85, W220, L129	1,284+	82	74-107	51-128	66	17	-
OK: H127, J25, J41, J33, O37	61+	111	94-132	58-135	17+	14	5-21

		TL			Weight		
	No.	Mean of means	Central 50%	Range	No.	Mean	Range
Age I (cont.)							
S13 TN Reelfoot L	16	127	-	109-150	16	51	28-71
Combined	1,364+	97	74-127	51-150	99+	21	5-71
Age II							
OR: G75, B233	6	103	-	79-161	-	-	-
IA, IN, IL: B86, G80, H200, L32, L86, R34, W220, L129	705+	121	120-131	71-178	56	60	-
NC: D68, R217	4	123	-	122-124	-	-	-
OK: H127, J25, J26, J33, J41, O37, W7	30+	149	112-173	104-206	15+	53	28-173
S13 TN Reelfoot L.	44	160	-	140-175	44	99	48-136
Combined	789+	131	120-147	71-206	115+	65	28-173
Age III							
B218 MD Redington L.	+	91	-	-	-	-	-
C213 MI Sugar Loaf L.	3	117	-	99-130	-	-	-
OR: B233, G75, O34	12+	125	-	108-170	-	-	-
IA, IN, IL: B86, B94, G80, L32, L86, H200, R34, W61, W105, W220	465+	145	132-163	91-190	24	105	-
NC: D68, R217	22	148	-	137-157	-	-	-
OK: J25, J33, H127, W7	23+	177	157-190	140-198	17	81	74-85
S13 TN Reelfoot L.	254	178	-	160-196	254	142	85-241
V44 LA Springhill L.	1	190	-	-	1	119	-
Combined	780+	151	132-168	91-198	296	109	74-241
Age IV							
OR: B233, G75	3	140	-	122-175	-	-	-
MI: C213, C215	12	146	-	114-163	-	-	-
IA, IL, IN: H200, G38, L32, L86, R34, R40, W105, W61, W220	354+	165	154-180	122-208	14	147	-
D68 NC	32	178	-	-	-	-	-
S13 TN Reelfoot L.	130	196	-	170-221	130	187	133-284
OK: H127, J25, J26, J33, W7	11+	202	-	188-234	9+	156	136-179
Combined	542+	172	154-188	114-234	153+	164	133-284
Age V							
MI: C213, C215	17	168	-	155-185	-	-	-
OR: B233, O34	4+	175	-	161-190	-	-	-
IA, IL, IN: G80, H200, L32, L86, R34, W105, W220, S183	99+	177	173-195	109-237	1	204	-
D68 NC	10	198	-	-	-	-	-
S13 TN Reelfoot L.	49	206	-	196-236	49	221	184-284
Combined	179+	180	168-197	109-237	50	218	184-284
Age VI							
MI: C213, C215	4	195	-	183-208	-	-	-
B233 OR L. of Woods	2	170	-	-	-	-	-
D68 NC	1	206	-	-	-	-	-
IA, IL, IN: L32, R34, W220, W105	14+	197	-	166-224	-	-	-
Combined	21	193	175-198	166-224	-	-	-
Age VII							
L32 IA Red Haw L.	1	213	-	-	-	-	-
Age VIII							
C215 MI Whitmore L.	1	170	-	-	-	-	-

	No.	1	2	3	4	5	6	7
L86 IL McKenzie L.	12	33	61	89	112	127	137	
L86 IL Park Pond Slough	242	38	66	91	117			
L157 NC Black L.	1	25	48	99	150			
M320 IA E. Osceola L.	37	51	76	99	140			
L157 NC Salters L.	10	33	69	107	157	196	234	251
L86 IL 40 & 8 L.	22	38	74	107	152	178		
R40 IN Grassy L.	1	34	86	113	146			
L86 IL Chautauqua L.	30	36	69	114	150	173		
L86 IL Fairmont quarries	100	43	76	114	142			
L86 IL Staunton L.	18	43	79	114	142			
L86 IL Mount Clare L.	16	33	71	117	145			
L157 NC Waccamum L.	2	41	86	117	135			
L86 IL Venard L.	334	43	81	117				
L157 NC Singletary L.	6	43	66	122	183	213	231	
G80 IL Horseshoe L.	19	65	102	123	147	157		
L157 NC Jones L.	8	36	86	124	180	206		
L157 NC White L.	18	38	89	124	142	165		
D68 NC	61	56	94	127	155	175	190	
W105 IL Crab Orchard L.	9	65	106	129	158	183	209	
J31 OK Grand L.	16	66	102	135	180			
S391 OK Rock C.	2	66	107					
H303 OK Latonka L.	1	76	117	137				
G77 OR 10 waters	+	58	108	137	155	180	221	278
G161 OR 10 lakes	34	61	97	128	153	163	273	287
F64 OK Little R.	97	46	102	140	168	201		
L86 IL Weldon Springs	28	43	89	145				
J41 OK Ft. Gibson L.	7	56	104	150	178			
J38 OK Rod and Gun Club L.	22	79	127	152	165	183		
F62 OK Sub Prison L.	2	51	107	155	183	206		
B86 IL Onized L.	101	36	102	155				
U4 IL Mississippi R.	26	46	122	157	180			
L86 IL Glendale L.	108	53	117	157				
J33 OK Franklin P.	42	76	122	157	190			
J44 OK statewide	1,131	84	132	160	183	203	231	274
slowest	-	38	76	117	132	160	206	
fastest	-	155	203	234	246	239	254	
15 reservoirs	333	84	132	155	185	211	254	274
8 large lakes	138	86	140	170	188	201	234	
33 small lakes	368	84	130	157	175	201	206	
8 ponds	44	84	130	165	188	196		
4 streams	248	79	132	170	196	218		
H424 OK Rod and Gun Club L.	52	81	137	163				
B380 TN Woods L.	89	74	124	165	178	173		
J95 OK Ardmore City L.	17	97	142	165	180	196		
H150 OK Illinois R.	118	46	99	165	196	211		
J45 OK Verdigris R.	3	66	135	170	183			
tributaries	24	69	107	140	147			
E61 OK Salt C.	2	84						
T54 KY streams	16	89	140	173				
J38 OR Chickasaw L.	81	104	157	178	196			
H150 OK Tenkiller L.	11	137	160	178				
J25 OK Illinois R.	12	61	114	180	198			

WARMOUTH

Calculated TL and increment each year

		1	2	3	4	5	6	7	8
L32 IA Red Haw L.		41	91	147	178	190	203	221	
	Incr.	41	52	61	38	21	15	9	
	No.	55	50	33	10	7	5	1	
W142 IL Little Grassy L.		43	92	143	175	196			
	Incr.	43	39	52	31	23			
	No.	11	11	10	4	1			
R217 NC Rhodhiss L.		114	122	132					
	Incr.	114	8	8					
	No.	8	8	6					
H200 IA Ahquabi L.		43	91	124	165	180			
	Incr.	43	48	41	46	30			
	No.	192	110	44	15	1			
E60 IL 2 ponds	Incr.	46	38	29	24				
	No.	41	38	22	7				
W61 IL Thompson's L.		59	77	108	125				
	Incr.	59	36	32	20				
	No.	25	5	5	2				
L86 IL Park P.		42	86	125	163	189	204	215	217
	Incr.	42	43	35	29	13	25	13	17
	No.	1,063	973	786	547	185	63	17	1
P71 MO Clearwater L.									
year of impoundment	Incr.	64	102	53					
later		43	46	41					

Calculated weight at each annulus

	No.	1	2	3	4	5	6	7	8
L32 IA Red Haw L.	55	1	15	68	129	175	214	242	
M320 IA E. Osceola L.	37	–	14	60	133				
H200 IA Ahquabi L.	192	1	15	42	103	137			
increments		1	14	30	67	61			
J44 OK statewide	1,131	10	47	87	135	189	286	495	
slowest	–	–	8	32	47	87	201		
fastest	–	79	189	296	350	318	387		
L86 IL Park Pond	1,063	2	12	40	91	153	196	232	239

Characteristics of annuli are described and pictured (L86). False annuli developed in August, when the shoreline was dredged and thus disturbed the habitat (L86). These false annuli would not have been detectable except for the regular monthly sampling. Annuli usually form about the first week in May in Illinois (L86). Growth was calculated on a direct proportion basis for most studies, but a Fraser-type modification was used in L32, L86, E60, W61, and W142. Growth was rapid in May through July in Park L., IL, but was slow thereafter, except for growth of the smaller fish, until October (L86). Compensatory growth was shown (L86).

The data tabulated above indicate faster growth in Oklahoma and Tennessee and slower growth in Michigan, a relationship which might be expected on the basis of lengths of the growing season. In Oklahoma there seemed to be little difference in the average growth rates in different types of water (J44). Growth was slower in the tributaries than in the Verdigris R., if the small samples were typical (J45). Growth was better in the year of impoundment than in later years in Clearwater Res., MO (P71). Growth was rapid in

L. Glendale, IL, the first 6 years after impoundment (L86), and in Tenkiller
Res., OK, the first year (H150). In Grand L., OK, growth declined during impoundment (J31). Reduction in the population of bullheads and catfish was followed by increased numbers, faster growth, and better condition of warmouth
in South Rod and Gun Club L., OK (H424). Artificial thinning of the population
in Park L., IL, also increased growth (L86). However, L86 and R285 point out
that warmouth do not tend to become dominant and stunted like many sunfish
but establish small broods each year without seriously restricting the reproduction and growth of other sunfishes. Even when their numbers were not reduced at draining as were bluegills, warmouth never constituted over 5% of the
population in Ridge L., IL (B372).

Males were consistently but slightly larger than females at each age (L86),
but no sex differences in growth were noted in S13 or L32.

Growth seems to be slower in turbid waters (H438, E60).

Warmouth mature sexually at 75-100 mm (L86). In Vernard L., they matured at age I and 79-86 mm; but in Park P., where growth was slower, they
matured at age II and 89 mm. Size is probably more important than age in at-
taining maturity (L86).

In central Illinois, nesting begins the second week of May, peaks in early
June, and may extend into August (L86). The earliest spawning in Red Haw L.,
IA, was in the last week of June (L32). In Texas, one pair spawned 3 times
from April to October; and warmouth, which probably had spawned in May or
June, spawned in August when put into a new pond in Illinois (L86). Larger
warmouth spawn over a longer period than smaller females (L86). The first
hatch is when the water at 15 cm is 26.7°C (S334).

Number of eggs per female

		TL	Weight	No.	Mean	Range
L86 IL Park L., Jan.-June		89-91	14	2	5,350	4,500-6,200
		102-124	23-41	6	10,100	8,100-11,200
		127-150	41-73	15	16,900	11,200-26,000
		152-175	77-127	7	27,400	23,400-32,700
		180	132	1	37,500	
	July-Aug.	135-145	54-68	3	13,100	11,000-15,100
		157-170	73-109	4	15,850	11,300-19,800
Vernard L.	May	94	18	1	17,200	
		104-117	27-36	6	29,200	26,000-31,600
		130-137	50-64	3	53,500	46,400-63,200

The number of eggs per female was higher in Vernard L. where the warmouth grew more rapidly than in Park L. (L86). Waterhardened eggs averaged
1.01±0.049 mm (M529).

Nests are usually built near a stump, clump of vegetation, or other cover
and not on clean sand, such as selected by bluegills and pumpkinseed (L86).
The nests may be clumped when nesting sites are limited but will be separated
otherwise. They may be at depths of 5 cm to 1.5 m (L86, R27). Nesting may
be necessary for ripening of the testes, as only aggressive males which established territories ripened in tanks where only a limited number of nests could
be built (L86). Eggs hatched in 44-53 hours at 25.5°C (M544) and in 33-36
hours at 25°-26.4°C (L86), and embryological and larval development were described. Caudal curvature accounted for most of the 5% deformity of fry noted
(M544). Sudden drops in water temperature promote fungus growth and egg

mortality. Many fish also eat fry, and 19 mm warmouth were observed eating 5 mm warmouth fry (L86). Survival of late broods is frequently higher than early broods because of denser aquatic vegetation and less danger of water temperature drops (L86). Annual mortality of adult warmouth in Spear Lake, IN, was estimated at 60% (R152).

Warmouth feed largely by sight (L86). Insects, crayfish, and fish are the main food items of all but the smallest fingerlings, which eat entomostraca (F155, L86, L32, B233, M336, F148, P205, and R310). The relative proportions of insects, etc., differs from season to season in relation to the availability (L86). The large size of the mouth is related to a more piscivorous diet than most sunfishes (F155, L86). It took warmouth about 24 days to consume food equal to their body weight, compared to 15 days for largemouth bass (H244).

Warmouth transport well, withstand low dissolved oxygen levels, and are good animals for bioassay; but they need live food and are difficult to collect because they remain in weedy areas (G156). The sizes taken in gillnets of different mesh size were (L216):

Bar measure	No.	Mean TL	Range
25	18	185	140-229
38	21	190	152-216
51	2	226	203-241

The red blood cell count of 3 warmouth was $1,880,000 \pm 477,703$ per cu. mm. and the hemoglobin reading was 38.0 (S151). The number of chromosomes is 48 (R286).

ORANGESPOTTED SUNFISH, *Lepomis humilis* Girard.

This small sunfish is found from North Dakota to western Ohio and south to Texas and northern Alabama (E119), mostly in slow moving streams (M215), but also in weedy lakes and ponds. The orangespot is quite tolerant of silt and mud and some pollution, but it is disappearing where man's influence is strong even though the seemingly preferred habitat is increased (S582). Orangespot sunfish have been found in water of 0.74 ppt salinity (C368). They are less tolerant of pH change than green sunfish but tolerated changes of pH 7.9 to 9.2 and 8.1 to 6.0 at 8 ppm O_2 or changes of 7.9 to 9.6 at higher oxygen levels (W237). They spawned at pH 9.3.

C30 IA		67 fish		FL = 1.15 SL	TL = 1.22 SL
		Weight			
TL	No.	Mean	Range	Citation	
25-50	3,500	1.2	0.5-1.2	S472 AL (probably too high a weight)	
41	1	0.9	-	J44 OK	
46	1	1	-	C30 IA	
59	8	2	1-3	C30 IA	
51-75	3,833	1.6	0.9-2.3	S472 AL	
	33	4.1	-	J44 OK	
68	29	5	2-7	C30 IA	
78	31	7	4-10	C30 IA	
85	3	11	10-11	C30 IA	

TL	No.	Weight Mean	Range	Citation
76-101	1,303	5	-	A40, S472 AL
	39	12	-	J44 OK
104	3	19	17-23	C30 IA
115	1	30	-	C30 IA
102-126	125	13	6-45	S472 AL
	2	27	25-30	J44 OK
127-151	10	35	27-40	S472 AL
152-177	6	50	45-59	S472 AL

S462 AL	5,277 fish	25-175 mm	Log W = -5.908 + 2.53 Log TL
J44 OK	75 fish	44-114 mm	Log W = -5.547 + 3.271 Log TL

The slope of the S462 regression is too low because the K(TL) of the 25-50 mm fish was 7.07 compared to 1.24 for the larger fish. Other condition factors are:

	TL	No.	K(SL)	Range
C36 AL L. Laverne	38-39	2	3.48	-
W105 IL	-	31	3.08	2.1-4.0

	TL	No.	K(TL)	Range
J44 OK	41-114	75	-	1.38-1.97
S472 AL	51-175	5,277	1.24	1.12-1.71
S225 SD Ft. Randall L.	102-126	3	2.02	-

	No.	TL	Range
Age 0-June			
B19 LA	26	22	-
J44 OK	2	46	-
Age 0-July			
C134 OK Stillwater R.	+	-	22-38
B19 LA	22	24	-
B19 IA	4	26	-
J44 OK	2	56	-
Age 0-Aug.			
B19 IA	97	23	-
B19 LA	75	28	-
J44 OK	4	71	-
Age 0-Sept.			
B19 IA	3	29	-
B19 LA	32	32	-
M308 WI	+	49	-
S225 SD Ft. Randall L.	+	53	-
Age 0-Oct.			
T113 OH	+	-	25-51
Age 0-Nov.			
B19 LA	8	43	-

	No.	TL	Range	No.	Weight	Range
Age I						
B19 IA Fairport	120	37	31-55	-	-	-
M215 KS Big Blue R.	15	43	-	-	-	-
B19 LA	16	46	-	-	-	-
T113 OH	+	-	25-64	-	-	-
W105 IL Crab Orchard L.	1	59	-	1	3	-
E126 OK Rocket Plant L.	8	61	-	-	-	-
L129 IL	+	71	-	-	-	-
J42 OK Ft. Gibson L.	23	89	-	-	-	-
P207 CO Carbody L.	1	90	-	1	20	-
M308 WI pond	+	96	-	-	-	-
Age II						
B19 IA Fairport	120	60	48-67	-	-	-
B19 LA	1	62	-	-	-	-
M215 KS Big Blue R.	10	61	-	-	-	-
W105 IL Crab Orchard L.	12	76	-	-	-	-
L129 IL	+	89	-	-	-	-
Age III						
B19 IA Fairport	29	74	60-90	-	-	-
W105 IL Crab Orchard L.	9	79	-	-	-	-
L129 IL	+	94	-	-	-	-
G158 CO Carbody L.	2	86	-	2	10	-
B134 CO Loveland L.	1	137	-	-	-	-
Age IV						
B19 IA Fairport	10	109	-	-	-	-
L129 IL	+	137	-	-	-	-
G158 CO Carbody L.	4	94	89-96	4	13	10-20

	Average Calculated TL at each annulus			
	No.	1	2	3
C270 MN Mississippi R.	18	28	66	
J45 OK Verdigris R.	2	43		
tributaries	63	41	69	79
W105 IL Crab Orchard L.	19	45	71	79
F62 OK Sub Prison L.	42	61		
B134 CO Loveland L.	1	43	107	127
J44 OK statewide	550	53	81	99
slowest	+	23	56	91
fastest	+	86	97	117
10 reservoirs	105	51	79	
11 large lakes	82	61	84	
14 small lakes	325	51	76	97
6 ponds	16	58	94	102
1 stream	22	43		
S225 SD Ft. Randall L.	22	41	94	79
increments before	-	33	23	
no.	-	18	1	
after impoundment	-	84	64	18
no.	-	3	18	1

Calculated weight at each annulus

		1	2	3
J44	OK	2	9	18
	slowest	-	3	14
	fastest	11	16	31

Little can be said about the difference in growth rates. The OK data (J44) indicate little difference in growth in various types of water.

Males matured at 48 and females at 61 mm (C271). Age III females less than 63 mm were full of eggs (H342). They nest May-July in Illinois (R27), May-August in Iowa (B19), and April-September in Louisiana (B19). Nests are in water about 45 cm deep, and parents stay close to the nest (R27). The males make grunting sounds during courting (G191).

	SL	Number	Eggs per female
B19 LA	30-49	13	50-752
	50-66	7	300-2,680
	80-88	6	520-3,000
	90-105	6	1,340-4,700

Insects constitute most of the food of orangespotted sunfish (K73, R310, W221).

The standing crop was 2.9 kg/ha in an Oklahoma pond and a net annual production of 1.6 kg/ha (W247).

Orangespotted sunfish make good bioassay animals because they transport and hold well, are not particularly excitable, and withstand low oxygen conditions to 1.7 ppm (G156). They do require live food however.

The chromosome number was reported as 46, whereas most sunfish have 48 (R286).

BLUEGILL, *Lepomis macrochirus* Rafinesque

Bluegill originally ranged from southern Ontario and south through the Great Lakes and Mississippi drainages to the Gulf of Mexico, to northeastern Mexico and Florida, and up the coastal area to the Carolinas (E138). Widespread introductions have greatly extended the range in North America, Europe, and South Africa. Bluegill are most abundant in ponds, lakes, and sluggish streams. They have been an important species in pond management. In streams of the Sierra Nevada foothills, California, abundance of bluegills was positively correlated with depth of stream, % rooted vegetation, % pools, and number of fish species and negatively correlated with elevation, % riffles, shade, and % native fish (M566). Bluegill are recommended for stocking in slightly alkaline lakes with less than 900 ppm total alkalinity, 250 ppm carbonate alkalinity, and 200 ppm K + Na (M552). Bluegills have been taken in water with 4.5 ppt salinity (B399).

Breathing rates of bluegill increased after handling, and this response did not decrease when bluegills were handled frequently (C392).

			Length	No.	$\frac{FL}{SL}$	$\frac{TL}{SL}$	$\frac{TL}{FL}$
R33	IN	SL to last scale	-	-	1.17	-	-
		SL to hypural	-	-	1.22	-	-
C37	MN		SL 10-99	136	1.178	1.247	-
			SL 100-159	155	1.164	1.225	-
			SL 160-189	45	1.152	1.205	-
			SL 190-219	17	1.137	1.193	-
C30	IA		-	321	1.20	1.28	-
C33	IA	ponds	SL 40-154	195	1.208	1.271	-
L29	IA	Red Haw L.	SL 27-212	107	1.191	-	-
			SL 19-212	142	-	1.265	-
L29	IA	East L.	SL 20-169	98	1.204	-	-
			SL 10-169	144	-	1.270	-
C38	IA	Little Wall L.	SL 71-90	88	-	1.291	-
			SL 91-105	34	-	1.271	-
			SL 106-120	5	-	1.223	-
B30	MI		SL <102	2,335	-	1.278	-
			SL 102-163	3,712	-	1.261	-
			SL >163	1,253	-	1.246	-
R33	IN		FL <102	-	-	-	1.064
			FL 103-150	-	-	-	1.053
			FL 150-180	-	-	-	1.042
R125	IA	Ike L.	TL 43-185	214	-	1.276	1.053
L78	MO	Clearwater L.	TL 25-210	-	-	1.279	-
W105	IL	Crab Orchard L.	SL 66-155	111	-	1.274	-
S277	IA	McFarland L.	TL 58-234	985	1.217	1.300	1.068
C157	IA	Little Wall L.	SL 70-89	93	-	1.291	-
			SL 90-119	61	-	1.267	-
			SL 160-199	12	-	1.238	-
S289	MA		SL 40-199	797	-	1.285	-
D72	IA	Clear L.	TL 50-267	905	1.220	1.277	1.036
C257	PA		SL 98-200	47	-	1.252	-
M311	MO		TL 30-69	1,480	-	1.303	-
			TL 70-190	-	-	1.284	-
M302	IA	lakes	TL 79-255	63	-	1.27	-
R147	NY		TL 33-224	412	-	-	1.054
S584	TX	L. Bastrop	SL 26-120	264	-	1.282	-
		L. Nasworthy	SL 20-120	640	1.194	1.268	-
		North L.	SL 30-145	252	1.214	1.293	-
		L. Colorado City	SL 33-130	143	-	1.309	-

R116 VA Claytor L.   TL = 0.307 mm + 1.265 SL

H200 IA Ahquabi L.   241 fish TL 41-201   SL = 3.8 mm + 0.8071 TL

R147 NY 28 ponds, 1955   TL 79-216   277 fish   TL = 1.5 mm + 1.041 FL

         1 pond, 1956   TL 33-224   135 fish   TL = 0.25 mm + 1.052 FL

The 1.5 mm intercept was significantly different from 0 (R147), indicating that a ratio estimator is not equally valid as fish increase in length. However, in comparing the FL to TL conversions, Regier did not believe the discrepancies were large enough to worry about.

Where it has been necessary to convert SL or FL to TL the following table has been used (interpolating in between):

TL, inches	0.5	1.0	1.5	2.0	2.5	3.0	3.5	4.0	4.5	5.0	5.5
mm	13	25	38	51	64	76	89	102	114	127	140
FL	12	24	36	48	61	72	84	97	108	121	133
SL	10	20	30	40	50	60	70	80	90	100	110

TL, inches	6.0	6.5	7.0	7.5	8.0	8.5	9.0	9.5	10.0	10.5	11.0
mm	152	165	178	190	203	216	229	241	254	267	279
FL	144	157	169	180	193	205	216	229	241	253	265
SL	120	130	141	151	161	171	182	192	202	212	222

TL, inches	11.5	12.0	12.5	13.0	13.5	14.0	14.5	15.0
mm	292	305	318	330	343	356	368	381
FL	277	290	302	214	326	338	350	362
SL	232	242	252	262	272	283	293	303

TL	No.	Weight Mean of means	Central 50%	Range	Citations
15-24	2	0.12	-	-	OH: M536
	1	0.2	-	-	TX: S584
25-50	+	0.3	-	-	MO: L68
	30	0.47	-	0.1-1.1	OH: M536
	53,327	0.4	0.3-0.4	0.2-4.0	AL: A39, A40, L97, L189, S472
	1	1.1	-	-	AR: C202, T92
	153+	1.4	0.9-1.4	0.9-2.5	MI: B16, B30, B77
	+	1.4	-	-	TN: K71
	11	1.4	-	-	OK: J44
	219	1.5	1.0-1.7	0.5-4	IA: C30, L29, M191
	9	1.8	-	1.2-2.8	PA: C257
	94	2.6	2.3-2.7	0.4-3.8	TX: S584
	53,845+	1.33	0.5-2.2	0.2-4.0	Combined
51-75	34,635	2.0	2.0-2.1	0.7-10	AL: A39, A40, L97, L189, S472
	1+	3.0	-	1.7-5.7	MO: L68, L78
	+	4.0	-	-	TN: K71
	18	4.0	-	-	OR: K80
	62+	4.4	-	3.6-6	OK: H392, J44
	88	4.5	-	2-10	PA: C257
	40+	4.9	-	1-10	IN, OH: K23, R79, M536
	+	5.0	-	-	CA: T132
	+	5.0	-	-	SD: S225
	996	5.1	-	4-7	MI: B30, S348
	793	5.3	4.5-6.0	1-14	IA: C30, C33, C106, F31, D72, L29, S277
	8	5.4	-	4-7	MD: D103
	37	6.5	-	2-8	AR: C202, T92
	36,678+	4.4	2.5-5.7	0.7-10	Combined
76-101	21,991	8	7-11	1-22	AL: A39, A40, L97, L189, S472
	200	9	8-9	5-13	TX: S584
	39	9	-	-	OR: K80
	41+	10	-	7-12	MO: L68, L78

TL	No.	Weight Mean of means	Central 50%	Range	Citation
	36	11	-	8-17	PA: C257
	42	12	-	9-14	IL: M499
	+	14	-	8-23	LA: M31
	12	14	-	10-20	MD: D103
	24	14	-	7-20	AR: C202, T92
	1,279	14	11-17	5-28	IA: C30, C33, C38, C105, C106, C157, F31, L29, M191, S277, D72, G163, I10, M302
	+	14	-	-	SD: S225
	297	15	11-18	6-23	MI, WI: B30, F96, M20, S348
	133+	15	-	7-17	IN, OH: K42, R79, M536
	185+	16	-	12-36	OK: H156, H392, J44
	+	17	-	9-24	TN: K71
	42+	18	-	8-26	CA: M140, T132, W71
	1	28	-	-	NY: G25
	24,322+	14	10-17	1-36	Combined
102-126	13,630	18	-	9-45	AL: A39, A40, L97, L189, S472
	91+	25	-	18-28	MO: L68, L78
	2	26	-	24-28	MD: D103
	444+	27	-	15-45	OK: H156, H392, J44
	27	28	-	14-45	AR: C202, T92
	767	28	22-31	14-41	TX: S584
	3	28	-	17-34	LA: V44
	+	29	27-32	23-32	SD: N94, N115, S225-9
	220	29	24-33	18-54	IL: M499, S348, W142
	+	29	-	-	TN: K71
	1,159	30	27-33	14-100	IA: C30, C33, C38, C105, C106, C157, F31, L29, M191, M246, S277, D72, G163, I10, M302
	219+	32	31-33	28-37	IN, OH: K42, O7, R79, M134, M536
	1,539+	32	25-37	16-51	MI, MN, WI: B30, C20, F96, G42, M20, M136, R24, S348
	59+	32	-	17-51	CA: M140, T132, W71
	11	33	-	26-40	PA: C257
	45	34	-	24-48	OR: K80, B317
	18,206+	29	25-32	9-100	Combined
127-151	8,675	34	33-37	6-64	AL: A39, A40, L97, L189, S472
	29+	42	-	32-45	NC, GA: D68, P189
	114+	44	-	37-57	MO: L68, L78, P112
	475+	48	47-50	31-68	OK: H156, H392, J44, W7
	32	49	-	37-62	AR: C202, T92
	9+	50	-	37-102	LA: M31, V44

TL	No.	Weight Mean of means	Central 50%	Range	Citations
	846	50	49-53	37-63	TX: S584
	2,640+	50	46-59	37-116	MI, MN, WI: B30, F96, G42, M20, R24, S348
	9	51	-	40-68	MD: D103
	597	52	44-62	27-91	IL: M499, S348, W142
	20	54	-	40-85	PA: C257
	+	55	54-59	50-59	SD: N94, M115, S225-9
	1,250	55	47-63	27-104	IA: B150, C33, C38, C106, C157, F31, M191, M246, L29, S277, D72, G162, G163, I10, M302
	288+	56	51-62	35-65	IN, OH: K42, M134, O7, R79, M536
	15	62	-	54-77	FL: H202
	39+	62	-	45-99	CA: M140, T132, W71
	59	67	-	46-88	OR: B317, K80
	15+	71	-	45-102	TN: K71, S11, S19, W132
	15,112+	52	46-59	6-116	Combined
152-177	37	68	-	45-136	AR: C202, T92
	5,300	69	62-64	29-104	AL: A39, A40, L97, L189, S472
	553	80	74-92	50-122	IL: M499, S348, W142
	54+	82	-	57-91	GA, NC: D68, P189
	174+	84	71-96	68-105	MO: L68, L78, P112
	4	85	-	74-91	LA: V44
	+	88	-	-	TN: K71
	55	88	83-94	72-134	TX: S584
	166+	90	85-90	68-105	IN, OH: K42, M134, O7, R79, M536
	2,018+	90	86-95	54-132	OK: H156, H392, J44
	3,133+	93	72-100	51-150	MI, MN, WI: B30, C20, F96, G42, M20, M136, S348
	+	95	84-95	82-141	SD: F100, N94, N115, S225-9
	1,064+	96	88-105	45-156	IA: B150, C30, C33, C106, F31, L29, M191, S227, D72, I10, G163, M246, M302
	30	96	-	51-133	MD: D130
	40	99	-	68-142	PA: C257
	19+	100	85-112	62-153	CA: M140, T132, W71
	24	110	-	65-105	FL: H202, M242
	65	113	-	80-136	OR: B317, K80
	12,736+	90	79-99	29-156	Combined
178-202	1,962	98	91-106	68-181	AL: A39, A40, L97, L189, S472
	6	113	-	91-136	AR: C202, T92
	1+	121	-	91-147	TN: K71, W132
	6	122	-	113-139	MD: D103

TL	No.	Weight Mean of mean	Central 50%	Range	Citation
	218+	124	-	113-153	MO: L68, L78, P112
	628+	126	123-129	104-154	OK: H156, H392, J44
	44	136	-	91-227	FL: H202, M242
	1	136	-	-	MT: B231
	236	140	127-153	82-196	IL: M499, S348
	75+	143	113-155	105-225	IN, OH: K42, O7, M134, M536
	2,923+	143	121-173	90-247	MI, MN, WI: B30, C20, F96, G42, M20, M136, S348
	22	145	-	113-198	PA: C257
	+	146	145-159	122-163	SD: N94, S225-9
	105+	146	-	102-159	GA, NC: D68, P189
	4	147	-	122-173	LA: V44
	375	150	133-165	100-230	IA: C30, C33, C106, D72, F31, G163, L29, M191, M246, M302
	15+	182	181-193	125-224	CA: M140, T132, W71
	10	233	-	215-312	OR: B317, K80
	6,631+	141	129-159	68-312	Combined
203-228	5	136	-	-	AR: C202, T92
	382	156	147-167	91-286	AL: A40, L97, L189, S472
	17+	172	150-176	139-227	IN: OH: M134, O7, R79, M536
	4+	178	-	176-181	MO: L68, L78, P112
	6	181	-	162-216	PA: C257
	42	190	-	136-272	FL: H202, M242
	72	197	-	195-201	OK: J25, J44, W7
	2,150+	205	173-244	139-354	MI, MN, WI: B30, C20, F96, G42, M20, M136, R24, S348, N52
	156	213	196-231	138-318	IA: C106, C157, D72, C163, M191, M246, M302, S277, L29
	86	217	195-220	128-272	IL: H278, M499, S348
	+	245	-	-	SD: S226
	55	245	-	-	MD: D68
	42+	265	-	125-351	CA: M140, T132
	1	336	-	-	NY: R237
	19	351	-	255-482	OR: B317, K80
	3,037+	208	176-241	91-482	Combined
229-253	2	233	-	227-240	AR: C202, T92
	5+	277	-	216-369	OH: M134, R79
	9	263	-	227-310	IL: M499, S348
	27	290	263-313	213-425	IA: C30, D72, L29, M191, M302, S277
	623+	294	264-321	247-462	MI, MN, WI: B30, C20, G42, M20, M136, S348
	1	304	-	-	NY: R237
	2	308	-	304-312	OK: J44

TL	No.	Weight Mean of means	Central 50%	Range	Citation
	19	315	-	181-413	AL: A40, S472
	9+	316	-	260-502	CA: T132, M140
	6	318	-	227-408	FL: M242
	10	340	-	-	MD: D68
	1	580	-	-	OR: B317
	714+	299	263-340	181-580	Combined
254-278	32+	418	366-445	312-576	MI, MN, WI: C20, M20, M136, R24, S348
	1	425	-	-	South Africa: P93
	15	427	-	288-510	IA: C30, D72, L29
	7	435	-	376-508	AL: S472
	4	454	-	-	FL: M242
	1	499	-	-	IL: B408
	+	510	-	-	OH: R79
	60+	433	386-468	288-576	Combined
279-304	3+	545	-	454-624	MI, WI: M136, S323, S348
	2	790	-	459-925	AL: S472
	1	906	-	-	MD: E105
305-329	6+	622	-	303-1191	MI, MN: C20, G36, M136
330-335	+	726	-	-	MI: M136
356-380	1	765	-	-	MD: S398
	1	1,588	-	-	AL: S472
381-405	2	851	-	-	MN: C20
	+	1,318	-	-	MI: M136
	2	2,056	-	1,956-2,155	AL: F11, M99

No consistent regional differences in the weight-length relationship are evident. The average weights from AL, MO, and AR were often low, and those from CA and OR were often high but probably do not indicate regional differences.

The mean maximum depth at each total length was determined to estimate the sizes which could serve as forage for bass of various sizes (L189).

TL	25	51	76	102	127	152	178	203
Mean depth	7.1	14.0	24	34	44	54	65	78
No.	141	195	189	146	180	176	180	50

The relationships were described as (L226).

under 100 mm $\quad$ TL = 15.89 mm + 2.466 depth
100-199 $\quad$ = 42.26 mm + 1.982 depth
over 200 mm $\quad$ = 114.2 mm + 1.1919 depth

			No.	Lengths	Standard lengths
L29	IA	East L	-	-	log W = -4.238 + 2.920 log SL
M191	IA	West Okoboji L.	228	26-191	log W = -4.168 + 2.935 log SL
E60	IL	Upper P.	50	30-159	log W = -4.323 + 2.940 log SL
S584	TX	L. Colorado City	420	24-130	log W = -4.338 + 3.005 log SL
S584	TX	L. Bastrop	259	20-130	log W = -4.463 + 3.018 log SL
M311	MO	L. of the Ozarks	1,446	-	log W = -4.585 + 3.086 log SL
L29	IA	Red Haw L.	-	-	log W = -4.572 + 3.090 log SL
B30	MI		-	-	log W = -4.651 + 3.110 log SL
E60	IL	Lower P.	188	30-159	log W = -4.650 + 3.112 log SL
S584	TX	L. Nasworthy	671	25-125	log W = -4.647 + 3.133 log SL
S584	TX	North L.	228	25-130	log W = -4.712 + 3.165 log SL
W105	IL	Crab Orchard L.	111	65-155	log W = -5.098 + 3.325 log SL
D72	IA	Clear L.	844	44-221	log W = -5.305 + 3.447 log SL

			No.	Lengths	Fork lengths
T132	CA	Folson L	338	30-190	log W = -4.905 + 3.129 log FL
L227	CA	Sutherland L.	1,198	-	log W = -5.518 + 3.400 log FL

			No.	Lengths	Total lengths
P208	IN	White R. Sec. B, 1970	-	-	log W = -2.083 + 1.863 log TL
		A, 1970	-	-	log W = -2.564 + 2.072 log TL
		C, 1970	-	-	log W = -2.709 + 2.173 log TL
		C, 1969	-	-	log W = -3.201 + 2.335 log TL
		B, 1969	-	-	log W = -3.542 + 2.504 log TL
		A, 1969	-	-	log W = -4.095 + 2.773 log TL
G163	IA	McLain P.	16	99-130	log W = -4.217 + 2.69 log TL
H156	OK	Canton L.	266	99-185	log W = -4.245 + 2.798 log TL
H202	FL	Canal 34	50	127-251	log W = -5.833 + 2.859 log TL
S188	RI		50	-	log W = -4.517 + 2.924 log TL
H202	FL	Blue Cypress L.	40	142-218	log W = -6.011 + 2.940 log TL
J37	OK	Rod & Gun Club L.	214	102-210	log W = -4.643 + 2.950 log TL
S472	AL		2,150	171-225	log W = -4.608 + 2.95 log TL
P122	WI	Flora L. 1952	191	-	log W = -4.757 + 2.989 log TL
P122	WI	Flora L. 1956	191	-	log W = -5.236 + 3.183 log TL
G163	IA	Huffacker P.	16	117-170	log W = -4.867 + 3.06 log TL
H202	FL	Blue Cypress L.	39	142-218	log W = -6.291 + 3.063 log TL
S225	SD	Ft. Randall L.	-	64-180	log W = -4.814 + 3.064 log TL
S472	AL		82,212	51-170	log W = -4.887 + 3.07 log TL
H200	IA	Ahquabi L.	241	41-201	log W = -4.874 + 3.076 log TL
P193	MI	Jewett L.	371	102-254	log W = -4.984 + 3.102 log TL
B403	SC	Par P.—central area	56	-	log W = -4.937 + 3.109 log TL
H202	FL	Canal and Blue Cypress L.	71	142-218	log W = -6.400 + 3.112 log TL
R125	IA	Ike L.	214	43-185	log W = -5.081 + 3.156 log TL
H202	FL	Canal 34	32	142-218	log W = -6.526 + 3.169 log TL
G163	IA	Sparks P.	16	102-178	log W = -5.193 + 3.20 log TL
F31	IA	ponds	193	51-201	log W = -5.156 + 3.209 log TL
B403	SC	Par P.—heated area	124	-	log W = -5.091 + 3.256 log TL
J37	OK	Chickasaw L.	109	127-235	log W = -5.322 + 3.281 log TL
M246	IA	ponds	1,435	89-218	log W = -5.971 + 3.282 log TL
L78	MO	Clearwater L	-	-	log W = -5.382 + 3.306 log TL

		No.	Lengths	Total lengths
J44	OK	3,617	51-280	log W = -5.450 + 3.333 log TL
C355	PA Alanconnie L.	1,146	28-254	log W = -5.515 + 3.371 log TL
K144	KY ponds	567	38-208	log W = -5.240 + 3.250 log TL
H419	OK	-	-	log W = -5.667 + 3.431 log TL
J95	OK Ardmore City L.	24	99-160	log W = -5.590 + 3.436 log TL
G163	IA Ross P.	16	114-168	log W = -5.716 + 3.47 log TL
G163	IA Kimberly P.	16	112-193	log W = -5.809 + 3.51 log TL
C106	IA ponds	1,146	64-216	log W = -5.850 + 3.534 log TL
G42	Upper Mississippi R.	1,009	102-251	log W = -6.098 + 3.64 log TL
G163	IA Link P.	16	135-170	log W = -7.151 + 4.11 log TL

H439 used dry weight (which was uniformly 26% of preserved wet weight with liquid drained from coelom and air bladder) in the following regressions: The ponds were all near Cornell University, NY.

Nutrient level	Pond no.	No. fish	
low	26	124	log W = -5.58 + 3.28 log SL
	31	60	log W = -5.65 + 3.31 log SL
	40	185	log W = -5.45 + 3.20 log SL
medium	25	105	log W = -5.61 + 3.30 log SL
	29	69	log W = -5.69 + 3.35 log SL
	39	93	log W = -5.54 + 3.26 log SL
high	24	74	log W = -5.66 + 3.34 log SL
	33	70	log W = -5.77 + 3.40 log SL
	35	74	log W = -5.62 + 3.32 log SL

The slopes were higher in ponds (H439) with good nutrition than those with low nutrition; but in all these cases the weight increased faster than the cube of the length, indicating that the fish improved in condition with increase in length.

The slopes were above 3.0 in most populations. Tests of significance indicated with 95% confidence that the slope differed from 3.0 in Clear L. (D72) and in two cases in H202 where the slopes were 2.859 and 3.112, but tests were not made in the majority of the studies. The weight-length relationship in T92 was obviously not a good fit and was not listed above.

				K(SL)	
		SL	No.	Mean	Range
M19	WI	108-255	3,528	1.83	-
S348	IN Warsaw L.	-	136	2.92	-
S348	WI Wildcat L.	-	124	3.03	-
H62	IN	-	618	3.17	-
M536	OH balanced ponds	20-190	264	3.15	1.02-4.28
M536	OH unbalanced ponds	20-216	100	3.25	0.62-5.67
S348	IN Indian Valley L.	-	197	3.23	-
S348	IN Barbec L.	-	68	3.29	-
L230	IL L. Murphysboro, Oct.	113-144	25	3.3	-
S348	IN Wawasee L.	-	130	3.33	-
S348	WI Big Bass L. females	-	28	3.42	-
	males	-	49	4.09	-
S348	WI Muskellunge L. females	52-154	938	3.43	2.57-3.96
	males	52-167	920	3.54	2.4 -4.0

			SL	No.	Mean	K(SL) Range
F31	IA	ponds	178-202	193	3.43	2.61-5.10
C33	IA	ponds	40-154	194	3.45	-
L230	IL	L. Murphysboro, Aug.	107-140	25	3.5	-
L230	IL	Campus L., Nov.	96-140	25	3.5	-
M308	WI		Age 0	327	3.5	-
			older	138	3.92	-
E60	IL	ponds	30-59	18	3.53	-
			60-109	85	3.70	-
			110-159	148	3.84	-
C202	AR	Ft. Smith L.	36-198	170	3.57	-
W105	IL	Crab Orchard L. Spring	-	51	3.59	-
		May-June	-	42	3.96	-
		Summer	-	14	3.86	-
B114	MO	ponds	-	461	3.59	2.25-5.27
L230	IL	Campus L., July	91-120	25	3.6	-
C257	PA	Upper Pymatuning	Age 0	5	3.64	-
		Lower Pymatuning	Age 0	72	3.50	-
		Sanctuary	Age 0	66	3.58	-
		Upper Pymatuning	older	37	4.51	-
		Lower Pymatuning	older	42	3.97	-
		Sanctuary	older	11	4.34	-
S348	WI	Allequash L.	-	150	3.64	-
W142	IL	Little Grassy L.	95-139	359	3.68	-
L230	IL	Izaak Walton L.	91-139	20	3.7	-
L230	IL	Midland Hills L.	102-146	25	3.7	-
C34	MN	Leech L.	-	2	3.72	-
C23	MN	1941-43	50-229	498	3.72	-
C36	IA		34-221	136	3.72	2.80-5.67
B30	MI		29-202	+	-	3.23-4.00
S584	TX	L. Bastrop	34-123	212	3.78	3.44-4.43
L230	IL	Crab Orchard L.	89-132	25	3.8	-
W61	IL	Thompson's L.	45-154	400	3.86	-
D172	TX		63-173	556	3.86	1.73-6.10
S348	WI	Mud L.	-	90	3.89	-
C157	IA	Little Wall L.	70-165	133	3.92	3.54-4.89
L29	IA	East & Red Haw L.	45-212	179	3.95	-
C38	IA	Little Wall L.	71-120	127	3.95	-
S584	TX	North L.	25-143	230	3.98	3.00-4.71
M311	MO	L. of the Ozarks	-	1,446	4.0	3.2 -4.3
S348	IL	Horseshoe L.	-	85	4.00	-
R116	VA	Claytor L.	-	149	4.02	2.21-4.50
S372	NY	pond	81-127	93	4.1	3.5 -4.5
S348	IL	Chautauqua L.	-	115	4.15	-
S584	TX	L. Nasworthy	19-131	1,111	4.16	2.96-6.65
E28	TN	Chicamauga L.	-	114	4.2	-
L230	IL	Horseshoe L.	106-160	25	4.2	-
S584	TX	L. Colorado City	34-131	420	4.22	3.75-5.25
S348	WI	Ripley L.	-	200	4.31	-
K80	OR	ponds	43-179	211	4.47	4.09-6.49
D72	IA	Clear L.	44-221	844	4.52	2.23-5.51
S348	WI	Wingra L. males	-	26	4.59	-
		females	-	48	4.65	-

83

			SL	No.	Mean	K(SL) Range
F97	MI	Wintergreen L. netting	-	514	4.69	4.38-4.88
		angling	-	220	4.72	4.45-5.11
S348	WI	Chetoc L. females	-	106	4.76	-
		males	-	121	5.07	-
S348	IL	Pistakee L.	-	78	4.99	-
S348	IL	Grass L.	-	133	4.99	-
M191	IA	West Okoboji L.	26-191	228	5.04	-
S372	NY	pond	51-147	139	5.3	3.0 -10.6
E3	MN	1936-41	76-329	655	5.39	-
S348	WI	Kegonsa L.	-	72	5.65	-
C23	MN	Standards	poor	--	<3.0	-
			average	--	-	3.4 -4.0
			excellent	--	>4.5	-
Mean of means				--	3.91	-
Central 50%						3.50-4.29

			TL	No.	Mean	K(TL) Range
B403	SC	Par P.	40-120	+	-	0.87-1.6
			170-280	+	-	1.5 -2.3
F96	MI	Ford L.	80-140	406	1.27	1.21-1.34
			141-226	227	1.39	1.23-1.77
B42	IL	Homewood L.	51-75	+	1.11	-
			76-101	+	1.39	-
			102-126	+	1.66	-
			127-151	+	1.94	-
M573	TX	North L.	61-172	+	1.49	0.66-2.54
C106	IA	ponds	51-75	136	1.55	-
M306	MD	Snowden P., Oct.-Dec.	104-125	3	1.58	-
			140-157	178	2.01	-
C202	AR	Ft. Smith L.	46-249	170	1.64	1.38-1.91
R44	OH		-	+	1.66	-
T92	AR	Ft. Smith L.	46-240	172	1.71	1.03-2.08
H424	OK	Rod and Gun Club L.	76-216	+	1.75	1.63-1.86
J96	OK	2 ponds	-	+	1.78	-
K81	MI	Ford L.	81-216	433	1.78	1.03-2.88
J37	OK	statewide	100-230	+	-	1.63-1.94
J37	OK	Rod and Gun Club L.	100-210	214	-	1.74-1.83
R125	IA	Ike L.	76-213	404	1.78	1.69-2.67
C105, C106	IA	ponds	76-100	327	1.79	-
			101-125	152	1.91	-
D200	WI	Cox Hollow L.	>125	+	1.82	-
B30	MI		36-254	+	-	1.55-2.08
H392	OK	before treatment	74-135	+	1.85	1.55-1.99
H392	OK	after treatment with silvex	66-135	+	1.53	1.33-1.72
		before treatment	136-190	+	1.98	1.86-2.14
		after treatment with silvex	136-190	+	1.70	1.55-1.91
H200	IA	Ahquabi L.	41-201	497	1.94	-
M246	IA	ponds	89-218	1,435	1.94	1.00-2.91
B373	IL	3 experimental ponds	25-150	+	1.94	1.15-2.40

BLUEGILL

	TL	No.	Mean	K(TL) Range
C106 IA ponds	126-150	226	1.94	-
	151-177	212	2.02	-
	178-228	104	2.14	-
B408 IL Ridge L., summer				
stable water	145-200	974	1.98	1.71-2.04
drawdown & feeding	145-221	3,603	2.19	2.06-2.35
B41 IL Fork L.	127-210	+	-	1.41-2.52
J37 OK Chickasaw L.	125-230	109	-	1.86-2.19
H156 OK Canton L.	99-185	266	2.03	1.30-3.46
G137 TN pond A	125-192	500	2.04	-
pond B	152-213	500	2.57	-
pond C	150-202	500	2.60	-
A88 IA Little Wall L.	165-193	32	2.05	-
N94, N115, S229, S276, S277 SD				
Gavins Pt. L.	100-190	344	2.05	1.87-2.31
H58 South Africa	-	+	2.08	-
S225-6, S228, S393 SD Ft. Randall L.	58-210	1,460+	2.11	2.05-2.16
tailwaters	-	42+	2.11	2.03-2.19
D72 IA Clear L.	56-278	844	2.16	1.08-2.66
L127 MN Linwood L.	-	+	2.26	2.21-2.36
D168 NC ponds	125-235	243	2.27	0.91-3.05
B42 IL Fork L.	-	830	2.27	-
F100, F105, F125 SD Oahe L.	150-220	13+	2.29	1.33-2.82
B105 IA Little Wall L.	140-164	6	2.30	2.10-2.41
G164 NM Navajo L.	101-200	26	2.31	2.28-2.32
M575 IA Williamson P., May, 1972	-	83	2.33	2.19-2.44
P208, B416, B417 IN White R. 1969	81-180	85+	2.56	1.81-3.5
1970	80-190	178+	3.27	2.5 -5.6
B41 IL standards	poor		<1.39	
	average			1.39-2.22
	good		>2.22	
C23 MN standards	poor		<1.66	
	average			1.83-2.24
	excellent		>2.52	
Mean of means			1.89	-
Central 50%				1.78-2.05

As the regression slopes were mostly above 3.0, the K factors increased with increase in length in many populations. This situation was noted in B30, B42, B105, B416, B417, C106, D72, E60, F96, H392, L230, M246, M536, M306, M311, P208, and S348. However, no trend with size was evident in K80, S229, T92, and W142.

No sex difference in K was reported by M308, S229, S276, S584, and W142. Males had higher average K values in Muskellunge, Big Bass, and Chetoc Lakes; but females had the higher average in Lake Wingra (S348) and in Alabama ponds (S536).

K factors averaged higher in May and June in Illinois (W105) and in June in Minnesota (L127) than in the months preceding and following. A decrease in condition as summer progressed was reported in South Dakota (S229), but no seasonal trend was noted another summer (F97). Higher condition may be related to gonad development prior to spawning (B373, K81, M311). After

spawning, condition was low in winter and in June in Lake of the Ozarks, and it
peaked in May and August (M311). Condition declined in fall and winter in Felt
L., CA, (W123) as the availability of bottom fauna decreased. In a heated
Texas reservoir (M573), K factors averaged 1.38 in spring compared to 1.54
and 1.55 in summer and from fall to winter. Low average condition in this
reservoir was thought to be related to general thermal effects. Condition of
bluegills in the White R. was better in 1970 when the river was warmer, lower,
and less turbid than in 1969 (P208).

Age 0 bluegills with lymphocystis were heavier for their lengths than uninfected bluegills, perhaps because of the tendency for lymphocystis to infect
the more robust fish (P212).

The following relationships were given in M574:

	r
Percent dry weight = 20.001 + 4.725 (K factor)	0.866
Natural log ash weight in g = -16.309 + 3.459 natural log TL	0.997
Percent fat = 13.286 + 5.929 (K factor) - 3.693 (natural log TL)	0.796
Percent protein = 190.099 - 8.222 (K factor) - 43.631 (natural log TL) + 13.830 Nat. log W + 0.22 (Nat. log TL multiplied by Nat. log W)	0.768

Condition of bluegills increased after drawdown of Bear Camp L., GA
(P189). Artificial circulation of Cox Hollow L., WI, did not increase the average K of bluegills but was followed by continued decline of K (W222). A
severe overpopulation of bluegills was already present at the start of the experiment. Introduction of hard water into an infertile reservoir increased the
average K of bluegills (R289). Average condition of bluegills was better two
years after 90% removal of the population in Franklin P., OK, but not in Rod
and Gun Club L. (J96). K factors on bluegill 44-80 mm SL was better in fertilized than unfertilized ponds, but the reverse was true for bluegills over 100
mm (K80). A higher population density resulted in a lower condition factor in
one of two ponds where the bluegills were fed (S536). A parasite (*Tricodina*)
was also a factor in the denser population. Average condition was better in
two ponds with many turtles and other predators than in a pond lacking predators. The two ponds were also in more fertile soil (G167). The K factors
averaged higher at 5 of 7 equivalent lengths in balanced ponds than they did in
ponds with unbalanced populations (M536).

Average condition of bluegills and other species decreased after carp became abundant in Alanconnie Lake, PA (C355).

Condition factors of yearling bluegills fluctuated more than those of older
fish, perhaps because the food of the older fish was more diversified and more
stable in abundance; whereas, the yearlings were largely dependent upon zooplankton (B373). Average K of bluegills was positively correlated with abundance of zooplankton in a series of Ohio ponds (M536).

Fish with lower condition factors had higher deposits of hemosiderin,
which may be related to stress, in their spleens (G162). Stunting was more
evident in subnormal weight than in subnormal lengths in Ford L., MI (K81).

Parasitism with *Posthodiplostomum minimum* did not seem correlated
with K factors (L230).

Condition factors and growth in length for short periods of time may be
poorly correlated (B373, C355). A general correlation between K and growth
was reported in S348.

		TL	
	No.	Mean	Range
W223 IN ponds			
Age 16-19 days	32	13.2	12-14
20 days	76	13.8	13-15
21 days	77	14.5	13-15
22 days	149	15.1	13-16
23 days	134	15.8	12-18
24-26 days	19	16.6	15-19
27-29 days	9	18.7	18-20
30-39 days	24	22.1	20-26
40 days	11	24.2	21-27
43-46 days	8	25.8	23-29
C372 at 30°C with 0.07-0.08			
mg Zn/l.			
30 days	32	15	12-22
60 days	30	21	19-29
90 days	26	32	27-39
with 0.2-0.4 mg Zn/l.			
30 days	13	11	7-15
60 days	11	19	10-20
90 days	9	30	27-31

		TL			Weight		
		Mean of	Central				
	No.	means	50 %	Range	No.	Mean	Range
Age 0-May							
MI: K24	+	-	-	5-57	-	-	-
Age 0-June							
MN: L132	9	10	-	-	-	-	-
IA, IL, OH, WV: B86, R82,							
S110, S227, W105	307+	44	39-64	8-76	-	-	-
South Africa: H20							
(2nd month)	+	-	-	38-51	-	-	-
OK: J33, J44	21	65	-	53-69	-	-	-
Combined	337+	46	39-64	8-76	-	-	-
Age 0-July							
MI, MN, WI: B16, L132,							
N141, S323	191+	24	18-19	7-64	+	-	2-3
IA: B3, C101, C102, C109,							
H200, L29, R125, S277	411	26	22-23	15-47	-	-	-
SD: S225	+	28	-	-	-	-	-
OH, IN, IL: C284, G32,							
R83, W105	234+	32	20-47	10-84	-	-	-
MO: M311	1	41	-	-	1	-	1
CA: K59, M140	22+	48	-	21-129	-	-	-
NY: S372	30	58	-	25-81	30	14	7-21
OK: J44, J95	67	68	-	49-76	-	-	-
Combined	764+	35	22-56	10-129	31+	9	1-21
Age 0-Aug.							
WV: S124	+	21	-	16-38	-	-	-
MI, MN, WI: L56, L132,							
N141, S323	159+	29	28-30	11-49	+	-	0.3-0.6
IA: B3, C91, C101, C102,							
C109, H200, L29, M372,							
R125	1,387	32	26-38	10-74	-	-	-
OH, IL: G32, R83	+	36	-	28-46	-	-	-

	No.	Mean of means	Central 50%	Range	No.	Mean	Range
						Weight	
Age 0-Aug. (cont.)							
SD: S225	+	38	-	-	-	-	-
MO: M311	27	49	-	33-58	27	2	-
OK: C353, J25, J44	17+	62	-	26-76	-	-	-
Combined	1,590+	37	28-38	10-74	27+	1.5	0.3-2
Age 0-Sept.							
IA: C101-3, H200, I10, S277	393	50	41-57	25-102	2	10	9-11
MO: M311	1	43	-	-	-	-	-
OH, IL: B42, G32, R83	3+	61	-	28-114	-	-	-
MI, MN, WI: C13, L132, M265, N141, S323	303+	41	36-46	15-69	-	-	-
OK: C353, J26, J44	11+	76	-	36-45	-	-	-
CA: B221	5	80	-	-	-	-	-
SD: S225-6	+	46	-	43-49	-	-	-
WV: S124	+	70	-	26-118	-	-	-
NY: H439	+	-	-	15-66	-	-	-
AL: S54	-	-	-	-	+	0.3	-
Combined	716+	53	38-64	15-118	2+	5.0	0.3-11
Age 0-Oct.							
AL: E146 pond	+	30	-	20-60	-	-	-
MI, MN, WI: K24, L132, M308, M418, N141	460+	43	36-44	24-127	327+	5	0.2-17
MO: M311 L. of the Ozarks	10	46	-	33-58	10	1	-
IA, IL, IN: B42, B407, C284, B4, K23, L29, S277	47+	50	36-59	29-127	+	2	0.1-6
WV: S110, S124	+	60	-	15-120	-	-	-
AR: C202	14	61	-	-	-	-	-
NY: S372	48	74	-	42-107	48	20	14-36
OK: C353, B182, J42, J44, L63	418+	75	52-114	5-142	-	-	-
KS: T79	+	76	-	64-89	-	-	-
CA: B221	61	85	-	-	-	-	-
LA: V44	9	147	-	102-190	9	79	17-173
Combined	1,067+	62	38-80	5-190	394+	8	0.1-173
Age 0-Nov.-Dec.							
MT: B108	2	25	-	-	-	-	-
MD: M306	80	38	-	25-51	-	-	-
UT: W137	1	38	-	-	-	-	-
IA, IL: B42, B373, S277	567+	48	25-59	19-127	-	-	-
MO: M311	8	51	-	41-58	8	1	-
Cuba: H441	+	58	-	33-84	+	1.2	0.9-1.5
AL: S53, S54, E146	+	40	-	15-65	+	2.3	0.9-3.7
MI, WI: H87, W222	+	60	-	28-76	-	-	-
WV: S124	+	61	-	24-123	-	-	-
CA: B221	26	80	-	-	-	-	-
Combined	684+	50	29-67	19-127	8+	1.5	0.9-3.7
Age 0							
MI, WI: B31, H250, M94, S348, S414	85+	40	36-43	27-57	-	-	-
LA: M31	+	-	-	32-76	+	-	0.3-2
MD: B218	183	-	-	43-83	-	-	-
PA, NY: E9, M310	9+	66	-	28-94	9	0.6	-
KS: H332	2	69	-	-	-	-	-
IA, IL, IN: B416, G80, M323, P12	32+	70	61-81	22-114	23	24	-

		TL			Weight	
	No.	Mean of means	Central 50%	Range	No.	Mean	Range
Age 0 (cont.)							
AR: H442	32	72	-	61-81	-	-	-
MA: S289	11	74	-	-	-	-	-
KY: S266, K144	313	77	-	53-145	-	-	-
OK: A10, J95	+	122	-	-	+	13.0	-
TN: S18	2	147	-	-	-	-	-
Combined	669+	65	45-81	27-147	32+	9	0.3-24
Age I							
GA: O50	315	70	-	69-70	-	-	-
MD: B218, D103, E76, M306	110+	72	55-83	43-118	10	6	3-11
PA: M310 Pymatuning L.	13	74	-	64-84	13	6	-
MT, UT: B108, W137	16	75	-	38-132	-	-	-
OH: L14, L18, M134, O8, R83, R200, M536, T145	444+	75	48-97	20-135	13	14	5-28
TX: B61, B109, S584	221+	76	61-77	13-203	+	128	65-227
MN: C270, S358	58	83	-	46-127	-	-	-
MI: B31, B75, B77, B418, C13, H87, H94, K24, K23, P28, C213, H272, C260, K81	1,350+	83	64-105	28-191	555+	11	0.3-66
MA: S289	66	84	-	-	-	-	-
IN: B6, B416, H62, K23, R31, R33, R34, R40, W220	422+	85	75-106	45-160	86	16	1-45
IA, D72, H177, H200, L29, P12, R125, M191, M323, M372, I10, S535, M503	1,369+	94	69-112	61-183	269	28	14-104
WI: C117, F2, H34, M19, M20, M97, T76, U8, S348, S414, M308, H250, W222, C356, S534	589+	97	69-124	28-163	256+	23	8-45
CA, OR: B221, M140, O24, T132	1,301	101	94-109	70-158	-	-	-
AR, MO: C202, T92, M131, H442	889	102	86-109	41-142	732	9	2-26
KS: C226, S374, H332, T79, T133	143+	102	74-128	58-160	56+	35	7-119
IL: B41-2, B86, B373, E60, G80, L129, U8, W105	5,466+	102	91-119	51-178	102	34	-
NC, SC: H97, R217, S395, T103, N100	194+	103	99-112	79-137	-	-	-
AL: S53, S54, S536, E146	+	104	-	20-180	+	80	1-150
NY: B7, E9, E59, E112, R237, S184, S372	106+	108	91-130	46-157	44+	53	9-79
WV, KY: S124, S266, K144	782+	112	114-121	26-196	-	-	-
OK: J25, J26, T44, W7, H127, J42, B183, H156, J33, J95, B182, C271, O37	928+	116	105-132	56-155	40+	17	5-51
TN, other: E22, E28, G167, K71	363+	125	116-150	79-152	336	61	37-75
LA: M31, V44	7+	126	-	85-165	7+	33	8-91
TN, Reelfoot L.: S11, S18, S19, S136, S205, S537, S538	187+	152	146-155	122-170	56	81	48-108

		TL				Weight	
	No.	Mean of means	Central 50%	Range	No.	Mean	Range
Age I (cont.)							
Cuba: H441	+	166	-	160-173	+	126	125-127
FL: F74	2	183	-	-	-	-	-
Combined	15,341+	99	75-119	13-203	25-75	39	0.3-227
Age II							
GA: O50	68	99	-	90-105	-	-	-
TX: B61, S584	391+	104	102-106	84-140	-	-	-
OH: L14, M134, M536, O8, R53, R57, R200, S541, T145	847+	108	85-137	51-203	3	113	108-119
MI: B31, B75, B77, B418, C13, C77, C213-4, C232, F96, H94, H272, K81, M531, S588	1,406+	110	92-112	54-221	1,309+	35	3-214
MD: B218, D103, E76, M306	66+	112	104-120	83-188	19	40	8-128
OR: C279, O24, O34	13+	114	-	78-129	-	-	-
PA: M130	16	117	-	104-130	16	37	-
MO: M311 L. of the Ozarks	453	117	-	61-168	453	34	11-48
MN: C270, E3, G110, S358, S384	176	118	81-153	61-225	-	-	-
MA, S289 ponds	142	122	-	-	-	-	-
WI: C117, F2, H34, M20, M97, S348, T76, U8, S414, C356, D200, H250, M308, S534, W222	1,270+	125	97-145	52-213	741+	52	17-198
IA: B150, D72, H200, H177, I10, L29, M191, M323, M372, P12, R125, S277, S535, G163, M503	1,507+	126	107-147	81-208	126	49	13-127
IN: B60, B416, G39, G193, H62, R31, R33-4, R39-40, S183, U1, R152, W220	2,478+	127	115-143	80-190	293	48	43-51
KS: C226, S374, H332, T79, T133	134+	132	124-150	90-175	16+	73	12-139
NC: D68, H97, M368, R217, S395, T103	303+	133	127-140	99-152	-	-	-
IL: B41-2, B86, B373, E60, G80, L129, S348, U8, W105	4,221+	134	109-147	102-190	134	64	45-88
AL: S54, E146	117	138	-	60-215	43+	92	11-170
TN, other: E22, E28, G167, K71	572+	141	128-165	81-168	386	88	64-103
UT: M340, W137	9	143	-	61-203	-	-	-
OK: B183, H127, H156, J25-6, J33, O37, T44, W7, C271	259+	145	142-155	71-178	83	41	9-71
WV, KY: S124, S266, K144	854+	148	142-149	89-208	-	-	-
CA: B221, M140, T132	1,786	148	134-174	118-217	-	-	-
MO ponds: D201	49	150	-	140-160	-	-	-
AR: C202, T92, H442	108	154	142-166	140-180	-	-	-
NY: E59, E112, G25, R237, S184	2+	154	-	122-183	2+	112	32-200
LA: M31, V44	3+	155	-	127-190	3+	85	43-167
TN Reelfoot L.: S11, S18, S19, S136, S205, S537	916	166	162-170	140-190	411	107	80-198

	TL				Weight		
	No.	Mean of means	Central 50%	Range	No.	Mean	Range
Age II (cont.)							
FL: F74	6	188	-	-	-	-	-
Combined	18,172+	130	107-149	52-234	4,038+	66	3-278
Age III							
GA: O50	15	120	-	112-122	-	-	-
TX: S584	506	127	125-130	124-130	-	-	-
OH, L14, M134, O8, R58, R200, S541, M536, T145	4,175+	131	105-165	81-229	3	147	133-162
OR: C197, C279, O24, O34	28+	132	139-155	73-155	-	-	-
MN: C270, E3, G110, S358, S384	878	133	97-163	86-288	-	-	-
MI: B31, B77, B75, B93, B418, C13, C213-5, C232, F96, L95, H272, K81, M351, S588	3,155+	134	119-140	94-216	2,677+	44	11-218
LA: V44	1	140	-	-	1	102	-
MT: B108	14	142	-	-	-	-	-
MO: M311 L. of the Ozarks	146	142	-	112-168	146	57	37-71
WI: C117, F2, H34, H250, M19-20, S348, S414, T76, U8, C356, D200, S534, W222	2,423+	144	119-163	57-231	1,725+	94	14-417
MA: S289 ponds	147	145	-	-	-	-	-
NY: B7, E59, E112, G25, R237	23+	148	94-185	84-206	7+	153	50-249
KS: C226, S374, H332, T133	146+	148	137-160	102-190	+	66	17-120
IA: B150, D72, H200, L29, M191, M323, P12, R125, S277, M372, M503, G163, S535	1,022+	149	135-163	102-224	103	51	17-118
NC: D68, D167, H97, M368, R217, S395, T103	252+	150	137-165	130-178	-	-	-
PA: M130	52	150	-	140-170	52	77	-
MD: B218, D103, E76, M306	123+	151	137-168	107-206	36	91	42-139
TN, other: E22, K71, G167	600+	155	125-177	107-182	452	122	82-139
IN: B60, B416, G39, G166, G193, H62, R31, R33-4, R152, U1, W220	4,543	156	140-168	89-233	690	95	53-147
IL: B41-2, B86, B373, E60, G80, L129, S348, U8, W105, B94	966+	158	143-172	124-208	191	106	94-125
OK: B183, H127, H156, J25-6, J33, O37, T44, C271, W7	227+	166	160-173	89-190	103+	72	59-96
MO ponds: D201	191	171	-	168-175	-	-	-
WV, KY: S26, S266, K144	357+	175	160-163	107-236	-	-	-
CA: B221, M140, T132	872	180	160-217	147-249	-	-	-
AR: C202, T92, H442	77	181	163-203	163-203	-	-	-
FL: F74	69	185	-	-	-	-	-
TN Reelfoot L.: S11, S18-9, S136, S205, S537	1,391	188	180-196	145-216	1,127	155	71-198
Cuba: H441	+	259	-	-	+	355	-
Combined	22,399+	151	132-170	73-288	7,313+	95	11-417

		TL			Weight		
	No.	Mean of means	Central 50%	Range	No.	Mean	Range
Age IV							
NY: B7, E112, G25	14+	134	-	97-198	+	113	28-198
TX: S584	81	149	141-150	141-166	-	-	-
MI: B31, B77, B93, B418, C13, C213-5, C232, F96, H94, L95, H272, M531, P124, S588	5,368+	153	140-168	76-239	2,136+	74	31-235
MA: S289	135	155	-	-	-	-	-
OH: L14, M134, O8, R53, R58, M536, S541, R200, T145	1,163+	156	127-185	122-218	3+	110	60-181
KS: C226, H332, S374, T133	66+	158	157-170	125-193	+	69	34-145
MN: C270, E3, G110, S358, S384, S464	510+	160	137-179	112-320	-	-	-
IL: B41-2, B86, E60, G80, L129, S348, U8, W61, W105, W142	381+	162	119-185	119-203	133	160	105-218
WI: C117, F2, H34, H250, M19-20, S348, S414, T76, U8, S534, W222, C356, D200	1,947+	163	136-186	85-246	1,469+	121	28-567
MO: M311	36	163	-	140-188	36	82	71-105
IA: D72, L29, H200, M191, M323, M372, P12, R125, S277, S535, M503, G163	623+	172	145-184	107-228	285	66	18-159
TN, other: E22, G167, K72	224+	173	164-190	124-190	220	151	105-176
NC: D68, H97, M368, R217, S395, T103	143+	174	160-185	142-213	-	-	-
IN: B60, B416, G39, G193, H62, R31, R33-4, R40, R152, U1, W220	1,558	177	165-195	94-235	183	146	108-198
MT: B108	16	178	-	-	-	-	-
OK: B183, C271, H127, H156, J25, J33, M107, O37, W7	88+	178	168-196	137-206	32	98	68-162
KY: S266, K144	36	178	-	145-211	-	-	-
MD: D103, E76, M306	256	180	170-196	140-203	3	88	51-108
PA: M130	104	183	-	173-190	104	136	-
FL: F74	142	190	-	-	-	-	-
OR: C279, O24, O34	45+	194	-	165-268	-	-	-
AR: C202, T92, H442	24	200	-	185-218	-	-	-
TN Reelfoot L.: S11, S18-19, S136, S205, S357	658	201	196-213	168-272	629	191	113-284
CA: B221, M140, T132	134	203	171-238	165-263	-	-	-
Combined	13,784+	168	149-190	76-272	5,233+	124	18-567
Age V							
OR: C198, C279, O24	19+	162	-	134-174	-	-	-
TX: S584	16	163	-	162-169	-	-	-
IL: B42, G80, L129, S348, U8, W61, W105, W142	75+	164	132-190	127-211	69	203	119-235
MT: B108	2	165	-	-	-	-	-
MA: S289	88	170	-	-	-	-	-
KS: C226, H332	18	172	-	160-178	-	-	-
WI: C117, F2, H34, H250, M19-20, S348, S414, S534, U8	1,016+	175	142-203	87-251	716+	158	28-267

	No.	TL Mean of means	Central 50%	Range	Weight No.	Mean	Range
Age V (cont.)							
MO: M311	5	178	-	170-188	5	113	108-119
MI: B31, B77, B93, B418, C13, C213-5, C232, F96, H272, L95, M531, S588	2,502+	177	163-188	117-246	1,426+	118	37-284
NY: B7, E112, G25, R237	8+	181	-	123-229	7+	189	40-336
MN: E3, G110, S345, S358	338	184	137-216	137-320	-	-	-
IA: G163, D72, H200, L29, M323, M372, M191, P12, S535	138+	184	167-198	127-229	32	90	34-150
NC: D68, D167, H97, M368, R217, T103	88+	191	170-206	163-216	-	-	-
MD: E76, M306	170	194	183-213	155-213	-	-	-
OH: L14, M134, O8, R58, S541, T145	102	188	165-203	147-241	-	-	-
IN: B60, B416, G39, G193, H62, R31, R33-4, R40, R152, W220	486	194	174-208	110-276	42	181	148-210
OK: B183, H127, H156, J25, W7	12+	197	185-218	178-221	2	200	198-201
KY: S266, K144	6	198	-	178-234	-	-	-
TN, other: G167, K71	106+	198	192-213	147-213	106	187	119-225
PA: M130	105	198	-	193-208	105	179	-
FL: F74	67	208	-	-	-	-	-
TN Reelfoot L.: S11, S18-9, S136, S205, S537-8	160	218	208-224	180-269	159	249	222-397
AR: C202, T92	5	220	-	216-229	-	-	-
CA: K58, M140, T132	16+	231	-	217-254	-	-	-
Combined	5,551+	186	174-206	87-320	2,669+	174	28-397
Age VI							
KS: C226, H332	2	170	-	155-185	-	-	-
MA: S289	79	175	-	-	-	-	-
OR: C198, C279, O24	5+	183	-	150-230	-	-	-
OK: W7	1	184	-	-	1	201	-
NY: G25	3	188	-	178-206	3	145	119-184
OH: M134, R58	69+	189	185-196	168-224	-	-	-
TN: K71	+	190	-	178-196	-	-	-
WI: C117, F2, H34, H250, M19-20, S348, S414, S534, U8	438+	191	168-221	96-251	390+	188	28-290
TX: S584	1	193	-	-	-	-	-
MI: B31, B77, B418, C13, C213-5, C232, H272, L95, M531, S588	1,368+	194	171-213	122-251	1,061+	168	74-369
MT: B108	2	198	-	-	-	-	-
MN: E3, G110, S348, S358	117	200	157-225	157-384	-	-	-
IL: L129, S348, W105	1+	208	-	196-213	42	225	201-351
IA: D72, L29, M191, M323	41+	211	201-220	152-251	-	-	-
NC: D68, H97, R217	13+	212	-	178-234	-	-	-
IN: B416, H62, R33-4, R152	172	212	193-229	160-257	33	203	187-215
PA: M130	8	213	-	211-218	8	224	-
MD: E76	2	218	-	-	-	-	-
FL: F74	8	229	-	-	-	-	-
CA: T132	2	238	-	-	-	-	-
TN Reelfoot L.: S11	22	259	-	254-264	22	510	454-567

	No.	TL Mean of means	TL Central 50%	Range	No.	Weight Mean	Weight Range
Age VI (cont.)							
South Africa: P93	1	268	-	-	1	425	-
Combined	2,355+	200	184-221	96-384	1,561+	203	28-567
Age VII							
NY: G25	3	185	-	165-197	3	145	94-198
MA: S289	51	193	-	-	-	-	-
OH: M134, R58	26+	194	-	185-196	-	-	-
WI: C117, F2, H34,							
M19-20, S348, S414, S534	356+	198	180-224	101-251	416+	196	63-318
MN: E3, G110, S358	47	204	140-249	132-384	-	-	-
NC: R217	1	208	-	-	-	-	-
MI: B31, B77, C13, C213-5,							
C232, L95, H272, M531,							
S588	646+	209	199-221	137-259	538+	204	97-417
OR: C279, O24	8+	217	-	208-257	-	-	-
IN: B60, H62, R33-4, R152	59	227	196-241	179-263	1	218	-
IL: S348	-	-	-	-	7	239	221-258
Combined	1,197+	205	190-239	101-384	965+	200	62-417
Age VIII							
OR: O24	2	193	-	-	-	-	-
NY: G25	3	203	-	196-216	3	193	164-238
WI: C117, M19, S348, S534	172+	203	126-249	105-257	358	187	62-400
MA: S289	24	206	-	-	-	-	-
MN: E3, G110, S358	25	210	150-272	150-384	-	-	-
MI: B31, B77, C13, C213-5,							
C232, H272, M531, S588	451+	222	213-226	168-287	252+	205	111-477
OH, IN, IL, IA: L29, M134,							
P12, R33-5, S348	24	228	203-238	172-267	1	250	-
Combined	701+	215	206-238	105-384	614+	201	62-400
Age IX							
MA: S289	10	208	-	-	-	-	-
MI, MN, WI: B31, B77,							
C117, C213, C215, C232,							
G110, M19, S348, S358,							
S588	213+	218	221-241	107-267	231+	199	68-460
IN, OH: M134, R33	6	236	-	224-271	-	-	-
Combined	229+	217	214-241	107-271	231+	199	68-460
Age X							
MA: S289	8	206	-	-	-	-	-
MI, WI: B31, B77, C213,							
C215, C232, S348	40	210	141-234	122-262	35	158	88-224
Age XI							
WI: S348	1	148	-	-	-	-	-

In the following tables, **F** in front of the number examined indicates that lengths were calculated using a Fraser-type correction and **E** indicates that an empirical curved body-scale relationship was used. Data which include annual increments are put in a separate section, and thus in checking growth data for any region both sections should be examined. The data are listed by regions or states from north to south, and generally within each region the slower growing populations are listed first.

	No.	1	2	3	4	5	6	7	8	9	10
				Calculated TL at each annulus							
M343 PA Duck Harbor	107	25	51	81	114	147	175	193	185	198	
M343 PA N. Jersey L.	62	28	64	97	119	150	163				
M343 PA Tingley L.	47	20	53	99	127						
M343 PA Cowan's Gap L.	92	46	71	99	130	175	206	229			
M343 PA Brady's L.	177	28	66	104	137	163	175	201			
M343 PA Winola L.	64	18	61	107	150	188	208				
M343 PA Peck's Pond	10	28	76	130	178	183	201				
S188 RI	F273	49	86	135	170	188	198	206			

	1	2	3	4	5	6	7	8	9	10	11
			Calculated TL and increments at each annulus								
S372 NY pond	81	123	155								
Incr.	81	32	23								
No.	F23	11	4								
Northeast											
Unweighted mean	36	72	112	141	171	189	207	185	198		

	No.	1	2	3	4	5	6	7	8	9	10
				Calculated TL at each annulus							
H94 MI Crystal L.	183	33	53								
B96 MI Battesse L.	195	37	84	130	172	203	228				
K81 MI Ford L.	F433	99	145	160							
C357 MI average	+	74	109	140	165	185	198	203	216	216	234
S348 WI Island L.	88	29	53	71	90	103	123	145			
S348 WI Wildcat L.	124	33	56	72	88	103	123	131			
S348 WI Mud L.	91	29	53	76	104	133	160	178	165	168	
S348 WI Big Bass L.	81	30	58	88	107	135	146				
S348 WI Ripley L.	225	29	65	89	138	193	210	235			
P122 WI Flora L.	F1,452	51	74	91	104	123	137	151	162	167	177
S348 WI Allequash L.	150	35	61	91	115	132	166	173			
S348 WI Chetec L.	227	32	64	120	164	189	200	206	203	208	
S323 WI Murphy Flowage	+	38	76	114	152						
D89 WI Murphy Flowage	+	56	122	142	168	188	198	218	229	234	236
S348 WI Wingra L.	74	36	102	144	175	189	196				
S348 WI Kegonsa L.	119	37	82	170	212	231	229	241			
C270 MN Miller L.	F101	41	97	175							
K33, M258 MN state average	2,248	49	86	124	155	180	198	211	218	231	244
E3 MN stage average	1,609	58	103	142	177	207	244	286	311		

	1	2	3	4	5	6	7	8	9	10	11
			Calculated TL and increments at each annulus								
F96 MI Ford L. 1952	43	86	117	150	173						
Incr.	43	43	30	23	18						
No.	F631	630	552	95	4						
S348 WI Muskellunge L.											
females	42	74	97	117	131	137	147	150	154	168	
Incr.	42	31	23	20	14	11	10	9	9	6	
No.	E965	957	829	680	560	387	274	133	39	7	
males	43	75	98	119	137	147	155	157	168	179	185
Incr.	43	32	23	23	19	14	13	10	9	8	3
No.	E963	958	874	707	513	307	206	93	39	6	1
C270 MN Mississippi R.											
backwaters	51	112	163	213							
Incr.	51	64	56	25							
No.	F169	95	12	2							

		Calculated TL and increments at each annulus										
		1	2	3	4	5	6	7	8	9	10	11
S229, S276, N94, N115, SD												
Gavins Point L.		52	99	117	138	158	180					
Incr.		52	51	28	24	23	25					
No.		234	185	80	24	3	1					
S225-6, S230, S393, SD												
Ft. Randall L.		50	122	146	156	188						
Incr.		50	75	43	35	22						
No.		596	451	107	18	3						
S225-6 SD Ft. Randall L.												
tailwaters		52	113	146	139	150						
Incr.		52	67	47	23	18						
No.		79	49	24	5	3						
F125 SD Oahe L.		56	107	147	173	188	196					
Incr.		56	51	40	23	25	13					
No.		12	12	12	9	5	1					
MI, WI, MN, SD mean		45	86	115	145	164	180	191	201	193	214	185

		Calculated TL at each annulus										
	No.	1	2	3	4	5	6	7	8	9	10	11
B232 MT 3 Forks ponds	119	33	61	91	112	127	137	155	175			
P159 MT 8 lakes	201	35	66	97	122	145	160	175	193			
B108 MT ponds	41	28	76	135	157	168	193					
MT mean	+	32	68	108	130	147	163	165	184			
K80 OR ponds, west	F77	37	81	124	147	165						
south	F53	35	85	133	173							
central	F99	35	117	189								
G99 OR	14	35	89	134	139	160						
G161 OR 1953-62	1,110	38	94	137	158	174	193	278				
G77 OR 35 waters	+	49	104	144	168	184	212	278				
O36 OR	+	89	128	160	174	177	179	195	198			
E138 CA Millerton L.	+	43	83	115	142	168						
L227 CA L. Havasu	+	55	134									
L227 CA Sutherland L	1,198	57	139	195								

		Calculated TL and increments at each annulus										
		1	2	3	4	5	6	7	8	9	10	11
O36 OR Coast lakes		91	125	150	174	184	198	208				
Incr.		91	33	26	16	9	14	10				
No.		F22	19	15	11	8	7	7				
O36 OR Williamette Valley		89	123	160	179	182						
Incr.		89	35	24	16	9						
No.		F138	122	71	14	2						
O36 OR West, pcnds & lakes		91	125	158	177	184	198	208				
Incr.		91	32	24	14	11	11	10				
No.		F160	141	86	25	10	7	7				
T132 CA Folsom L.		38	83	129	188	212	233					
Incr.		38	49	52	41	24	11					
No.		389	248	115	43	17	2					
B221 CA Felt L.												
Incr.		80	45	30	16							
CA, OR mean		57	110	148	165	180	202	233	198			

		Calculated TL at each annulus										
	No.	1	2	3	4	5	6	7	8	9	10	11
C284 OH St. Mary's L.	9	36	64	102	130	152	170	183	216	229		
R83 OH L. Vesuvius	+	46	79	102	127	152						
M134 OH Buckeye L.	300	41	74	104	132	152	180	188	196	213		

BLUEGILL

					Calculated TL at each annulus									
			No.	1	2	3	4	5	6	7	8	9	10	11
R55	OH	Kiser L.	+	28	74	112	132							
R78	OH	L. Meander 1947	+	33	76	114	145	168	183	203				
R82	OH	L. Alma	+	41	91	140	178	203						
B102	OH		+	58	102	140								
R44	OH		+	51	99	145	180							
R78	OH	L. Meander 1938	+	102	132	168								
E30	OH	poorest lake	+	36	69	81	99	117	127	132				
E30	OH	average	1,000	41	91	130	152	173	190	193	216			
E30	OH	best lake	+	49	132	168	193							
R31	IN	gravel pit	47	32	54									
R31	IN	Springwood L.	117	32	54	70	85	108						
R152	IN	Spear L.	F+	38	78	134	179	208	222	236	241			
R40	IN	Foots P.	12	41	104	151	175	175						
R34	IN	Muskellunge L.	+	49	99	153	179	195	206					
R40	IN	Grassy P.	1	49	104									
R33	IN	poorest lake	10	38	59	80	99	125	137					
R33	IN	average of 56 lakes	+	40	80	128	174	198	217	246	238			
R33	IN	best lake	10	54	120	184								
L121	IL	Murphysboro L.	212	41	74	97	130	147	157	193	236	236		
B86	IL	5 lakes	592	33	89	124	152	173	203	208	203			
G80	IL	Horseshoe L.	74	61	98	132	155	174						
B42	IL	Homewood L.	152	71	114	135	142	147						
B86	IL	Onized L.	329	49	117	165	185							
U4	IL	Mississippi R.	+	76	135	168	185	198						
B142	IL	Ridge L.	1,947	119	157	180	193	198						
L129, L228	IL	state average	9,059	81	117	145	168	188	213					
B408	IL	Ridge L., stable water levels	326	51	119	146	161	169						
		drawdown	112	48	137	170	188							
		drawdown, feed	316	83	147	185	205	221						
M320	IA	E. Osceola L.	154	46	64	102	127	140	163					
R125	IA	Ike L.	308	36	81	107	163							
M201	IA	L. Thayer	866	51	94	127	145	175	190					
M373	IA	Williamson L.	+	70	106	128	170	181						
J104	IA	Center L.	108	53	114	135	157	173	180					
M575	IA	Williamson P.	83	48	100	137	157	180						
L29	IA	East L.	F145	43	92	142	177	189	199					
M198	IA	Keomah L.	404	51	89	142	173	190	211					
M246	IA	Balanced ponds	374	43	104	155	178							
		ponds, too many bass	40	38	112	155	180							
		ponds, too many bluegill	153	30	76	127	147							
B150	IA	Little Wall L.	6	53	135	152								
M197	IA	L. McBride	112	86	119	152								
L29	IA	Red Haw L.	F133	36	87	156	183	206	217	245	259			

		Calculated TL and increments at each annulus							
		1	2	3	4	5	6	7	8
M536 OH	8 ponds	46	90	129	153				
	Incr.	46	45	31	20				
	No.	303	222	144	38				
R40 IN	Foot's P.	45	104	153	179	179			
	Incr.	45	64	45	25	28			
	No.	12	9	7	4	1			
G193 IN	Wyland L.	43	73	106	125	136			
	Incr.	43	30	34	25	18			
	No.	F1,106	1,106	981	185	8			

Calculated TL and increments at each annulus

			1	2	3	4	5	6	7	8	9
W61	IL	Thompson's L.	52	100	125	148	156	191			
		Incr.	52	42	27	20	18	10			
		No.	F536	414	346	147	28	2			
W142	IL	L. Grassy L.	51	105	135	152	166				
		Incr.	51	54	27	19	14				
		No.	F363	358	319	129	28				
W105	IL	Crab Orchard L.	79	112	132	152	175	198			
		Incr.	F79	28	20	15	15	8			
E60	IL	Upper P.	44	94	130	158					
		Incr.	44	50	36	30					
		No.	F50								
		Lower P.	42	120	156	166					
		Incr.	42	79	43	25					
		No.	F188								
S535	IA	McLain P.	38	74	97	112	119				
		Incr.	38	36	23	15	10				
		No.	117	117	117	107	17				
S277	IA	McFarland L.	53	80	107	207					
		Incr.	53	29	23	48					
		No.	F1,003	705	101	5					
S535	IA	Kimberly P.	46	79	119	157					
		Incr.	46	33	36	38					
		No.	121	121	35	18					
H200	IA	Ahquabi L.	49	94	119	142	160				
		Incr.	49	43	28	25	25				
		No.	F635	312	229	167	29				
S535	IA	Link P.	58	97	122	140	150				
		Incr.	58	39	25	20	20				
		No.	114	113	100	24	1				
S535	IA	Sparks P.	49	97	127	152					
		Incr.	49	48	33	25					
		No.	163	163	125	12					
S535	IA	Huffacker P.	61	99	130	152	168				
		Incr.	61	38	33	20	18				
		No.	123	123	91	40	3				
M191	IA	W. Okoboji L.	40	88	137	170	189	210	230		
		Incr.	40	48	53	30	20	15	10		
		No.	F228	221	130	96	73	27	4		
D72	IA	Clear L.	61	107	142	157	198	208			
		Incr.	61	46	39	28	25	20			
		No.	F1,166	823	497	59	12	3			
D172	IA	Ventura Marsh	61	107	147						
		Incr.	61	46	43						
		No.	F23	23	20						
I10	IA	DeSoto Bend	56	118							
		Incr.	56	53							
		No.	40	20							
C157	IA	Little Wall L.	81	119	173	216	226				
		Incr.	81	49	57	43	47				
		No.	F164	42	17	8	1				
OH, IN, IL, IA mean			51	98	134	158	171	190	205	226	226

Calculated TL at each annulus

		No.	1	2	3	4	5	6	7	8	9	10
T54	KY streams	15	43	79								
S266	KY ponds	F1,396	79	119	145	163	185					

		No.	1	2	3	4	5	6	7	8	9	10
K71	TN White Oak L.	591	36	76	117	140	168					
	slowest		13	41	76	112	157					
	fastest		97	137	160	175	183					
T127, F151	TN East	48	53	94	127	157	188					
T127	TN Kentucky Res.	13	74	119	152							
E28	TN Chickamauga L.	138	53	122								
B380	TN Woods L.	190	81	130								

		Calculated TL and increments at each annulus							
		1	2	3	4	5	6	7	8
K144	KY 260 ponds	86	126	149	172	183			
	Incr.	86	41	26	19	11			
	No.	F695	430	156	14	3			
TN, KY mean		63	108	138	158	182			

			Calculated TL at each annulus								
		No.	1	2	3	4	5	6	7	8	9	10
W27	MO L. of the Ozarks	21	25	46	66							
P73	MO Wappapello L.	F97	43	76	112	142	173					
L68	MO Clearwater L.	F830	64	107	142	168						
B114	MO ponds	F1,010	56	117	145	157	163					
P113	MO statewide	565	38	79	112	137	157	173				
	headwaters	189	38	74	107	140	160	157				
	middle	166	38	74	104	132	157	157				
	lower	210	41	86	127	150	157	190				
	poorest station	+	36	69	94	109	137	145	170	185		
	best station	+	46	89	137	163	180	190				

		Calculated TL and increments at each annulus								
		1	2	3	4	5	6	7	8	9
M311	MO L. of the Ozarks	43	94	132	155	175				
	Incr.	43	49	36	25	20				
	No.	1,372	640	187	41	5				
L78	MO Clearwater L.	64	107	145	168					
	Incr.	64	43	38	28					
	No.	F830								
MO mean		48	89	122	154	167	173			

			Calculated TL at each annulus							
		No.	1	2	3	4	5	6	7	8	9
A87	AR Bull Shoals L.	124	49	84	102	122	150			
H442	AR Catherine L.	42	64	114	152	180	208			
H442	AR Hamilton L.	45	76	117	155	175				
H442	AR Ouchita L.	45	84	147	188	206				

		Calculated TL and increments at each annulus							
		1	2	3	4	5	6	7	8
T92, C202 AR Ft. Smith L.		61	104	135	178	193			
	Incr.	61	43	36	38	15			
	No.	162	106	41	5	1			

			Calculated TL at each annulus								
		No.	1	2	3	4	5	6	7	8	9	10
J42	OK Ft. Jenkins L.	23	56	109	135	150						
E61	OK Salt C.	10	58									
J95	OK Ardmore City L.	30	69	109	135	160	185					
K65, 66 OK Hiwassee L.		100	76	122	137	157						

			No.	1	2	3	4	5	6	7	8	9	10
						Calculated TL at each annulus							
O37	OK	Redwine L.	+	76	112	137	160	180					
J101	OK	L. Eucha	1,209	56	102	140	170	190					
E126	OK	Rocket Plant L.	3	81	102								
S391	OK	Rock C.	6	84	119	145							
F64	OK	Little R.	587	51	102	145	170	198					
J31	OK	Grand L.	75	79	124	147	170						
J27	OK	Spavinaw L.	30	89	127	150	170	183					
J25	OK	Poteau R.	+	91	130	152	163	175					
J25	OK	Illinois R.	94	58	114	155	185	201					
F62	OK	Sub Prison L.	41	58	119	155							
		restocked	67	97	170								
J45	OK	Verdigris R.	3	94	119	155							
		tributaries	66	69	109	142	150						
H150	OK	Illinois R.	337	66	135	157	188	201					
		Tenkiller L.	186	122	168								
J101	OK	Spavinaw L.	188	69	122	157	175	193					
J37	OK	Rod and Gun Club L.	62	94	137	163	178	193					
F87	OK	Little R.	98	56	119	165	180						
		tributaries	87	46	99	137							
		cutoff lakes	402	49	119	140	165	198					
O37	OK	Wister L.	+	99	137	170	180	201					
H424	OK	Rod and Gun Club L.	215	99	147	185							
J37	OK	Chickasaw L.	109	117	170	188	203	221					
J44, H419	OK		5,464	81	127	152	175	185	185				
		slowest	-	38	76	109	137	163					
		fastest	-	157	196	213	239	259					
		reservoirs	2,462	81	132	160	183	241					
		large lakes	427	79	117	140	160	193					
		small lakes	1,695	81	127	155	178	190					
		ponds	507	84	127	150	170	183	185				
		streams	373	71	117	152	185	201					
		10 new waters	+	97	147	173	201	259					
		59 clear waters	+	81	127	155	178	196					
		24 turbid waters	+	76	114	142	165	185					

			1	2	3	4	5	6	7	8	9
				Calculated TL and increments at each annulus							
O37	OK	Heyburn L.	49	94	122	135					
		Incr.	49	49	15	20					
		No.	148	128	88	18					
H156	OK	Canton L.	58	99	132	152	165				
		Incr.	58	49	33	28	23				
		No.	F151	88	64	19	2				
T44	OK	Grand L.	81	112	137						
		Incr.	81	30	23						
		No.	50	31	14						
H303	OK	Lawtonka L.	71	130	165	180					
		Incr.	71	59	41	20					
		No.	64	64	13	5					
C271, B183	OK	Canton L.	102	130	165	185					
		Incr.	102	46	30	25					
		No.	F280	172	17	2					

	No.	1	2	3	4	5	6	7	8	9	10
				Calculated TL at each annulus							
AR, OK mean		75	122	150	170	190	185				

	No.	1	2	3	4	5	6	7	8	9	10
			Calculated TL at each annulus									
H158 DE	+	102	147	175	193	213	226				
S192 MD Potomac R.	+	56	104	122	147						
B218 MD Cash L.	+	83	108								
R116-7 VA Claytor L.	F150	53	102	157	189	220	241				
D68 NC	243	49	89	127	160	180	206				
L157 NC Waccamow L.	14	43	94	135	165	185					
L157 NC White L.	3	49	119	178	213	236	236				
M187 GA	23	81	142	193	224	254	279				

		1	2	3	4	5	6	7	8	9
			Calculated TL and increment at each annulus							
M306 MD Snowden Pond		38	74	109	150					
	Incr.	38	38	36	41	30				
	No.	465	351	278	240	149				
D167 NC Kitty Hawk L.		56	84	112	135	198				
	Incr.	56	28	28	20	63				
	No.	8	8	8	1	1				
R217 NC Lookout Shoals L.		69	97	122	142	152	170	198		
	Incr.	69	28	23	18	18	20	13		
	No.	F84	62	44	30	12	4	1		
T103 NC Mt. Island L.		64	102	124	155					
	Incr.	64	38	30	28					
	No.	F50	36	16	5					
R127 NC Hickory L.		71	104	127	145	163				
	Incr.	71	33	25	20	20				
	No.	F59	54	33	12	3				
M368 NC Glenville L.		114	122	130	152	175				
	Incr.	114	8	8	10	10				
	No.	F19	19	16	6	1				
R217 NC L. Lure		58	97	130	173					
	Incr.	58	38	33	23					
	No.	F33	26	19	1					
R217 NC Rhodhiss L.		69	107	132	168					
	Incr.	69	33	23	18					
	No.	F59	48	33	4					
T103 NC Tillery L.		69	102	137	160	175				
	Incr.	69	30	28	18	15				
	No.	F75	56	21	5	2				
T103 NC Blewett Falls L.		71	107							
	Incr.	71	41							
	No.	F34	17							
T103 NC Badin L.		69	107	140						
	Incr.	69	41	33						
	No.	F79	48	8						
T103 NC Catawba L.		79	114	145	165					
	Incr.	79	41	33	23					
	No.	F118	81	34	6					
S395 NC Roanoke Rapids L.		69	112	155						
	Incr.	69	53	43						
	No.	F31	14	2						
M368 NC Cedar Cliff L.		76	122	157	185					
	Incr.	76	46	30	25					
	No.	F17	17	8	2					
M368 NC Bear Creek L.		89	140	150	160					
	Incr.	89	51	10	8					
	No.	F24	24	8	2					

	Calculated TL and increment at each annulus								
	1	2	3	4	5	6	7	8	9
S395 NC Kerr L.	99	132	170	203					
Incr.	99	33	20	13					
No.	F44	41	5	1					
Southeast, mean	70	110	142	169	196	226	198		
S584 TX North L. F236	39	77	111	141	159	190			
L. Bastrop F244	57	91	116	153					
L. Colorado City F123	53	91	122	141	166				
L. Nasworthy F613	62	97	122	140					
O50 GA L. Sinclair	34	69	100						
Incr.	34	34	37						
No.	398	83	15						

	1	2	3	4	5	6	7	8	9	10	11
TX, GA mean	49	85	114	144	163	190					
Mean of regional means	53	95	128	153	173	189	200	199	206	214	185

Although there is great variation within regions, the regional averages are usually higher where growing seasons are longer.

		Calculated weight at each annulus							
	No.	1	2	3	4	5	6	7	8
R152 IN Spear L.	+	0.6	6	41	122	194	242	271	300
L228 IL	+	14	36	73	86	141	213		
S348 IL Grass L.	133	26	67	116	162	185	201	207	
S348 IL Pistakee L.	78	29	75	123	166	177	207	212	206
S348 IL Chautauqua L.	115	43	96	133	141	170			
S348 IL Senachawine L.	96	41	94	134	145	151			
S348 IL Horseshoe L.	85	40	107	145	166				
E60 IL Lower P., Incr.	+	1.2	31	51	47				
Upper P., Incr.	+	1.7	13	23	33				
W105 IL Crab Orchard L., Incr.	8	13	26	30	25	16			
R125 IA Ike L.	308	1	7	19	64				
H200 IA Ahquabi L.	764	2	16	33	56	81			
Incr.	-	2	13	18	26	33			
M201 IA Thayer L.	866	14	28	40	79	113	170		
M320 IA East Osceola L.	154	-	14	43	57	62	85		
D72 IA Clear L.	1,166	3	21	56	81	189	209		
Incr.	-	3	18	37	39	86	66		
M198 IA Keoma L.	404	23	37	71	85	119	173		
J44 OK slowest		0.5	6	22	47	83			
statewide	5,464	9	35	68	105	127	128		
fastest		75	153	205	298	392			
H419 OK slowest mean	+	0.5	6	21	45	82			
statewide	+	8	35	68	106	130			
fastest mean	+	75	156	211	310	411			

Growth from 12-25 mm was 0.1 mm/day in ponds with poor growth and 0.6 mm/day in ponds with good growth in IN (K24), 0.2-0.5 mm/day in MN (L132), 0.5 mm/day in WI (N141), 0.16-0.36 mm/day in NY (H439), and 0.4 mm/day in IN (W223). In aquaria, 1.85 g fingerlings increased 0.028-0.041 g/day (C260). Instantaneous growth rates were reported as 1.78 for age I,

1.36 for age II, 0.55 for age III, 0.23 for age IV, and 0.08 for age V in a slow-growing population, Wyland L., IN (G127, G193). Instantaneous growth rates for Mill L., MI, were 1.174, 0.820, 0.588, 0.470, 0.298, 0.135, and 0.301 for age groups II through VIII respectively (S588).

Good pictures depicting characteristics of annuli are given in B42, S11, · P124, R147, and R231; and the last 3 papers have extensive analyses of the scale characteristics with respect to their use in interpreting age and growth. Difficulties in recognizing annuli are discussed in R33 as well as problems in collecting proper samples. Annuli are characterized by contrasting growth that may or may not be accompanied by a clear space between them (M308). Anastomosis of a few circuli on the lateral fields may occur many times a year (M308). False annuli have been detected or suspected in several studies (B42, B41, B114, C157, R31, R33, R147, S11, S277, C358, M539, and R33). False annuli were observed in 80% of known-age bluegills in Louisiana (M539). Many bluegills showed a false annulus in their third summer corresponding to the time of thermal stratification in Red Haw Hill L. (M376, M532). False annuli in a new population were associated with rapid growth (B41). False annuli seem to be evident in new impoundments with relatively unstable populations (R147). Annulus formation was induced by starvation (von Limbach in B41, C358), by injection of alizarin dye (P124), and by handling (C358). Failure to form annuli some years was observed in Louisiana (M539).

The scale must measure at least 0.25 mm from the focus to form a recognizable annulus (P124), and small bluegills at the first winter may not form a recognizable annulus (R33, B114, P124, and R147). Scales first appear on the caudal peduncle when the bluegills are about 14 mm TL; and therefore, the first annulus is more apt to be on caudal peduncle scales than on scales further forward (P124). The small scales on the head, breast, and nape and some of those above the lateral line usually lack the first annulus. The first ctenii on scales were found on 21 mm bluegill (P124) or on 24 mm bluegills (C353). The pattern of scale formation was similar to that on green and longear sunfish but begins when the bluegills are larger (P124).

In examination of 264 known-age bluegills (age I-III), 23.9% were incorrectly read and 18.9% did not have the correct number of annuli even when the number to be found was known (P194). Some bluegills in slow-growing populations failed to form annuli, and some even grew so little that the 2 annuli were missing (R147). Marginal resorption of scales even took place in some populations (B114, R147). Annuli could not be detected on scales from bluegill in a thermal outflow that never fell below 22.3°C (S584).

The completion of annulus formation was listed (R147) as at:

Latitude 38°N	late April	B114
40°N	early May	B11, M134, R79
42°N	late May	R33, S277, B28
44°N	early June	B28
46°N	late June	B28

However, the range in annulus formation is greater as indicated by other studies:

Latitude 32°N	November-July, with peak February-April, TX (S584);
36°N	May, OK (C271);
38°N	February-August, CA (B221); early April to late June, L. of the Ozarks, MO (M311);
40°N	late April to early May, MD (M306); mid-April to end of May, IL (B41);

Latitude 42°N   early May to mid-June, NY (R147); April to late June,
                 IA (S277); late April, OH (M536); February-June,
                 northern IN (R33); mid-March to early June, IN
                 (G166, G168); April, 20-August; Ford L., MI (F96);
                 May, IN (G193);
         44°N   mid-May, SD (S226);
         46°N   late June to early July, WI (S348);

Annulus formation occurred when water temperatures reached 55°F (A47).
Annuli usually formed earlier on younger bluegills (M311, F96, B221, R33,
B41, and B11), perhaps because of a shortage of food for older bluegills (B41).
The differences with age are often not statistically significant (G166), but the
consistency of the observations suggests that they are valid. Larger individuals in a year class formed annuli earlier than the smaller ones (F96), but no
such difference was evident in another lake (S277). Annuli formed earlier in
years with warm springs than usual (F96, S277). The lag between the time that
environmental factors are suitable and the time of annulus formation was suggested (S348) as indicating that the fish compensate for lost condition and build
reproductive products before growing in length. Various physical and chemical factors, including temperature, photoperiod, and oxygen distribution, were
not significantly related to time of annulus formation in Indiana lakes (G166).
Bluegills showed little or no growth from November to February in one Texas
reservoir and from October through February in another (S584).

In Newcomb P., with many predators, 60% of the bluegill scales collected
were regenerative but in Duke's L., with few predators, only 8% were regenerative (P124). The percentage increased with size up to 100 mm in Newcomb
P. but did not increase thereafter. Regeneration appeared to have no effect on
the size of neighboring scales.

In most studies growth was calculated upon a direct proportion basis, although many used a straight line relationship with an intercept other than zero
(marked F in the above tables). A curvilinear relationship was used in the
Muskellung L., WI, study (S348). Populations in some ponds also showed
curvilinear body-scale relationships; but it was believed (R147) that this was
the result of combining data from the original stock with their progeny, because the original stock and the progeny did not have the same environmental
conditions at the time of scale formation. Scales should come from the same
area of the body throughout the study, since the body-scale relationship differs
with area of body (P124, R231). Scales above the lateral line gave lower calculated lengths than scales from below, using a correction value of 15 mm,
based upon the size of fry forming scales (K80). In calculating lengths a better correction factor can usually be derived from the body-scale relationship
than from study of the size at scale formation, because the early growth of the
scale may compensate for some of the lateness in scale formation. No relationships between rate of growth or sex were found with the body-scale relationship (P124). Numbers of fish required to estimate the body-scale relationship within 2% of the scale size were computed as 23 fish per size group in
Newcomb P. and 33 in Duke's L. (P124).

The intercept values in 23 ponds varied from 6 to 65 mm, but use of an
average value was not thought to introduce much error (R231). Elimination of
data from 7 ponds with few data or data poorly distributed over the usual
range of lengths left intercept values of 10 to 21 mm. Some difficulties in try-
ing to compute correction factors for each population are evident in the calculated lengths reported from North Carolina (M368, R217, T103, and S395).

Intercept	20	23	28	43	46	61	71	97
Calculated length								
at first annulus	64	58	69	69	76	89	89	114
Number	50	33	59	84	17	49	24	19

It is very likely that the higher calculated lengths at the first annulus are at least partly the result of a high intercept value, which may not be very accurate.

In general, errors in aging are more serious than those arising from body-scale relationships (P124). Lee's phenomenon has been noted in some populations, probably the result of selective sampling (R33, G193). Selectivity of sampling methods may significantly affect the average sizes taken in some age groups (H200).

The oldest age reported from scale reading was age XI (S348), and F6 reports the longest a bluegill was kept in captivity was 8 years.

Both slow and rapid growth have been reported from most regions, indicating that population and edaphic conditions may have a greater effect on average growth rates than do growing season or latitude. Review of the tabulated data does however indicate more rapid growth in the southern part of the range than the north. The life span seems greater in the north. Growth was faster in southern than northern Wisconsin (S534), but the good growth in L. Chetac in northern Wisconsin indicated that local factors may overshadow the climatic ones (S348). The slower growth in ponds in southern Michigan compared to farther south was thought to be due to the shorter growing season (B92). In Minnesota, latitudinal differences in growth of bluegill were not evident, though they were for 11 other species (E94). It was thought that overpopulation of bluegills was more common in southern Minnesota due to reduction in larger predatory fish. Growth was faster in the MN-WI portion of the Mississippi R. than in the IA-IL portion to the south (U8). Growth in moderate to very fertile ponds in New York (R237) are considerably slower than minimum acceptable rates in Alabama.

No correlation between mean air temperatures and annual increments in bluegill growth was found in L. Muskellunge, WI (S348), but there was some indication of more abundant year classes in warmer years. There was little feeding and no growth below $10°$-$13°C$ (A47). At $20.5°C$, growth was more rapid on equivalent daily rations than at $25.5°C$ (R291). Little growth was found at temperatures over $26.7°C$, because most energy then went to spawning (S550).

Significant differences in growth were neither found in heated water and in other areas of White River (B417), nor in 4 Texas reservoirs (S584), nor in Georgia (O50). An inverse relationship between megawatts per hectare and first year growth of bluegill was believed to be the result of zooplankton destruction by intake of water for cooling (S584).

Most of the growth is completed in spring and early summer in some populations (B11, C157, D72, J44, W123, A47, L229, and G168). Growth continued into September in Buckeye L., OH (M134). A study quoted by W123 indicated that bottom fauna suitable for bluegill over 130 mm was sparse by late summer and fall and that growth and condition decreased as the bluegills eat molluscs and plant remains. The length of the growing season varied from 98 to 189 days in northern Indiana (G168, G195), but no physical or chemical characteristics could be identified as possible causes. I suggest that growth stops when the carrying capacity is reached, i.e., when any factor becomes limiting. Standing crop in relation to carrying capacity may thus be the major consideration in length of growing season (C348). Growth slowed after thermal stratification and oxygen depletion concentrated the bluegills near the surface (M376, M532). Average size of young bluegills increased over winter in Oklahoma, but this was due to selective predation, eliminating the smaller fish (C353).

Males grew more rapidly than females in some populations (S277, S536, L78, M308, M306, S348, S543, S344, D72, M311, K24, and E3), but the differences were usually small. In other lakes no sexual difference in growth was detected (S534, S393, B31, S228, F96, and E3), and in a few, females grew larger than males (S348, S584, M134, and R33). The percentage of females decreased with age, and all bluegills over Age V were males in L. of the Ozarks, MO (M311). In Ford L., MI (F96), there were only 33 females to 100 males in age groups IV and V, but there were 121 and 164 per 100 in age groups II and III, respectively. Males constituted 50% of the bluegills over 150 mm—but only about 30% of those under 100 mm (S543). In Muskellunge L., WI, however, females predominated in ages over VI (S348). In 9 years of collection in Lake Muskellunge there were 963 males to 965 females (S348).

Females outnumbered males in 4 Alabama populations, with 61.6%, 68.1%, 72.5%, and 79.5% of the populations being females (S543). The estimate for the population with 68.1% female was based on the total population, while the others were based upon the 75-125 mm fish.

In all seasons females had higher metabolic rates than males, but the differences were not statistically significant (W123). M134 suggested that marked decline in increments in the 2nd and 4th year indicated spawning activity. Spawning checks were more frequent on scales of males than females (S348).

Growth compensation was demonstrated (M11, S348, G166). On the other hand, a positive correlation between 1st and 2nd year growth was found (H90). Earlier annulus formation by the smaller fish was suggested as the mechanism of growth compensation (G166). Growth increments were better at the 2nd and 3rd annuli in lakes with poor growth the 1st year in 4 Texas reservoirs (S584).

Considerable variation in the 1st year's growth can be expected because of the prolonged spawning season (L29, M134). Reduced variability in sizes in age groups beyond V were noted and thought to be the result of slower growth at larger sizes (R33).

Growth of bluegills stocked at equal rates and of their offspring were both correlated with nutrient levels in NY ponds (H439).

Growth was generally better in clear than in turbid ponds (H332, B182). Growth improved in one pond when turbidity decreased after cattle were fenced out (S535).

In southern Iowa growth was better in reservoirs with wooded shoreline and deep basins than in shallow reservoirs with exposed shorelines, and it was poorest in water-supply reservoirs with fluctuating water level and with copper sulfate control of aquatic growth (M323, M530).

First year growth was correlated with fertility of 4 Texas reservoirs (S584). Average growth was correlated with total carbonates, total dissolved solids, and plankton abundance in a series of Minnesota lakes (E94). Fertilizing experiments have shown the growth of bluegill to increase with increase in phytoplankton in ponds (S118) and in a river (P196). No association was evident between growth rate and size of lake, average depth, transparency of water, abundance of weeds, or abundance of predatory or of competing fish in a series of Indiana lakes (R33). Growth was slower in a strip mine lake at pH 6.2-7.4 than at pH 7.0-7.8 (S374). Artificial circulation of a lake did not result in more rapid growth of bluegills (W222).

Growth increased in a marl lake when the water level raised (M531). In-creased water levels in Reelfoot L., TN, also may have been related to in-creased average growth of bluegills (S19).

Slow growth of bluegills associated with overpopulation has often been recorded (E3, R33, R31). In a series of 6 similar lakes in Minnesota (E94), 3 lakes showed low populations (60-125 fish per hectare) and rapid growth (taking 2.8 to 3.5 years to reach a length of 127 mm), and 3 lakes had high populations (790-1,240 fish per hectare) and slow growth (4 to 5 years to reach the same length). In another pair of MN lakes, growth was better and stomachs were fuller in the lake with the lower population density (S464). Growth was best in ponds of lowest population density in MT ponds, where dredging of the shallow areas interferred with bluegill spawning and kept populations low (B232). Crowding has been shown to decrease growth of bluegills in aquaria and other controlled experiments (C260) and in pond experiments (B373, E146, R237). In Illinois ponds, growth was limited in direct proportion to population density of all species (B400).

Thinning of the population has sometimes increased growth (P122, M107, B93). Growth was better only after 60% removal of standing crop in Williamson P., IA (M373). Annual removal of 33 to 38 percent of the standing crop per year did not increase bluegill growth in Michigan ponds (B418). No change in growth was noted in a 4 year period of 60%, 40%, 20% and 0% annual harvest in small Missouri ponds (G190). Annual cropping of 49 to 181 kg of bluegills/ha did not maintain rapid growth in Fork L., IL (B41). Removal of 1 to 24 kg of bluegills/ha had no effect on bluegill growth in a Minnesota lake (S358). Reduction of bluegill and bullhead populations by up to 47 kg/ha did not increase bluegill growth rate, but the average growth rate decreased because of increased numbers of small fish (H424). Restrictions on commercial removal of bluegills from Reelfoot L., TN, was followed by decreased growth rates (S538, S205). Intensive removal of over half the bluegills in Alanconnie L., PA, resulted in an increase of the mean weight of age IV bluegills from 76 g to 157 g. Good reproduction again brought density almost up to the previous level, and the average age IV bluegill weighed 104 g (C355). Increased harvest, in Murphy Flowage, WI, resulted in a decrease in average growth rate, which was perhaps due to greater survival of the young fish (S414). When improvement of growth followed winterkill in southern Michigan lakes, the benefit was usually temporary (B96); and benefits were slight in another thinning study (P193), although periodic partial winterkills were related to more rapid growth. Growth was better in a lake following partial winterkill (R33). After chemical treatment of a winterkill lake, stunted bluegills grew to 229 mm in 7 months (S548).

Treatment of bluegill nests with copper sulfate was not effective in controlling bluegill populations in Michigan lakes (B375). The little improvement in average size of young bluegills, which was noticed, may have been due primarily to elimination of the smallest fry. Removal of many rough fish was followed by improved growth of bluegills in Bass L., IN (R33), but not in 2 Minnesota lakes (S384). Competition with carp for bottom fauna was believed to be the cause of slow growth of bluegills in a CA reservoir (W71). Reclamation of lakes by eliminating the fish population and restocking often results in more rapid growth of bluegills (F62).

Exposure of bluegills to 0.0 to 0.025, 0.0375, and 0.05 ppm heptachlor resulted in successively increased mortality, but the remaining fish grew more rapidly where more heavily treated, because of smaller numbers remaining (A89). A similar situation was reported with 0, 10, 20, and 40 ppm herbicide dichlobenil (C359) and in ponds with 0, 0.1, 0.5, 1.0, 5.0, and 10.0 ppm 2, 4-D (C360). Spawning of bluegills in 5 in 10 ppm 2, 4-D was delayed about 2 weeks, and pathological conditions involved the liver, vascular system, and brain.

Growth and condition of bluegills was better in two ponds with many turtles and other predators than in two ponds with few predators (G167). Introduction of northern pike did not result in increased bluegill growth (B374). Stocking of northern pike in Murphy Flowage failed to reduce the numbers of bluegill, which increased in numbers and decreased in average growth rate (S590). Age IV bluegills declined from 180 mm in 1954 to 125 mm in 1964-68. Growth is better in ponds with balanced populations, (maintained by adequate predation) than in ponds with too many bluegill or other forage fish (M246, S266, S535). Growth of bluegills was better in ponds with regular fall drawdown to permit bass to control numbers of bluegills (B372).

In newly stocked waters, the growth rate of the original stock depends upon the population density which is dependent upon the number stocked and the rate of survival (K24, G165, R237). The original stock may grow much faster than the progeny, which may be much more abundant (H440, M536). Stocking with bass in the summer before the bluegills in the fall resulted in more rapid growth of bluegills because of better control by the bass (D201).

Growth is often good in new impoundments with slower growth is characteristic of later years (C271, G180, H150, S229, L78, B221, and H442). Rapid growth the 1st year has been associated with influx of nutrients, and zooplankton, with rising water levels (B221) and with low population density. Factors related to the reduced growth rate were listed as 1. less food than when lake first filled, 2. water fluctuation delaying spawning, and 3. increased population density (L78). Failure to grow rapidly in fluctuating reservoirs was attributed to lack of insect life (E22). In a small artificial lake, increase of pondweed, *Potamogeton foliosus*, was related to decreased growth rates of bluegills; but dieoff of the pondweed was not followed by increased growth of the bluegills (B41). Bass grew more rapidly, however, when the bluegills were no longer hidden in the plants. Shortage of food organisms was believed to be the cause of slow bluegill growth. Growth was more rapid in L. of the Ozarks, MO, in 1951 when the water level was high, and thus washing food in, than in 1953 when water levels were low (M311).

Jaw tagging slowed growth of the tagged bluegills (S372).

Exposure to 0.0146 mg/l of H_2S for 126 days reduced growth, but H_2S levels below 0.0067 mg/l showed no effect (O51).

Diquat up to 3 ppm seemed to have no effect upon bluegill growth (G165). Growth was reduced about 50% in 13 ppm detergent (alkyl benzene sulfonate) in one experiment, but no clear effects were evident in others (L174). Although mirex, an insecticide used in fire ant control, had no effect on mortality of bluegills, the growth rate was reduced when the bluegills were fed 5 mg mirex/kg food (V87). The mirex levels in the bodies also accumulated. Concentrations of 0.01 and 0.04 ppm methoxychlor for 13 weeks did not affect bluegill mortality or growth but did cause pathological changes in liver and circulatory systems (K142). Exposure to over 4 ppm sodium arsenite reduced growth rates of bluegills (G169), despite lower survival rates. Heptachlor fed to bluegills resulted in slower growth approximately in proportion to the concentrations in the body (A89).

Growth of fingerling bluegill was not appreciably reduced over a 6-week period, when the fish were infected with an average of 102 metacercaria, *Posthodiplostomum minimum;* but when infection rates were 353 and 547 per fish, growth was significantly reduced over a 32-week period (S539). Mortality rate was also increased.

Growth indexes and relative condition of bluegill were not well correlated in Alanconnie L. (C355) or in Illinois ponds (B373).

After sexual maturity, males can be distinguished from females at any season of the year because the male does not have a small, swollen, doughnut-like ring around the urigenital pore (M538). In males the urigenital opening is wedge shaped and pigmented, and in females it is round and little pigmented (M538); but the use of opercular, lobe shape, and coloration of gular area, dorsal fin, and body increased the accuracy of sex determination (B419). Sex could not be accurately determined in fall or winter on the basis of color, fins, or opercular flaps (C365). Methyltestosterone failed to reverse sex of blue-gills as it has done in some other fish and thus gives no promise as a method of preventing over population (C398).

Although some bluegill may reproduce during their first summer in Ala-bama (S542), most mature at age I (AL, S542; OK, C271; CA, M140; IL, J115; KS, T133; MO, L68; OH, M248; NY, R237). Most did not mature until age II in L. of the Ozarks, MO, (M311) or in Muskellung L., WI (S348). If bluegills are not at least 14 g at age I, they will not spawn then (S542). Smallest mature in-dividuals have been reported as 76-90 mm (M248), 112 mm (C271), or 23 g (L189). The smallest mature males were 58 mm (S348, R125) and females, 89 mm (R125). Although males were mature at 112 mm in an Ohio pond (S541), none under 125 mm were able to maintain territories. In this crowded popu-lation, females successfully spawned a year or two earlier than the males. Differences in size of bluegills, 25-75 mm TL, at stocking in the fall in New York ponds did not affect the probability that they would spawn at age I (R237).

			No. of eggs per female		
		Size of females	No.	Mean	Range
U1, IN		2 year olds	7	3,820	2,360-5,066
		3 year olds	7	9,264	6,518-13,137
		4 year olds	2	19,169	16,220-22,119
C54	IN	177 mm	1	11,257	-
L13	OH	140 mm	1	11,267	-
O8	OH	-	+	-	8,000-12,000
C77		184 g	1	32,950	-
J25	OK	113 g	+	-	10,000-13,000
M134	OH Buckeye L.	102-126 mm	15	6,787	-
		127-151 mm	101	12,398	-
		152-177 mm	60	16,402	-
		178-202 mm	57	22,779	-
		203 mm	4	29,769	-
		216-241 mm	16	48,846	-
E98	TX	119 mm, 34 g	1	7,410	-
		130 mm, 45 g	1	9,273	-
		165-175 mm, 88-108 g	3	14,641	13,705-15,917
		196 mm, 136 g	1	9,463	-
S323	WI Murphy Flowage	114 mm	1	2,500	-
		165 mm	1	12,000	-
		large	1	38,000	-
L189		-	+	22,000	-
S348	WI Kegonsa L.	149-194 mm, 201-374 g	21	58,000	38,897-81,104

There was no increase in egg numbers with increase in size of female over the range covered in S348 above. Females spawned an average of 5 times a year in Texas, with a 120 mm female spawning about 80,000 eggs per year (E98). Unfertilized ova averaged 1.04 mm (T136). Eggs averaged 1.09 ± 0.048 mm, and egg size showed no correlation with length of the female (M529). Peripheral ova in the ovary are less mature than those located in the median section (T136), and ova nearest the urigenital sinus showed the greatest maturity. Although they are not usually stripped, eggs artificially stripped from a ripe female and fertilized with sperm (taken by dissection) of a male will give good hatches (T136, H443). When ripe females were held in the laboratory 2 days, the eggs were not fertile and their pale watery appearance indicated that the yolk had been resorbed (T136).

Bluegills may spawn almost throughout the growing season: February-October in FL (C361), March-fall in FL (M242), March-September in TX (E98, S571), April-October in AL (S542), May-September in IL (B41), late May-August in IL (J115), June-September in MI (S467), and May-August in WI (J105, S323). Peak spawning was in early June in Murphy Flowage, WI (S323), but it was in mid-July in Muskellung L., WI (S348). Peak numbers of pelagic fry occurred in mid-May to early June in Rathbun Res., IA, but a second peak occurred in mid-July, indicating 2 spawning periods, 3 years in a row (M561). Successful spawning may be delayed until August or September, even in Alabama, if the population is crowded (S542). When food is scarce, bluegills commonly eat their own eggs, and the earliest eggs are usually eaten by the male bluegills, so that the early spawns are rarely successful. This may also prevent hatching of small fish when food is scarce (S542). In Michigan, spawning may be delayed 4-5 weeks by low temperatures, and some summers the temperature may never reach suitable levels (B92). Water temperatures were not thought to affect average reproductive success in New York ponds (R237). Spawning failed in some Illinois ponds in 1962 when surface-water temperatures rarely reached 27°C and frequently dropped to 20°C (B373). Late spawning in Ohio in 1950 was associated with a cold spring and summer (M134). Spawning was earlier where males predominated than in populations where females were more abundant (M308). Males matured earlier than females and large males than small males (B42). Bluegills nest about 3 weeks after pumpkinseeds (C365). Nesting started June 7 in 1970; first eggs, June 15, 1970 and June 24, 1969; and first hatch, June 22, 1970 and July 2, 1969 in L. Opinicon, ON, about 2 weeks after pumpkinseeds (A102). The water warmed earlier in 1970. During the spawning season, the ratio of body weight to gonad weight is about 10:1 in mature females compared to ratios of 40-80:1 the rest of the year (M541).

The first hatch occurs when water at 150 mm reaches 26.5°C (L189, S334, S550). Spawning may occur at 22°-26°C (S336) or about 24°C (D178), or when water temperature is rising and near 27°C (S541). The first nests were found when surface-water temperatures were 17°C one year and 23°C another (C365). A drop in temperature followed by a temperature rise stimulated renesting. In May 1967, spawning occurred in 55 nests in Ohio at temperatures of 17°C (S541), Spawning even occurred at 32°C (C361). Threshold temperatures for spawning were given as 17°-26°C (K161).

In Felt L., CA (B221), a fish kill in early June mostly affected the older, larger bluegills which were spawning on the shoals later than the younger fish.

Males build nests in water usually 15-120 cm (S542), or 30-100 cm (M134) and over a variety of substrates with fine gravel preferred (S542). When vegetation was abundant in shallow water, bluegills nested at 2.5-3.3 m (S542). In other cases most nests were at depths of 30-85 cm (C365). Increased aquatic vegetation in Lehman's P., OH (S541), one summer resulted in more spawning on coarser gravel than in other years. The finer substrates produced more fry per nest (S541).

	No.	Fry per nest	
	nests	Mean	Range
Sand and fine gravel	4	64,025	56,457-73,755
1967 Medium gravel	5	63,848	33,345-71,557
1967 Large gravel	4	21,150	10,676-28,736
1967 Mud, gravel, and detritus	5	48,285	7,599-86,700
1967 Mud and detritus	4	19,932	17,034-24,010
1965 Sand	4	19,263	6,576-36,383
1965 Mud and gravel	3	3,597	3,106-4,537
1966 Gravel	4	3,279	1,372-5,655

Nests were usually in areas exposed to the sun (C354) but often in the shade of trees (C365). Successful reproduction took place in the large embayments rather than the littoral areas of the main reservoir in Rathbun, IA (M561). Nesting and breeding behavior were described in C54, M134, B41, K140, M528, C354, and S534. Courting males make grunting sounds and may be attracted to areas of spawning by odor (G191). Density of adults did not affect the numbers of fry produced in various years in an Ohio pond (S541). There were 3 times as many nests in 1965 as in 1966 with about the same number of adults (S541). There were also more fry per nest in 1965. Swingle (S390) reported a repressive factor, probably a hormone, in the water which increases with population density and retards reproduction, but not sufficiently to prevent crowding and slow growth. High populations delayed spawning until August or September in Alabama (S452). Shortage of other food also resulted in bluegills eating most or all eggs and fry produced. Density of the bluegill population did not affect reproductive success in New York ponds (R237) or in Ohio (S541). Bluegills spawned in September in a portion of Big P., IL, which was separated off and treated with rotenone but did not spawn in areas where the population was not reduced (C354). Reproduction was poorer from high populations of adults than from low (A90). Levels of feeding which affected growth rate did not seem to affect reproductive success (H439). Density of vegetation was not related to abundance of bluegill fingerlings in Michigan ponds (B92), although S117 reports unusually high survival of young when vegetation is heavy.

	No.	Fry per nest	
	nests	Mean	Range
C54 IN Winona L.	4	86,631	11,257-224,900
C6 MI	17	17,914	4,670-61,815
C115 WI	-	-	22,000

There were an estimated 914 nests per hectare in Murphy Flowage, WI (S323).

Sperm viability was 8-12 minutes under a microscope in water (M134). Less than half the eggs hatched if sperm were in water over 1.25 minutes and none if the sperm were over 4 minutes old (C354). Eggs fertilized 30 minutes after discharge still gave over 50% hatch, and some hatched if not fertilized

until 3 hours old (C354). Average functional life span of sperm was estimated
at 1.0 minutes and of eggs, 67 minutes. Incubation takes 32-62 hours (M428),
71 hours at 22.6°C, 34 hours at 26.9°C, and 32.5 hours at 27.3°C (C354). At
24°C eggs from one female hatched in 43 to 72 hours with a mean of 50 hours
(H443), and at 25.5°C they hatched in 43 to 56 hours with a mean of 46 hours
(M544). The mean hatching time at 23.5°C was 41.87 hours in light and 45.75
hours in dark (T136).

Embryological development was described (M134, H443, T136, and M525).
Fry averaged 4.28 ± 0.17 mm when incubated at 22.6°C, 4.98 ± 0.11 mm at
26.9°C, 4.69 ± 0.08 mm at 27.3°C (C354), and 3.26-3.72 mm at 23.5°C (T136).
Hatching success was 56% at 22.6°C, 83% at 26.9°C, and 90% at 27.3°C (C354).
Eggs stripped and fertilized gave a 51.7% hatch at 24°C (H443). Treatment
with 0.001 ppm malachite green gave a 71% hatch compared to 44% in untreated
eggs and 50% at 0.01 ppm and 47% at 0.1 ppm (M527). A few even hatched when
treated at 0.35 ppm. Maximum survival rate of fry was greatest at 0.01 ppm.
There was no surviving spawn at salinities of 0.5% (S550), but hatches were as
good in 19% sea water as in the control (T141). The 11 day TL_m of fry was
13% sea water and 14-21 mm fingerlings had 96 hour TL_m of 29%-30% sea
water (T141). No fry survived at 0.23 mg/l zinc (C372). Bluegill gonads developed at concentrations of 0.25 mg/l zinc, but spawning was inhibited (C393).

Fry do not become free swimming until about 10 days after fertilization
(H439) or 4 days after hatching, and hydra killed many in this time in a Michigan study (C362). After absorbing the yolk sac, the 6 mm fry abandon the
nest and move into littoral vegetation. At 10-12 mm they move to the limnetic
zone in the upper 3 m depth for 6-7 weeks, returning to the littoral when 22-25
mm (W223, W224). In some New York ponds they were taken in the plankton
samples only when from 4-8 mm long (H439). Pumpkinseed, redear, and
warmouth did not appear to have a limnetic stage. Larval bluegills appeared
in the limnetic zones in late June through July in northern Wisconsin (F156).

Survival of yolk fry was subject to food supply, space, and competition
with other species (K24). Fry starved if not provided with food by 8.5 to 9
days after fertilization of the eggs or by 6.5-7 days after hatching (T136).
Finclipping, except possibly of the anal fin, did not affect predation loss of
59-67 mm bluegills to largemouth bass (C389).

When bluegill fry were stocked alone, survival rates to the first fall were
75%-100% if food was abundant (S244). When stocked with bass fingerlings, the
survival rates were 76%-85%. No correlation occurred between calculated egg
production and number of young bluegills surviving to fall (B418). Instantaneous mortality rate of age 0 bluegills were 0.68, 0.42, and 0.43 in 1971, 1972,
and 1973 in Rathbun L., IA (M561). The average annual mortality rate of bluegills for 4 years in ponds when stocked with bass was about 30%; but if the
bluegills were stocked the year following bass, there was a mortality rate of
74% the first 15 months and about 30% annually thereafter (R237). At 19
months in ponds where bluegills were fed and unfed, there was 84.7%-87.2%
survival of the stocked bluegills, even though growth stopped at 14 months in
the ponds where the bluegills were not fed (S536). Annual mortality rates of
bluegills over age III-IV were estimated at 57%-99% (P193), 81.6% (G48), 60-
76% (R34), 71-88% (R152), 77% (R35), 73% (G49), and 89% (G38). Mortality due
to pH changes was similar for bluegill and largemouth bass fingerlings, so
that the manipulation of pH is not a feasible method of selectively thinning
bluegill populations (C363).

Adult bluegills feed mainly on aquatic insects, small crayfish, and small
fish (A87, B41, B368, D72, D103, D161, E135, F148, G127, G193, H229, H250,

J25, K81, K125, L29, M134, M310, M336, M404, M523, M524, N100, R125,
R200, R212, R285, R288, R310, S357, S372, S467, T135, W221, and W246).
Aquatic vegetation is sometimes a major food item (B15, H229, L127, M134,
P205, R24, R200, S348, S372, and S464). Aquatic vegetation may be eaten primarily as roughage (G193). Bluegill were shown to get energy from algae and
to grow more when algae were added to the diet (K145). In streams, terrestrial arthropods were a significant portion of the food (M404). Terrestrial
organisms also were most of the food of larger bluegill during the winter-
spring rise of water levels in Beaver Res. (M523) and in spring, summer, and
fall in White Oak L., TN (N135). In denser populations, bluegills ate more
small fish (S464). Alewives were the major food of 200-250 mm bluegills in
L. Hopatcong, NJ (G97). Increase of plant material in the food as the summer
progresses was indicated as a cause of reduced catchability of bluegills by
anglers (L127). The fish louse, *Argulus*, has been found in bluegill stomachs,
and there is evidence that bluegills may perform a "cleaning" function on infected fish (S545). Planktonic crustacea are taken mostly during the day when
the crustacea are near the bottom and only rarely in open water at night, as
are some other plankton feeders (K135). Sediments taken in with the food
were believed to result in a lower assimilation of 137 Cs than shown by other
fish which did not take in as such sediment material (N135).

Young bluegills feed on smaller crustacea and aquatic insects (R285,
T135, S467, E121, N93, N94, R212, S372, K81, K153, and H439) and sometimes
eat plants (R285). Feeding started at 5 mm on *Polyarthra* rotifers and copepod nauplii; at 7 mm, other rotifers and *Cyclops* were taken; and cladocerans
became the major food at 8 mm-15 mm (S586). An inverse relation was noted
between size of bluegill and percentage of microcrustacea in the food (T135).
While in the limnetic stage, 10-25 mm TL, bluegills feed almost exclusively on
planktonic crustacea (W223). Electivity indexes indicated that these small
fingerlings selected copepods and *Bosmina* over *Daphnia* (W223). The mouth
gape was 0.20 mm at 6 days after fertilization and increased to 0.27 mm by 9
days (T136). Particle size of prey was a major factor in food selection of
bluegills as they increased in size (H439). The mouth gape = 0.217 mm + 0.093
SL with a correlation coefficient, r = 0.99 (W249). The fact that bluegills at 50
mm can handle the whole range of foods of larger bluegills was cited as one
cause of stunted population (W249). The handling time (time between seizure
of prey and swallowing) increases as a bluegill nears satiation and may be 2.5
to 3 times that of a hungry bluegill (W249). As a bluegill becomes satiated,
feeding behavior becomes more complex and indicates more selective response
to the prey (C394). Size selection of prey increases as prey density increases
(W253). Bluegills are able to move the eyelens, primarily along the roatral
and caudal axis, and to use their eyes in feeding more than do bullheads, suckers, and goldfish (S580). Young bluegills fed mostly on *Hyalella* in littoral
zones but ate mostly zooplankton in heated outflow water where zooplankton
were abundant (N141).

Feeding occurred in the winter in Buckeye L., OH (M134), with fat ac-
cumulating in the tissue. The fat disappeared in April-May as the fish became
more active. In aquaria no feeding occurred below 3.3°C (C260), and feeding
in ponds has been found at 3.9°C (S550). In ponds there was no growth and
little feeding at temperatures below 10°C (A47). In a Michigan lake, bluegills
fed little in the winter, and cladocera were the major food in mid-winter
(M540). Most active feeding was in the spawning season (M134) or at least
the early part of the summer (S464). Optimum temperature for feeding was
given as 27°C with a maximum temperature of 31°C (K161).

Peak catch of bluegills in traps was from 9 P.M.-5 A.M. (M134). Peak activity was just before dawn (D211, D212). Bluegills are largely diurnal in feeding but also feed in the evening (K135). Although bluegills were more diurnal than nocturnal, they were most active from 4-10 P.M. and 8-10 A.M. in the Mississippi R. (R288). In Clear L., IA (C247), they were mostly diurnal. Limnetic fry were most active at dusk with peak activity about one hour after sunset. At that time most were near the surface, suggesting a vertical migration. Small bluegills did eat in the dark when concentrations of zooplankton were high, but it is unlikely that they can eat at the light intensities found one hour after sunset with plankton concentrations usually found in the lake (W223). Bluegills tended to be limnetic during the daytime, feeding largely on plankton, but moved to littoral areas about sundown and fed on benthic and aufwuchs organisms (B421). Larger bluegills remained in the littoral areas more than did those 70-129 mm TL. The onshore and offshore movements were more pronounced in September and October. Activity patterns were typically diurnal (B423).

At 10°C, 14-19 g bluegills ate 0.7%-1.9% of their body weight per day, and at 20°C, 4.7%-6.2% (H288). During the growing season bluegills ate 336% of the body weight but only 13% during the rest of the year in Missouri (A47). Daily rations have been estimated at 1%-2% of body weight (S464), 2.0%-2.8% (W225), 1.16%-3.59% (G171), and 3% (G49). Daily rations were less in dense populations and less for large than small bluegills (S464). Basal food requirements, without growth, were estimated at 1.1%-1.2% of body weight, when bluegills were fed beef liver (M542) and 1.8% at 20.5°C and 2.5% at 25.5°C, when bluegills were fed earthworms (R291). Bluegills averaging 29.7 g needed 7.2 mg of nitrogen (G171) or 5.8 mg (G172) per day for maintenance. At 25.5°C bluegills consumed less food than at 20.5°C. Food consumption rates were positively correlated with water temperature except at temperatures above 25°C and in autumn after a hot summer. The daily meal was 0.8% of body weight in February and 3.2% in June, averaging 1.75% for the year in White Oak L., TN (K146, K162). Food is utilized less efficiently as age increases (G49, G170, G171). Bluegills fed at 1.16% of body weight per day showed no increase in protein but 140% increase in fat in 27 days, while those fed 3.59% increased 38% in protein and 383% in fat (G171).

Bluegills fed mayfly nymphs to satiation ate 1.16 to 3.43% of their weight per day, and intake was greater at 25°C than at 15° or 20°C (P210). Energy expenditure for food utilization ranged from 4.8% to 24.4% of the caloric content of the daily ration (P210).

Growth efficiency (utilization of protein) decreases with body weight (G195, G190).

Mean weight	No.	
13.9	11	Nitrogen retention = -2.25 + 0.439 Nitrogen consumption
27.0	18	Nitrogen retention = -3.17 + 0.419 Nitrogen consumption
41.3	16	Nitrogen retention = -5.93 + 0.361 Nitrogen consumption
68.9	6	Nitrogen retention = -2.54 + 0.290 Nitrogen consumption
85.2	9	Nitrogen retention = -3.21 + 0.141 Nitrogen consumption

Starvation experiments (S544) indicated that nitrogen excretion rates of starved fish were lower than of controls, suggesting use of fats rather than proteins for energy requirements. Over 29 days of starvation, 70-90 g bluegills lost an average of 1.73 g protein and 1.61 g fat. The protein provides 5.65 kilocalories per gram and the fat 9.45. The daily requirements were estimated at 0.86 kilocalories per day. An average of 36.5 kg of bluegills/ha was estimated to consume annually 142 kg of food organisms/ha (G49). About 30% of the protein consumed went to growth. Protein maintenance levels were highest at $29°$-$32°C$ and decreased at $23.9°C$ and $15.6°C$ but did not decrease further at $7.2°C$ (S585). Seasonal changes in protein content of pituitary gland was reported (L229). In March-May the pituitaries were 24%-26% protein; in June-July, 36%-37%; and in August-September, 55%-57%.

Ultraviolet lights attracted insects to provide 40% of the growth of caged bluegills (H473).

Meals of natural food organisms required about 18 hours at $21°C$ for complete digestion, but 50% was digested in the first 6 hours (W225). Force feeding resulted in considerable variation in digestion rate. Rate of digestion decreased after bluegills were starved 7, 14, and 25 days, with rates after 25 days being less than half those of control fish. Size of the meal had little effect upon the time needed to empty the stomach, and when the meal was increased 2.7 times, the rate of digestion per unit of time increased 2.2 times. In Minnesota lakes at about $22°C$, it took 20-26 hours for complete digestion, with 50% digestion in 4-8 hours (S464). If more food than 1% of the volume of the fish is eaten at a time, digestive enzymes become exhausted before the food is digested and some nutrients are not assimilated (N143).

Metabolic rates of small bluegills were less affected by a change of temperature from $25°C$ to $30°C$ than larger bluegills. The oxygen consumption, in mg/hr, was described by the following formulas (O49):

at $25°C$, 64 bluegills, 1.4-116 g $\quad$ OC = -0.966 + 0.717 W (in g)
at $30°C$, 19 bluegills, 3.7-36.5 g $\quad$ OC = -0.785 + 0.749 W (in g)

Metabolic rates ranged from 30.7 to 160.9 mg O_2/kg/hr and were 26% higher at night than in daytime (P210). Metabolic rates increased with temperature ($15°$, $20°$, and $25°C$) but declined with increase in weight (P210). Oxygen consumption of isolated bluegills (7-11 g) was 8.8 mg O_2/hr compared to 4.4-4.6 mg O_2/hr for schooled bluegills (P201). Optimum temperature for respiration was given as $30°C$ with a maximum of $34°C$ (K161). At $2.5°$-$4.0°C$, bluegill ventilation rates increased as O_2 was reduced from 4.0 to 1.0 mg/l but did not increase further at 0.5 mg/l and all died at 0.26 mg/l (P211). Bluegills did not come up to the ice surface at low oxygen as did northern pike and yellow perch, which are less subject to winterkill. Critical dissolved oxygen levels in winter are 0.5 mg/l (P211), 0.6 mg/l (C391), and 0.8-3.6 mg/l (M543). At $21.3°$-$21.8°C$ fingerling bluegills died in 20 hours at 0.8 ppm O_2, quit eating at 1 ppm, and showed increased breathing movements at 5 ppm (P202). Concentrations of 0.25 mg/l zinc increased the respiration rate of adult bluegills, but 1/10 of that concentration killed fry and inhabited growth (C393).

The Q_{10} values for bluegills were from 1.6 to 1.8 for the size ranges covered compared to 2.1-3.0 for pumpkinseeds, indicating that the bluegills are better adapted to high temperature (O49).

Using elimination of 137 Cs at various temperatures, the Q_{10} values for small bluegills (under 70 g) ranged from 1.3 to 2.1 and for large bluegills ranged from 1.8 to 2.6 (U12). The bluegill toleration limits of the pH range, using a single acid or base in a continuous-flow method were pH 4.00 ± 0.15 to pH 10.35 ± 0.15 (T137). Bluegills had been reported to withstand transfers from pH of 8.2-8.7 to pH 4.4-6.4, but at 5 ppm O_2 bluegills showed some mortality when transferred from pH 7.9 to pH 9.6 (W237). At higher O_2 concentrations pH effects were less. Maximum salinities at which bluegills were found were 2 ppt (C368) or 5.6 ppt (K151).

The final preferred temperature of bluegills was 32.5°C (**F161**). Preferred temperatures in L. Monona were 28°-33°C and in laboratory tests were 28°-32°C (**N141**). Bluegills acclimated at 29°C died in 8 minutes in 36.5°C water, but resistance time increased until bluegills acclimated at 33°C survived an average of 340 minutes at 36.5°C (**N141**). Bluegills were often more abundant in the heated outflow than elsewhere in L. Monona. Preferred temperatures in L. Wingra were 31.2°C and did not differ when exposed 30 minutes at 21°C but were 0.6°C higher for survivors after 30 minutes exposure to 36.1°C (**B423**). Activity was reduced immediately after treatment at 21°C or 36.1°C (**B423**).

Acclimation temperature (°C)	Lower lethal (°C)	Upper lethal (°C)	Citation
10	-	28	H255
15	2.5	30.7	S333, H141
20	0.5	31.5	S333, H141
21.5	-	31	H461
25	7.5	-	S333, H141
30	11.0	33.8	S333, H141
30	-	35	H255
-	-	37-39	T102

Florida bluegills showed a different temperature tolerance than bluegills from Tennessee (H141).

Survival occurred even at body temperatures up to 35.6°C (T102). The ultimate incipient upper lethal temperature was 35.5°C, but a few fish survived to 41.5°C (H474).

Breathing and swimming activity rates increased when the diurnal temperature cycle was changed from 24.8°-26.0°C to 24.8°-29.2°C (C390).

For juvenile bluegill the 96 hour TL_{50} were (B422):

Acclimated (°C)	Lower TL_{50} (°C)	Upper TL_{50} (°C)
12.1	3.2 ± 0.56	27.5 ± 0.58
19.0	6.3 ± 0.41	33.0 ± 0.26
26.0	9.8 ± 1.00	36.1 ± 1.10
32.9	15.3 ± 0.22	37.3 ± 0.19
Eggs spawned at 26	21.9	33.8
fry	11	34

The preferred temperatures for bluegills acclimated at various temperatures were (C396):

Acclimated (°C)	6	9	12	15	18	21	24	27	30
Preferred (°C)	18.7	19.6	23.9	25.9	29.2	30.1	31.2	31.4	31.7
95% C.I.	17.3	19.5	21.6	23.7	25.5	27.2	38.7	30.1	31.5
	22.3	23.6	25.0	26.5	28.2	30.1	32.1	34.2	35.4

The minimum dissolved oxygen concentration at which bluegills survived (M351):

Temperature	Immediate exposure	Acclimated
25°C	0.75 ppm	0.70 ppm
30°C	1.00 ppm	0.80 ppm
35°C	1.23 ppm	0.90 ppm

The lowest observed oxygen tensions at which all bluegills survived 24 hours in summer were 3.4 ppm at 23°C or for 48 hours in winter were 3.6 ppm, but the highest tensions which killed all bluegills were 3.1 ppm at 15°C in summer and 0.8 ppm in winter (M543). Small bluegills were less tolerant of oxygen deficiency than larger ones. Bluegills at temperatures of 17°-24°C in the summer avoided water with 1.5 ppm O_2 and to some extent at 3 ppm O_2 but not 4.5 or 6.0 ppm (W226). At 30-32 ppm dissolved oxygen, bluegills died from gas emboli in gill capillaries (B382). Feeding sac fry were more sensitive to LAS toxicity than other early life stages (H472). Rise in temperature from 20°C to 30°C reduced the survival time of bluegills to lethal and sublethal concentrations of zinc (B420). Bluegills exposed to 0.004 mg/l H_2S had increased endurance at slow speeds but reduced endurance at higher speeds than did control fish. Resistance to copper was increased by exposure to H_2S, but resistance to malathion was not affected (O51).

When weights were put in bluegill stomachs, the bluegill rested on the bottom but gradually increased the volume of the air bladder to approach neutral buoyancy (G194).

Tagging studies indicated that bluegills did not wander around the lake during the summer (K139), but R280 reported an average of 1.8 km movement (up to 5.5 km) for tagged bluegills in Folsom L., CA. No territory or homing was noted in a 40 acre Michigan lake, May to August (F97). In winter bluegills were near the bottom of the lake at 4-6 m (M376). Dominant bluegills survived longer in 32 mg/l zinc than submissive bluegills, but when shelters were provided in the bioassay tanks the dominance-submissive relationships were not sufficient to affect mortality rates (S587).

Carrying capacity for bluegills in Illinois ponds was estimated at 56-84 kg/ha in forest lands, 112-225 kg/ha on light colored soils, and 225-450 kg/ha on black soils (L228).

			Standing crops (kg/ha)	Yield or production (kg/ha/yr)
G193	IN	Wyland L.		
		June	98	
		year round average	72	y = 91
B417	IN	White R.	74, 80	
		thermally enhanced areas	42, 16	
B418	MI	pond	83-480, mean 344	
		ponds with 33-38% annual renewal	197, 227	
S588	MI	Mill R.	49	p = 23
M575	IA	Williamson P.	221	
B407	IL	end of 1 season	68, 79, 99	
		end of 2 seasons	117, 138, 141, 188, 204, 233, 246, 302	
E146	AL	pond, consecutive years		y = 142, 83, 64
B373	IL	ponds, only bluegills	112, 113, 142, 194	
	OR	ponds, only bluegills	101, 260	
	KY	ponds, only bluegills	353, 368, 744	
P193	MI			p = 2.5, 8.2, 28, 33
G127	IN			p = 91

Standing crop (no/ha)

S584 TX reservoirs, spring   35, 834, 2284
S588 MI Mill L.   2678

Standing crops of bluegills were higher in ponds with largemouth bass and channel catfish than in ponds with largemouth bass (B407).

There have been several detailed studies of production rates in bluegill (G49, G127, G166, G193, G195, C355, P193, H439, and R237). Annual production averaged 244 kg/ha in a Michigan pond and 312 and 330 kg/ha in two similar ponds where 33% to 38% annual removal took place (B418). The production by bluegills over a year old was 179 kg/ha compared to 112 and 158 kg/ha in the ponds with annual removal. Net production was 4.4 kg/ha in an Oklahoma pond with a standing crop of 8.4 kg/ha (W247). Cropping more than doubled the production rate in Fork L., IL (B373), and a cropping of 50% was suggested as producing an optimum production rate in MI (P193). When spring standing crops were high, annual production was low in PA (C355). Biological productivity of bluegills was thought (W123) to be lower in Felt L., CA, because the bluegill has not developed mechanisms for suppressing maintenance metabolic requirements after the growing season or for maintaining adequate food supplies to allow for potential growth over the entire growing season in a fashion suitable for the situation where temperatures remain high most of the year. A model of biomass dynamics was developed with data on bluegills (K161).

When Mill L., MI, was opened after 5 years of no fishing, there were 513 bluegills/ha and 13% of them were caught in 3 days with 96 hrs of fishing per hectare (S595). There was no evidence that catchability decreased over the 3 days.

Hematocrit values of 29%-50% (V87) and 35% with a range of 18%-55% (H435) were reported. Hematocrits of bluegills did not differ upon exposure to diuron (M533) but increased on exposure to 0.05 ppm heptachlor (A89) and to 0.04 ppm methoxychlor (K142). Serum protein was recorded as 2.4-5.96 g% (V87); and total plasma protein, at 6.8 g% (3.0-11.5 g%) (H435). Hemoglobin was 7.9 g% (6.0-11.9 g%) (H435). Hemoglobin was also reported at 45.2 ± 9.7 g% (S151). Red blood cells per mm^3 of 27 bluegills averaged $2,605,555 \pm 654,670$ (S151).

The number of chromosomes is 48 (R286) but reported at about 44 in one study (B381). Bluegills hybridize with *Lepomis auritus*, *L. cyanellus*, *L. gibbosus*, *L. gulosus*, *L. humilis*, *L. megalotis*, *L. microlophus*, and *L. punctatus* (C354, M428). Laboratory crosses have been made with *Pomoxis annularis* (M428) and with *P. nigromaculatus* (M529). The significant difference in egg size with *P. nigromaculatus* is not believed to be the cause of poor success of the hybrids (M529).

Good management discussions of bluegills in ponds are available in B373, R237, and B382. In cage culture 5 to 27 bluegills per m^3 were sufficient to eliminate fighting (H473).

Sizes of bluegill taken in gillnets of various mesh sizes (L216):

Bar measure	No. fish	TL	Range
25	40	137	89-190
38	63	170	127-203

Bar measure	No. fish	TL	Range
51	25	201	152-216
64	3	218	203-229
89	1	196	

Models of yield of bluegills in Mill L., MI, were constructed to show effects of exploitation rates and minimum size limits (S588) and increased growth rates (S589). At the observed growth rate maximum biomass was reached when bluegills were 152 mm (S588).

DOLLAR SUNFISH, *Lepomis marginatus* (Holbrook)
    This sunfish, which reaches a maximum of about 180 mm, is found in freshwaters from Oklahoma to South Carolina and Florida to Texas (E119, B399). It has 48 chromosomes (R286).

LONGEAR SUNFISH, *Lepomis megalotis* (Rafinesque)
    The longear sunfish ranges from southern Minnesota to Ontario and western New York, southward through the Mississippi basin to the Gulf states and Mexico (M420, R285) and is usually found in sluggish streams. It appears to be intolerant of large amounts of silt (R285) and of salinity (B399).

L110 IL	161 fish	46-137 mm SL	TL = 1.22 SL
W105 IL	17 fish	54-119 mm SL	TL = 1.242 SL

TL	No.	Mean weight	Range	Citation
25-50	1,292	1.3	0.1-7	S472 AL
51-75	4,539	2.5	0.8-5	S472 AL
	14	6	-	J44 OK
76-101	3,210	9	1-31	S472 AL
	43	18	-	J44 OK
102-126	2,873	20	8-45	S472 AL
	53	34	-	J44 OK
127-151	867	40	18-86	S472 AL
	62	53	-	J44 OK
152-177	153	59	15-100	S472 AL
	5	87	-	J44 OK
178-202	8	100	68-200	S472 AL
	2	134	127-142	D179 TX

L110 IL	Hutchins & Clear Creeks	164 fish, 47-137 mm,		$\log W = -4.77 + 3.16 \log SL$
W105 IL	Crab Orchard L.	19 fish, 54-119 mm,		$\log W = -5.28 + 3.431 \log SL$
P208 IN	White R. Sec. B, 1970	-	-	$\log W = -2.371 + 2.023 \log TL$
	Sec. C, 1970	-	-	$\log W = -2.449 + 2.059 \log TL$
	Sec. A, 1970	-	-	$\log W = -2.892 + 2.265 \log TL$
	Sec. B, 1969	-	-	$\log W = -3.117 + 2.332 \log TL$
	Sec. A, 1969	-	-	$\log W = -4.066 + 2.781 \log TL$
	Sec. C, 1969	-	-	$\log W = -4.766 + 3.083 \log TL$
S472 AL		352 fish, 52-180 mm,		$\log W = -4.47 + 2.88 \log TL$
J40 OK		177 fish, 52-160 mm,		$\log W = -4.05 + 3.197 \log TL$

The weight-length regression given in B376 with a slope of 0.2675 is obviously in error and possibly should be 2.675.

	Length	No.	K(SL)	Range
L110 IL Hutchins & Clear Creeks	47-137	164	3.56	
D179 TX Medina L.	151-160	2	3.61	3.10-4.12
W142 IL Little Grassy L.	85-135	21	4.15	
			K(TL)	
S472 AL	52-190	11,649	1.93	1.72-2.05
J40 OK Little R.	52-160	177	-	1.77-2.44
P208, B416, B417 White R., 1969	80-170	128+	3.02	2.5-5.1
1970	80-170	101+	4.25	2.9-6.6

	TL				Weight		
	No.	Mean of means	Central 50%	Range	No.	Mean	Range
Age 0-July							
OK: J44	2	58	-	-	-	-	-
Age 0-Aug.							
AR King's R.: B378	+	-	-	35-45	-	-	-
IL Big C.: L108	8	46	-	-	-	-	-
OK: J44	3	58	-	-	-	-	-
Age 0-Sept.							
OK: J44	1	76	-	-	-	-	-
Age 0							
IL Jordan C.: D88	3	51	-	-	-	-	-
OK Illinois R.: J25	3	61	-	-	-	-	-
IN White R.: B416	16	61	-	-	16	29	-
AR 2 lakes: H442	7	62	-	58-66	-	-	-
Age I							
MI north: H90	490	38	-	25-64	-	-	-
MI south: H90	59	53	-	25-81	-	-	-
AR King's R.: B378	12	56	-	-	-	-	-
IL rivers: D88, L108, L110	334	64	56-74	56-74	36	6	-
WV Elk R.: H218	1	71	-	-	-	-	-
MO Black R.: P74	42	71	-	38-114	-	-	-
KS: C348	+	78	-	-	-	-	-
IN: H62	1	79	-	-	1	9	-
IL Horseshoe L.: G80	7	79	-	-	-	-	-
IN White R.: B416	33	82	-	-	33	48	-
OK Illinois R.: J25	13	84	-	76-107	-	-	-
AR 3 lakes: H442	9	91	-	79-94	-	-	-
OK 3 lakes: J26, J42, H150	25+	104	-	97-107	+	17	-
Rivers combined	435	69	56-82	56-114	36	27	6-48
Lakes combined	41+	95	79-107	79-107	+	17	-
Combined	1,026+	73	56-82	25-114	70+	24	6-48
Age II							
IN Muscatutuck R.: M317	2	53	-	51-56	2	3	1-6
MI north: H90	180	61	-	38-94	-	-	-
MI south: H90	142	69	-	48-107	-	-	-
AR King's R.: B378	14	74	-	-	-	-	-
KS Farlington L.: C226	14	79	-	-	-	-	-
IL 4 streams: D88, L108, L110	333	87	79-104	79-104	36	20	-
KS: C349	+	89	-	-	-	-	-
MO Black R.: P74	128	97	-	58-152	-	-	-
IN: H62	18	97	-	-	20	17	-
IL 2 lakes: G80, W105	15	105	-	104-107	-	-	-
WV Elk R.: H218	17	107	-	-	-	-	-
IN White R.: B416	51	109	-	-	51	87	-

	No.	TL Mean of means	Central 50%	Range	Weight No.	Mean	Range
Age II (cont.)							
OK Illinois R.: J25	119	112	-	94-132	119	31	-
OK Tenkiller L.: H150	2	127	-	-	-	-	-
AR 3 lakes: H442	10	146	-	127-165	-	-	-
Rivers combined	664	93	79-107	51-152	208	40	1-87
Lakes combined	41	117	104-135	79-165	-	-	-
Combined	994+	95	79-107	38-165	228	35	1-87
Age III							
MI north: H90	80	69	-	51-99	-	-	-
IN Muscatutuck R.: M317	20	76	-	56-104	20	6	1-20
MI south: H90	53	89	-	69-112	-	-	-
MO Black R.: B378	5	90	-	-	-	-	-
KS: C349	+	103	-	-	-	-	-
IL 2 lakes: D88, W105	149	108	-	102-112	-	-	-
KS Farlington L.: C226	13	112	-	-	-	-	-
MO Clearwater L.: P71	+	112	-	-	-	-	-
IL 3 streams: L108, L110	69	116	-	102-130	30	45	-
IN: H62	9	117	-	-	8	34	-
MO Black R.: P74	148	119	-	74-168	-	-	-
IN White R.: B416	188	128	-	-	188	95	-
OK Illinois R.: J25	23	130	-	114-142	23	48	-
WV Elk R.: H218	7	137	-	-	-	-	-
AR 2 lakes: H442	9	154	-	142-173	-	-	-
OK L. Duncan: W1	1	190	-	-	1	227	-
Rivers combined	460	115	102-130	56-168	261	51	1-95
Lakes combined	172+	125	110-142	102-190	1	227	-
Combined	782+	113	102-130	51-190	270	57	1-227
Age IV							
MI north: H90	80	76	-	64-94	-	-	-
MI south: H90	28	97	-	76-124	-	-	-
IN Muscatutuck R.: M317	64	99	-	76-112	64	20	6-31
AR King's R.: B378	11	111	-	-	-	-	-
KS Farlington L.: C226	1	119	-	-	-	-	-
IN: H62	3	119	-	-	3	40	-
IL 4 streams: D88, L108, L110	268	124	107-145	107-145	16	62	-
IL 2 lakes: G80, W105	14	124	-	119-128	-	-	-
KS: C349	+	133	-	-	-	-	-
WV Elk R.: H218	4	135	-	-	-	-	-
OK Illinois R.: J25	2	135	-	132-137	2	65	-
MO Black R.: P74	100	137	-	89-201	-	-	-
IN White R.: B416	25	140	-	-	25	102	-
Rivers combined	474	125	111-137	76-201	107	62	6-102
Lakes combined	15	123	-	119-128	-	-	-
Combined	603+	119	107-137	64-201	110	59	6-102
Age V							
MI north: H90	14	76	-	56-130	-	-	-
IN Muscatutuck R.: M317	86	122	-	99-140	86	34	14-65
AR King's R.: B378	6	124	-	-	-	-	-
IN: H62	3	127	-	-	3	45	-
MI south: H90	1	135	-	-	-	-	-
IL 4 streams: D88, L108, L110	110	140	129-160	124-160	9	85	-
IN White R.: B416	13	149	-	-	13	124	-
OK Illinois R.: J25	1	150	-	-	1	71	-
MO Black R.: P74	45	152	-	102-183	-	-	-

	No.	TL Mean of means	Central 50%	Range	Weight No.	Mean	Range
Age V (cont.)							
IL Crab Orchard L.: W105	1	173	-	-	-	-	-
Rivers combined	261	139	124-152	99-183	109	80	14-124
Combined	280	132	124-150	56-183	112	75	14-124
Age VI							
MI north: H90	4	109	-	94-130	-	-	-
MI south: H90	2	127	-	117-137	-	-	-
IN Muscatutuck R.: M317	36	135	-	127-142	36	45	31-68
IL Jordan C.: D88	28	142	-	137-150	-	-	-
AR King's R.: B378	4	146	-	-155	-	-	-
IN White R.: B416	2	152	-	-	2	143	-
MO Black R.: P74	7	157	-	137-178	-	-	-
Combined	83	139	135-152	94-178	38	78	31-143
Age VII							
MI north: H90	2	99	-	94-102	-	-	-
IL Jordan C.: D88	5	132	-	-	-	-	-
IN Muscatutuck R.: M317	8	142	-	127-150	8	51	40-71
MO Black R.: P74	3	165	-	152-168	-	-	-
Combined	18	135	-	94-168	8	51	40-71
Age IX							
MI north: H90	1	89	-	-	-	-	-

		Calculated TL at each annulus								
	No.	1	2	3	4	5	6	7	8	9
S241 IN Brummetl's R.	195	21	33	53	71	89	100			
S241 IN Jack's Defeat R.	223	21	33	58	79	102	112			
D88 IL Jordan C., upper	353	28	61	89	114	127	142			
lower	462	28	58	86	109	122	130	132		
B376 IL Big Walnut C.	55	61	87	106	117	121				
G80 IL Horseshoe L.	15	43	50	107	120					
P113 MO statewide	3,009	33	64	91	109	122	127			
headwaters	1,209	28	58	84	102	119	127			
middle stream	1,003	33	66	94	112	122	124			
downstream	887	36	71	99	114	127	130			
poorest station	+	25	48	74	84	104	102	122	130	119
best station	+	38	79	117	137	150	152	168	142	147
P74 MO Black R.	473	30	61	89	112	130	142	152		
B116 MO	+	36	81	119	137	165	170			
B378 AR King's R., females	+	50	75	93	108	121				
males	+	49	75	94	111	126	139			
Beaver L., females	+	55	79	98	113	124	134			
males	+	56	79	100	117	129	137			
A87 AR Bull Shoals L.	150	43	81	104	114					
T54 KY streams	57	66	107	135	150					
J41 OK Ft. Gibson L.	5	36	69	104	122					
J27 OK Spavinaw L.	3	43	89	104						
H303 OK Lawtonka L.	1	43	94							
E126 OK Rocket Plant L.	11	64	86	104						
J25 OK Illinois R.	161	46	86	112	124	135				
H150 OK Illinois R.	195	58	89	112	127	135				
J45 OK Verdigris R.	11	64	91	117	109	117				
tributaries	78	48	76	94	102					
S391 OK Rock Creek	20	64	112							
E61 OK Salt Creek	4	66	99							
F62 OK Sub Prison L.	8	71								
F64 OK Little R.	261	51	97	130	152					

		No.	1	2	3	4	5	6	7	8	9
			Calculated TL at each annulus								
J44	OK	656	69	102	117	132	135				
	slowest average	-	33	69	84	117					
	fastest average	-	114	140	150	152					
	12 reservoirs	109	76	102	124	137					
	5 large lakes	38	76	102							
	21 small lakes	247	66	102	114	130					
	7 ponds	19	61	104	124						
	7 streams	243	61	94	114	127	135				

	Calculated TL and increments at each annulus				
	1	2	3	4	5
L108 IL Big C.	30	56	79	94	117
Incr.	30	33	22	18	16
No.	333	203	103	64	17
W142 IL Little Grassy L.	41	94	128	165	
Incr.	41	53	33	33	
No.	21	21	17	1	
L110 IL Hutchins and Clear C.	41	79	107	132	147
Incr.	41	39	28	26	15
No.	133	97	55	25	9
W105 IL Crab Orchard L.	69	102	122	142	178
Incr.	69	33	18	15	8
No.	40				
P71 MO Clearwater L. Incr.	46	41	30		

	Increments in grams at each annulus				
	1	2	3	4	5
W105 IL Crab Orchard L. Incr.	5	14	12	14	9
L110 IL Hutchins and Clear C. Incr.	1	8	15	23	22

Calculation of growth data from scale measurements was usually on a direct proportion basis, but W105, P74, and B378 used Fraser-type correction factors. The body-scale relationship did not differ by sex or between Beaver Res. and King's River (B378).

In general, growth in rivers and streams is slower than in lakes of the same region. Growth was also better downstream than in headwaters (P113, J45, D88). Growth was more rapid in southern than in northern Michigan (H90) and generally was more rapid in the southern part of the range than in the north.

No growth compensation was noted by H90 but rather a positive correlation between growth in the 1st and in the 2nd season. Males grow faster than females (H90, B378).

Von Bertalanffy equations were computed (B378):

Beaver L., AR, males $L_t = 220 (1 - e^{-0.14(t+1.2)})$
females $L_t = 169 (1 - e^{-0.24(t+0.63)})$
King's R., AR, males $L_t = 203 (1 - e^{-0.17(t+0.65)})$
females $L_t = 165 (1 - e^{-0.23(t+0.60)})$

The males had greater ultimate lengths, 203 and 220 mm.

Both males and females matured at age II in Michigan, and a few larger age I fish were mature (H90). In New York they mature at age II and about 75 mm (R285). Spawning leaves a double annulus mark on the scales (H90).

Longear sunfish spawn in late June to early August in Michigan (H90), from early May to late July, at 24°-30°C, in Kansas (C349), and from late May to mid-August in Illinois (H8) and in Louisiana (H444).

Males defend territories and build nests on gravel areas (H444, H8, K140, B386) at water depths of 20-60 cm (H444) or 0.2-3.4 m (B383, B386). Long-ears may have spawned at greater depths in the Arkansas reservoirs (B383, B386) to take advantage, at higher water levels, of substrate normally utilized. The nests were usually in brush-free areas. A typical nesting cycle lasted almost 2 weeks, and most eggs in a colony hatched within a week. Occasionally, old nests were utilized, and another batch of larvae hatched (B383, B386). Water temperature at the beginning of spawning was from 21.6° to 22.8°C, but the temperature at the start of a second spawning in late July was 28.3°C. When temperature rose to 28.9°C, nests were abandoned (B384). Availability of food for guarding males may influence colony location (B384). Occasionally, smaller males may nest individually, away from the colonies. Nests are usually closely spaced in colonies (S583); and other males sometimes intrude to try to fertilize eggs, and females intrude to eat eggs (K163). Eggs are also eaten by hognose and white suckers, redhorse, and minnows (K163). A male guarding a nest paid no attention to topminnows but drove off other species, except for a largemouth bass capable of eating the male (W256). Males showed high levels of aggression in aquaria at 25°C, with either a short or long photoperiod, but not at 11°-13°C (S531). A long photoperiod (16 hour light and 8 hour dark) induced nesting at both 11°-13° and 25°C, but a short photoperiod inhibited nesting (S531). Injection of human chorionic gonadotrophin induces nest digging under a short photoperiod at 25°C but not at 11°-13°C. Although aggression may occur at other than the breeding season in confinement, it rarely occurs in nature, except on spawning territories (H444). Males make grunting sounds during courtship (G191).

The number of ova stripped from 5 female longear, 83-98 mm TL and ages II-III, averaged 414 and ranged from 177 to 717. After counting the ova retained in the female, the total number of developed ova averaged 548 and ranged from 236 to 940 (B384). In the spring many young oocytes are formed, but most are resorbed prior to spawning. Females under 101 mm had 5,500 fewer oocytes in May than April, those 101-129 mm had lost 4,500 oocytes, and the larger females had 15,000 fewer. In another population the oocyte production and loss was much less. The number of spawned ova (May to July) was estimated at 1,417 for females under 101 mm, 3,440 for those 101-129 mm, and 4,213 for the large females in Beaver Res. (B384). In Bull Shoals L. the figures were 3,600 for those under 101 mm and 4,136 for those 101-129 mm. Postspawning resorption of ova and regeneration of oocytes occurred simultaneously.

Collection of longear eggs and fry from 12 nests using an underwater vacuum device yielded 608-2,756 eggs or fry per nest, indicating that several females spawned in each nest (B384, B386). Spawning females often deposited eggs in one nest and shortly thereafter in another nest (B384). The average gonadalsomatic index for prespawning males was 1.2%, and females spawned at indexes of 2% to 9% (B383). The percentage of successful nests was 35% and 12% in Beaver Res. in 1967 and 1968 and 64% and 39% in Bull Shoals L. (B384). Larvae were 6.9-11 mm at the time of leaving the nest (B383), and males often guarded the nest even after the last larvae emerged (H444).

Aquatic insects and entomostracans were the major food of fish under 50 mm (A87). In Beaver Res., however, aquatic insects were the major food as entomostracans were scarce (M523, M524). Fish eggs and terrestrial insects became more important than the entomostracans as food of the larger longears (A87, M524, M523, P205, and W246). *Hyallella,* snails, and small crayfish were also eaten by longears in Kansas strip mines (M336) and oligochaetes in Beaver Res. (M524, M523). Fish was the principal food of longear sunfish in one study (L249). The use of terrestrial organisms was greatest in the spring when water levels were highest in Beaver Res. (M523). Longear sunfish were observed on two occasions to follow hogsuckers, *Hypentelium,* and feed on organisms stirred up by the suckers (H444). In a Kentucky stream longear sunfish frequented eddy currents, feeding on mayflies and on terrestrial insects and occasional scuds and crayfish (M404). Males guarding nests continued to feed but were limited to food available in the nesting territory (B384, B386). *Argulus* have been found in stomachs of longear sunfish, suggesting that the sunfish may sometimes serve a "cleaning" function (S545). In feeding experiments a 10 g longear sunfish utilized 33% of the protein consumed for growth, but a 105 g fish utilized only 5% for growth. A 75 mm sunfish needed 0.076 g food to add 1 mm TL, and 102-154 mm sunfish needed 0.16-0.285 g (G170). Longear sunfish ate less in aquaria than pumpkinseed or redbreast sunfish and often only ate on alternate days (M528). Longear sunfish are easily handled as test animals (W214).

In the dark longear sunfish rest on the bottom, apparently sleeping (B384), but they may sometimes feed at the surface in bright moonlight. At temperatures below 7°C longear sunfish are relatively inactive and were rarely taken in trap nets (B383). They tend to remain in home territories in reservoirs (B383) and in streams (G173-6). Although repopulation by longear sunfish of stream areas decimated of fish has usually been slower than most other species, longear sunfish repopulated Bayou Locombe II quickly to about 3.5 times the original population, probably because of the lack of predators (B385). Longear sunfish were one of the first fishes to repopulate an Arkansas stream after a fish kill (O52).

Standing crops in White R., IN, were 21.4 and 53.9 kg/ha compared to 19.1 and 14.9 kg/ha in heated effluent areas (B417).

Longear sunfish were taken at water temperatures up to 37.8°C (P208).

The two longear sunfish taken in 25 mm bar measure gill net were 114 and 127 mm TL (L216).

The number of chromosomes is 48 (R286).

REDEAR SUNFISH, *Lepomis microlophus* (Günther)

Redear sunfish, or shellcrackers, were native from North Carolina to Florida west to Texas and southern Missouri and Ohio (T113). They have been introduced in a number of other areas, including Oklahoma in 1935 (J44), Illinois in 1946 (C354), and California in 1948 or 1949 (E139). They are most common in large warm rivers, bayous, and lakes, occasionally in brackish water. They were frequently in 4.5 ppt salinity (B399), at 4.1 ppt (C368), or at 12.3 ppt (K151). In introduced areas they seem to require relatively clear water and some vegetation (T113, C354). Turbidity was not considered an important factor in a Louisiana study (C368).

Since no length conversion data were found, the bluegill conversion factors were used in the few cases where conversion to TL was necessary (K42, R24, W7, G48, R152, T132, and W254).

TL	No.	Weight Mean of means	Central 50%	Range	Citation
25-50	80	0.6	-	0.3-0.7	A39, S472: AL
51-75	106	2.3	-	1.1-4	A39, A40, L189, J44: AL, OK
	1,993	2.9	-	0.5-5.9	S472: AL
	2,089	2.5	2.0-2.9	0.5-5.9	Combined
76-101	290	6.8	-	5-18	S472: AL
	23	9.8	-	5-14	A39, A40, L189: AL
	93	14.5	-	12-17	J44, K42: IN, OK
	406	10.9	6.8-14	5-18	Combined
102-126	502	20	16-22	7-31	A39, S472, L189: AL
	228	26	-	-	J44: OK
	99+	44	-	26-48	C285, K44, S12: IL, IN, TN
	819	26	20-31	7-48	Combined
127-151	594	32	27-35	14-91	A39, A40, L189, S472: AL
	1+	34	-	27-43	H202, P189: FL, GA
	345	47	-	-	J44: OK
	62	58	-	43-74	C285, K42: IL, IN
	1,002+	41	32-56	14-91	Combined
152-177	278+	58	54-64	45-113	A39, A40, L189, S472: AL
	+	68	-	-	P189: GA
	341	76	-	-	J44: OK
	83	90	-	60-102	C285, K42: IL, IN
	+	99	-	-	S12: TN
	39	109	-	77-118	F74, H202, M242: FL
	741+	81	64-105	45-118	Combined
178-202	+	95	-	-	P189: GA
	250	105	100-111	77-136	S472, L189: AL
	372	115	-	-	J44: OK
	886	142	136-150	96-173	F74, H202, M242: FL
	36+	148	-	147-150	K42, S12: IN, TN
	1,544+	125	100-147	77-173	Combined
203-228	149	159	145-176	91-186	S472, L189: AL
	146	166	-	-	J44: OK
	+	175	-	150-201	P189: GA
	+	180	-	-	S12: TN
	1,310	216	190-227	118-272	F74, H202, M242: FL
	3	218	-	-	K42: IN
	1,608+	187	166-227	91-272	Combined
229-253	34	209	-	91-240	S472: AL
	+	227	-	199-255	P189: GA
	53	262	-	-	J44: OK
	789	292	267-328	227-363	F74, H202, M242: FL
	1	306	-	-	K42: IN
	877+	265	255-272	91-363	Combined
254-278	26	299	-	227-318	S472: AL
	20	355	-	-	J44: OK
	274	420	377-449	296-500	F74, H202, M242: FL
	320	377	346-428	227-500	Combined

TL	No.	Weight Mean of means	Central 50%	Range	Citation
279-304	3	558	-	-	J44: OK
	91	626	617-635	363-1088	F74, M242: FL
	94	612	-	363-1088	Combined
305-329	1	624	-	-	J44: OK
	76	891	831-952	635-1043	F74, M242: FL
330-355	17	1013	953-1043	817-1270	F74, M242: FL

		TL	No.	
J34	OK Ardmore L.	102-145	79	$\log W = -4.576 + 2.9357 \log TL$
J37	OK Rod and Gun Club L.	102-254	247	$\log W = -4.762 + 3.0065 \log TL$
	Chickasaw L.	127-279	73	$\log W = -4.919 + 3.1103 \log TL$
J44	OK	61-306	1,567	$\log W = -5.324 + 3.2580 \log TL$
H202	FL Blue Cypress L.	165-257	40	$\log W = -4.914 + 3.089 \log TL$
	Canal area	135-254	28	$\log W = -5.061 + 3.154 \log TL$
	Combined	135-257	68	$\log W = -4.982 + 3.119 \log TL$
S472	AL	51-254	3,937	$\log W = -4.622 + 2.96 \log TL$
T132	CA Folsom L. FL	55-230	+	$\log W = -4.004 + 2.712 \log FL$

The regressions given for Blue Cypress and the canal area in H202 did not differ significantly in slope nor adjusted mean, and the combined slope 3.119 did not differ significantly at the 95% level from 3.0.

	SL	No.	K(SL) Mean	Range
D179 TX ponds	90-165	28	3.85	2.27-4.67
W254 FL Griffin L.	175-224	130	4.90	4.5-5.7
Weir L.	175-224	94	3.67	3.45-4.2

	TL	No.	K(TL) Mean	Range
J40 OK	61-254	1,567	-	1.41-2.08
J37 OK Rod and Gun Club L.	102-254	247	-	1.77-1.80
Chickasaw L.	127-279	73	-	2.05-2.25
M539 LA 2 ponds	-	+	-	2.05-2.19
J96 OK Rod and Gun Club L.	-	-	1.80	-
H424 OK Rod and Gun Club L.				
summer	114-140	+	1.72	-
winter	114-140	+	1.83	

No differences in K of males and females were significant (W254). Condition factors increased with length and were highest in March and April (W254).

Drawdown of water levels resulted in higher condition factors of redears in Houston L., GA (P189). Reduction of bullhead and bluegill populations did not improve average condition of redears in Rod and Gun Club L., OK (H424, J96), although the growth rate did improve.

The following equation may be used to estimate total length of redear sunfish from the maximum body depth (L226):

$$TL = 6.06 \text{ mm} + 2.8917 \text{ depth of body.}$$

	No.	TL Mean of means	TL Central 50%	Range	No.	Mean Weight	Range
Age 0							
M524 CA							
5 days	1	5.5	-	-	-	-	-
30 days	1	10.2	-	-	-	-	-
57 days	1	14.0	-	-	-	-	-
124 days	1	32.0	-	-	-	-	-
Age 0							
W254 FL							
1 month	+	26	-	-	-	-	-
5 months	+	75	-	-	-	-	-
7 months	+	147	-	-	-	-	-
Age 0-June							
J44 OK	14	79	-	-	-	-	-
Age 0-July							
J33, J44 OK	13	69	-	58-71	-	-	-
Age 0-Aug.							
J44 OK	5	89	-	-	-	-	-
Age 0-Sept.							
C260 MI	+	-	-	15-25	-	-	-
J44 OK	3	91	-	-	-	-	-
Age 0-Oct.							
J44, H150 OK	32	144	-	142-145	-	-	-
Age 0-Fall							
B182 OK turbid ponds	-	-	-	-	+	43	22-54
intermediate	-	-	-	-	+	62	40-107
clear ponds	-	-	-	-	+	74	45-113
Age 0							
H442 AR 2 lakes	11	57	-	56-58	-	-	-
L151 IL	243	76	-	-	243	14	-
Age I							
C260 MI May	+	-	-	25-46	-	-	-
E146 AL	+	88	56-120	38-150	-	-	-
R34 IN Shoe L.	2	96	-	-	-	-	-
S12 TN Reelfoot L.	2	109	-	-	2	37	28-43
T132 CA Folsom L.	79	110	-	-	-	-	-
C285, L151 IL	910	111	-	46-132	710	59	-
W220 IN 4 lakes	125	112	105-141	99-155	-	-	-
C260 MI Sept. 4 ponds	+	113	96-130	79-157	+	37	8-96
W254 FL ponds	+	123	116-140	89-160	+	54	22-93
J25, J26, J41, B182 OK	10+	125	109-143	97-155	5+	70	20-164
H442 AR 3 lakes	32	126	-	114-147	-	-	-
B109, V44 LA TX	1+	133	-	130-135	1+	60	43-68
F74 FL L. Harris	12	198	-	-	-	-	-
Combined	1,173+	118	105-132	25-198	718+	55	8-164
Age II							
G48, R34, R152, W220 IN	482	137	125-149	96-194	-	-	-
E146, W254 FL AL	+	140	-	127-155	+	79	66-91
C285, L151 IL	676	151	138-163	109-170	676	83	26-109

	No.	TL Mean of means	Central 50%	Range	No.	Mean Weight	Range
Age II (cont.)							
S12 TN Reelfoot L.	31	152	-	147-157	31	74	62-85
J25, J26 OK	11+	158	143-168	140-180	11+	75	65-85
T132 CA Folsom L.	75	158	-	-	-	-	-
H442 AR 3 lakes	33	190	173-216	173-216	-	-	-
F74 FL L. Harris	43	203	-	-	-	-	-
Combined	1,351+	154	138-163	96-216	718+	79	26-109
Age III							
W254 FL	+	133	-	132-134	+	71	66-76
G48, R34, R152, W220 IN	1,008	163	151-177	115-224	-	-	-
S12 TN Reelfoot L.	389	175	-	152-196	389	125	62-181
L151 IL	410	180	-	-	410	74	-
J25, J26, H150, O37, W7 OK	30+	181	162-188	150-221	29+	92	68-113
T132 CA Folsom L.	21	206	-	-	-	-	-
E146 AL	+	210	-	-	-	-	-
F74 FL Harris L.	44	218	-	-	-	-	-
H442 AR 3 lakes	24	242	-	208-269	-	-	-
Combined	1,926+	183	162-206	115-269	828+	92	62-181
Age IV							
G48, R34, R152, W220 IN	298	189	171-202	137-253	-	-	-
S12 TN Reelfoot L.	124	190	-	165-211	124	150	96-213
L151 IL	135	198	-	-	135	191	-
J25, M107, W7 OK	7+	222	-	211-244	7+	134	60-170
F74 FL Harris L.	53	226	-	-	-	-	-
H442 AR 2 lakes	7	233	-	211-254	-	-	-
T132 CA Folsom L.	12	241	-	-	-	-	-
E146 AL	+	246	-	-	-	-	-
Combined	636+	205	187-220	137-254	266+	154	60-213
Age V							
S12 TN Reelfoot L.	17	206	-	198-221	17	181	156-221
G48, R34, R152, W220 IN	54	216	214-220	168-258	-	-	-
L121 IL	65	234	-	-	65	286	-
F74 FL Harris L.	15	236	-	-	-	-	-
W7 OK Duncan L.	3	247	-	-	3	272	-
Combined	154	221	206-236	168-258	85	241	156-286
Age VI							
R34, R152 IN	6	230	-	184-257	-	-	-
L151 IL	31	236	-	-	31	290	-
W7 OK Duncan L.	2	251	-	-	-	-	-
F74 FL Harris L.	15	269	-	-	-	-	-
Combined	54	246	236-269	184-269	31	290	-
Age VII							
L151 IL	8	241	-	-	8	304	-
R152 IN Spear L.	2	254	-	246-259	-	-	-
F74 FL Harris L.	6	262	-	-	-	-	-
Age VIII							
R34 IN Shoe L.	1	179	-	-	-	-	-
F74 FL Harris L.	1	318	-	-	-	-	-

		No.	Average calculated TL at each annulus							
			1	2	3	4	5	6	7	8
F62	OK Sub Prison L.	23	49	89	119	155				
	restocked	19	49	127	175					
L121	IL Murphysboro L.	216	46	94	132	165	188	216	259	
R152	IN Spear L.	117	43	91	147	182	208			
H424	OK Rod and Gun Club L.	91	97	132	147	193				
F64	OK Little R.	49	58	122	155	185	201			
S291	OK Rock C.	8	66							
E61	OK Salt C.	1	69							
J45	OK Verdigris R.	3	104	142	160					
H150	OK Illinois R.	26	89	132	168	206				
J25	OK Illinois R.	26	76	130	170	206				
J101	OK Spavinaw L.	23	71	135	180	198	226			
J34	OK Ardmore L.	63	112	152	175	188	203			
J27	OK Spavinaw L.	6	84	130	180	203	213			
F74	FL Harris L.	189	112	147	183	198	224	251	257	216
J38	OK Rod and Gun Club L.	54	109	168	190	206	216	221		
K65 & 66	OK Hiwassee L.	86	117	170	208	240				
M539	LA pond	+	30	185	191					
J38	OK Chickasaw L.	73	142	193	218	234	246	257		
B380	TN Woods L.	74	74	152	224					
J44	OK	2,046	94	147	180	208	239	257		
	slowest	-	38	79	117	163	203	218		
	fastest	-	185	221	251	277	284	306		
	16 reservoirs	377	99	135	178	211	241			
	10 large lakes	219	102	155	183	206				
	48 small lakes	1,128	91	150	183	208	241	272		
	21 ponds	294	91	145	178	206	226	234		
	2 streams	28	74	130	170	206				

		Calculated TL and increments at each annulus		
		1	2	3
H303	OK Latonka L.	74	124	165
	Incr.	74	50	61
	No.	4	4	2

		No.	Calculated weight in grams at each annulus					
			1	2	3	4	5	6
J44	OK	2,046	13	54	102	171	261	336
	slowest	-	1	7	26	76	157	200
	fastest	-	116	206	314	430	500	513

Growth was usually calculated by the direct proportion method, but J44, R152, and T132, used Fraser-type corrections. In one pond in Louisiana (M539) fall drawdown resulted in annulus-type marks on the scales. In another pond it was estimated that 3 of 49 redear had false annuli. Scales were not considered interpretable in Florida (W254). In general, growth was slower in the northern part of the range, Michigan, moderate in Illinois and Indiana, and faster to the south. Growth in streams appears to be slower, at least for the first 2 years, than in lakes (J44). Growth was better in clear than in turbid ponds (B182). In aquaria 2.8 g fingerlings grew 0.013-0.018 g/day, more slowly than did bluegills under the same conditions (C260). In

Illinois (L228) and Oklahoma (J25) redear growth averages faster than that of the bluegill. Crowding decreased the growth in the aquaria and in ponds (C260). The most rapid growth in these ponds was when the redears were stocked with bass and probably kept at a lower population density. Reduction of bullhead and bluegill populations was followed by increased growth rates of redears (H424). Age IV redear increased from 60 to 170 g after population thinning (M107). Abundance of bluegill had a greater effect on depressing the growth rate of redears than did the abundance of the redears (E146). Growth was more rapid in newly renovated canals and slowed after reproduction occurred (W254). Finclipping had no discernable effect on growth or survival of redear sunfish (R39). Males averaged a little longer and heavier than females at age II (C285), but no sex difference in growth was noted at Reelfoot L. (S12).

In Florida and Texas, redears may spawn late in their first year of life (W254, B109) but not until the second summer in Tennessee (S12), Illinois (L151), and Michigan (C260). In Florida lakes, redear probably do not spawn until age II because of competition (W254). Maturity is reached at 134-147 mm, where growth is rapid, but not until about 188 mm in two lakes studied (W254). In Florida spawning may start in late February as water approaches 21°C and may continue until about October 1st (C361, W254), or may occur in any month (K151). Spawning may even continue at 32°C. In Alabama spawning occurs in the spring when surface-water temperatures reach 24°C and again in the fall, but only sparingly in summer (S550). In Texas redears spawn from early May to early July (S571). In Tennessee spawning begins in May and continues to September (S12). In Illinois redears spawned May and June at 20°-21°C (L151) with no fall spawning reported (C354). In Michigan spawning was in July when temperatures had been over 21°C for several days (C260). Five synchronous spawning peaks were observed in two Florida lakes, 2 occurring at new moon and 3 at full moon (W254). The nests are in 45-90 cm of water (C361), or as deep as 2 m (W254). The nests tend to be grouped (L189) usually in submerged vegetation (L228). Areas of water lilies, *Nuphar*, seemed to be preferred (W254), but a variety of spawning sites have been used. Courting males made a popping sound quite different from other sunfish (G191). Elimination of the red portion of the opercular tabs of nesting males increased hybridization with female bluegills (C354). Males could usually be distinguished from females by the greater amount of red border even when not spawning (W254).

Females had about 5,000 eggs (L189) or 2,000-10,000 eggs (L151). The average number of mature eggs in 2 Florida lakes were (W254):

TL	No. females	Mean
190-200	8	16,196
201-210	20	20,263
211-220	17	21,352
221-231	17	21,903
323-242	13	25,965

Eggs were 1.3-1.6 mm in diameter (M525). The alpha threshold temperature for hatching was estimated as 18.3°C, and 280 degree hours were needed for hatching (C354). At 23.6°C, hatching occurred in 50.3±5.4 hours; and at 28.7°C, in 26.6±1.9 or 28.1±2.2 hours (C354). At hatching, fry averaged 4.8, and 5.1 mm respectively in the experiments at 23.6°C and 28.7°C. At 23.6°C, 34%-53% of the eggs hatched and 24%-46% were normal fry; and at 28.7°C, 29-66% hatched and 20%-41% were normal fry (C354). The eggs hatch in 6-10 days, and the fry remain in the nest about a week with the male guarding (L228).

Redears failed to spawn successfully in several ponds in Oklahoma where bluegills and largemouth bass were successful (G177). There was evidence that the sexes were segregated at some seasons in Louisiana (L173).

Reproduction is not adequate to maintain good fishing in combination with bass (S550), and a buffer species is needed. Sudden cold spells may kill an entire population in a shallow pond in Alabama (S550).

Annual mortality rates were estimated at 55% (R34); (G48); 77%, 82%, and 82% (R152). The herbicide, Hydrothol 191, caused pathological changes in blood, gills, liver, and testes of redear (E141).

Redear were reported as good test animals, which handle easy and are not nervous (W214).

Redear eat insect larvae, snails, and cladocerans (L151, C260, H229, C364, and W254). Snails were the major food items (H229, P205), and redears practically eliminated snails in experimental ponds (C364). The gill arches bear rounded molarlike teeth suited for crushing snails (W254). Copepods were a major food of redears from 30-69 mm and less up to 129 mm, in L. Griffin, FL, but not in L. Weir (W254). In these lakes Tendipedids, *Hexagenia*, and *Hyalella* were more important foods than snails. Food changed seasonally, primarily on the basis of availability, and no diurnal cycle in feeding was evident (W254). Males on spawning beds ate less but still fed (W254). Redears seldom feed on surface insects like bluegills (L151). In aquaria with bluegills, bluegills fed nearer the surface and got more food than the redears (C260). Redears chewed most of their good (C260). They were inactive and did not feed at temperatures below 6.5°C (C260).

In Oklahoma ponds the standing crop was 5.3 kg/ha, with an annual net production of 2.2 kg/ha (W247). Annual yield in an Alabama pond was 42.6, 15.1, and 27.5 kg/ha in consecutive years (E146).

In gill nets with the following bar measure mesh sizes the total lengths of redear caught increased with mesh size (L216).

Mesh size	No.	Mean TL	Range
25	13	147	114-241
38	57	188	165-254
51	33	231	178-254
64	2	273	267-279

Redear have 48 (24 pairs) chromosomes (C354, R286).

SPOTTED SUNFISH, *Lepomis punctatus* (Valenciennes)

The spotted sunfish is found on the east coast as far north as North Carolina and in the Mississippi Basin as far north as southern Indiana and Oklahoma (M420). It tolerates saltier water than other *Lepomis* (H460) and is frequent in water up to 4.5 ppt salinity (B399) or 11.8 ppt (K151). Spotted sunfish were found in more diverse habitats in the delta in Louisiana than other centarchids (C368).

Since no length conversion data were found, the available data are recorded in the length types given in the original papers.

Standard length-weight—C148—Florida

SL	No.	Weight	Range	SL	No.	Weight	Range
12.9	7	0.12	0.1-0.2	81.4	38	26.0	22-31
17.5	28	0.26	0.1-0.4	87.0	28	31.2	28-39
21.9	36	0.46	0.3-0.6	91.8	14	37.5	33-47
27.4	33	0.9	0.5-1.5	97.1	9	46.4	43-52
31.6	28	1.4	1.0-2.1	104.0	2	60.0	58-62
37.1	32	2.4	1.7-3.3	107.2	6	58.7	56-63
42.2	49	3.5	2.7-4.6	111.5	4	66.2	57-75
47.0	50	5.1	4.0-7.3	116.0	1	80.0	-
52.3	49	7.1	5.9-8.2	121.8	5	85.1	75-96
56.9	72	9.1	7.1-12.1	127.3	3	95.5	79-109
62.2	76	11.6	9.6-13.6	132.5	2	112.4	105-120
67.1	53	14.5	11.6-17.3	137.7	3	130.9	120-142
71.9	68	17.8	15.1-22.5	142.0	1	127.1	-
76.7	59	21.6	18.0-27.3				

$\log W = -4.32 + 3.002 \log SL$

Average K(SL) = 4.82

In a Florida constant-temperature spring (C148), growth was estimated at 0.12 mm per day. At 9 days of age spotted sunfish were 6.5-7.0 mm SL in Florida (C41).

		Mean TL at each annulus			
	No.	1	2	3	4
F64 OK Little R.	15	33	74	114	150

Spotted sunfish mature by the time they are 55 mm SL (C148). In Florida they spawn from early spring into November (C361) or throughout the year (H460, K151). Some spawned when midday water surface temperatures were as low as 18°C and as high as 33°C.

Peak spawning occurred at 27°-29°C in late August (C361). Males were more pugnacious in guarding the nest than other sunfishes (C361). Courting males make grunting sounds (G191). Spotted sunfish showed little evidence of schooling (L189).

Six spotted sunfish taken in a 25 mm bar measure gill net averaged 124 mm and ranged from 102-140 (L216).

CENTRARCHID HYBRIDS

All centrarchid hybrids are considered in this section because while most involve *Lepomis* spp., some are intergeneric. Some reported to be intergeneric are no longer considered such, since *Chaenobryttus* has been put into *Lepomis*. David Starr Jordan (J94) had several genera for the sunfishes: *Allotis* for the orangespotted, *Xenotis* for longeared, *Helioperca* for bluegill and dollar, *Eupomotis* for pumpkinseed and redear, *Lepomis* for redbreast, and *Apomotis* for green, spotted, and bantam.

Where possible, hybrids are listed with the female parent species first (following H445, but not C354); but in some cases of natural hybridization, the particular cross is not known. Morphological characters of the reciprocal hybrids are not such that the particular cross can be recognized from the F_1 generation (H446). The female parent is placed first, since many of the comparisons are made upon the basis of survival of eggs of hybrids compared with eggs of parent stock.

Hybrid sunfish are more vulnerable to angling than the parent species (C366, C354).

Most viable crosses of sunfish produce fertile F_1 generations (B389).

ROCK BASS X WARMOUTH, *Ambloplites rupestris* X *Lepomis gulosus*
Rock bass eggs fertilized with warmouth sperm failed to hatch (T147).

ROCK BASS X BLUEGILL, *Ambloplites rupestris* X *Lepomis macrochirus*
Rock bass eggs fertilized with bluegill sperm gave 26.2% as good a hatch as when fertilized with rock bass sperm (T147).

ROCK BASS X LARGEMOUTH BASS, *Ambloplites rupestris* X *Micropterus salmoides*
Rock bass eggs fertilized with largemouth bass sperm failed to hatch (T147).

ROCK BASS X BLACK CRAPPIE, *Ambloplites rupestris* X *Pomoxis nigromaculatus*
Rock bass eggs fertilized with black crappie sperm gave 40.7% as good a hatch as when fertilized with rock bass sperm (T147).

BANDED SUNFISH X ROCK BASS, *Enneacanthus obesus* X *Ambloplites rupestris*
Banded sunfish eggs fertilized with rock bass sperm gave 14.2% as good a hatch as when fertilized with banded sunfish sperm (T147).

REDBREAST SUNFISH X BLUEGILL, *Lepomis auritus* X *L. macrochirus*
Only 5% of 20 redbreast eggs were fertilized with bluegill sperm and none hatched (S591).

REDBREAST X REDEAR SUNFISH, *Lepomis auritus* X *L. microlophus*
Only 17% of 60 redbreast eggs were fertilized with redear sperm and only one hatched (S591), but better results were indicated in other studies.

GREEN X REDBREAST SUNFISH, *Lepomis cyanellus* X *L. auritus*
Green sunfish eggs fertilized with redbreast sperm gave 96% fertilization and 57% hatch (S591).

GREEN X PUMPKINSEED, *Lepomis cyanellus* X *L. gibbosus*
Most natural hybrids of these two species are assumed to be of this combination, since the male pumpkinseed is more aggressive in spawning (H446). Green sunfish X pumpkinseed F_1 hybrids have a bright orange margin on the opercular lobes, quite pronounced blue reticulations on the cheeks, and brown flecking of the posterior part of the dorsal fin (E142).

	No.	SL	TL	Range
Age 0				
H93 MI Wiard's P.	14	35	-	28-45
Age I				
H94 MI Willow R.	44	37	-	22-49
B7 NY	48	44	-	34-54

	No.	SL	TL	Range
Age I (cont.)				
H446 MI Laboratory	+	54	-	35-95
C13 MI	6	-	56	51-61
H93 MI Wiard's P.	14	63	-	50-75
H94 MI Middle Range R.	55	75	-	63-92
Age II				
B7 NY	29	56	-	49-71
H94 MI Willow R.	8	70	-	58-85
H94 MI Range R.	27	94	-	78-112
H93 MI Wiard's P.	1	107	-	-
Age III				
B7 NY	10	75	-	67-86
H94 MI Range R.	5	113	-	98-138
Age IV				
B7 NY	8	85	-	68-92
C13 MI	1	-	170	-
Age V				
B7 NY	17	99	-	33-111
Age VI				
B7 NY	1	115	-	-

	Calculated SL at annulus		
	No.	1	2
H93 MI Wiard's P.	14	29	63

Spawning occurred in aquaria in February even in Michigan (H446). Most hybrids were males (H94, E142). Hybrid males guarded nests in Squaw L., MN (E142). Growth of hybrids was no more variable than that of bluegills or pumpkinseeds in Sieverson L., MN (E142). Growth of hybrids was faster than either parent species in Middle Rouge R., MI (H94).

Hybrids ate larger food items than did pumpkinseeds, probably because their mouths were larger (E135). Fish were more frequently eaten by the hybrids than by either parent species, but even then they were not a very significant part of the food. The hybrids ate more *Hyalella* and snails than did the green sunfish or green X bluegill hybrids, thus showing more affinity with the pumpkinseeds in their feeding.

GREEN X WARMOUTH, *Lepomis cyanellus* X *L. gulosus*

A color illustration of this hybrid is given in C354. There was no significant difference in the percentage hatched at 24.4°C, 27.3°C, and 28.1°C or between the hybrids and the green sunfish matings (C354). The average hatch of the hybrids was 78% (62%-89%) and that of the green sunfish was 78% (65%-86%). The mean hatching time of the hybrid eggs was 48.7 hours (48.4-48.9) at 23.8°C, 29.8 hours (29.0-30.0) at 27°C, and 29.0 hours (28.4-29.6) at 27.6°C. The mean hatching time of the green sunfish was 50.0 hours (49.2-50.8), 31.5 hours (30.7-32.8), and 29.1 hours (28.8-29.8) for the same temperatures (C354). The percentages of the fry which were normal did not differ with temperature and were 56% (0.79%) for the hybrids and 75% (63%-82%) for the green sunfish. The hybrid fry averaged 4.60 mm at hatching, and the green fry averaged 4.71 mm (C354). The hybrids took 258 degree-hours above 18.6°C to hatch compared to 266 degree-hours above 18.5°C for the green sunfish (C354). F_1

hybrids were 97% male (L250). These hybrids are recommended as predators in some pond stocking (L250).

F_2 and F_3 hybrids were readily obtained (C354).

GREEN X BLUEGILL, *Lepomis cyanellus* X *L. macrochirus*

Green sunfish X bluegill F_1 hybrids have a dark gray margin on the opercular lobe with possibly a trace of orange. The large black spot, characteristic of both parents, is on the posterior part of the second dorsal fin (E142). The color illustration in C354 does not show such a spot however. Although the reciprocal crosses both give good results in experiments (C354), the male bluegill X female green experiments in ponds were unsuccessful because the ponds were contaminated with male green sunfish and produced only green sunfish (C354). The male green X female bluegill ponds produced many hybrids (C354). With no other information available, the data on natural hybrids of these two species are recorded as bluegill X green sunfish.

C285 IL	TL	No.	Weight	Range	TL	No.	Weight	Range
	71-81	3	7	7-8	121-133	11	56	37-63
	82-95	6	10	8-13	134-146	80	63	52-81
	96-108	5	20	14-26	147-157	29	80	79-92
	109-120	8	29	23-39				

Only 4 of the above fish were females, and their weights were not consistently different from those of the males.

		No.	SL	Range
Age I				
H446 MI experimental cross		+	65	46-95
	experimental cross	+	70	48-119
Age—22 months				
H94 MI Crowded aquaria		51	97	50-165

		No.	TL	Range	Weight	Range
Age I—May						
C285 IL females		4	124	109-140	41	28-59
	males	138	135	71-157	59	7-92
Age III						
C286 IL		10	218	183-234	336	227-390

The growth of the hybrids collected in May, Age I was a little faster than that of green sunfish under the same conditions.

Green sunfish eggs fertilized by bluegill sperm gave 92% fertilization and 89% hatch (S591). There was little or no difference in the percentage hatched at 24.4°, 27.3° and 28.1°C or between the hybrid and green sunfish matings (C354). The average hatch of the hybrids was 73% (50%-84%) and of the green sunfish was 78% (65%-86%). The mean hatching time of the hybrids was 49.5 hours (48.6-51.4) at 24.4°C, 31.4 hours (30.8-31.8) at 27.3°C, and 28.0 hours (27.1-28.4) at 28.1°C compared to 50.0 hours (49.2-50.8), 31.5 hours (30.7-32.8), and 29.1 hours (28.8-29.8) for green sunfish at the same temperatures (C354). The percentage of fry which were normal was 66% (62%-68%) for hybrids at 28.1°C compared to 78% (67%-82%) for green sunfish; but at 24.4° and 27.3°C, the percentages did not differ, hybrids 72% (50%-84%) and green sunfish 74% (63%-80%). The hybrid fry averaged 4.76 mm at hatching, and the green sunfish fry averaged 4.72 mm (C354). The hybrids took 258 degree-hours above 18.6°C to hatch compared to 266 degree-hours above 18.5°C for green sunfish (C354). Males constituted 81% of the F_1 generation (H94) or 97% (L250). An experiment to produce F_2 generation failed (C285). These hybrids readily utilize artificial feeds and shallow water areas (L250).

GREEN SUNFISH X (BLUEGILL X REDBREAST) *Lepomis cyanellus* X (*L. macrochirus* X *L. auritus*)

Green sunfish eggs fertilized by bluegill X redbreast F_1 hybrids sperm gave 79% fertilization and 79% hatch (S591).

GREEN X LONGEAR, *Lepomis cyanellus* X *L. megalotis*

An experimental crossing of a green sunfish female by a male longear produced eggs but no fry (H446).

GREEN X REDEAR, *Lepomis cyanellus* X *L. microlophus*

Hybrids were readily produced in ponds, although the reciprocal cross failed in pond experiments (C354).

	No.	TL	Range	No.	Weight	Range
Age 0--Fall						
H466 IL	-	-	-	1,778	31	5-59
Age I						
C285 IL June	79	127	104-147	-	-	-
H466 IL Sept.-Nov.	-	-	-	360	114	64-159
C366 IL Sept.-Oct.	42	165	140-178	47	118	-
Age II						
C285 IL May	79	173	132-206	-	-	-
H466 IL Sept.-Nov.	-	-	-	227	225	132-318
C336 IL possibly F_2	1	211	-	-	-	-
Age III						
C366 IL possibly 2 were F_2	7	244	-	-	-	-
H466 IL Sept.-Nov.	-	-	-	46	387	227-635
Age IV						
C366 IL	9	236	-	-	-	-
Age V						
C366 IL	7	239	-	-	-	-
Age VI						
C366 IL	2	264	-	-	-	-

In southern Illinois, hybrids held under crowded conditions averaged 2.9 g at Age I March (L231). The approximate weights (taken from graphs) at Age I October were:

	Mean	Range	Percent over 110 g
Not fed, stocked at 7,426 fish per ha	35	10-110	0
stocked at 3,713 fish per ha	50	30-100	0
Fed, stocked at 7,426 fish per ha	60	10-170	13-19
stocked at 3,713 fish per ha	70	10-160	8-20

When Age II March hybrids, averaging 60-68 g, were stocked at 7,426 per hectare, the average weight of the stocked fish after feeding to October was about 190 g with some reaching 300 g and with F_2 fish averaging about 65 g

(L231). Survival of the fingerlings in the first ponds ranged from 75%-91% and in the second year, 84%-97%. The standing crops at harvest were in kg/ha.

	Total	Harvestable
Not fed, stocked at 3,713 fish per ha	137	0
stocked at 7,426 fish per ha	249	0
Fed, stocked at 3,713 fish per ha	169-172	11-83
stocked at 7,426 fish per ha	414-426	71-106
The second year, fed, stocked at 7,426	872-898	708-748

Green sunfish eggs fertilized with redear sperm gave 82% fertilization and 73% hatch (S591).

Green sunfish eggs fertilized with redear sperm gave 86% hatch (83%-89%) compared to 76% hatch (71%-80%) of eggs fertilized with green sunfish sperm at 24.4°C; 81% hatch (65%-92%) compared to 76% hatch (65%-83%) at 27.3°C; and 77% hatch (75%-79%) compared to 82% (74%-86%) at 28.1°C (C354). The mean hatching times, 48 hours (47.7-48.4) at 23.8°C, 31.7 hours (31.4-32.5) at 27.1°C, and 27.4 hours (26.7-28.9) at 27.6°C, did not differ from those of green sunfish. Incubation took 267 degree-hours above 18.5°C compared to 266 degree-hours above 18.5°C for green sunfish (C354). Practically all the fry which hatched were normal and averaged 4.62 mm at hatching at 24.4°C, 4.84 mm at 27.3°C, and 4.53 mm at 28.1°C (C354).

In one experiment only 1 of over 2000 hybrids was a female (L231); but in others, 69%-70% were males (C285, C354) or 99% were males (L250). F_2 hybrids were readily produced (C354), but largemouth bass and grass pickerel greatly reduced the production of F_2 hybrids (C366). No F_2 hybrids were seen in other ponds (H466). Hybrids were produced with a male bluegill and a female green X redear F_1 (C354). Hybrids were readily taken by anglers and perhaps also increased the ease with which bass could be caught (C366). They utilized deeper areas of the pond than do bluegill X green sunfish hybrids (L250).

PUMPKINSEED X REDBREAST, *Lepomis gibbosus* X *L. auritus*
Three males and one female hybrids of these species (but particular cross not known) were reported in the University of Michigan Museum's collections (H94).

PUMPKINSEED X GREEN SUNFISH, *Lepomis gibbosus* X *L. cyanellus*
While female pumpkinseed X male green sunfish hybrids have been produced in the laboratory, most natural hybrids of these two species probably have pumpkinseeds as the males since the male pumpkinseed is the more aggressive (H446).

	No.	SL	Range
Age 0			
H93, H446 MI aquarium	41	54	39-95
H94 MI aquarium, 9 months	1	99	-
Age I			
H93 MI aquarium	12	67	50-90
H94 MI aquarium, 16 months	1	122	-

One hybrid started nest building at 9 months when 99 mm SL (H94). Of the hybrids reared under control, 95% were male (H94).

PUMPKINSEED X WARMOUTH, *Lepomis gibbosus* X *L. gulosus*
    The one pumpkinseed X warmouth hybrid (which cross, not known) in the Michigan collections was a female (H94).

PUMPKINSEED X BLUEGILL, *Lepomis gibbosus* X *L. macrochirus*
    Experimental crosses both ways were unsuccessful (H446), but ponds with pumpkinseed males and bluegill females and with bluegill males and pumpkinseed females both produced abundant hybrids (L232). All field data, where the cross was not known, are listed as bluegill X pumpkinseed.

PUMPKINSEED X LONGEAR SUNFISH, *Lepomis gibbosus* X *L. megalotus*
    It was suggested that there was more probability of this combination rather than the reciprocal because the male longears are too busy in aggressive activity in the colony to distinguish females (S583).

WARMOUTH X ROCK BASS, *Lepomis gulosus* X *Ambloplites rupestris*
    An experimental cross showed 5% hatch compared to the control warmouth hatch (H445), and this was interpreted as indicating a more distant relationship than between the warmouth and bluegill or largemouth bass. Another experiment gave 5.6% hatch compared to the control (T147).

WARMOUTH X GREEN SUNFISH, *Lepomis gulosus* X *L. cyanellus*
    No pond produced hybrids were found; but hybrids of this cross were produced in the laboratory, and a color photo is given (C354). Only 16% of the F_1 generation were males, the reverse of most Centrarchid crosses. The average hatch was 62% (51%-80%), not significantly different than 58% (48%-77%) of the warmouth controls, and 47% (35%-62%) of the fry were normal compared to 49% (45%-56%) of the controls. The incubation period was longer than that of the controls—30.9 hours (30.4-31.3) compared to 29.4 hours (28.8-30.1) of the controls at 27.3°C and 29.7 hours (29.6-29.8) compared to 28.9 hours (28.4-29.6) of the controls at 27.6°C.
    There was evidence that the gas transport properties of the blood were better in the hybrids than in either parent stock (C354).
    F_2 hybrids were abundant in two ponds stocked with F_1 hybrids (C354). Green sunfish X warmouth F_1 females were successfully crossed with male *Lepomis cyanellus*, *L. microlophus*, *L. microlophus* X *L. macrochirus*, and *L. macrochirus* X *L. gulosus*.
    The two warmouth X green sunfish hybrids in the Michigan collections were females (H94).

WARMOUTH X BLUEGILL, *Lepomis gulosus* X *L. macrochirus*

Descriptions and a photograph of this hybrid are given in W229; and a color photo, in C354. Hybridization by stripping, fertilization, and incubation gave success with female warmouth X male bluegill, but not the other way (M544). Hatching was described as normal, but 25% of the fry showed caudal curvature or cephalic swelling (M544), compared to 5% in warmouth X warmouth.

A natural population of hybrids, probably less than 1% of the parental stocks, was found in a Florida pond, which had few bluegill nesting sites (B389). Hybrid vigor was evident in greater body depth, caudal peduncle depth, and color intensity. No hybrids were produced in ponds with only female warmouth and male bluegills (C354).

The hatch was reported as 64% as good as warmouth eggs (H445) or 85.8% with 98.7% as good fertility (M544). The hybrids gave 77.5% hatch compared to 90.4% (M544) or 45.3% compared to 77.6% (W229). In the latter case, 7.2% were raised to fingerlings. In an Illinois study (C354), the hatch was 53% (52%-54%) at 27.2°C and 68% (61%-73%) at 28.1°C, slightly higher than the hatching percentage of warmouth in the same conditions. The hatching time, 30.8 hours (30.5-31.2) at 27.3°C and 28.9 hours (28.9-29.0) at 27.6°C, was also about equal of that of the warmouth (C354). The fry averaged 4.77 mm at hatching.

The F_1 generation had 69% males (C354) and 89% males (H389). F_2 hybrids were produced (C354, W228). Warmouth X bluegill males were backcrossed with bluegill females; warmouth X bluegill females and male redears gave successful outcrossing; and a warmouth X bluegill male was crossed with a warmouth X green female (C354). A warmouth X bluegill female was also backcrossed with male bluegills (26% hatch, 22.7% survival), a male warmouth (25.8% hatch, 5.2% survival), and a male warmouth X bluegill (4.7% hatch, 3.9% survival) (W228).

Five F_1 females (218-229 mm, 311-402 g) had an average of 18,200 ova (9,883-30,334) (W228).

WARMOUTH X REDEAR SUNFISH, *Lepomis gulosus* X *L. microlophus*

A warmouth X redear hybrid is illustrated in color in C354. Warmouth eggs fertilized with redear sperm gave a 47% (39%-53%) hatch at 27.3°C and an 80%(77%-83%) at 27.6°C compared to 51% and 69% for warmouth X warmouth. The hatching time was also similar, 29.4 hours (29.0-29.8) at 27.3°C and 29.2 hours (29.0-29.6) at 27.6°C (C354). The percentage of fry, which were normal, was the same as with the warmouth fry, and the average length at hatching was 4.61 mm (C354). The F_1 generation was 55% male. Two ponds failed to give natural hybrids when stocked with female warmouth and male redears, but F_2 hybrids were produced in a pond stocked with warmouth X redear F_1. Warmouth X redear females were successfully backcrossed with male warmouth and male redears and outcrossed with male bluegills and male green sunfish (C354).

WARMOUTH X LARGEMOUTH BASS, *Lepomis gulosus* X *Micropterus*
   salmoides
   A photograph and description of this hybrid are given in W229. Experimental crosses of this combination and its reciprocal have given good results, indicating close phylogenetic relationships (H445).
   The hatching rate of warmouth eggs fertilized with largemouth bass sperm was 71% (H445), 62% (W229), and 96.2% (M544) as high as that of eggs fertilized by warmouth sperm, and the percentage which were fertilized was 94% (W229) or better (M544). The percentage fertile was 83% (W229) and 89.3% (M544) with 48% (W229) and 86.9% (62.7%-91.9%) (M544) hatch. Survival to 60 mm fingerlings was 4.8% (W229). Hatching was normal, and the fry showed 10% with spinal curvature or cephalic swelling, slightly higher than warmouth fry (M544). The first hatch was at 40.8 hours compared to 44.1 hours for warmouth, and the hatching period lasted 18 hours compared to 8.5 hours (M544).
   No F_2 or backcrosses resulted when F_1 hybrids were stocked with one of the parents (W228). Study of oogenesis indicated that the hybrids were not capable of producing viable eggs and that the testes of the males did not produce viable sperm (W228).

WARMOUTH X BLACK CRAPPIE, *Lepomis gulosus* X *Pomoxis nigro-*
   maculatus
   Warmouth eggs fertilized with crappie sperm gave only a 1% (H445) or 4.3% (M544) as good a hatch as when fertilized with warmouth sperm. In another study, a 0.1% hatch was secured although 91.1% of the eggs started embryonic development (W229). In yet another study, a 3.9% (2.7%-6.0%) hatch was secured with 87.7% embryonic development (M544). Hatching was abnormal with most dying at emergence from the chorion (M544). Hatching started at 55 hours, and some continued for 25 hours compared to 44.1 and 8.5 hours for warmouth X warmouth (M544).

ORANGESPOTTED X GREEN SUNFISH, *Lepomis humilis* X *L. cyanellus*
   One male and one female natural hybrids (particular cross not known) of these two species were recorded by H94.

BLUEGILL X ROCK BASS, *Lepomis macrochirus* X *Ambloplites rupestris*
   Bluegill eggs fertilized with rock bass sperm gave 5% (4.6% (T147)) as good a hatch as when fertilized by bluegill sperm (H445).

BLUEGILL X FLIER, *Lepomis macrochirus* X *Centrarchus macropterus*
   Bluegill eggs fertilized with flier sperm gave 12% as good a hatch as when fertilized by bluegill sperm (H445).

BLUEGILL X BLUESPOTTED SUNFISH, *Lepomis macrochirus* X
   Enneacanthus gloriosus
   Bluegill eggs fertilized with bluespotted sunfish sperm gave 22% as good a hatch as when fertilized by bluegill sperm (H445).

BLUEGILL X REDBREAST SUNFISH, *Lepomis macrochirus* X *L. auritus*
   Bluegill eggs fertilized with redbreast sunfish sperm gave 80% fertilization and 66% hatch (S591), which was better than when fertilized by bluegill sperm (H445).

BLUEGILL X GREEN SUNFISH, *Lepomis macrochirus* X *L. cyanellus*

Ponds stocked with female bluegills and male green sunfish produced many hybrids (C354), and hybrids have been found in nature (B7, C13, H94, and E142). An experimental cross gave 79% as good a hatch as the bluegill control (H445) and almost the same hatch (82%-91%) as the control (84%-90%) in other studies (C354).

The average K(TL) of 40 hybrids from South Dakota reservoirs was 2.05 (S228, S276).

	No.	SL	TL	Range
Age 0--Aug.				
C285 IL	322	-	38	-
Age I				
B7 NY	18	46	-	36-49
C13 MI	1	-	51	-
Age II				
B7 NY	6	60	-	49-68
H94 MI	41	97	-	-
C285 IL Oct.	44	-	175	132-190
Age III				
B7 NY	9	70	-	59-81
C13 MI	1	-	109	-
Age VI				
C354 IL (965 g)	1	-	310	-

	Calculated TL at annulus		
	1	2	3
S228 SD Ft. Randall L.	51	122	147

In southern Illinois, hybrids held under crowded conditions averaged 7.2 g at age I, March (L231). The approximate weights (taken from graphs) at age I, October were:

		Mean	Range
Not fed, stocked at 7,426 fish per ha	females	25	10-50
	males	35	10-80
stocked at 3,713 fish per ha	females	35	10-70
	males	45	10-80
Fed, stocked at 7,426 fish per ha	females	40	10-110
	males	110	20-180
stocked at 3,713 fish per ha	females	60	20-120
	males	130	20-210

When stocked at 1,238 fish per ha and permitted to subsist on natural foods, there was no sexual difference in growth (L231). The faster growth of males where competition was greater was believed to be due to the greater acceptance of artificial foods by males.

Hatching averaged 32.3 hours (32.2-32.4) compared to 32.5 hours (32.3-32.9) for bluegills at 27.0°C; 34.3 hours (32.8-35.7) compared to 33.9 hours (33.4-34.2) at 26.8°C; and 73.4 hours (72.7-74.8) compared to 71.1 hours (69.3-71.8) at 22.3°C (C354). At 27°C, 91% (88%-93%) of the hybrid fry were normal compared to 90% (71%-96%) of the bluegill fry; at 26.8°C, 88% (78%-93%) were normal compared to 83% (79%-85%); and at 22.3°C, 41% (6%-87%) were normal compared to 56% (8%-89%).

At hatching, the hybrids averaged 4.61, 5.00, and 4.48 mm at 27.0°, 26.8°, and 22.3°C compared to 4.69, 4.98, and 4.28 mm for bluegills (C354). It took 279 degree-hours above 18.5°C to hatch hybrid eggs compared to 287 degree-hours above 18.3°C for bluegill eggs (C354). F_1 hybrids were 70% males (C285) and 64% and 70% males in two other ponds (C354).

Hybrid males guarded nests in 3 Minnesota lakes (E142).

Hybrids ate larger food items than did bluegills (E135).

Hybrids had 24 pairs of chromosomes (C354), the same as both parent species. When hybrids were stocked, 50% were caught by anglers the first day, and 82%, in 3 days (C354).

BLUEGILL X PUMPKINSEED, *Lepomis macrochirus* X *L. gibbosus*

Hybrids of these species have been observed in nature (B7, C13, H94, B93, and N52). In aquaria, attempts at crossing both directions failed (H446). Ponds with male pumpkinseed and female bluegills and with male bluegills and female pumpkinseeds both produced abundant hybrids (L232). When equal numbers of male and female bluegills and pumpkinseeds were put in uncrowded ponds, there was no evidence of hybridization (C365). Visual recognition of conspecific mates through behavior was believed to be the main barrier to hybridization, since nesting overlapped in time and habitat.

P122 WI Flora L.   71 hybrids, 1952   log W = -11.764 + 3.177 log TL (base n)
                    71 hybrids, 1956   log W = -12.91 + 3.417 log TL

A 152 mm TL hybrid weighed 85 g in a Wisconsin lake (N52).

The average K(SL) of hybrids taken from May to August from Wintergreen L., MI, was 5.32 for 63 caught in nets and 5.68 for 19 caught by angling (F97).

	No.	SL	TL	Range	Weight
Age 0					
C13 MI Sept.	11	-	41	36-48	-
H94 MI Crystal L.	135	49	-	34-69	-
Age I					
C13 MI	269	-	69	43-97	-
H94 MI Crystal L.	84	99	-	66-121	-
Age II					
C13 MI	56	-	104	79-135	25
Age III					
H94 MI Third Sister L.	1	60	-	-	-
B7 NY	5	71	-	62-77	-
C13 MI	40	-	130	109-163	42
B93 MI North Twin L.	50	-	155	-	-
After thinning					
Age IV					
B7 NY	2	85	-	68-92	-
H94 MI Third Sister L.	6	85	-	56-111	-
C13 MI	25	-	152	114-206	65
H94 MI Sunken L.	4	149	-	120-160	-
Age V					
B7 NY	2	94	-	89-98	-
C13 MI	5	-	180	155-226	136

	No.	SL	TL	Range	Weight
Age VI					
C13 MI	3	-	213	211-216	230

	Calculated SL at annulus		
	No.	1	2
H94 MI Crystal L.	83	32	67

	Calculated TL at annulus						
	No.	1	2	3	4	5	6	7
P122 WI Flora L.	152	53	76	99	122	145	155	173
after thinning	224	51	81	107	112	130	150	157

The growth (P122) was calculated using a Fraser-type correction of 30 mm. Some improvement in growth was noted at first after the population was thinned (P122, B93).

Bluegill eggs fertilized with pumpkinseed sperm gave a 63.4% hatch compared to a 51.7% hatch of eggs fertilized with bluegill sperm (H443). The incubation time at 24°C for the hybrid eggs was 47.3 ± 5.3 hours compared to 50.4 ± 7.5 hours for the bluegills. Embryonic development of the hybrid and bluegill were similar in most respects (H443).

About 80% of the hybrids were males (H94). Ponds stocked only with hybrids produced an abundant F_2 generation (L232).

BLUEGILL X WARMOUTH, *Lepomis macrochirus* X *L. gulosus*

Crosses with female bluegill X male warmouth gave abnormal hatching. Over 99% of the fry were deformed, and they died at or right after emergence (M544). First hatching was at 39.8 hours at 25.5°C, and hatching continued 16.2 hours with a median hatching time of 46.3 hours. This time is not very different from bluegill X bluegill eggs. Only 3.6% hatched (M544) although 79% were fertile. In another study (W229), 16.6% hatched with 69.3% fertile, and in another study (S591) 19% hatched with 95% fertilized. The hatching success was only 25% that of bluegills (H445).

In another series of experiments (C354), 22% (16%-25%) hatched at 22.6°C, and 92% (86%-98%) at 26.9° and 27.3°C, but none of the fry were normal. Hatching time was less than that of bluegill X bluegill eggs, 59.8 hours at 22.3°C, and 29.4 and 27.9 hours at 26.8°C compared to 71.1, 33.1, and 32.5 hours respectively.

BLUEGILL X (BLUEGILL X REDBREAST), *Lepomis macrochirus* X (*L. macrochirus* X *L. auritus*)

Bluegill eggs fertilized with sperm from F_1 hybrid bluegill X redbreast gave 88% fertility and 69% hatch (S591).

BLUEGILL X REDEAR, *Lepomis macrochirus* X *L. microlophus*

Female bluegills and male redears in ponds failed to produce hybrids (K42), although the reciprocal cross was successful. However, 11 young believed to be hybrids but not positively identified as such were found in 1 of 4 ponds with female bluegills and male redears (C354).

	No.	TL	Range
C285 IL			
Age I—June	463	102	86-112
Age II—July	92	135	97-150

	No.	SL	SD
W255 IL			
Age 0	1	28	-
I	24	72	4.7
II	9	73	2.6
III	48	80	5.5
IV	20	93	4.1
V	216	99	6.5
VI	62	113	20.3

The age VI fish were F_1 generation, the age IV and V fish were F_2, and the younger fish were F_{2+} (W255).

Bluegill eggs fertilized by redear sperm gave 68% as good a hatch as when fertilized by bluegill sperm (H445). Bluegill eggs and redear sperm gave 97% fertility and 38% hatch (S591).

In other experiments, the percent hatched was 81% (76%-88%) at 22.3°C, 94% (91%-98%) at 26.8°C, and 88% (83%-90%) at 27°C compared to 87%, 84%, and 90% respectively for bluegill eggs fertilized with bluegill sperm (C354). The hatching time was 67.9 ± 2.80 hours at 22.3°C, $34.6 + 1.66$ hours at 26.8°C, and 31.8 ± 0.83 hours at 27.0°C compared to 71.1, 33.9, and 32.5 hours for bluegill fry. Incubation took 299 degree-hours above 17.9°C for the hybrids compared to 287 degree-hours above 18.3°C for the bluegill fry. Variations in incubation times were much greater with the hybrid (C354). The percentages of eggs giving normal fry were 76% (68%-88%) at 22.6°C, 93% (90%-98%) at 26.9°C, and 85% (78%-90%) at 27.3°C; and the fry at hatching averaged 4.42 ± 0.108 mm, 4.98 ± 0.151 mm, and 4.63 ± 0.069 mm at these temperatures (C354). The hybrids were 97% males (C354), 99.4% males (W255), or 100% male (C285). F_2 generation hybrids were also almost all males, but F_3 and later generations were 44%-46% males (W255). Sperm from male hybrids successfully fertilized eggs of a warmouth and of a warmouth X green sunfish female (C354). One F_2 hybrid was secured in 1 of 3 ponds stocked with F_1 generation (C354), and many were found for 5 generations (W255); but none were secured in other studies (C285). At age I, F_2 hybrids showed all 3 isozyme phenotypes at most loci in the approximate ratio of 1:2:1 expected; but at age V, F_2 hybrids were all morphologically like the F_1 and were heterozygous for the 3 enzyme loci, thus suggesting differential survival of mature heterozygous individuals (W255).

BLUEGILL X LARGEMOUTH BASS, Lepomis macrochirus X Micropterus salmoides

Bluegill eggs fertilized with largemouth bass sperm gave only 9% as good a hatch as when fertilized with bluegill sperm (H445). The hatch was 4% (W229) or 3.5% (M544), although 85% (W229) and 95.6% (M544) of the eggs started development. Deficient embryonic development is described, and the critical period seems to be epiboly (22 hours) (M544). No survival of fry was reported (W229).

BLUEGILL X BLACK CRAPPIE, Lepomis macrochirus X Pomoxis nigromaculatus

Bluegill eggs fertilized with black crappie sperm gave only 11% as good a hatch as when fertilized with bluegill sperm (H445). The hatch was 7.7% (W229) or 3.7% (M544), although 73.5% (W229) and 63.0% (M544) of the eggs started development. The critical periods in development were at epiboly and at emergence (M544). Almost 100% were deformed (M544), and no fry survived (W229).

LONGEAR X GREEN SUNFISH, *Lepomis megalotis* X *L. cyanellus*
Experimental cross of longear female X green sunfish male showed some success (H94). Most of the hybrids were males and averaged larger than either parent species (H94). Hybrids from Michigan lakes were 64-147 mm SL.

LONGEAR X PUMPKINSEED SUNFISH, *Lepomis megalotis* X *L. gibbosus*
Museum specimens of hybrids of these two species were 42% male (H94). Aquarium attempts with female longears and male pumpkinseed were not successful, and the reverse was not tried (H446). Hybrids were larger than longear sunfish, 55-160 mm SL (H94).

REDEAR X REDBREAST SUNFISH, *Lepomis microlophus* X *L. auritus*
Redear eggs fertilized with redbreast sunfish sperm gave 64% as good a hatch as when fertilized with redear sperm (H445), or 47% fertility and 27% hatch (S591).

REDEAR X GREEN SUNFISH, *Lepomis microlophus* X *L. cyanellus*
No hybrids were produced in ponds with male green and female redear sunfish (C354), but hybrids were produced by artificial spawning (C285, C354). A color photo is given in C354.
The following length-weight data were recorded for the hybrids in Illinois (C285).

	No.	Mean Weight	Range
135-145	2	79	74-85
147-157	28	82	62-94
165	74	100	76-113
178	19	119	111-133

	No.	TL	Range	Weight	Range
Age I - July-Aug.	528	117	97-140	-	-
Age II - May females	56	160	140-183	90	71-139
males	67	168	147-183	102	82-130
Age II - July	15	196	185-208	184	136-216

When stocked as fry in May at 10,000 per surface acre, 106 hybrids on August 30 averaged 94 mm, the slowest growth found in that Illinois study of hybrids (C354). In another pond, redear X green hybrids averaged 117 mm in early August of their first year (C354). When stocked with the parent species at relatively low population densities, no differences in growth between the hybrids and the parent species were noted (C354).
The growth was about the same as that of the parent species in the same ponds (C385).
Redear eggs and green sunfish sperm gave 91% fertilization and 47% hatch (S591).
Redear eggs fertilized with green sunfish sperm gave a 57% hatch (29%-70%) compared to a hatch of 45% (34%-53%) of eggs with redear sperm at 23.6°C and a 44% hatch (35%-64%) compared to 40% (29%-66%) at 28.7°C (C354). It took 49.6 hours (48.2-52.1) for the hybrids to hatch compared to 50.3 hours

(49.3-52.1) for the redears at 23.8°C; and 28.0 hours (27.2-28.6) compared to 27.4 hours (26.2-28.4) at 28.6°C. At 23.8°C, 50% (29%-60%) of the hybrid fry were normal compared to 38% (24%-46%) of the redear; and at 28.6°C, 34% (23%-56%) of the hybrids and 24% (15%-41%) of the redear. The hybrids averaged 4.98 mm at hatching compared to the redears at 4.78 mm at 23.8°C and 5.37 mm compared to 5.14 mm at 28.6°C. Hatching time was 47.61 hours at 21°C which did not differ from bluegill X green or bluegill X redear hybrids (C285). Of the hybrids, 48% were male (C285, C354). Hybrids produced only a few F_2 generation when the F_1 hybrids were abundant (C285, C354) but when only a few F_1 hybrids remained there was good production of an F_2 generation (C354).

REDEAR X BLUEGILL, *Lepomis microlophus* X *L. macrochirus*

A color photograph is given in C354. Hybrids were readily produced by placing female redears with male bluegills in some ponds (R38, K42) but not in another pond (C354). This hybrid has been recommended for stocking in farm ponds because of their rapid growth and low fecundity (R38, K42).

K42 IN, Fork length-weight

FL	90	95	100	105	110	115	120	125	130	135	140		
W	13	19	20	23	28	31	37	42	48	56	64		
No.	2	2	12	26	22	23	46	26	40	35	55		
FL	145	150	155	160	165	170	175	180	185	190	195	200	210
W	70	81	91	103	122	140	156	169	190	204	223	192	328
No.	73	51	49	33	31	40	49	45	36	13	11	1	1

Total length-weight

TL	95	100	105	110	115	120	125	130	135	140	145	150	155
W	11	16	18	23	26	30	35	39	45	51	58	67	72
No.	1	2	9	21	21	24	36	31	35	30	45	59	60
TL	160	165	170	175	180	185	190	195	200	205	215	220	-
W	84	94	105	124	140	160	173	191	201	224	192	328	-
No.	50	48	26	30	42	58	38	29	13	12	1	1	-

The hybrids were heavier for their lengths than the average bluegills or redears from the same ponds (K42). The hybrids were also heavier at most lengths when in ponds with no competition (B in next table) than those in ponds (A) with other species (K42):

FL	105	110	115	120	125	130	135	140	145	150	155	160	165
A	23	28	31	37	43	47	56	61	68	79	90	103	113
No.	17	8	10	37	19	27	20	32	52	31	29	11	8
B	24	28	31	37	41	49	55	67	76	84	93	103	124
No.	9	14	13	9	7	13	15	23	21	20	20	22	23
FL	170	175	180	185	190	195	200	210					
A	138	152	160	178	191	198	192	-					
No.	22	24	20	12	6	3	1	-					
B	143	160	176	196	215	232	-	328					
No.	18	25	25	24	7	8	-	1					

R38 IN ponds		No.	FL	Range	Weight
Age 0 - July 6,	late spawn	35	15	13-17	0.06
	early spawn	141	24	22-27	0.26

R38 IN ponds (cont.)		No.	FL	Range		Weight	
Age 0 - July 21,	late spawn	95	21	17-25		0.14	
	early spawn	53	31	27-33		0.56	
Age 0 - Aug. 4,	late spawn	57	25	23-28		0.32	
	early spawn	17	36	34-38		1.0	
Age 0 - Aug. 20,	late spawn	37	30	26-33		0.55	
	early spawn	15	41	37-45		1.20	
Age 0 - Sept. 9,	late spawn	62	31	28-35		0.55	
	early spawn	23	41	38-45		1.13	
Age 0 - Oct. 20,	late spawn	48	36	28-40		0.76	
	early spawn	101	47	41-57		1.96	
					No.		Range
Age 0 - Oct. - F_2 generation		39	54	22-114	36	8.5	0.7-42
Age I April 29,	late spawn	94	32	28-40		0.60	
	early spawn	60	51	43-64		2.60	
Age I July,	late spawn	38	77	61-87		7.3	5-13
	early spawn	12	121	107-135		49.0	28-71

	No.	FL	Range	No.	Weight	Range
Age I - Oct.-Nov.	874	97	57-195	1,324	15.7	13-198
Age II F generation	2	197	189-204	-	-	-
Age II	15	134	83-202	15	78	12-223
Age III	1	238	-	1	435	-
Age IV	2	243	237-248	2	461	455-467
K42 IN Age I - Oct.-Nov.	+	-	50-70	+	2.3	-

C285 IL	No.	TL	Range	Weight
Age 0 - Oct.	40	119	91-137	37
Age I - July	67	163	137-175	-
Age I - Sept.	43	188	157-211	159

Redear eggs and bluegill sperm gave 82% fertilization and 4.7% hatch (S591).

Redear eggs fertilized by bluegill sperm gave a higher percentage hatch than when fertilized with redear sperm (H445) and gave a 58% (46%-75%) hatch compared to 45% and 37% when fertilized with redear sperm (C354). The incubation time averaged 52.1 hours (49.9-53.5) at 23.8°C and 28.3 hours (27.6-29.1) at 28.6°C compared to 50.3 and 27.4 hours for redear fry. Incubation took 290 degree-hours above 18.3°C compared to 280 degree-hours for redears (C354). About 50% of the eggs gave normal fry at 23.8°C and 27% at 28.6°C compared to 38% and 24% for redear. The fry averaged 4.86 ± 0.139 mm when hatched at 23.8°C and 5.13 ± 0.192 mm or 5.58 ± 0.137 mm at 28.6°C, slightly longer than redear fry (C354).

The F_1 generation was 97.9% male (R38) and 97% male (C285, C354). F_2 generations were reported by R38 but not C354 or C285.

REDEAR X (BLUEGILL X REDEAR), *Lepomis microlophus* X (*L. macro-chirus* X *L. microlophus*)

Redear eggs and sperm from F_1 hybrid bluegill X redear gave 44% fertilization and 17% hatch (S591).

LARGEMOUTH X ROCK BASS, *Micropterus salmoides* X *Ambloplites rupestris*

Largemouth bass eggs fertilized with rock bass sperm gave less than 1%

as high a hatch as when fertilized with largemouth bass sperm (H445), or 0.3%
(T147).

LARGEMOUTH BASS X WARMOUTH, *Micropterus salmoides* X *Lepomis gulosus*

A photograph and description are given in W229.

The hatching rate of largemouth bass eggs fertilized with warmouth sperm
was 97% (W229), 104% (H445), and 116% (M544) compared to eggs fertilized by
largemouth bass sperm. The percentage hatch was 53.7% (W229) and 72.2%
(64.2%-80.0%) (M544). Survival to 60 mm fingerlings was 34% (W229). Hatching was normal with 5%-75% showing caudal curvature, about the same as
largemouth bass fry (M544). The first hatch was 53.7 hours compared to 51.2
hours for largemouth bass, and the hatching period lasted 14.7 hours compared
to 11.3 hours (M544).

Successful hybrid reproduction did not occur in two ponds where they were
stocked alone or in a pond with the parental species (W228). Spermatogenesis
and oogenesis in the hybrids were not normal.

LARGEMOUTH BASS X BLUEGILL, *Micropterus salmoides* X *Lepomis macrochirus*

Descriptions and photographs are given in W229.

Largemouth bass eggs fertilized with bluegill sperm gave 66% (H445) or
9.5% (M544) as good a hatch as when fertilized with bass sperm. The percentage of eggs fertilized was 81.6% (W229) and 84% (M544), and the percentage of
hatch was 30.3% (W229) and 51.8% (M544). Hatching was reported as normal,
but 95+% of the fry had strong caudal flexure and abbreviated heads (M544).
Only 2% of the fry survived to 60 mm (W229). The first eggs hatched at 52.2
hours, and hatching lasted 20.2 hours compared to 51.2 and 11.3 hours for bass
fry.

LARGEMOUTH X SMALLMOUTH BASS, *Micropterus salmoides* X *M. dolomieu*

Backcrosses with female largemouth bass grew approximately twice as
fast as backcrosses with male largemouth bass, indicating that hybrid vigor
results when mixed genetic material interacts with pure species egg cytoplasm
but not when mixed genetic material interacts with mixed egg cytoplasm
(C400).

LARGEMOUTH BASS X BLACK CRAPPIE, *Micropterus salmoides* X *Pomoxis nigromaculatus*

Largemouth bass eggs fertilized with crappie sperm gave less than 1% as
good a hatch as when fertilized with bass sperm (H445). No hatch was reported
in experiments by M544, and the hatch was less than 0.1% (W229). Embryonic
development started in 91% (M544) and 59% (W229).

WHITE CRAPPIE X FLIER, *Pomoxis annularis* X *Centrarchus macropterus*

A female 119 mm hybrid presumably of this combination was collected in
Illinois (B426).

WHITE X BLACK CRAPPIE, *Pomoxis annularis* X *P. nigromaculatus*

Hybrid crappies did not become overcrowded but produced F_2 hybrids in
an Illinois pond with smallmouth bass (B429). The annual yield was from 0.46
to 13.8 kg/ha and averaged 4.8 kg/ha for 13 years, or was from 2.4 to 46.9

crappies/ha with an average of 14.8. The hybrids show fast growth and most are males (S574).

BLACK CRAPPIE X ROCK BASS, *Pomoxis nigromaculatus* X *Ambloplites rupestris*

Black crappie eggs fertilized with rock bass sperm gave 86% as good a hatch as when fertilized with crappie sperm, suggesting a close relationship between these genera (H445, T147).

BLACK CRAPPIE X WARMOUTH, *Pomoxis nigromaculatus* X *Lepomis gulosus*

Black crappie eggs fertilized with warmouth sperm gave 2% (H445) or 5.2% (M544) as good a hatch as when fertilized with crappie sperm. Another study failed to get any hatch, although about 30% of the embryos started development (W229); and another study got a 1.8% hatch (0%-16.5% in 4 trials), and 60.8% showed some embryo development (M544). Hatching was abnormal with most deformed if hatched (M544). The first hatch was at 47.5 hours, and the hatching period lasted 24 hours, not much different than crappie eggs (M544).

BLACK CRAPPIE X BLUEGILL, *Pomoxis nigromaculatus* X *Lepomis macrochirus*

Descriptions and a photograph of this hybrid are given in W229.

Black crappie eggs fertilized with bluegill sperm gave 39% (H445) or 47.9% (M544) as good a hatch as when fertilized with crappie sperm. The hatch was 18.2% (W229) and 16.5% (M544), but 91,8% (W229) and 60.2% (M544) of the eggs began embryonic development. Hatching was abnormal with 80%-95% of the fry deformed and dying on emergence (M544). About 7% survived to 60 mm (W229). The first eggs hatched at 49.2 hours, and hatching continued 19.7 hours, about the same as crappie fry (M544).

BLACK CRAPPIE X LARGEMOUTH BASS, *Pomoxis nigromaculatus* X *Micropterus salmoides*

Black crappie eggs fertilized with bass sperm gave less than 1% (H445) and 0.3% (M544) as good a hatch as when fertilized with crappie sperm. The hatch was 0.1% (M544) or completely failed (W229), but 74.6% (M544) and 38.1% (W229) of the eggs began development. Hatching was abnormal with about 67% dead at 20 hours after fertilization, and the first hatch was at 72 hours with hatching continuing 22 hours compared to 48.1 and 19.7 hours for crappie eggs (M544).

BLACK BASS, *Micropterus* spp.

Much additional information is available in the Proceedings of the First National Bass Symposium, 1975, published by the Sport Fishing Institute.

SHOAL BASS, *Micropterus* sp.

The shoal bass, previously confounded with the redeye bass, is an Apalachicola R. endemic in large piedmont and coastal plain tributaries of the Chipola, Chattahoochee, and Flint R. drainages of Florida, Georgia, and Alabama (R311). Adults grow larger (to 3000 g) than redeye bass (to 1000 g).

The following data probably refer to the shoal bass.

P70 FL Chipola R.

TL	Mean Weight	Range	No.
190-201	88	77-102	2
202-215	124	108-142	4
267-278	247	227-255	3
279-304	303	255-369	3
305-329	431	369-482	9
330-355	539	482-595	2
356-380	768	652-851	6
391	800	-	1
406-430	1,134	-	2
483	2,041	-	1

	TL	Mean Weight	Range	No.
D62, P70 AL	480-486	1,820	1,644-1,996	2
R312 AL	521	2,700	-	1

		TL	Weight	
Age V	1 fish	384	800	P70 FL Chipola R.
Age VI	1 fish	480	1996	P70 AL Chattahoochie R.
Age VIII	1 fish	521	2700	R312 AL Halawakee C.
Age X	1 fish	483	1644	D62 AL Chattahoochie R.

S576 AL pond - first year

	July	Aug.	Sept.	Dec.	
average weight	12	46	70	99	67% survival

	Average calculated TL and increments at each annulus					
	No.	1	2	3	4	5
P70 FL Chipola R.	21	97	206	290	353	388
Incr.		97	109	84	64	36
No.		21	20	12	5	1

The faster growth of shoal bass in the Chipola River than of redeye bass in Tennessee was thought to be the result of a longer growing season, larger stream habitat, clear water, and productive shoal areas (P70).

In Auburn AL ponds, they spawned successfully on May 19 at 25°C compared to Alabama R. redeye which spawned on April 14 at 22.8°C (S576). Eggs were 2.0 mm in diameter and the fry were 4.5 mm at hatching. The data for December follows:

No. measured	TL	W	% survival	Stocked at
76	142	44.5	81	7400/ha
68	160	49.5	89.9	2190/ha

Females 314 to 442 mm had 5,400 to 21,000 eggs (R312).

These bass have been taken in streams which reach 26.0° and 27.3°C in the summer (R312).

Aquatic insects, particularly mayfly nymphs, are the major food of shoal bass 40-80 mm, but they then turn to crayfish and fish (R312).

REDEYE BASS, *Micropterus coosae* Hubbs and Bailey

The redeye bass was originally found in the upland streams of Alabama, Georgia, Tennessee, and possibly Florida (probably shoal bass), and North and South Carolina (J67). They have been introduced into California with questionable success (G178) and into Puerto Rico (R312). They live mainly in streams too cold for other warmwater fish but too warm for trout. In streams with smallmouth bass, the redeye occupies the headwaters, with little overlap of distribution.

P67 TN TL = 1.21 SL, based on 157 fish.

TL	Mean of means	Weight range	No. of fish	Citations
51	0.2	0.09-0.23	54	S472 AL
76-101	6	0.3-10	613	S472 AL, P67 TN
102-126	11	8-20	255	S472, P67
127-151	27	17-50	32	S472, P67
152-177	52	32-74	37	S472, P67
178-202	83	50-112	40	S472, P67
203-228	124	104-150	31	S472, P67
229-253	170	149-202	22	S472, P67
254-278	221	179-227	7	S472, P67
279-304	259	227-272	4	S472
305-329	363	-	2	S472
330-355	499	-	1	S472

P67 TN Sheed's C.	157 fish 66-223 mm	$\log W = -4.749 + 3.0359 \log SL$	
S472 AL	944 fish 51-254 mm	$\log W = -5.18 + 3.12 \log TL$	
P67 TN Sheed's C.	157 fish 66-223 mm	average K(SL) 2.11	range 1.60-2.60
P67 TN Hatchery P.	311 fish 51-150 mm	2.16	1.77-2.38
	13 adults	2.30	2.08-2.60
S472 AL	54 fish 51 mm	average K(TL) 1.39	
	850 fish 76-127 mm	1.02	0.80-1.18
	40 fish 152-330 mm	1.35	1.20-1.50

The low average K indicates that redeye bass are more slender than the other black basses (P67).

The growth of redeye bass in Alabama ponds was (R307) as follows:

days	22	35	64	134
SL	19	37	60	87-106

In other Alabama ponds, growth the first year (S576) follows:

	July	Aug.	Dec.
Mean weight	8	37	58

	No.	Mean TL	Range
P67 TN creek Age 0 - June	30	23	15-32
P67 TN pond Age 0 - Sept.	311	112	51-150
K60 CA pond Age 0 - Oct.	10	-	64-94
P67 TN creek Age 0 - Nov.	+	54	-
S576 AL pond Age 0 - Nov. (19 g)	81	134	-
P28 TN Sheed's C. Age I	15	99	-
II	15	123	-
III	5	157	-
IV	15	171	-
V	14	183	-
VI	17	203	-

P28 TN Sheed's C. (cont.)	No.	Mean TL	Range
Age VII	11	221	-
VIII	5	227	-
IX	2	212	-
X	1	264	-

		Average calculated TL and increments at each annulus									
	No.	1	2	3	4	5	6	7	8	9	10
P67 TN Sheed's C.	100	48	83	114	143	169	191	211	215	217	255
Incr.		48	38	33	30	29	22	18	11	12	18
No.		100	85	70	65	50	36	19	8	3	1
P70 TN Spring C.	+	56	107	155							

Scales taken on May 15 showed evidence of annulus formation, and annuli were clear on scales collected in early June in Tennessee (P67). In Sheed's C. (P67), 82% of the scales examined were regenerated compared to 46%-67% in other centrarchids.

Redeye bass spawn in coarse gravel at the heads of pools in late May to early July at water temperatures of 16.7 -20.6 C (P67). Newly hatched young remain in schools only a short time (P67), and schools break up at 12-18 mm or 6 days (S576).

The smallest mature female was 120 mm, 16 g, age III, and the smallest mature male was 122 mm, 19 g, age IV in Sheed's C. (P67). In newly stocked streams, however, several males and females matured as yearlings when over 150 mm.

TL	Weight	Age	Eggs/female
145	35	V	2,084
205	98	-	2,334

The first spawn in Alabama ponds was on April 14 at 22.8 C (S576). The eggs averaged 3.5 mm in diameter, and the fry were 6 mm at hatching. Survival was 81.7% when stocked at 7,400/ha and was 100% from July to December (S576).

Insects taken from the surface were the major foods in 56 stomachs examined in May and August, but some fish, crayfish, salamander, and aquatic insects were also taken (P67). Insects, particularly chironomids, are the main food but they also eat crayfish, fish eggs, and salamanders (G178). Redeye bass moved out of the headwaters at the first heavy frost and returned in the spring (P67).

Redeye bass do not do well in ponds or reservoirs (P67).

SMALLMOUTH BASS, *Micropterus dolomieui* Lacépède

Smallmouth bass originally ranged from northern Minnesota and southern Quebec to the Tennessee River in Alabama and west to eastern Oklahoma (H450), but they have been widely introduced elsewhere (for an extensive summary see R312). They invaded the Hudson Valley with the opening of the Erie Canal about 1825 (H450) and were introduced in the Atlantic coastal states in the 1850's. They were taken to California as early as 1874 (E143). Although survival and even spawning have been successful in a few lakes in Saskatchewan no populations have become established, probably because of the low summer temperatures (M547, R295).

Smallmouth bass were suggested as suitable for stocking in slightly alkaline lakes, less than 900 ppm total alkalinity, 250 ppm carbonate alkalinity and 200 ppm K + Na (M552). Introduction of smallmouth bass into L. Opeongo, ON, may have resulted in decreased brook trout populations and introduced bass tapeworms, the larvae of which infected yellow perch, pumpkinseed, and brown bullheads, but probably had little effect on lake trout or the open lake habitat (M576).

	TL	No.	FL/SL	TL/FL	TL/SL
R210 NY	-	+	-	-	1.15
L142 L. Michigan	152-253	187	-	-	1.225
	254-355	184	-	-	1.238
	356-507	41	-	-	1.241
L123 MO	-	+	-	-	1.226
S299 NY Fall C.	30-279	488	-	-	1.160
F82 L. Ontario	-	-	-	1.05	-
S289 MS	60-419	108	-	-	1.140
T6 IA	103-299	104	1.17	-	1.216
C22 MN	54-419	8	1.121	-	-
B40 WI	90-420	188	-	-	1.18
S97 TN	50-400	254	-	-	1.216
B30 MI	to 64	+	-	-	1.250
	65-320	+	-	-	1.211
	over 320	+	-	-	1.220
E144 NY	-	+	1.14	1.04	1.18
R294 IA Des Moines R.	30-430	137	1.18	1.06	1.24
L178 L. Michigan	150-253	187	-	-	1.225
	254-355	184	-	-	1.238
	356-500	42	-	-	1.241
T138 ON Tadenac L.	90-412	164	-	-	1.212
R117 VA Claytor L.	-	83	TL = -1.3 mm + 1.2124 SL		

In compiling the table on length and weight it is assumed that total length equals 1.216 SL and 1.040 FL. The lengths given in the first three columns are not class centers but the lower length in the range covered in that line.

SL	FL	TL	Mean of means	Weight Central 50%	Range	No. of fish	Citations
21	24	25	1.6	-	-	17	S299 NY, Fall C.
42	50	52	1.9	-	1.7-3.5	130	S472 AL
			3.8	3-5	2-6	143	B30, L9, P18, S97, S299
			3.2	1.9-4	1.7-6	173	Combined
63	73	76	6	-	4-9	561	S472 AL
			8	6-10	6-12	99	B30, S97, S299
			8	-	-	322	L9 IN Greenwood L.
			7.5	6-8	4-12	982	Combined
84	98	102	13	-	2-23	546	S472 AL
			16	13-20	9-28	89	M499, B30, J15, P18, S97, W39
			20	-	-	322	L9 IN
			22	-	15-28	104	S299 NY
			17	13-20	2-28	1,061	Combined
104	122	127	25	-	12-59	255	S472 AL

SL	FL	TL	Mean of means	Weight Central 50%	Range	No. of fish	Citations
104	122	127	34	24-43	24-45	25	L142, L178, M499, P18, S97, T6
			36	34-43	20-57	97	B30, J15, W39, W144 North lakes
			37	37-39	28-44	115	B192, C182, S299 North streams
			40	-	-	367	L9 IN
			35	28-40	12-59	859	Combined
125	146	152	45	-	27-82	114	S472 AL
			53	45-54	45-71	45+	L9, P113, R24, S97, W132 South
			53	43-65	20-65	158	B30, J15, W39, W144 North lakes
			60	59-61	28-85	131	B192, C182, S299, T6 North streams
			63	-	36-86	19	L142, L178, M499
			56	45-62	27-86	467+	Combined
146	171	178	68	-	45-113	63	S472 AL
			85	-	28-142	101	J15, W39 WI
			91	85-93	57-113	128	B192, C182, S299, T6 North streams
			96	91-94	71-142	444	B30, L9, L142, L178, M20, M499, P18, P113, R47, S97, W132
			91	85-94	28-142	336+	Combined
167	195	203	113	-	68-181	41	S472 AL
			113	-	57-227	105	J15, W39 WI
			130	111-142	91-190	59+	L142, L178, P18, P113, M499, S97
			139	125-150	91-227	116+	B192, C182, L54, S299, T6 North streams
			144	142-147	99-170	64	B30, M20, W144 North lakes
			133	113-142	57-227	385+	Combined
188	220	229	145	-	113-200	32	S472 AL
			164	-	143-247	8+	E32, P113, S97, T44 South
			170	-	-	96	B192 OH, Miami R.
			170	-	113-227	87	J15, W39, WI
			195	186-201	142-255	68	C182, S299, T6 North streams
			196	-	-	104	B30 MI
			196	-	-	67	L142, L178 L. Michigan
			215	184-200	159-284	24+	M20, M499, R47, W144
			187	170-198	113-284	486+	Combined

SL	FL	TL	Mean of means	Weight Central 50%	Range	No. of fish	Citations
209	244	254	221	-	219-230	12+	P113, S97, T44 South
			222	-	195-286	19	S472 AL
			227	-	-	69	B192 OH Miami R.
			227	-	113-312	131	J15, W39 WI
			252	-	-	253	B30 MI
			252	227-272	170-340	27	C22, E32, M1, M20, M499, P18, R24, W144
			257	255-275	198-312	43+	C182, L54, S299, T6 North streams
			267	-	-	131	L142, L178 L. Michigan
			245	227-261	113-340	685+	Combined
230	268	279	303	-	-	38	B192 OH Miami R.
			312	-	170-397	161	J15, W39 WI
			321	-	284-347	42	S97 TN Norris L.
			322	-	200-445	15	S472 AL
			329	-	-	75	B30 MI
			335	312-351	298-425	57+	C22, C182, E32, M20, M499, P113, R47, S97, W144
			354	-	-	107	L142, L178 L. Michigan
			329	312-349	170-445	495+	Combined
251	293	305	411	-	-	64	B30 MI
			424	-	318-477	20+	P113, S97, S472 South
			425	-	170-680	124	J15, W39 WI
			441	411-445	363-567	77+	C22, C182, B192, E32, L54, M20, M499, R47, T6, W144
			477	-	-	84	L142, L178 L. Michigan
			437	411-477	170-680	369+	Combined
272	317	330	524	-	-	31	B30 MI
			539	-	312-709	92	J15, W39 WI
			542	471-601	400-709	35+	B192, C22, E32, H62, M20, M499, P18, P113, R47, S472, W144
			578	-	-	41	S97 TN Norris L.
			601	-	-	36	L142, L178 L. Michigan
			551	504-601	312-709	235+	Combined
292	342	356	644	-	-	44	B30 MI
			652	-	340-907	69	J15, W39 WI
			697	627-740	454-964	50+	B192, C22, C182, E32, M20, M1, M499, P113, P99, P18, L54, R24, R47, S97, S472, W144, T6

SL	FL	TL	Mean of means	Weight Central 50%	Range	No. of fish	Citations
292	342	356	700	-	-	21	S97 TN Norris L.
			772	-	-	22	L142, L178 L. Michigan
			695	652-740	340-964	206+	Combined
313	366	381	765	-	397-1,134	67	J15, W39 WI
			773	-	-	23	S97 TN Norris L.
			845	-	-	26	B30 MI
			885	808-987	709-1,021	39+	C182, B192, E32, L142, L178, M20, M499, P113, P99, R47, R25, T44, W25, W144
			854	773-896	397-1,134	155+	Combined
334	390	406	964	-	539-1,474	62	J15, W39 WI
			1,019	907-1,134	794-1,361	44+	B192, C22, C182, D21, L54, L142, L178, M20, M499, P113, P54, R47, S97, S472, W144
			1,049	-	-	31	B30 MI
			1,016	945-1,049	539-1,474	137+	Combined
335	415	432	1,117	-	-	13	S97 TN Norris L.
			1,134	-	624-1,814	42	J15, W39 WI
			1,176	-	-	18	B30 MI
			1,280	1,137-1,332	1,021-1,514	18+	C22, D21, E32, L142, L178, M20, M499, P54, P113, R24, R47, W25
			1,223	1,134-1,297	624-1,814	91+	Combined
376	439	457	1,276	-	680-2,268	36	J15, W39 WI
			1,449	1,298-1,758	862-1,837	33+	B30, C22, C182, D21, E32, J13, L54, L142, L178, H62, M20, P113, R24, R47, S97, S472
			1,420	1,276-1,559	680-2,268	69+	Combined
397	464	483	1,446	-	765-2,381	23	J15, W39 WI
			1,723	1,588-1,817	1,446-2,155	19+	B30, C22, E32, F22, H52, H62, L142, L178, M20, R39, P54, P113, R47, R24, S472
			1,672	1,559-1,817	765-2,381	42+	Combined
418	488	508	1,895	-	1,361-2,523	4+	E32, H52, L54, P113
			1,899	-	1,276-2,750	21	J15, W39 WI
			1,896	1,724-1,899	1,276-2,750	25+	Combined
439	513	533	1,831	-	907-2,722	21	C22, H52, J15, W39, M20
460	538	559	1,985	-	1,418-3,005	18	J15, W39 WI
			2,268	-	-	+	L54 PA

SL	FL	TL	Mean of means	Weight Central 50%	Range	No. of fish	Citations
460	538	559	2,353	-	2,041-2,835	7	H52 IN
			4,763	-	-	1	M99 AL
			2,430	-	1,418-4,763	26+	Combined
480	562	584	1,304	-	-	1	C22 MN L. of Woods
			1,985	-	1,814-2,155	5	J15, W39, WI
			2,991	-	-	1	W65 WV
			3,113	-	2,438-3,970	7	H52 IN
501	587	610	1,503	-	-	1	P42 S. Africa
			1,928	-	1,814-2,041	2	J15, W39 WI
			3,573	-	3,176-3,970	2	H52 IN
543	635	660	2,495	-	-	1	J15, W39 WI
			4,540	-	4,422-5,216	3	H52 IN
-	-	686	5,419	-	-	1	T66 TN

In general there was little consistency in differences in weights at various lengths in different parts of the country.

	No. of fish	SL	
R116 VA Claytor L.	83		$\log W = -4.6145 + 2.995 \log SL$
B30 MI			$\log W = -4.725 + 3.0515 \log SL$
S97 TN			$\log W = -4.8441 + 3.0804 \log SL$
T6 IA			$\log W = -4.828 + 3.0935 \log SL$
L142, L178 L. Michigan	537	127-488	$\log W = -5.3980 + 3.3561 \log SL$

T132 CA Folsom L.		$\log W = -5.1227 + 3.125 \log FL$

	No. of fish	TL	
B192 OH Little Miami R.	491	127-411	$\log W = -4.6861 + 2.9172 \log TL$
S472 AL	1,742	51-260	$\log W = -4.798 + 2.95 \log TL$
	26	270-485	$\log W = -4.878 + 3.00 \log TL$
P74 MO Black R.			$\log W = -4.888 + 2.9964 \log TL$
P113 MO			$\log W = -5.0536 + 3.077 \log TL$
R293 IA Des Moines R.	220	76-429	$\log W = -4.868 + 3.0096 \log TL$
T138 ON Tadenac L.			
young of year			$\log W = -4.177 + 2.701 \log TL$
older fish			$\log W = -5.122 + 3.2047 \log TL$
A104 IA Turkey R.	104	100-440	$\log W = -4.201 + 3.13 \log TL$
S188 RI			$\log W = -5.8412 + 3.372 \log TL$
F166 MO Big Piney R.			$\log W = -4.921 + 3.016 \log TL$
L123 MO	934		$\log W = -5.0560 + 3.077 \log TL$
upper river stations	500		$\log W = -4.1948 + 3.0163 \log TL$
substation	194		$\log W = -5.2747 + 3.1644 \log TL$
substation	159		$\log W = -4.9034 + 3.0087 \log TL$
middle river			
stations	290		$\log W = -5.0540 + 3.0765 \log TL$
substation	44		$\log W = -5.5076 + 3.2633 \log TL$
substation	47		$\log W = -4.9546 + 3.0244 \log TL$
lower river station	144		$\log W = -5.11975 + 3.1047 \log TL$
substation	51		$\log W = -4.7666 + 2.9638 \log TL$

	No. of fish	TL
substation	68	log W = -5.01536 + 3.0479 log TL
substation	154	log W = -4.6920 + 2.9239 log TL
substation	52	log W = -4.9118 + 3.0304 log TL

The regression with a slope of 0.3184 given by B376 is obviously in error and probably should be 3.184. The regression slopes for each year 1956-62 on the Des Moines R. were 3.032, 2.919, 3.260, 2.909, 3.014, and 3.062, respectively (R293). None of these differed significantly from 3.0, but the intercept values differed.

	SL	No.	K(SL)	Range
E3 MN	203-507	11	1.94	-
B40 WI	Age III	90	2.08	-
S100 TN Cherokee L.	-	5	2.16	-
S110 NC Hiwassee L.	-	21	2.16	-
B40 WI Muskellunge L.	100-259	171	2.20	2.00-2.56
S97 TN Norris L.	-	250	2.23	-
S299 NY Fall C.	30-280	488	2.41	-
R116 VA Claytor L.	-	83	2.42	1.82-2.59
B30 MI	44-401	+	2.43	2.29-2.56
T6 IA streams	-	104	2.49	2.39-2.69
C23 MN Hubert L.	250-390	4	2.53	-
W25 MI Potogannissing Bay	-	23	2.64	-
L142 L. Michigan	150-480	542	2.74	2.27-3.35
C22 MN L. of Woods	203-507	13	2.80	-

	FL	No.	K(FL)	Range
H21, H162, H163, P99, P92, P42, P53, P54 S. Africa	305-610	23+	1.59	0.66-2.08

	TL	No.	K(TL)	Range
A104 IA Turkey R.	100-440	104	1.27	1.08-1.44
S472 AL	52-483	1,786	1.29	1.20-1.49
B30 MI	53-483	-	1.30	1.16-1.41
R293 IA Des Moines R.	76-429	271	1.45	1.29-1.69
M499 IL	114-445	77	1.50	1.22-1.94
H21 South Africa	-	+	1.61	-
B373 IL added stock ponds		+	1.33	
central ponds		+	1.36	
cropped ponds		±	1.88	

There was a tendency for condition factors to decrease with increase in length in M499 and for it to increase with increase in length in B39, L142, and L178. In Lake Michigan (L142, L178) the slope of the weight-length regression, 3.356, also indicates an increase of condition factors with increase in length. Most of the regression slopes were close to 3.0, suggesting no trend in K with increase in length (see R293). Mandible tags did not significantly affect the condition factors, although they did slow the growth of smallmouth bass (Y13). In general, condition factors and growth rates were correlated in Illinois ponds (B373) but there were some inconsistencies. The best condition was found in the same streams as the best growth in Iowa (T6), but the slower growing populations showed higher condition factors than the faster growing in a Wisconsin study (B40). Females over 400 mm averaged heavier than males at the same lengths (H198). Condition was highest in July in 4 or 5 years of Des Moines R. collections (R293).

| | | TL | | | | Weight | |
	No.	Mean of means	Central 50%	Range	No.	Mean of means	Range
Age 0-May							
WV streams: S106, S108	183+	11	-	10-11	-	-	-
Age 0-June							
SA L. Waskesiu: R15	+	6	-	5-8	-	-	-
ON lakes: D20, D21, M17, T8	62+	13	12-15	8-17	-	-	-
NY lakes: A1, M243	128	15	-	11-18	-	-	-
IL Fork L.: T19	46	23	-	-	-	-	-
OH: R71	+	30	-	25-35	-	-	-
OK Grand L.: H150	227	38	-	28-51	-	-	-
TN, WV streams: D43, S106, S108	217+	46	36-49	23-65	-	-	-
AL 29 and 37 days: R307	+	55	-	36-74	-	-	-
WV ponds: S171	+	71	-	-	-	-	-
Combined	680+	29	12-38	5-74	-	-	-
Age 0-July							
L. Erie: F5	7	15	-	10-20	-	-	-
SA Waskesiu L.: R15	+	16	-	8-24	-	-	-
NY George L.: M243	42	16	-	13-48	-	-	-
ID: K147	100+	-	-	10-28	-	-	-
CA ponds: B176	-	-	-	-	+	2.5	1-6
NY ponds: M243	33	28	-	-	-	-	-
ON: D21, M17, T8	244+	33	29-43	16-52	-	-	-
OH: L11, R71	+	48	-	43-57	-	-	-
OK Little R.: F64	44	51	-	33-71	-	-	-
IA, IL, IN lakes: B3, E32, T19	11+	58	52-64	33-71	-	-	-
TN, WV streams: D43, H76, S106, S108	216+	62	37-64	36-98	-	-	-
NY hatcheries: T27	+	74	-	51-97	-	-	-
OK Grand L.: H150	1,292	86	-	33-178	-	-	-
AL: R307, S576	+	97	-	-	+	19	-
Combined	1,989+	45	28-64	8-178	+	11	1-19
Age 0-Aug.							
SA Waskesiu L.: R15	+	39	-	26-52	-	-	-
L. Ontario: S182	+	-	-	38-53	-	-	-
MO ponds: C370	90	53	-	51-54	-	-	-
OK lakes: H150, J42	7+	56	-	53-58	-	-	-
IN Maxinkuckee L.: E32	+	56	-	43-69	-	-	-
ON: D21, M17, T138	18+	61	-	35-92	-	-	-
NY streams: L48	248	63	-	39-95	-	-	-
ID rivers: K147	148	-	-	37-96	-	-	-
OH: L11, L40, R71	+	77	-	51-152	-	-	-
TN, WV streams: D43, S108	226	83	-	51-173	-	-	-
WI lake: C117	2	84	-	-	-	-	-
WV ponds: S298	+	-	-	74-127	-	-	-
IL Fork L.: T19	59	114	-	-152	-	-	-
AL: R207, S576	+	183	-	-	+	72	-
Combined	834+	75	58-88	26-173	+	72	-

		TL				Weight	
	No.	Mean of means	Central 50%	Range	No.	Mean of means	Range
Age 0-Sept.							
SA Waskesiu L.: R15	+	55	-	42-82	-	-	-
MO ponds: C370	69	60	-	58-61	-	-	-
NY George L.: M243	4	62	-	58-64	-	-	-
ID: K147	38	-	-	51-97	-	-	-
NJ Hatcheries: M14	+	-	-	-	+	9	-
TN, WV streams: D43,							
S106, S108	202+	91	84-96	48-146	-	-	-
NY ponds: M243	7	99	-	-	-	-	-
WV ponds: S298	+	-	-	74-137	-	-	-
OH: L11, L40, L79, R71	+	118	94-135	94-152	+	38	-
NJ cannibals: M14	30	129	-	-	30	17	-
TN Norris L.: J10	20	135	-	-	-	-	-
IL ponds: B144, T19	45+	149	-	127-229	-	-	-
AL: S576	-	-	-	-	+	105	-
Combined	415+	103	81-129	42-229	30+	37	9-105
Age 0-Oct.							
ID: K147	76	-	-	53-120	-	-	-
TN streams: D43	3	76	-	41-124	-	-	-
IA hatchery: D5	+	77	-	65-89	+	20	-
IL ponds: B373	4,735	95	86-104	69-119	-	-	-
OK ponds: B427	289	125	99-156	118-183	-	-	-
TN Norris L.: J10	23	135	-	-	-	-	-
OH: L11, L14	600+	142	-	64-229	-	-	-
IL Fork L.: T19	74	165	-	-	-	-	-
Combined	5,800+	107	86-135	41-229	+	20	-
Age 0-Nov., Dec.							
NJ: S258	12	-	-	102-127	-	-	-
AL: S576	116	174	-	171-177	56+	66	55-108
Age 0							
NK Potters L.: S66	104	39	-	-	104	-	-
L. Erie: W32	313	42	-	10-74	-	-	-
ON Perch L.: T9	2	51	-	-	-	-	-
AR: P197	323	51	-	30-70	-	-	-
OK Illinois R.: Neosho							
subsp.: L106, J25	126	53	-	25-112	-	-	-
IL: D88, V2	2+	72	-	64-81	-	-	-
NY ponds: E9	+	72	-	65-79	-	-	-
NY Fall Creek L.: S299	99	73	-	-	-	-	-
MO streams: L123, P74	811	76	-	20-190	-	-	-
MI: B41	33	84	-	-	33	9	-
VA Claytor L.: R116	1	95	-	-	-	-	-
IA Des Moines R.: R293	+	+	-	38-165	-	-	-
WV streams: H218, S217	23	103	-	84-135	-	-	-
TN Norris L.: E26, S97	36	123	-	61-170	-	-	-
WI: J15	4	124	-	117-132	-	-	-
South Africa: H19	+	127	-	-	-	-	-
OK Grand L.: T44	1	241	-	-	1	150	-
Combined	1,878+	79	51-94	10-241	138	-	0.9-150
Age I							
OH streams: B192	550	83	-	41-145	-	-	-
IN Muscatatik R.: M317	10	89	-	64-114	10	9	3-17
ID early spring: K147	27	-	-	73-120	-	-	-
MD Deep Creek L.: E76	16	99	-	-	-	-	-
ON: D21, H15, T9, T138	44+	115	-	66-153	40	45	14-60

		TL				Weight	
	No.	Mean of means	Central 50%	Range	No.	Mean of means	Range
Age I (cont.)							
NY ponds: E9	+	119	-	106-132	-	-	-
MI, MN, WI:E3, J15, M20, N37, B31, J16, G152	325	124	93-150	89-231	55	61	48-88
MA lakes: M1, M87, S289	42	124	-	117-135	29	27	26-28
NK Potters L.: S66	5	128	-	-	5	28	-
ME lakes: H280	15	131	-	112-140	15	33	28-35
NY Fall Creek L.: S299	272	132	-	-	-	-	-
NY: B54, G19, G25, G27	47	137	-	71-193	38	40	14-88
MO streams: L123, P74	734	141	-	36-274	-	-	-
IA streams: T6	18	142	-	-	-	-	-
OH ponds: L14	+	150	-	-	-	-	-
OK Little R.: F64	37	155	-	107-193	-	-	-
AR: P197	363	155	-	91-213	-	-	-
L. Erie: R200	17	157	–	104-201	17	51	17-79
NY Cayuga L.: W92	291	163	-	s.d. 18	291	62	-
OK Illinois R. Neosho subsp.: J25, L106	185	163	-	99-218	185	57	-
TN, WV streams: S133, S217, V48, H218, D43	94+	166	165-170	118-193	-	-	-
IL streams: D88, L108	131+	169	-	145-178	-	-	-
VA Claytor R.: R116	15	169	-	109-231	-	-	-
PA streams: B247, H448	+	170	-	102-244	-	-	-
OH Sandusky R.: C182	14	178	-	-	-	-	-
OK reservoirs: H150, J42, T44	14+	181	152-218	114-188	13	48	
IL lakes, ponds: L84, B144, L129, V2, B373	1,675+	183	173-193	89-251	-	-	-
NJ Steenykill L.: S258	26	-	-	152-218	-	-	-
AR Ouachita L.: H442	7	188	-	-	-	-	-
South Africa: H19, H21, H162	2+	188	-	132-252	-	-	-
NC reservoirs: M368, R217	60	204	178-206	178-241	-	-	-
CA Colorado R.: T132	191	216	-	-	-	-	-
IA Des Moines R.: R293	138	218	-	64-292	-	-	-
TN lakes: B380, E22, S97, E26	155	221	178-284	73-292	-	-	-
VA streams after reclamation: H177, H304	+	222	-	178-279	-	-	-
MO ponds: A91, C270	365+	245	-	135-343	348	92	44-238
CA, OR: B122, S363	2	313	-	303-324	1	408	-
Combined	5,889+	166	137-185	41-343	1,047	53	3-408
Age II							
IN Muscatatak R.: M317	7	132	-	89-170	7	28	14-37
OH streams: B192	538	155	-	94-236	-	-	-
ON lakes: H15, T9, D21, W144, T138	173+	178	173-191	123-224	137+	75	28-170
MD Deep Creek L.: E76	15	180	-	-	-	-	-
ME lakes: H280	51	181	-	178-183	50	85	71-99
WI lakes: M20, J15, N37, G152	228	182	156-183	122-356	12	147	-
NY Fall Creek L.: S299	59	185	-	-	-	-	-

| | | TL | | | | Weight | |
	No.	Mean of means	Central 50%	Range	No.	Mean of means	Range
Age II (cont.)							
NY ponds: E9	+	190	-	178-203	-	-	-
NY streams and lakes: B53-4, G19-23, G25, G27, P72	325	190	-	114-279	103	99	20-286
MA lakes: M1, M87, M146, S289	87	201	183-216	183-216	59	102	91-122
AR: P197	439	203	-	122-388	-	-	-
MO streams: L123, P74	434	203	-	114-376	-	-	-
NK Potters L.: S66	14	205	-	-	14	128	-
OH ponds: L14	+	206	-	-	-	-	-
TN streams: D43	4	206	-	145-244	-	-	-
OK Little R.: F64	41	211	-	165-250	-	-	-
IL Jordan Creek: D88	138	213	-	206-234	-	-	-
NY Cayuga L.: W92	794	213	-	s.d. 20	794	142	-
Great Lakes: F73, R200	104	221	-	211-251	5	136	74-176
OK Illinois R.: Neosho subsp. J25, L106	88	221	-	168-267	88	130	-
WV streams: S133, S217, V48, H218	450+	224	216-229	118-366	-	-	-
PA streams: L54, B247	+	224	-	183-254	-	-	-
MI: B31	95	229	-	-	95	164	-
IL: V2	+	229	-	-	+	284	227-340
IA streams: T6, H177, R293	82+	238	-	190-368	-	-	-
NC, VA reservoirs: M368, R116, R217	87	238	221-268	216-328	-	-	-
IL ponds: L84, B144, L129, B373	1,283+	241	229-259	152-394	-	-	-
OK reservoirs: J42, T44, H150	16+	242	198-284	175-287	16	197	111-312
OH Sandusky R.: C182	101	246	-	-	-	-	-
NJ Steenykill L.: S258	3	-	-	254-269	-	-	-
MN: E3	5	278	-	246-308	-	-	-
CA Colorado R.: T132	237	290	-	-	-	-	-
AR Ouaichita L.: H442	9	292	-	-	-	-	-
TN reservoirs: B380, E22, E26, J10, S97	299	323	296-325	146-413	-	-	-
South Africa: H19, H162, H164, P53-4, P93, P96	12+	328	-	254-422	8+	716	284-1,361
Combined	6,218+	227	191-254	94-422	1,378+	142	14-1,361
Age III							
IN Muscatatik R.: M317	16	183	-	157-236	16	79	43-150
OH streams: B192	195	214	-	145-290	-	-	-
NY Fall Creek L.: S299	27	214	-	-	-	-	-
ME lakes: F26, H280	56	217	181-251	133-297	55	138	28-298
QE: P17	-	-	-	-	+	142	-
NJ lakes: N50, S258	1+	230	-	229-231	-	-	-
WI lakes: G152, C117, W144, M20, B40, J15, N37	512	232	219-253	171-417	64	198	
ON lakes: H15, T9, D21, F82, T138	441+	224	189-254	179-267	300+	160	62-233
AR streams: P197	191	235	-	152-328	-	-	-

| | | TL | | | | Weight | |
	No.	Mean of means	Central 50%	Range	No.	Mean of means	Range
Age III (cont.)							
TN, WV streams: S133, D43, S217, V48, L105	167+	242	190-282	53-340	-	-	-
MA lakes: M87, M146, S289	83	243	-	229-259	50	211	184-238
PA streams: L54, B247	+	248	-	218-282	-	-	-
NY: B53-4, G19-23, G25, G27, P72	523	251	-	145-381	198	204	45-811
NK Potters L.	6	253	-	-	6	233	-
Great Lakes: F73, R200	490	254	-	239-312	185	249	201-392
OK Little R.: F64	21	259	-	229-297	-	-	-
NY Cauyga L.: W92	1,022	262	-	s.d. 23	1,022	261	-
MO streams: L123, P74	212	264	-	155-455	-	-	-
OK Illinois R.: Neosho subsp. J25, L106	29	267	-	208-312	29	241	-
OH ponds: L14	+	269	-	-	-	-	-
MD Deep Creek L.: E76	13	269	-	-	-	-	-
OH Sandusky R.: C182	35	277	-	-	-	-	-
IL Jordan C.: D88	108	279	-	264-306	-	-	-
NY Oneida L.: F103	56	280	-	262-295	-	-	-
IL ponds: B144, L129, B373	277+	282	254-295	239-366	-	-	-
MI lakes: M31	158	284	-	-	158	315	-
NC lakes: M368	5	288	-	249-328	-	-	-
IA streams: T6, R293	106	291	-	241-400	-	-	-
OK lakes: H150, J42, T44	18+	293	-	211-424	18	521	213-1,137
WA Columbia R.: H198	15	295	-	229-330	15	425	198-567
AR Ouaichita L.: H443	1	318	-	-	-	-	-
MN: E3	12	319	-	246-433	-	-	-
VA Claytor L.: R116	9	339	-	318-377	-	-	-
CA Colorado R.: T132	31	348	-	-	-	-	-
South Africa: H21, H62, P54, P99	10	363	-	317-505	10	712	468-1,616
TN lakes: E22, E26, J10, S97, B380	377	367	328-404	268-438	-	-	-
Combined	5,242+	267	231-287	53-505	2,126+	256	28-1,616
Age IV							
IN Muscatatik R.: M317	22	226	-	203-264	22	125	88-230
NY Fall Creek L.: S299	17	240	-	-	-	-	-
ME lakes: F26, H280	210	240	194-252	145-450	209	234	40-1,389
OH streams: B192	110	267	-	208-348	-	-	-
IA streams: T6, R293	30	302	-	272-371	-	-	-
NY: B53-4, G19-23, G25, G27, P72, S182	370	273	262-277	140-394	116	255	91-635
ON lakes: H15, T9, D21, F82, W144	819+	273	361-302	232-346	241+	302	156-397
AR streams: P197	68	274	-	213-346	-	-	-
PA streams: L54, B247	+	282	-	236-320	-	-	-
Great Lakes: G25, F73, S182, R200	617	286	262-305	206-351	445	403	329-652
NK Potters L.: S66	55	287	-	-	15	346	-
WI lakes: N37, C117, J15, M20, G152	334	288	254-318	211-460	64	519	-

		TL			Weight	
	No.	Mean of means	Central 50%	Range	No.	Mean of means	Range

	No.	Mean of means	Central 50%	Range	No.	Mean of means	Range
Age IV (cont.)							
WV, TN streams: S133, D43, S217, V48, H219, L105	248+	293	274-322	165-424	-	-	-
NJ lakes: N50, S258	1+	299	-	292-312	-	-	-
OK Little R.: F64	8	302	-	274-328	-	-	-
IL Jordan C.: D88	18	307	-	279-318	-	-	-
NY Cayuga L.: W92	497	307	-	s.d. 24	497	445	-
MA lakes: M87, M1, M146, S289	48	308	284-323	284-330	35	389	343-474
OH Sandusky R.: C182	3	312	-	-	-	-	-
MO streams: L123, P74	92	316	-	216-455	-	-	-
WA Columbia R.: H198	20	320	-	305-340	20	482	425-680
NY Oneida L.: F105	399	320	305-328	297-338	-	-	-
OH ponds: L14	+	323	-	-	-	-	-
MN: E3	11	324	-	278-339	-	-	-
MI lakes: B31, C215	129	328	-	277-338	128	539	-
OK lakes: H150, J42, T44	4+	328	-	284-401	4	534	340-923
OK Illinois R. Neosho subsp.: J25, L106	7	333	-	310-371	7	442	-
MD Deep Creek L.: E76	9	340	-	-	-	-	-
IL lakes: B144, L129	15+	369	-	323-417	-	-	-
South Africa: H162, P42, P54, P93, P96, P99	8	371	-	264-475	6	1,304	794-1,814
IN lakes: H62, G39	5	381	-	333-435	3	420	-
NC lake: M368	1	401	-	-	-	-	-
TN lakes: E22, J10, S97, B380	83	404	-	340-387	-	-	-
AR Ouaichita L.: H442	5	409	-	-	-	-	-
CA Colorado R.: T132	3	409	-	-	-	-	-
Combined	4,203+	302	269-325	140-475	1,812+	393	40-1,814
Age V							
ME lakes: F26, H280	52	263	244-297	213-356	52	230	113-448
NY: B53-4, G19-23, G25, G27, P72, S182, S299	376	284	277-300	193-419	74	383	116-914
TN streams: D43, L105	26	286	-	198-388	-	-	-
IN Muscatatik R.: M317	7	290	-	244-351	7	312	162-669
QE: P17	-	-	-	-	+	241	-
AR: P197	39	302	-	219-395	-	-	-
ON lakes: W144, T9, H15, D21, F82, T138	877+	306	291-327	264-366	165+	414	300-548
NK Potters L.: S66	16	313	-	-	16	403	-
Great Lakes: W25, F73, S182, R200	635	313	277-363	251-371	332	563	471-709
WI lakes: G152, C117, M20, J15	166	327	295-351	267-508	23	865	-
TN Wawassee L.: H62	1	331	-	-	1	400	-
PA streams: L54, B247	+	331	-	287-394	-	-	-
NJ ponds: N50	+	333	-	-	-	-	-
WA Columbia R.: H198	18	335	-	305-366	18	624	340-879
OK lakes: J42, T44	5	337	-	292-320	5	614	397-1,049
IL Jordan C.: D88	33	340	-	335-348	-	-	-
NY Oneida L.: F103	422	341	-	333-358	-	-	-

	No.	TL Mean of means	Central 50%	Range	No.	Weight Mean of means	Range
Age V (cont.)						-	-
NY Cayuga L.: W92	220	348	-	s.d. 22	220	712	-
MO streams: L123, P74	49	349	-	244-439	-	-	-
MA lakes: M1, M87, M146	17	350	-	345-353	17	659	633-691
WV streams: S133, S217, V48, H218	95+	351	301-380	197-519	-	-	-
OK Little R.: F64	6	353	-	351-356	-	-	-
IA streams: T6, R293	6	358	-	297-429	-	-	-
South Africa: H21, H162, P93	4	362	-	290-441	2	1,162	1,148-1,176
OH ponds: L14	+	368	-	-	-	-	-
WV ponds: S289	8	388	-	-	-	-	-
MI lakes: B31, C215	80	390	-	381-424	79	760	-
MD, VA, NC lakes: R116, F76, M368	8	397	-	384-425	-	-	-
IL ponds: B144, L129	15+	418	-	368-467	-	-	-
MN: E3	3	422	-	339-494	-	-	-
OH Sandusky R.: C182	3	424	-	-	-	-	-
TN lakes: S97, B380	19	440	-	388-487	-	-	-
Combined	3,206+	330	297-353	193-519	1,011+	522	113-1,176
Age VI							
IN Muscatatik R.: M317	2	305	-	290-320	2	343	293-394
QE: P17	-	-	-	-	+	340	-
ME lakes: H280, F26	40	317	293-330	278-386	40	420	216-760
NY: B53-4, G19, G21, G23, G25, G27, P72	97	323	-	213-470	24	556	201-803
Great Lakes: F73, S182, W25	766	326	302-337	241-394	330	623	621-624
NK Potters L.: S66	20	337	-	-	20	474	-
ON lakes: F82, W144, T9, H15, D21, T138, W258	734+	337	315-353	300-396	97+	525	384-709
AR: P197	18	338	-	255-406	-	-	-
OH streams: B192	13	349	-	297-386	-	-	-
TN, WV streams: S133, D43, S217, H218, L105	50	350	320-403	241-470	-	-	-
IL Jordan C.: D88	11	358	-	353-391	-	-	-
WI lakes: G152, J15, M20, C117	106	363	333-391	292-536	9	1,327	-
PA streams: L54, B247	+	365	-	310-427	-	-	-
WA Columbia R.: H198	24	368	-	330-381	24	822	454-964
NY Cayuga L.: W92	103	373	-	s.d. 17	103	879	-
NJ ponds: N50	+	373	-	-	-	-	-
NY Oneida L.: F103	265	374	-	358-386	-	-	-
OK Little R.: F64	4	381	-	378-384	-	-	-
MI: B31	45	388	-	-	45	817	-
IL: V2	-	-	-	-	+	822	652-907
IN Wawassee L.: H62	4	392	-	-	4	581	-
OH ponds: L14	+	394	-	-	-	-	-
MA lakes: M146, S289, M1, M87	21	397	-	361-429	18	849	652-916
MO streams: L123, P74	18	398	-	330-486	-	-	-
OH Sandusky R.: C182	1	406	-	-	-	-	-

		TL				Weight	
	No.	Mean of means	Central 50%	Range	No.	Mean of means	Range
Age VI (cont.)							
IA Des Moines R.: R293	1	411	-	-	-	-	-
IL ponds: L129	+	414	-	-	-	-	-
MD Deep Creek L.: E76	1	424	-	-	-	-	-
TN Norris L.: S97	1	462	-	-	-	-	-
South Africa: H162, P96	4	474	-	449-528	4	1,524	1,361-1,814
VA Claytor L,: R116	1	523	-	-	-	-	-
MN lakes: E3	2	556	-	526-588	-	-	-
Combined	2,352+	360	330-388	241-588	720+	676	201-1,814
Age VII							
QE: P17	-	-	-	-	+	510	-
Great lakes: W25, F73, S182	557	348	318-366	241-445	180	720	647-901
NK Potters L.: S66	5	356	-	-	5	578	-
IA streams: T6	3	358	-	-	-	-	-
ON lakes: T9, H15, D21, F82, W144, T138	260+	360	348-372	307-425	45+	606	397-794
OH streams: B192	9	371	-	335-411	-	-	-
NY: B54, G19-20, G22-23, G25, G27	27	373	-	221-445	4	803	680-867
AR: P197	6	374	-	318-438	-	-	-
ME lakes: F26, H280	14	377	-	273-433	14	682	247-918
NY Oneida L.: F103	55	390	-	388-409	-	-	-
NY Cayuga L.: W92	64	396	-	s.d. 19	64	1,044	-
WA Columbia R.: H198	39	396	-	356-437	39	1,106	765-1,418
WI lakes: J15, M20, C117, G152	42	399	376-432	297-607	11	1,611	-
NJ ponds: N50	+	401	-	-	-	-	-
PA streams: L54, B247	+	401	-	363-445	-	-	-
TN, WV streams: S133, S217, H218, L105	13	406	367-458	343-490	-	-	-
MO streams: L123	8	411	-	305-516	-	-	-
MI: B31	20	417	-	-	20	989	-
MA lakes: M146, S289	15	428	-	409-437	13	1,164	-
TN Norris L.: S97	3	471	-	412-487	-	-	-
VA Claytor L.: R116	2	572	-	-	-	-	-
Combined	1,162+	388	366-409	221-607	395+	862	247-1,611
Age VIII							
QE: P17	-	-	-	-	+	624	-
Great Lakes: W25, S169, F73, S182	448+	369	348-385	267-445	76	894	776-953
NY: G19, G21, G23, G25, G27	25	371	-	310-470	3	774	524-910
OH streams: B192	8	373	-	335-414	-	-	-
ON lakes: T9, H15, D21, F82, W144, W258	240+	374	359-385	338-456	18+	763	425-964
ME lakes: F26	27	377	-	336-442	27	807	340-1,134
NY Oneida L.: F103	31	401	-	391-424	-	-	-
AR, WV, TN streams: L105, P197, S133	17	403	-	330-488	-	-	-
NJ ponds: N50	+	414	-	-	-	-	-
MI lakes: R25, B31	10	417	-	388-427	10	1,078	1,021-1,097
NY Cayuga L.: W92	34	424	-	s.d. 19	34	1,261	-
WA Columbia R.: H198	69	427	-	386-470	69	1,304	964-1,701

		TL				Weight	
	No.	Mean of means	Central	Range	No.	Mean of means	Range
Age VIII (cont.)							
IA, MO streams: L123, T6	12	431	-	394-486	-	-	-
WI lakes: J15, M20, C117, G152, S593	69	435	394-476	305-487	6	1,653	-
MA lakes: M146, S289	4	456	-	432-480	2	1,500	-
IN Wawassee L.: H62	1	499	-	-	1	1,591	-
Combined	995+	400	373-425	305-499	246+	1,012	340-1,701
Age IX							
QE: P17	-	-	-	-	+	737	-
NY: G19, G22, G23, G27	27	384	-	325-521	9	779	567-1,348
Great Lakes: W25, F73, S182	344	390	366-414	292-445	20	1,139	1,094-1,162
MI lakes: R25	6	391	-	373-421	6	959	765-1,106
ON lakes: T9, H15, D21, F22, F82, T138	45+	405	384-414	356-483	10+	1,004	567-1,871
AR streams: P197	1	406	-	-	-	-	-
ME lakes: F26	19	411	-	353-486	19	931	595-1,446
NY Oneida L.: F103	29	417	-	409-478	-	-	-
NY Cayuga L.: W92	7	432	-	-	-	-	-
MO streams: L123	5	439	-	394-465	-	-	-
NJ ponds: N50	+	445	-	-	-	-	-
WA Columbia R.: H198	23	457	-	432-470	23	1,616	1,361-1,900
IN Wawassee L.: H62	1	460	-	-	1	1,001	-
WI lakes: G152, J15, M20, C117	55	464	430-490	343-673	2	1,724	-
MA lakes: M146	6	490	-	-	6	1,767	-
WV streams: S133	1	593	-	-	-	-	-
Combined	569+	424	394-439	325-673	96+	1,155	567-1,900
Age X							
MI lakes: R25	5	391	-	357-417	5	936	737-1,035
Great Lakes: W25, F73, S182, F82	137	405	391-420	292-470	9	1,077	907-1,162
ON lakes: D21, H15, T9	9	406	-	403-414	6	970	907-1,030
ME lakes: F26	15	420	-	367-460	15	978	907-1,276
NY: G19-20, G23, B54	17	421	-	340-533	1	882	-
NY Oneida L.: F103	51	425	419-434	419-455	-	-	-
MO streams: L123	3	447	-	417-465	-	-	-
NY Cayuga L.: W92	10	457	-	-	-	-	-
WI lakes: J15, M20, C117, G152, S593	32	468	445-472	394-623	1	2,155	-
WA Columbia R.: H198	13	475	-	457-483	13	1,814	1,531-2,155
Combined	292	429	408-448	292-623	50	1,228	709-2,155
Age XI							
NY Raquette R.: G19	6	386	-	371-434	-	-	-
MI lakes: R25	2	396	-	384-402	2	1,132	964-1,290
MO streams: L123	1	417	-	-	-	-	-
NY Oneida L.: F103	11	422	-	419-424	-	-	-
ME lakes: F26	7	423	-	373-470	7	1,066	765-1,389
ON lakes: D21, T9	6	445	-	401-466	4	1,547	1,261-1,758
WI lakes: J15	10	488	-	414-610	-	-	-
WA Columbia R.: H198	4	490	-	483-498	4	1,985	1,814-2,155
NY Chautauqua L.: G23	3	513	-	495-521	-	-	-
CT: W13	1	559	-	-	1	2,608	-
Combined	51	444	410-488	371-610	18	1,447	765-2,608

		TL				Weight	
	No.	Mean of means	Central 50%	Range	No.	Mean of means	Range
Age XII							
Great Lakes: G25, F73, S182, R200	184	416	384-450	292-495	2	1,985	-
ON lakes: T9, H15, D21	4	442	-	403-513	3	1,210	936-1,758
MI Loon L.: R25	1	452	-	-	1	1,332	-
ME lakes: F26	9	467	-	410-516	9	1,372	1,077-1,758
NY lakes: F103, G23	3	484	-	437-521	-	-	-
MO streams: L123	1	486	-	-	-	-	-
MA ponds: S289	1	488	-	-	-	-	-
WI lakes: J15	8	495	-	406-559	-	-	-
Combined	211	451	415-491	292-559	15	1,435	936-1,985
Age XIII							
L. Ontario: S182	22	411	-	318-472	-	-	-
ON lakes: D21	1	455	-	-	1	1,304	-
NY lakes: F103, G22	2	470	-	452-488	-	-	-
WI lakes: J15	3	503	-	465-536	-	-	-
TN Dale Hollow L.: T66	1	686	-	-	1	5,419	-
Combined	29	473	411-503	318-686	2	-	1,304-5,419
Age XIV							
L. Ontario: S182	3	429	-	396-483	-	-	-
ON lakes: T9	2	445	-	-	1	1,332	-
ME lakes: F26	3	480	-	470-492	3	1,304	1,106-1,446
MA ponds: S289	1	508	-	-	-	-	-
WI lakes: J15	1	533	-	-	-	-	-
Combined	10	469	445-480	396-533	4	1,313	1,106-1,446

Mean calculated TL at each annulus

	No.	1	2	3	4	5	6	7	8	9	10	11	12	13	14
D19 L. Erie	+	171	214	255	295	330	361	387							

Mean calculated TL and annual increment at each annulus

	No.	1	2	3	4	5	6	7	8	9	10	11	12	13	14
L142, L178 L. Michigan, Waugoshance Point		109	157	188	221	267	310	358	396	424	442	452	447		
Incr.		109	48	33	38	38	43	38	25	18	15	13	13		
No.		1,892	1,863	1,599	1,423	734	671	152	45	24	18	10	2		
L142, L178 L. Michigan, Hog Island		114	150	188	221	267	310	356	396	411	432	450	470		
Incr.		114	36	38	34	46	41	41	25	20	20	15	13		
No.		157	157	155	139	139	72	65	20	15	13	10	1		
L142, L178 L. Michigan, Cecil Bay		104	155	188	221	269	343	381	414	424	427	434			
Incr.		104	51	38	41	51	43	56	23	13	7	7			
No.		214	214	164	111	107	10	8	3	2	1	1			
W259 L. Huron, South Bay		111	154	234	267	293	326	356	–	–					
Incr.		111	43	80	33	28	36	31	–	–					
No.		3,119	3,119	3,117	3,001	2,477	1,320	271	56	5					
Unweighted mean Great lakes		122	166	211	245	285	330	368	402	420	434	445	458		

Mean calculated TL at each annulus

	No.	1	2	3	4	5	6	7	8	9	10	11	12	13	14
B40 WI Owen L.	11	49	112	177	234	273	311	325	365	417					
B40 WI	1,322	61	135	208	269	318	358	388	424	447	465	480			
E2 MN	17	79	169	238	348										
E3, S62, M258, B365 MN	54	99	183	254	310	462	521								
B40 WI Weber L.	41	82	180	267	358	437	499								
J16 WI Weber L.	41	81													
J16 WI Weber L.	110	112													

Mean calculated TL at each annulus (cont.)

	No.	1	2	3	4	5	6	7	8	9	10	11	12	13	14
P213 WI Red Cedar R.	544	100	190	274	329	383	407	424	444	432	465	480			
Unweighted mean MI,MN,WI		83	161	236	308	363	419	379	411						

Mean calculated TL at each annulus

	No.	1	2	3	4	5	6	7	8	9	10	11	12	13	14
S299 NY Fall C.	488	71	112	165	203	224									
B247 PA Upper Wood P.	21	56	112	183	262	312	363								
B247 PA Winola L.	9	53	147	213	297										
W89 ME Big L.	736	76	147	218	279	330	376	409	434						
S188 RI	113	84	170	229	300	356	399	437	457	480					
B247 I A Idlewild L.	36	61	152	231	307	363	411	457	493	508	523				
E93 ME	+	-	175	234	287	302	340	394	396	445	447				
W13, T23 CT	172	102	183	241	290	333	371	401	429						
F168 NY Oneida L.															
Constantia	2,699	94	171	248	309	347	374	395	412	425	437				
Shackleton	2,565	93	174	250	309	345	371	391	407	417	429				
F103 NY Oneida L.															
Constantia area	885	99	175	249	312	343	373	399	417	424	434				
Shackleton Shoals	252	81	152	236	290	333	366	378	396	406	417	427	442		
Shackleton Point	184	94	178	251	307	343	368	386	399	411	417	419			
B247 PA Clarke L.	123	81	173	259	307	335									
M292 MA Quabbin L.,															
1946-52	85	89	170	259	328	373	409	424	434	439	445				
1953-57, after smelt	24	84	193	279	348										

Mean calculated TL and annual increment at each annulus

	No.	1	2	3	4	5	6	7	8	9	10	11	12	13	14
F103 NY Oneida L.		99	175	249	312	343	373	399	417	424	434				
Incr.		99	76	74	63	41	25	25	15	10	10				
No.		885	885	885	859	533	254	96	51	27	13				
Unweighted mean Northeast		82	163	235	297	332	376	406	424	438	443	423	442	452	

Mean calculated TL at each annulus

	No.	1	2	3	4	5	6	7	8	9	10	11	12	13	14
T6 IA lakes	7	99	140	187	229	273	306	340	386						
R47 OH	+	102	152	190	224	254	279	305	325	335	355	370	385	395	410
B376 IL Big Walnut C.	65	96	149	193	235	278	326	356							

Mean calculated TL at each annulus (cont.)

	No.	1	2	3	4	5	6	7	8	9	10	11	12	13	14
T6 IA streams	109	92	149	208	246	270	300	339							
C122 IA streams	206	91	175	224	282	348	432								
B192 OH Sandusky R.	125	86	170	234	302	363	381								
C182 OH Sandusky C.	157	86	175	236	307	384	381								
B192 OH Stillwater R.	135	97	160	241	312	373	404		480						
D88 IL Jordan C.															
upper	153	86	180	259	292	335	366	427							
lower	288	84	163	236	284	320	345								
M546 NB McConaughy R.	92	74	169	263	327	318	385								
B373 IL Siloom Springs L.	+	107	178	267	335										
L129 IL	114	142	208	269	323	368	414								
B144 IL Malcomsen's P.	347	135	236	282	310	363									

Mean calculated TL and annual increments at each annulus

	No.	1	2	3	4	5	6	7	8	9	10	11	12	13	14
B192 OH Massie C.		76	145	216	274	315	351	373	384						
Incr.		76	71	61	38	28	20	15							
No.	718	718	288	139	56	19	7	4							
B192 OH Little Miami R.		74	150	224	279	323	353	368	384						
Incr.		74	76	76	61	46	30	20	15						
No.	213	213	104	57	25	15	11	5							
A104 IA Turkey		91	155	226	279	328	376	401	434						
Incr.		91	62	74	60	37	47	46	40						
No.	104	104	55	33	17	6	6	3	2						
R293 IA Des Moines R.		119	229	297	340	388	411								
Incr.		119	119	71	46	25	13								
No.	270	270	132	62	10	4	1								
Unweighted mean OH to NB		96	171	236	288	330	363	364	392	335	355	370	385	395	410

Mean calculated TL at each annulus

	No.	1	2	3	4	5	6	7	8	9	10	11	12	13	14
F166, F169 MO Big Piney R.	2,407	86	160	216	269	325	378	411	434	445					
L190 MO upper stations	+	89	163	224	277	333	363	388	406	419	417	429	445		
middle stations	+	97	173	246	302	340	373	399	424	437					
lower stations	+	102	196	284	366	417	457	470	488						

Mean calculated TL at each annulus (cont.)

	No.	1	2	3	4	5	6	7	8	9	10	11	12	13	14
P113 MO statewide	3,448	89	170	244	290	343	371								
headwaters	2,258	84	160	226	277	323	363								
middle reaches	873	91	168	241	284	351	353								
lower	317	97	201	290	353	394	439								
lowest station average	+	64	142	175	234	249	267	348	368	388	399	411	445		
highest station average	+	107	213	310	381	434	445	442	442	470	488	427			
P197* AR unweighted															
mean of 51 stations	1,147	87	181	257	329	400	500	534	594						
lowest mean	+	58	118	170	227	286	388	452	504						
highest mean	+	128	315	346	402	515	632	665	753						

*These data from P197 are not in agreement with the data on lengths at capture. The largest bass at capture was 467 mm TL. An erroneous method of calculating length was used.

	No.	1	2	3	4	5	6	7	8	9	10	11	12	13	14
A87 AR Bull Shoals L.	107	91	175												
J42 OK Ft. Gibson L.	46	64	140	203	257	284									
H150 OK Illinois R.	38	99	185	234	300										
J25 OK Illinois R.	417	86	173	246	302										
L106 OK Caney C.	130	97	180	244											
L106 OK Caney C. pools	40	89	165	236	312										
Ballard C.	50	69	165	254	325										
Illinois R.	25	99	183	241	335										
Snake R.	59	76	173	262	333										
Illinois R. system	309	86	173	246	328										
F87 OK Little R. trib.	107	99	188	241	290	338	371								
Little R.	9	102	206	274											
cutoff lakes	1	64													
J101 OK L. Eucha	17	99	226	302	373										

Mean calculated TL and annual increments at each annulus

	No.	1	2	3	4	5	6	7	8	9	10	11	12	13	14
P74 MO Black R.		68	157	236	302	363	406								
Incr.		68	81	76	61	61	23								
No.		181	91	40	35	8	1								

Mean calculated TL and annual increments at each annulus (cont.)

	No.	1	2	3	4	5	6	7	8	9	10	11	12	13	14
F64 OK Little L.		94	190	246	290	338	371								
Incr.		94	94	51	41	41	28								
No.		117	80	39	18	10	4								
L123 MO streams		81	175	251	300	358	391	406	414	419	417	419	445		
Incr.		81	97	81	53	46	30	28	25	18	18	18	33		
No.		1,386	742	359	162	87	45	28	20	10	5	2	1		
P71 MO Clearwater L.		117	112	81	69	61									
Incr.		117	74	46	18	17									
No.		87	74	46	18	17									
Unweighted mean MO,AR,OK		89	174	246	309	344	387	409	425	429	430	422	445		
Mean calculated TL at each annulus															
B192 WV Lost R.	+	79	140	188											
B192 WV 15 streams	+	81	155	213											
S192 MD Bennett C.	+	58	137	206	216	239									
S192 MD Catoctin C.	+	76	173	226	282	340									
S192 MD Israel C.	+	74	168	234	287										
S192 MD Potomac R., Allegany Co.	+	99	193	244	284	335	373								
S192 MD Potomac R., Fredrick Co.	61	107	188	249	295	333	386								
E73 MD Potomac R.	+	102	201	264	323	406	470								
M182 MD Potomac R.	+	99	196	262	328	404	457								
E73 MD Deep Creek L.	+	97	178	267	343	391	437								
M182 MD Deep Creek L.	+	99	175	267	338	378	427								
E73 MD Loch Raven	+	102	201	305	386	447	483								
M182 MD Loch Raven	+	99	201	305	386	445	475								
T54 KY streams	42	175	234	282	333	417									
R116 VA Claytor L. (Fraser correction)	83	91	198	300	384	439	423	559							
R117 VA Claytor L. (Dahl-Lea)	83	61	173	277	358	429	518	559							

Mean calculated TL at each annulus (cont.)

	No.	1	2	3	4	5	6	7	8	9	10	11	12	13	14
S99 NC Hiwassee L.	20	91	232	319	356	389									
R292 TN Dale Hollow L. (from Polloch, 1965) (from Nichols Turner, 1966)	+	155													
R292 TN Center Hill L.	+	114	244	274	348										
T127, F151 TN, E. TN	+	168	264	353	404										
Valley, 1944–62	192	107	241	328	386	401									
S99 TN Norris L.	5	85													
E22 TN Norris L.	97	141	272	393											
S97 TN Norris L.	599	118	258	358	411	445	457	473							
C169 TN Norris L.	+	79	226	338	401	442	457	472							
T54 KY lakes	23	130	279	388	518										

Mean calculated TL and annual increment at each annulus

		1	2	3	4	5	6	7	8	9	10	11	12	13	14
B380 TN Woods L.															
before impoundment		81	168	218	282										
Incr.		81	99	84	64										
No.		50	28	6	2										
after impoundment															
Incr.		193	175	119	114	124									
No.		12	22	22	4	2									
M368 NC Nantahala L.		102	170												
Incr.		102	64												
No.		26	13												
M368 NC Santeelah L.		109	188	229	302	358									
Incr.		109	81	38	41	56									
No.		20	10	3	2	2									
M368 NC Fontana L.		117	185	251											
Incr.		117	48	38											
No.		15	7	1											

Mean calculated TL and annual increment at each annulus (cont.)

	No.	1	2	3	4	5	6	7	8	9	10	11	12	13	14
R217 NC James L.		165	206	257											
Incr.		165	41	51											
No.		16	9	2											
M368 NC Hiwassee L.		109	213	262	363										
Incr.		109	104	49	101										
No.		26	4	2	1										
R292, R308 TN Dale Hollow L.		97	183	332											
Incr.		97	87	128											
No.		70	14	4											
After threadfin shad introduction		100	175	259											
Incr.		100	83	102											
No.		59	39	8											
E26 TN Norris L.		141	271	363											
Incr.		141	132	94											
No.		382	311	126											
Unweighted mean Southeast		107	202	282	346	391	455	501							

Mean calculated TL and annual increment at each annulus

	No.	1	2	3	4	5	6	7	8	9	10	11	12	13	14
G99 OR	1	60	150	238	311	324									
G161 OR 3 lakes	134	73	181	274	330	370	382	416	449						
E143 CA Pine Flat L.	+	137	222	312	367	415	446	457							
E143 CA Folsom L.	+	137	272	359	404	406									
H21 South Africa	10	109	211	318	406										
H162 South Africa	5	116	277	361	409	435	412								
T132 CA		137	272	358	403										
Incr.		137	135	86	45										
No.		462	271	34	3										
K147 ID Salmon R.		82	137	180	221	249	268	288	306						
Incr.		82	55	43	41	27	19	20	18						
No.		24	22	20	20	11	9	4	1						

Mean calculated TL and annual increment at each annulus (cont.)

	No.	1	2	3	4	5	6	7	8	9	10	11	12	13	14
K147 ID Clearwater R.		83	142	193	241	269	295	315	325	338					
Incr.		83	60	54	45	29	24	19	13	12					
No.		166	147	127	96	73	53	29	19	8					
K147 ID Lower Snake R.		84	145	206	239	267	292	310	322						
Incr.		84	62	59	34	26	24	18	15						
No.		155	149	136	126	101	76	45	19						
K147 ID Upper Snake R.		87	150	211	249	272	297	310	325						
Incr.		87	63	61	38	24	23	14	16						
No.		162	157	126	105	88	62	37	16						
Unweighted mean		100	196	274	297	334	342	349	345	338					
Unweighted mean of regional means		97	176	246	298	341	382	397	400	399	425	428	432	424	410

Calculated weight in grams at each annulus

	No.	1	2	3	4	5	6	7	8
T6 IA	116	43	71	181	289	414	567	791	
W259 L. Huron S. Bay		2.0	63	232	298	388	527	744	
Incr.		20	43	168	78	94	151	189	
No.		2,451	2,451	2,449	2,358	2,019	1,158	212	
D19 L. Erie	+	85	142	255	369	539	765	992	
R293 IA Des Moines R.	270	23	181	395	676	840	1,002		
Incr.		23	163	231	213	154	91		
No.		270	132	62	10	4	1		
P213 WI Red Cedar R.	544	12	91	291	520	844	1,020	1,170	1,350

Instantaneous growth in weight $G = \log_e (W_e/W_0)$

	1	2	3	4	5	6	7	8
P213 WI Red Cedar R.	–	2.026	1.162	0.581	0.484	0.189	0.137	0.143

The maximum life span as indicated by tag returns in L. Oneida is about 18 years (F168).

Consistency of data on growth increments and abundance of year classes gave evidence of the validity of annuli (B192, F103, R293). Back calculations of lengths and of weights were verified by the fact that they agreed with the expected trends on the basis of a minimum size limit in the capture technique (W259). Annuli were also verified from tagged fish (L142, B192). The tagging returns indicated a greater probability of missing annuli than of reading false annuli, particularly in older fish (B192). In Arkansas, P197 reported no difficulty with false annuli. Crowded and indistinct annuli on scales of smallmouth bass over 8 years old prevented accurate age determinations, and some large bass (457-521 mm) with 10 to 15 annuli were interpreted differently each time examined (K147). On a second reading of scales from the Des Moines R., 95% agreement was secured (R293). Scale pictures with the annuli marked are in R293 and H450.

In Ohio, smaller bass had new annuli by late April and early May and all had formed annuli by mid-June (B192). The same situation occurred in Norris L., TN (S97). Yearling bass in NY hatcheries had not formed annuli by mid-June, but 97% of age II and III bass in L. Oneida had, compared to 55% for age IV, and 15% for older smallmouths (F103). In L. Michigan, annuli formed in early June to late July (L142). Annuli were formed in May in the Des Moines R., IA (R293). In California, a new annulus was found on 19 of 25 bass collected in May (T132).

Most growth calculations were made on a direct proportion basis, but many used the Fraser modification (B192, C122, F103, G152, L142, L178, M292, M368, R116, F217, R293, S188, T6, and T132). A comparison is available in the tabulations for Claytor L. R116 and R117. In most cases a correction factor of 30-50 mm was used, but R217 used a correction factor of 117 mm, which unduly increases the first year's calculated growth. Some other workers (K147, L123, P74, P197, and S299) calculated lengths by referring to a body-scale chart developed for the population. This method was apparently not used correctly in P197 because the average calculated lengths were much greater than the lengths of the fish at capture.

Scales are first formed when the young bass are 21 mm and first appear along the lateral line on the caudal peduncle (E144). Although the body-scale relationship was curvilinear, it did not differ much from a straight line relationship beyond a body length of 156 mm (150 mm FL) (E52). The body-scale regression was represented by 2 straight lines with the inflection point between 145 and 150 mm (P213). The anterior radius gave a better indication of body length than the anterolateral radius and scale width. Key scales did not significantly increase the accuracy of back-calculation (E52). Annual increments of the scales were reported (P199) to be proportionately less than body length increments when temperatures were low or when individuals were old.

Weights at various annuli were calculated from scale measurements using a log-log regression (W259).

Lee's phenomenon was reported in B192 as related to selective sampling, possibly due to selective removal by angling, and possibly due to the effects of tagging. In Oneida L. (F103) earlier maturity of the faster growing individuals resulted in selective sampling and the appearance of Lee's phenomenon in age groups III and IV. RNA-DNA ratios were shown to be related to growth, and age I showed higher RNA-DNA ratios than age II smallmouth bass (H476). Caution should be expressed as to the use of RNA-DNA ratios as measures of long term growth as proposed by H476, however, since the data presented may

merely indicate that fish which were growing faster through the period were
also growing faster at the conclusion of the experiment.

Since the growth curve was sigmoid, a Von Bertalanffy equation fit to the
data overestimated the sizes from the 4th through the 6th year in the South Bay
population (W259).

Although growth rates varied within each region as tabulated above, there
was a tendency for faster growth in the southern part of the range and also for
fewer bass over age VI or VII in the South.

Faster growth in the South than the north was also pointed out by B192,
E143, T6, B40, P199, and H450. The length of the growing season was correlated with bass growth in Wisconsin (B40), but a 10-20 day difference in growing season gave no detectable effect on growth of bass in Arkansas (P197).
Progeny of stock from Green L., MN, grew faster than those from Elk R., MO,
when held under the same conditions (A91). The lake bass consumed more
food and used it more efficiently in that study. The above tabulated data do not
give an indication of whether growth is better in lakes or streams. Growth is
usually slower in headwaters or small tributary streams than in the larger
rivers (F87, L190, P113, P197, B192, and S192). In Iowa, no smallmouth bass
over age IV were found in small streams, indicating a movement to larger
rivers as they grow (T6). Slower growth in smaller streams was attributed to
lack of deeper water permitting free movement (B192). Generally, growth is
more rapid in lakes and reservoirs than in streams, according to R312.

Populations with different average growth rates were found in different
parts of L. Simcoe (W144, W231), L. Oneida (F103), L. Huron (F82), and L.
Ontario (S286); but average growth rates from three locations in L. Michigan
were quite similar (L178). Analysis of variance detected no significant differences in growth of smallmouth bass in various streams in 3 different basins in
Arkansas but showed differences between stations on the streams (P197).

A positive correlation between smallmouth bass growth and pH, conduc-
tivity, and higher sucker populations was reported from a study by Sullivan,
1956 (S552). Growth rate increased with fertilization of the Obed R., TN
(P196). Soil type of the watershed had no evident effect on growth of smallmouth in Arkansas streams (P197). Average TL of smallmouth in fall age I
was 210 mm in a pond with a Secchi disk reading of 15 cm, 277 mm in a pond
with disk reading of 25 cm, and 306-343 mm in ponds where the disk could be
seen in 60 cm of water (A91). The growth rate was inversely related to the
clay turbidity. No relationship between water transparency and growth of bass
was evident in some Missouri ponds, but there were indications that there was
a relationship masked by differences in population density (C370). Growth was
generally better in eutrophic waters of the Montreal Plain than in the oligotrophic waters of the Laurentides (P199). More rapid growth of bass in Center
Hill L. than in Dale Hollow L., TN, was attributed to the greater fertility and
warmer water of Center Hill L. (R292). Growth of smallmouth bass was
slower in fertilized ponds than in less fertile ponds perhaps because of increased diurnal dissolved oxygen flux, decreased transparency, or increased
ammonia (H475).

In the Des Moines R., IA, the year of best growth was the year with the
lowest average water levels; and correlations between water levels and
growth increments of age I, II, and III bass were all negative, but nonsignifi-
cant (R293). In Ohio streams the slowest growth was in years of high stream
flow and lower temperatures (B192). The low water stages of drought years
did not markedly decrease growth rates (B192), but no correlation was
detected between growth and rainfall.

Growth was rapid after impoundment of Norris L., TN, but decreased after the first few years and then increased again when water levels raised, flooding new areas (S97). Growth also was rapid after impoundment of Grand L., OK, (H150).

Temperatures, particularly July-August temperatures but also April temperatures, were positively correlated with annual increments of smallmouth bass in Ohio (B192). Maximum growth occurred at 28°-29°C (B427). With juvenile smallmouth (28-53 g), the percentage increase in weight per day for 50 days (except 23 days at 35°C) at various temperatures were (H474):

Temperature (°C)	16.1	23.1	26.4	28.9	32.2	35.1
% increase/day	0.15	0.70	1.23	1.02	0.76	-0.08

Fry, 11 days after hatching, averaged 12.9 mm (10.2-14.0 mm) at 25°C and 4.4 mg O_2/liter; 9.4 mm (9.0-9.9 mm) at 20°C and 4.4 mg O_2/liter; and 9.9 mm (8.4-10.1 mm) at 20°C and 8.7 mg O_2/liter (S594). A temperature drop from 23°C to 14.4°C in 3 days resulted in decreased growth of young bass (D20). Growth was slow in spring-fed ponds, probably because of lower temperatures (L11). Positive relationships between temperature and bass growth were reported by D21, L11, and P199. Annual growth increments did not vary in accordance with the annual sums of degree-days (over 10°C) in Idaho rivers (K147). Waters with smallmouth bass had 886-3100 degree-days (over 10°C) annually (C371). Annual increments of age III-V smallmouth bass in South Bay, L. Huron, and in several other waters were related with mean surface-water temperatures from July through September, but no such relationship was evident in L. Opeongo, nor for older bass in South Bay (C371). First year growth of smallmouth bass in South Bay was more rapid in warmer summers (C401). Slow growth in L. Michigan in 1950 and 1951 was related to cold, rainy summers (L178).

In L. Oneida (F168) increments for the years of life 1 through 4 and 6 were significantly correlated with mean air temperatures June through September, and the abundance of young yellow perch improved the prediction of annual increments for the 3rd year on. Growth of young smallmouth bass was negatively correlated with abundance of young perch, suggesting some competition.

Over 50% of the seasonal growth was completed by the end of June, 1956 in Ohio streams, with an abrupt leveling off in late July (B192).

No sex differences in growth were reported by L178, S229, W92, W144, T6, S182, B40, B31, W231, or L190. Growth is slower after sexual maturity is reached (H450).

Growth has generally been rapid where bass have been introduced, as the tabulated data for South Africa, California, and Oregon show or data in B122, S363, T132, E26, J10, and E22. Smallmouth bass introduced into rivers after reclamation showed good growth (H177, H304). There was some indication of increased growth of smallmouth bass in George L., WI, after muskellunge were added (G152).

Growth in a series of experimental ponds was inversely related to population density (C370). Growth was usually better in cropped ponds and slower in ponds where population density was artificially increased, but the relationships were not as clear as anticipated because of differences in the carrying capacities of the ponds (B373). Slow growth with high population density was noted (H450). The more abundant year classes showed slower growth in L. Michigan (L142). The abundant 1952 year class in Big Piney R. showed good growth, but the growth of the 1953 year class was believed to be slowed by the competition

(F166). In California (T132), little variation was noted in the growth of the various year classes.

Growth increments of smallmouth bass were better in L. Oneida in 1952, 1954, and 1955, than in 1951, 1953, and 1956, a pattern similar to that shown by walleyes (F103).

Growth of smallmouth bass increased somewhat when smelt were added as forage fish (M292) and when lake chubsuckers were introduced (B307). An abundance of minnows was noted in Arkansas streams where bass showed good growth (P197). Abundant forage was also associated with good growth in the Columbia R. (H198) and some Ontario lakes (D21). Cannibalistic bass grow most rapidly in rearing ponds (L12, M14). No effect on bass growth by adding threadfin shad to Dale Hollow L., TN, was noted (R292, R308).

Where rock bass and yellowbelly sunfish were abundant in Maryland streams, growth of smallmouth bass was slow (S192). Although growth was slowed where bass competed with rock bass and other fish, growth was rapid after the bass reached a size beyond competition with these fishes (L190). It was suggested (S182) that growth of smallmouth bass increased in L. Ontario because of removal of competitive fishes by the commercial fisheries. Growth of smallmouth bass, at 20 kg/ha, did not differ from that in ponds with 20 kg/ha bass plus 70 kg/ha carp (H475).

When the effect of smallmouth bass density was removed from some pond studies, growth showed a curvilinear regression with the minnow standing crop (C370). As standing crop of minnows increased, growth rates of bass increased up to a level of about 60 kg of minnows per hectare after which growth of bass did not increase further.

Young smallmouth bass infected with 193-708 metacercaria of *Crassiphiala ambloplites* grew more slowly than uninfected bass (H449).

Mandible tags reduced growth (Y13), but finclipping had no detectable effect on growth rates (C370). A spaghetti-type tag did not affect growth (W258).

For a few of the populations, I have computed the average annual increment as a percentage of the mean calculated length at the beginning of the year. The numbers of fish are given in parentheses.

TL	F103 NY Oneida L.	T6 IA Streams	F64 OK Illinois R.	E26 TN Norris L.	S97 TN Norris L.
51-75	-	-	-	-	174 (77)
76-101	85 (580)	55 (84)	114 (62)	206 (4)	200 (81)
102-126	62 (305)	46 (11)	75 (18)	103 (104)	239 (31)
127-151	-	45 (55)	-	102 (72)	94 (6)
152-177	42 (533)	43 (12)	-	81 (131)	-
178-202	43 (352)	25 (14)	27 (29)	-	68 (25)
203-228	-	17 (1)	26 (10)	-	71 (64)
229-253	26 (533)	25 (3)	17 (8)	40 (62)	47 (14)
254-278	25 (326)	11 (1)	16 (10)	45 (4)	50 (28)
279-304	13 (337)	13 (1)	15 (6)	29 (60)	27 (7)
305-329	13 (196)	-	12 (4)	-	26 (37)
330-355	8 (230)	-	8 (4)	-	18 (6)
356-380	7 (96)	-	-	-	14 (10)
381-405	4 (75)	-	-	-	9 (4)
406-431	2 (40)	-	-	-	13 (10)
432-456	-	-	-	-	4 (7)

This type of tabulation may be most helpful in answering such questions as: How much can a fish of a given size be expected to grow the next year? Some daily growth rates have been reported:

	No.	Size beginning	Daily increment	Daily instantaneous rates	Food conversion
W111	4	4-13 g	0.14-0.2 g for 98 days	0.0018-0.0095	2.6-4.1
	5	22-34 g	0.07-0.35 g for 98 days	0.0024-0.0097	4.0-7.9
	5	39-60 g	0.15-0.53 g for 98 days	0.0033-0.0079	3.6-6.5
	2	111-112 g	0.65-1.0 g for 33-65 days	0.0047-0.0079	3.7-4.8
	4	83-109 mm	0.4 mm for 99 days	-	-
	7	123-152 mm	0.1-0.2 mm for 98 days	-	-
	5	160-202 mm	0.2-0.4 mm for 98 days	-	-
L73	4	157-290 mm	0.1-0.7 mm -	-	3.3-4.1
H150	+	fingerlings	1.3 mm for 140 days	-	-

Smallmouth bass usually mature at 3-4 years, but extremes of 2-9 years have been reported (E143).

	Males Age	TL	Females Age	TL
MO: L190	II	-	II	-
IA Des Moines R.: R293	II-III	231-	II-III	229-
IA: T6	III	229	III	229
MI: L178	III-V	254	IV-VII	318
NH: N95	III	-	III	-
ME: E93	III	229	III	229
L. Huron: F73	50% at III	-	50% at IV	-
	78% at IV	-	75% at V	-
	all at VI	-	all at VI	-
NY: W92	III-IV		IV-V	
ID: K147	IV	215-	V	225-
IA: C369	IV	200-250	-	-
L. Ontario: S182, S169	IV-VI	198-305	V-IX	258-356
ON: D202	-	305	-	305
ON Tadenac L.: T138	33% at IV	187-260	9% at IV	187-260
	88% at V		82% at V	
	all at VI	264-366	all at VI	264-366
Northern states: H450	IV-VI	-	IV-VI	-

The number of eggs per female increases with the age and size of the female, except that old females may have fewer eggs and the eggs may have a lower rate of survival (L190). In the Des Moines R., IA, (R293) no correlation was noted between size of female (204-484 g) and egg number. The number of eggs is reported at 7,000 (N25) or 8,000 (W89) per pound of female, which is low compared to the following data:

	Size of females	No.	No. of eggs per female Mean	Range
H49 NY	250 mm	1	5,148	-
	395 mm	1	11,428	-

	Size of females	No.	No. of eggs per female Mean	Range
C77	482 g	1	20,825	-
	652 g	1	5,040	-
specimen with tapeworm	1,247 g	1	13,862	-
B65 MI	-	-	-	2,000-10,000
L14 OH	356 mm	1	5,440	-
R293 IA	229-320 mm, 204-484 g	8	7,708	4,964-15,203
C397 MI Katherine L.	228-278 mm	5	3,120	2,757-3,567
	279-328 mm	16	4,876	3,711-5,659
	329-379 mm	10	7,734	6,081-9,289
	380-430 mm	4	7,241	6,450-8,433

It was indicated that some of the females in R293 may have been partially spent and that the fecundity estimates were thus low.

Egg diameters, after fertilization, were 1.8-2.2 mm (M525), 2.5 mm (R297), and 3.5 mm (S576).

Smallmouth bass spawn earlier than other centrarchids in the same area (H450). They spawned on April 16 at 22.8°C in Alabama, but the first successful spawn was on April 25 at 23.9°C and hatched on May 4 (S576). April to early May is reported for the southeast (L190), mid-April to May for Ohio (B192), late April to May for Missouri (P195), May and June for Maryland (S192), Iowa (L190, H451, R293), New Hampshire (N95), Maine (N145), New York and Michigan (H450), May to July for Wisconsin (S593), and June for Ontario (T8, T138) and Saskatchewan (R295). In L. Ontario smallmouth bass do not spawn until late June or July, except for some that enter shallow bays or streams to spawn in late May or early June (S182). Spawning in the main basin of Cayuga L., NY, is not until late June to mid-July (W92), although it is completed by mid- to late June on shoal areas or in streams. Sometimes a second spawning period may take place, particularly if a temperature drop or high water interferes with the first spawn (P195, R295, N145, C369). The same males were often involved in the second nesting (P195). In the Toledo, Ohio, aquarium, spawning occurred in May even though water temperatures were fairly stable (B192).

Spawning activity occurs in rising water temperatures when the daily minimum is not less than 13°C (B192, N145), or at temperatures of 13°-21°C (N95), 15°-18°C (T138), 19°-21°C (R295), 15°-21°C (L190), 11.7°C (W92).

If the water warms gradually, spawning can be expected at about 15°C; but if it warms rapidly, temperatures may be about 18°C before spawning begins (H450). By holding breeders in sufficiently cold water, fish culturists have been able to delay spawning until the latter part of the summer (H450).

Nesting may occur at temperatures below that of egg laying. Nesting occurred at 9.4°-21°C (B192).

The male defends a territory and fans out a nest with his caudal fin (B390, R297, M534). Large objects may be removed by mouth. Nest construction requires from 4-48 hours or more, depending upon temperature and bottom type (H450). Nest sites are chosen with care, and 2 or 3 nests may be constructed before a final one is accepted (C369).

Nests are usually dished out down to coarse, egg-sized rubble or even to bedrock (C369). Nests may also be in sand areas with the bottom of the bowl covered with woody debris and broken clam shells (T138). Nests may even be on soft bottoms (B144). Nests were usually about 60 cm in diameter and 5-10 cm deep (P195); 61 cm in diameter (40-80 cm) in water 84 cm deep (66-100 cm) (N145); but some may be only 30 cm in diameter (L178). Most nests in L. Oneida were at depths of 3.7-5.5 m (F168). In streams, most nests were in less than 75 cm of water, often near overhead cover or stumps, logs, stones, or steep banks (C369, P195). The nests were usually in margins of the deeper pools to take advantage of minimum current, and lack of minimum velocity pools eliminated spawning (C369, S553).

In lakes, the nests are usually at 0.6-1.2 m depth and 0.3-2.4 m from the leeward shore (T138, H450, L178). Nests have been reported in water 3.5 m deep (B370) and up to 10 m from shore (S182).

Nesting was usually in the tributaries rather than the Des Moines R. or other rivers, particularly when water levels were high in the river (H451, R293, C369). High water levels in April to May 15 were important to let bass migrate to suitable tributary stream areas, and high water also rejuvenated gravel areas in the streams by flushing away silt (C369). Rearing enclosures can be used in ponds to permit nests to be closer together (R16, L234).

Egg deposition may occur shortly after nest completion or only after several days (P195). Before spawning the male may make only perfunctory attempts to guard the nest, but he guards the eggs and fry almost continuously (P195). Fry may be guarded 2-10 days or even 28 days after leaving the nest (H450, W89, L178). Males on nests were not observed to feed (P195). In an aquarium, 4 males spawned with one female over a nest (J106). Spawning behavior has been observed by SCUBA (S592).

Several females may spawn in the same nest (H450).

	No. of eggs per nest		No. of fry per nest	
	Mean	Range	Mean	Range
P195 MO streams	2,517	1,092-4,235	2,363	1,651-3,952
S553 WV	-	-	-	1,998-2,210
N145 ME S. Branch L.	4,270	474-8,447	3,943	451-7,856
L178	7,437	5,281-9,593	2,054	1,355-2,819
S554	-	2,000-10,000	-	-
L8 OH hatcheries	-	-	3,062	92-11,329
R295 SA	-	-	-	1,500-3,700
C397 MI Katherine L.	2,149	453-5,858	736	103-2,608

In experiments with constant temperatures (W232), incubation periods were as follows:

Temperature ($^\circ$C)	25.0	23.9	21.7	21.1	19.4	18.3	15.6	15.0	12.8
days	2.17	2.25	2.92	3.25	3.75	4.08	6.25	6.96	9.83

		Days after spawning		
Citation	Temperature ($^\circ$C)	Hatching	Rise from nest	Disperse
B73	17+	3-4	-	-
N145	16-20	4-5	11 (after hatch)	-
	19	3.9	-	-
B390	21+	7	-	-
	17.8-21.1	14	-	-
	15.0-17.2	21	-	-
W232 (from Embody)	23.3	4	-	-
	15.6	10-12	-	-

| | | Days after spawning | | |
| | Temperature | | | |
Citation	($^\circ$C)	Hatching	Rise from nest	Disperse
L234	15.6-26.7	7-16	-	-
R16	18.3-21.1	4	-	-
P195	12.2-18.9+	3	12-13	14-15
	17.8-26.7	-	6-8	10-12
T138	15.2-18.2	4	-	-
	17.2-19.5	-	8-11	19-28
T8	12.2-22.8	-	12	0-28
			(after hatching)	(after rising)
M525	-	3	9	-
E143 (from Sigler)	25.6	2.5	-	-
	21.7	3.5	-	-
	19.4	3.75	-	-
	15.0	7	-	-
	12.8	9.5	-	-

Embryology is described by R297 and H450 and early fry development, by B390, F5, L8, T8, M525, and H450. Mean numbers of anal spines, and of dorsal soft rays decreased as incubation temperatures increased from 17° to 29°C (W257). A sigmoid relationship was found between temperature and number of lateral line scales, with greater numbers at 20° and 29°C than at 17°, 23°, and 26°C.

Egg mortality averaged 5.9% (0%-44.7%) in 17 nests (P195). The annual percentage of potential eggs deposited in nests were 15.2%, 30.9%, and 34.1%; and of the eggs deposited, 25.8%, 27.6%, and 32.5% survived to postlarvae; and of the surviving postlarvae, 3.0%, 5.7%, and 14.7% survived to fall fingerlings (C397). The survival of potential egg to postlarvae was 3.9%, 9.4%, and 10.0%; to fall fingerling, was 0.3%, 0.5%, and 0.6% (C397).

Egg mortality was highest in nests of males under 400 mm with large egg deposits, and nest fanning did not increase egg survival but did reduce mortality of sac fry (N145).

Egg mortality of those eggs collected at the 16-cell stage at 19°C, and transferred to water at 17°, 20°, 23°, 26°, and 29°C ranged from 35% at 23°C to 84% at 17°C and 95% at 29°C (W257). Survival to 14 days at 20°C was 50% at 8.7 mg O_2/l, 39%-42% at 4.4 mg/l, and 29%-31% at 4.4 mg/l at 25°C, and there was no survival at 2.5 mg/l and practically no hatch at 1.2 mg/l (S594).

		No. of nests	Percent producing fry
T138 ON Tadenac L.	1966	20	45
P195 MO Ozark streams, early		23	78
second period		18	83
L178 L. Michigan	1954	20	55
	1955	28	54
W92 NY Taughannock C.	1945	38	79
	1946	72	68
	1947	51	82
N145 ME S. Branch L.	1971	15	67
	1972	30	77

Nesting success was greater in cropped ponds than in ponds in which the brood stock density was increased (B373). There was a suggestion that accumulation of waste products had an inhibitory effect on reproduction in ponds receiving water from crowded ponds (B373).

Since nests without guarding males may be overlooked, the percentage success may be a high estimate (W92). When males were removed from the nest, there was no hatch (S593).

Many factors are involved with egg and fry loss.

Floods. Sudden increases in stream level and turbidity did not seem to damage nests with eggs, sac, or black fry unless flood proportions were reached (W92). When over 18 cm of rainfall occurred between May 15 and June 30, survival of eggs and fry was poor in Iowa streams (C369). High water destroyed nests (B192). Floods also destroy the microcrustaceans needed by the fry (C369).

Advanced fry, newly risen from the nest, may be washed downstream by slight rises in flow, but these fry may just be distributed to survive elsewhere (W92). The swimming speed of fry increased as acclimation temperatures increased from 5° to 30°C and was slightly slower at 35° than at 30°C (L233). Water temperatures, when flow increases, may thus affect the degree to which the fry are displaced downstream. The fry selected a flow of 9 cm/second in preference to quiet water but could not maintain themselves in flows of 27 cm/second (L233).

Water level recession. Drop of water levels resulted in leaving 12 nests exposed in Cayuga L. in 1947 (W92). Eggs are laid in shallower turbid water than when the water is clear, which affects the probability of being dried (W92). Water level drop was considered the major cause for lower nest success in Maine (N145).

Predators. Carp and common suckers may destroy nests and eat eggs and fry (W92). Male bass are not able to successfully fend off mass intrusions. White perch also destroy nests (B390). Predation appeared to be the major cause of mortality in L. Michigan (L178), with few dead or fungused eggs being found in the nests. Sunfishes prey on eggs and fry when nests are not guarded (P195).

Fungus. Most of the nests classified as lost by W92 were heavily infested with fungus, but the fungus generally invades only after some of the eggs are dead. Male bass usually desert after the eggs become heavily fungused (W92, M548, R297, and B390). Males have been observed guarding nests where fungus completely covered the eggs (W92).

Temperature. A rapid rise in temperature from 16.1° to 23.0°C just before hatching resulted in heavy mortality (T8), although newly hatched fry were not affected. Any drop of temperature below 15.6°C, particularly if for an extended period, may result in nest desertion, death of eggs, and to a lesser extent, death of fry which have not yet risen from the nests (R295). Experimental studies (W232) indicated that eggs in various stages of development were not affected by relatively rapid changes within the range of 10°-25°C. Refrigerated eggs gave almost as good a hatch as unrefrigerated eggs, but more of the fry were curved and abnormal (B192). In streams bass eggs are normally subject to considerable temperature change (W92). In two periods when temperatures dropped below 15.5°C, many nests were deserted and the eggs were lost to predators (L178). Rapid temperature drops to below 10°C, if for longer than a day, caused the eggs to lose their adhesiveness and to become fungused (C369). Egg survival was uniformly high at 15° to 25°C, but dropped with lower or higher temperatures. No survival was secured at 10°C or 30°C and only 45% at 12.5°C (C401).

Low temperatures were correlated with low production of bass fry in Ohio (B192). June and June-July temperatures were correlated with abundance of year classes in a northern Michigan lake (C397). Floods and low temperatures were related to low year class success in Big Piney R., MO (F166). The largest year class came from a year without floods. June temperatures, but not May or July temperatures, were positively correlated with abundance of young bass in Oneida L. (F168). Strong year classes were generally produced in years of above normal June temperatures (r = 0.75). In southwest Wisconsin streams, strong year classes depended upon presence of 3-year-old adults, sparse yearling bass population, lack of flooding, and favorable temperatures (S593). An extensive discussion of the effect of temperature on year class strength of smallmouth bass (C401) cites several examples. The correlations between year class strength and July to October temperatures in South Bay and May to October temperatures or May to August temperatures in L. Opeongo were related to the fact that L. Opeongo warmed earlier than South Bay.

Nest desertion. Male bass may desert the nest if there are sharp or prolonged drops in temperature, external disturbances such as man, dropping of water level, clearing of water, and fungusing of eggs (W92). Several instances of nest desertion on temperature drop have been given (L178). High temperatures appeared to weaken the nest guarding stimulus (B192).

Males guarding nests in late May mostly deserted when temperatures dropped from 18.3° to slightly below 10°C (W92). Angling took more bass over spawning areas than elsewhere (L178), but no differences were noted between a year when nesting was completed prior to the angling season and a year when angling overlapped the nesting period.

No differences in success of unguarded and guarded nests were noted (W92). Nests with guarding males become about as silted as unguarded nests (W92, R297, L178).

Spawning by other species. Common shiners were found to spawn on smallmouth nests. Although the nesting success of bass with common shiners in the same nest was lower than those without common shiners, the difference was not significant according to a Chi-square test (L178). Carp spawning increased the turbidity so much in one area that bass nests were unsuccessful (L178).

Survival of fry may be low if entomostracans are scarce (L11). Backswimmers, dragonfly larvae, cannibalism, and Ichthyopthirius may also cause losses of fry (L11).

Good year classes in Wisconsin were associated with an adequate number of spawners, small yearling crop, and dry, warm April-June (B388). No relationship between number of brood fish and success of year classes could be demonstrated at Oneida L. (F168) nor at South Bay (W259), but there was a suggestion of a positive correlation in L. Opeongo (C401).

Survival from hatching to early September was estimated at 0.15% in Little Saline C., MO (P195). Survival from 17-20 mm fingerlings for 211 days was 69.3% when stocked at 7400/ha and for 197 days was 58% when stocked at 2470/ha (S576). Overwinter survival of small age 0 bass was less than that of the larger age 0 (C401).

Tagging studies indicated an annual mortality of 58% of adult bass in L. Michigan (L178) with an angler harvest of 22% and annual mortality rates of 52%, 58%, and 18% in Oneida L. (F103) with an angler harvest of 18%, 21%, and 4.6%. The drop in exploitation rate from 21% to 4.6% from 1956 to 1957 was

believed to indicate a change in vulnerability. Growth increments were much
better in 1957 than in 1956, indicating better feeding (F103).

Annual mortality rate over a 14 year period in L. Oneida was 43% with a
natural mortality rate estimated at 12.5% (F166). Annual minimum exploitation rates (as determined from tag returns) ranged from about 5%-21% (F168).
The annual mortality rate 56.4% was such that increased exploitation would increase the yield in South Bay (W259).

Catch-curve analysis indicated average annual mortality rates of 55% for
ages II-VIII in Red Cedar R., 65% for ages I-VI in Plover R., 55% for ages 0-
III in Livingston Branch, 66% for age II-VII in Huzzah C., and 65% for ages II-
VIII in Courtois C. (!213). Exploitation rates were 29% and 34% in Red Cedar
R. (P213).

Mortality estimates from scale data indicated a 69% mortality from age
II-III, 53% from III-IV, and 65% from IV-V in Massie C., OH (B192).

Of 75-100 mm fingerlings stocked in Big Piney R., 4% were later caught
by anglers (F88).

Natural mortalities were less in cropped ponds and greater in ponds where
population density was artificially increased (B373).

Monel metal strap tags on the premaxillary and maxillary bones did not
increase the mortality rate (L178, E23, F103). Finclipping reduced survival of
fingerling bass in ponds to about 1/2 or 1/3 that of control fish (C370). No difference in catchability was shown between tagged and finclipped bass (L178).
Some tagged bass were recaptured after 6 years (F103).

Tag returns indicated little movement of smallmouth bass in L. Michigan
(L178), Cayuga L. (W92), L. Erie (D93), L. Ontario (F73, S286, S182), Oneida
L. (F103), L. Huron (F82) and other lakes (H450), and in streams (L84, G174,
R293, M549, F158, F167, and G179). Bass released in Ohio streams were
taken as much as 205 miles downstream, however (W234), but this may have
been because they were introduced. Hatchery reared bass move more, up to
113 km, than did marked native bass in Massie C., OH (B252). Most of the
natives remained within 1 km but a few moved as much as 30 km. A positive
correlation was found between the distance travelled between captures and the
number of times the fish was caught (L187, W233). Disturbance of the stream
by gravel operations caused desertion of the home range (F167). Fifteen tagged
bass voluntarily moved from 100 to 2,845 feet to other pools and later homed
(F167). A migration to a winter aggregation area was noted in Cayuga L. (W92).
One bass was captured 95 km from where it had been marked 13 months earlier
in L. Ontario (S182), and another traveled 150 km in L. Michigan (L142). A
bass in the Des Moines R., Iowa (R293, B351), was caught 65 km upstream after
2 years.

Bass fingerlings stocked in Missouri ponds in August-September and collected one year later gave survival rates of 2%, 3%, 15%, 22%, 28%, 33%, 57%,
and 62%, but no correlation with turbidity, food, or predation was detected
(C370). The average survival was 24% but was 48% for the fish not finclipped.
In another study in the same ponds, smallmouth averaging 95 mm when
stocked showed survival rates of 43%, 88%, 89%, and 90%. Hatchery reared
smallmouth bass could be distinguished from native bass by the larger distance
from focus to first annulus (B427).

Smallmouth bass are usually found in cool, flowing streams and large,
clear lakes (H45). Because of the temperature requirements, they are usually
in streams in the southern part of the range and in lakes in the north (H450).

Ideal smallmouth lakes have the following characteristics (H450, B39):

1) over 20 hectares, preferably over 40 hectares,
2) over 9 meters deep, thermally stratified,
3) water clear and not dark colored,
4) vegetation present, but scanty,
5) large shoal areas of rock, gravel, or sand with gravel patches,
6) ample forage (small fish, crayfish, or insects), and
7) moderate summer temperatures.

Estimated standing crops in lakes range from 11 to 342 kg/ha with most between 20 to 40 kg/ha (H475). Below are data on standing crops (kg/ha) in Illinois ponds with only smallmouth bass (B373).

	Control ponds	Cropped ponds	(Harvest)	Added stock	Gain or loss
	72.5	81.8	78.2	201.8	+ 150.0
	29.6	63.1	96.6	194.4	+ 45.4
	189.4	10.4	142.2	107.0	- 76.4
	95.2				
	104.4				
mean	98.2	51.8	+ 105.7	167.7	

There was considerable variation in these ponds relative to their apparent carrying capacity, as indicated by their standing crops with no harvest or added stock. The carrying capacity in two of the better ponds was about 190 to 202 kg/ha (B373).

New York ponds with only smallmouth bass had standing crops of 51, 156, 160, and 163 kg/ha (R219). An Illinois pond with smallmouth bass, hybrid crappies, and lake chubsuckers gave a 13 year average yield of 68 kg/ha, with a range of 30-109 kg/ha or an average yield of 222 bass/ha, with a range of 77 to 455 (B429). The effort ranged from 121 to 575 and averaged 321 manhours/ ha, and the mean catch ranged from 0.39 to 0.69 bass/hour. This combination provided continued fishing success.

Streams should have good current, clear, cool water, good gravel or rock bottom (H450). Adult bass are rarely in streams less than 10.5 m wide. Standing crops and harvest in streams were listed as follows (P213):

	bass/ha	kg/ha	harvest, kg/ha
Wisconsin	45, 121, 132, 164	3.7, 14.6, 15.1, 29.2, 51.6, 62.8, 82.9	5.1
Ohio	16, 26, 29, 87	3.3, 5.8, 6.4, 13.5	4.4
Illinois	-	32.8	-
Iowa	-	-	1, 2
Missouri	56, 58	8.6, 8.9	0.2, 1.1, 1.3, 2.2, 6.4, 6.7, 6.7, 6.9, 11.4, 11.4
Maryland		17.9	3.6, 5.7
MO (F166)			2.1-6.0

Stream bass readily used artificial shelters (H448). A definite preference of broken rock substrate over solid rock or sand was shown in Middle Snake R. (M549).

In winter, smallmouth tend to be inactive, at least in the northern part of their range (H450). In Cayuga L. (W92), bass congregate along limestone ridges, often arriving in early September when temperatures had dropped only slightly. Larger individuals do not form as compact aggregations as bass 150-200 mm (W92). In the Middle Snake R., smallmouth were not found in shallow water in the late fall after water temperatures reached 15.5°C but only in quiet pools at least 3.6 m deep (M549). Most of the fish remained under rocks at temperatures below 6.7°C. Below 10°C smallmouth tend to be lethargic (T113).

In the spring, adult bass in some of the larger lakes enter tributaries to spawn (W92). In August adult bass were mostly in *Potamogeton* weed beds (W231). Stream bass prefer shelter as velocities increase, and when water levels decrease they seek darker areas for shelter (H448).

In Cayuga L. (W92) most bass in the summer were taken in the epilimnion, not deeper than 12 m, but a few were taken at 24 m. Smallmouth bass at 17-24 m in Cayuga L. were thought to be following alewives (G114). After storms smallmouth bass were taken at 23-24 m in L. Ontario (S169).

The temperatures (C°) selected in the laboratory by fingerling and adult smallmouth bass from L. Erie were as follows (B428):

	Fingerling	Adult
summer	29-31	30-31
fall	26-30	21-27
winter	24-28	13-26
spring	22-28	18-26

Preferred temperatures (C°) of smallmouth bass acclimated at various temperatures were (C396):

Acclimated	15	18	21	24	27	30
Preferred	20.2	22.9	26.5	29.8	30.1	31.3
95% CI	18.5-23.7	21.4-25.3	24.2-27.2	26.5-29.5	28.3-32.2	29.9-35.1

The final preferred temperature was reported as 28°C, but in the field smallmouth were usually at 20.3°-21.3°C (F161).

The LD50 temperature was determined as 29°-32°C, but smallmouth bass have been seen in water at 34°C (T102). The 96-hour LD50 temperatures for juvenile smallmouth bass (mean 29 g) when acclimated at various temperatures were (H477):

Acclimated (C°)	15	18	22	26
LD50 (C°)	1.6	3.7	6.7	10.1
90% survival (C°)	2.5-4.2	5.5-5.7	8.3-10.0	10.4-12.7

The last line lists the minimum temperatures at which at least 90% survived for 96 hours. At 21°C, lethal concentrations of dissolved oxygen were 0.87 and 0.96 ppm (B392). Fingerlings to 127 mm withstood 40 ppm O_2 with no distress (W237). Fry tolerated sudden changes of pH from 9.3-6.0 and from 7.7 to 9.7 (W237). No significant difference in activity of smallmouth bass was observed at pH of 7, 6, 5, and 4; but more testing and acclimation is needed (C385). Smallmouth bass tend to be inactive at night, and in midday they may remain quiet in shallow water, "sunning" (M549). At night, smallmouth bass move to deeper water, rest on the bottom, and often are readily approached by divers (E153). Smallmouth bass feed opportunistically during the day, but the peak of feeding occurs at dusk and early sunlight (E153).

The smallmouth bass is essentially carnivorous, with small amounts of vegetable matter taken probably accidentally (H450). Bass are dependent upon sight to feed (L11).

The first food of the young bass is usually copepods and cladocerans (H450, L11, C402, and E143). The first foods of 8mm bass, rising from the nest in a Missouri stream were dipterous larvae, but they soon took copepods and rotifers (P195). One at 12 mm had already eaten fish fry. Small chironomid larvae and small mayfly nymphs were the first food of bass in Taughannock C. (W92). No evidence of cannibalism was found in this study (W92), although larval suckers, cyprinids, and alewives were eaten. Most of the fish eaten by 50-100 mm smallmouth bass in Bull Shoals L. were young bass (M542).

By the time the bass are 40 mm, insects and small fish comprise the diet (H450, T6, W92, A87, D183, P199, S109, and B239), but soon fish and crayfish form most of the food (H450, W92, A87, R293, L249, M524, P213, W231, and D163). Fish of any species in the same waters as the bass may be eaten at the appropriate size (H450, W92). Alewives were a major forage in Oneida L. (F103), L. Michigan (W235), and some other lakes (H380). Insects may contribute a significant part of the diet, particularly in streams where forage fish are limited (H450, T6, R293, B144, and D21), and in ponds without forage fish (B373). *Chaoborus* and chironomids were the major food of juvenile and adult smallmouths in Clady L., MI (C402).

Where crayfish are abundant, they frequently comprise over 2/3 of the food (H450, M549, F159, K147, W144, and R200). Smallmouth bass took crayfish and tadpoles in preference to bluegills and golden shiners (L236). Bass of 830 g ate 0.8% to 3.2% of their body weight per day (L236). Average stomach contents were greater in cropped ponds and less in ponds where fish population density was increased (B373).

Although smallmouth bass eat many of the same items that walleyes do, the bass's major use of crayfish made competition minimal (F159). Smallmouth bass are not competitors of brook trout (P199), however, they will replace trout and salmon in marginal trout lakes (W89). In warm water ponds, smallmouth bass could not compete successfully with largemouth bass, bluegills, green sunfish, and bullheads (B144). Smallmouth bass reared from stock from a Minnesota lake ate more minnows and used the food more efficiently than did those from a Missouri stream stock (A91).

Little seasonal change of diet was noted (W92). In Bull Shoals L. AR, however, young shad were the major summer food; centrachids and crayfish, in late autumn and winter; and mayflies, (*Stenonema* sp.) in spring (A99). Bass were still feeding in late November although less than in September (W92). Bass stomachs examined in mid-April in Ontario were shrunken and full of mucous, indicating little or no winter feeding (K137). Bass do not feed in winter (P199) but begin active feeding when temperatures near 15°C (B382).

The number of chromosomes is 46 (R286).

Propagation of smallmouth bass is described in B390, L234, R297, and S554. Smallmouth bass, at 60-70 mm, were more successfully trained to artificial feeding than at 25-35 mm (A103).

In the Red Cedar R., WI, length limits of 274 or 329 mm would increase the harvest by 36% and 44% respectively over the current catch with no size limit (P213). A length limit of 306 mm would increase the yield in Oneida L. over that secured with the 254 mm size limit (F168), and fingerling stocking added little to the harvest in this lake. Stocking of fingerlings, averaging 75 mm, contributed little (3.3%) to the harvest, and the return of stocked fish to the creek was about 2.3% (F166). The quality of the bass fishery in Big Piney R., MO, declined under continued fishing pressure of over 250 hours/ha per year (F166).

Holding of smallmouth bass in nets was linked with increased incidence of furunculosis and mortality (L251).

SUWANNEE BASS, *Micropterus notius* Bailey and Hubbs
  The Suwannee bass is in the Suwannee R. system in Florida. They live in swifter water than the largemouth bass (H460). Brood stock brought to Auburn spawned in ponds in early April at 20°C (S576). Eggs were 2.0 mm in diameter; and fry at hatching were 5.5 mm, at one week, 6.5-7.5 mm, at 25 days, 12 mm, and at 42 days, 25 mm TL (S576). At 42 days they were 23 mm SL, at 61 days, 44 mm, and at 96 days, 69 mm (R307).

H460 FL Santa Fe R.

	No.	TL	Range
Age 0 - April	5	20	10-35
June	20	34	20-45
July	43	40	20-66
August	17	45	25-65
September	23	51	30-86
October	22	57	45-85

A 342 mm Suwannee bass (680 g) was probably 7 or 8 years old (R312).

SPOTTED BASS, *Micropterus punctulatus* (Rafinesque)
  The spotted bass was native to the Gulf of Mexico drainage from Florida to Texas, north to Kansas, southern Ohio, and West Virginia. They do not enter brackish water (B399). It has been introduced into other areas, though less than the smallmouth and largemouth bass. It was introduced into California in 1933 (M550) and into South Africa (P99, H164). The Alabama spotted bass, *M. p. henshalli*, is restricted to the Mobile Bay drainage, above the Fall line particularly thriving in deep oligotrophic reservoirs (G197).
  Spotted bass are predominately stream fish but are also in many reservoirs. In streams it is intermediate in habitat preference to smallmouth and largemouth (T113). In Ohio they were found in streams where river birch, *Betula nigra*, grew but not where there was smooth alder, *Alnus rugosa* (H273), or water willow (H452). Spotted bass were found in areas without aquatic vegetation. They fed mostly at night and spent the day in deeper pools (H273). Adults tend to school. Spotted bass are less tolerant of heat and mud than largemouth (B393). Spotted bass were taken at water temperatures of 36°C, however (P208).
  Pyloric caecae of fry and fingerlings ranged from 10-13 compared to 20-33 for largemouth bass and proved to be the best character for separating the fry (A92).

S97 TN Norris L.	306 fish, 30-370 mm SL	TL = 1.209 SL
R116 TN Norris L.	163-353 mm SL	TL = 2.284 mm + 1.1975 SL
R116 VA Claytor L.	357 fish, 45-396 mm SL	TL = 1.578 mm + 1.203 SL

	Weight (g) Mean of means	Central 50%	Range	No.	Citation
TL					
25	0.5	-	0.09-0.9	33	S472 AL
51	1.4	-	0.5-2.7	6,420	S472 AL
76	4.2	-	0.8-11.8	4,214	S472 AL

TL	Weight (g) Mean of means	Central 50%	Range	No.	Citation
85	5	-	-	1	S97 TN Norris L.
102	9.5	-	4-23	1,665	S472 AL
127	23	-	3-43	483	S472 AL
127-151	36	32-44	27-60	32	C373, V44 KY, LA
152	40	-	-	356	S472 AL
152-177	55	50-62	40-79	28	C373, V44
178	64	-	40-113	254	S472
178-202	85	77-91	45-108	28	C373, S97, V44, W132
203	100	-	59-159	173	S472
203-228	121	109-133	95-196	52	C373, S97, V44, V94, W132
229	136	-	91-213	147	S472
229-253	193	150-216	141-272	126	C373, S97, V44, W132
254	195	-	141-286	84	S472
254-278	257	221-273	200-329	124	C373, S97, V44, V94, W132
279	259	-	181-363	71	S472
279-304	344	301-372	295-432	53	C373, P98, S97, V94
305	363	-	272-454	53	S472
305-329	464	431-510	318-602	87	C373, S97, V94, W132
330	459	-	340-590	43	S472
330-355	598	550-624	480-712	152	C373, S97, V94
356	549	-	408-772	22	S472
356-380	711	676-801	567-834	26	C373, P98, S97, V94, W132
381	694	-	545-907	27	S472
381-405	907	883-920	794-964	29	C373, S97, V94
406	871	-	545-998	15	S472
406-431	1,051	1,032-1,070	907-1,135	8	S97, V94
432	1,084	-	726-1,270	8	S472
432-457	1,377	-	1,225-1,700	4	P99, R312, S97, V94
457	1,270	-	952-1,406	7	S472
635	3,900	-	-	1	R312

S97 TN Norris L.	301 bass,	70-372 mm	$\log W = -5.0529 + 3.2032 \log SL$
R116 VA Claytor L.	357 bass,	45-396 mm	$\log W = -5.1318 + 3.224 \log SL$
P208 IN White R.			
Sec. C, 1970	-	-	$\log W = -1.888 + 1.761 \log TL$
Sec. A, 1970	-	-	$\log W = -2.051 + 1.803 \log TL$
Sec. B, 1970	-	-	$\log W = -3.211 + 1.880 \log TL$
Sec. B, 1969	-	-	$\log W = -3.109 + 2.253 \log TL$
Sec. A, 1969	-	-	$\log W = -3.842 + 2.561 \log TL$
Sec. C, 1969	-	-	$\log W = -4.676 + 2.904 \log TL$
S472 AL	13,867 bass,	51-260 mm	$\log W = -5.15 + 3.09 \log TL$
	177 bass,	270-460 mm	$\log W = -5.05 + 3.06 \log TL$
C373 KY Cumberland L.	371 bass,	127-404 mm	$\log W = -5.554 + 3.271 \log TL$

Mean weights of females (V94) did not differ from the general weights from other waters.

	Length (mm)	No.	K(SL) Mean	Range
C256 TX L. Crockett	145-234	20	1.98	1.73-2.32
S100 NC Hiwassee L.		8	2.16	-
C256 TX L. Park	45-419	50	2.26	1.52-2.83
L110 IL Hutchins C.		8	2.46	
S100 TN Douglass L.		21	2.48	-
C256 TX Timber L.	71-440	48	2.54	2.07-3.62
R115 VA Claytor L.	45-396	357	2.62	1.72-2.78
S100 TN Cherokee L.		43	2.62	-
S97 TN Norris L.	70-372	300	2.76	-
			K(FL)	
H163, P98 South Africa	296-370	3	1.52	1.44-1.66
			K(TL)	
V44 LA Springhill L.	25-295	98	1.47	0.78-1.86
S472 AL	25	33	3.10	
	51-130	12,782	1.01	0.91-1.11
	150-280	1,085	1.15	1.12-1.19
P208, B416, B417 IN White R.				
1969	80-300	97+	1.64	1.2-4.6
1970	50-300	248+	2.28	1.1-8.1

The average K was higher in June, July, and October than other months in a Texas lake (C256).

		TL Mean of means	Central 50%	Range		Weight	
	No.				No.	Mean	Range
Age 0-April							
AL: S576	+	-	-	11-12	-	-	-
Age 0-May							
AL: S576	+	-	-	19-26	-	-	-
AL *M. p. henshalli:* R307	+	-	-	25-34	-	-	-
Age 0-June							
AL: S576	+	-	-	21-28	-	-	-
OH: R42	+	30	-	-	-	-	-
TN streams: D43	4	48	-	41-53	-	-	-
TX lake: C256	1	48	-	-	1	3	-
AL *M. p. henshalli:* R307	+	63	-	-	-	-	-
Age 0-July							
AL: S576, R307	+	63	-	-	+	9	-
AL Martin L.: B268	1,867	64	-	-	-	-	-
TX lake: C256	26	81	-	61-124	26	9	-
AL *M. p. henshalli:* R307	+	92	-	-	-	-	-
LA Springhill L.: V44	2	155	-	147-163	2	70	60-80
Age 0-Aug.							
IL Big C.: L108	2	56	-	-	-	-	-
OH: R42	+	-	-	30-97	-	-	-
TN streams: D43	2	74	-	51-97	-	-	-
LA Springhill L.: V44	6	185	-	170-203	6	94	71-133
Age 0-Sept.							
TN streams: D43	6	61	-	48-71	-	-	-
LA Springhill L.: V44	42	231	-	137-259	42	207	28-315
Age 0-Dec.							
AL *M. p. henshalli:* S576	58	136	-	-	58	19	-
AL *M. p. p.:* S576	44	189	-	-	44	67	-

		TL				Weight	
	No.	Mean of means	Central 50%	Range	No.	Mean	Range
Age 0							
IN White R.: B416	137	76	-	-	137	33	-
WV Elk R.: H218	11	86	-	-	-	-	-
KS: C349	+	-	-	64-127	-	-	-
AR, OK reservoirs: H442, T44	29	97	89-109	84-137	10	17	6-34
TN, VA reservoirs: E26, S97, R117, R292	110	98	-	60-155	-	-	-
KY ponds: S266	2	150	-	-	-	-	-
Combined	289+	96	86-109	60-155	147	26	6-34
Age I							
IN Muscatatuck R.: M317	6	56	-	53-61	6	3	-
OH Vesuvius L.: R83	9	79	-	-	-	-	-
OH: R42	+	102	-	-	+	28	-
IN White R.: B416	93	102	-	-	93	39	-
KY Cumberland L.: C373	13	135	-	-	13	27	-
VA Claytor L.: R117	33	147	-	121-192	-	-	-
IL Big C.: L108	8	154	-	-	-	-	-
KY, WV rivers: C175-6, H218	28	155	-	142-173	-	-	-
OK Illinois R.: J25, L106	105	170	-	107-224	-	-	-
AR, KS, OK reservoirs: C349, H127, H442, T44, V94	67	174	163-175	147-260	11	133	71-244
AL, TX reservoirs: B268, C256	167	177	173-178	114-236	25	86	26-111
TN Dale Hollow L.: R292	30	183	-	79-228	-	-	-
KY ponds: S266	36	213	-	-	-	-	-
TN Norris L.: E22, E26, S97	44	278	-	155-339	-	-	-
Combined	640+	163	147-178	53-339	54+	65	3-244
Age II							
IN Muscatatuck R.: M317	25	104	-	61-150	25	17	3-40
IN White R.: B416	48	153	-	-	48	66	-
OH: R42	+	165	-	-	+	57	-
KY Cumberland L.: C373	88	185	-	140-226	88	77	36-122
IL Big C.: L108	5	190	-	-	-	-	-
KY, WV rivers: C175-6, H218	76	194	-	183-206	-	-	-
TX reservoirs: C256	34	217	206-236	175-264	35	128	99-190
OK Illinois R.: J25, L106	30	226	-	188-274	-	-	-
VA Claytor L.: R117	86	236	-	133-326	-	-	-
AR, KS, OK reservoirs: C349, H127, H442	66	245	218-277	188-356	10	344	116-624
TN Dale Hollow L.: R292	79	254	-	214-268	-	-	-
KY ponds: S266	30	279	-	-	-	-	-
TN Norris L.: E22, E26, S97	309	323	-	193-387	-	-	-
South Africa P98	2	338	-	295-370	2	482	340-624
Combined	878+	226	190-277	61-387	208+	151	3-624
Age III							
IN Muscatatuck R.: M317	37	160	-	117-211	37	54	20-108
IN White R.: B416	139	195	-	-	139	114	-
IL Big C.: L108	5	229	-	-	-	-	-
OH: R42	+	229	-	-	+	142	-
KY, WV rivers: C175-6, H218	63	248	-	246-249	-	-	-

	No.	TL Mean of mean	Central 50%	Range	No.	Weight Mean	Range
Age III (cont.)							
KY Cumberland L.: C373	174	259	–	229-302	174	213	136-495
TX reservoirs: C256	16	271	246-307	234-424	16	287	125-638
VA Claytor L.: R117	132	280	–	230-339	–	–	–
OK Illinois R.: J25, L106	26	279	–	234-310	–	–	–
TN Dale Hollow L.: R292	35	280	–	–	–	–	–
AR, OK reservoirs: H127, T44, V94	38	294	–	244-402	25	578	577-822
South Africa: H163	1	317	–	–	1	454	–
TN Norris L.: E22, E26, S97	153	355	–	193-393	–	–	–
Combined	819	263	244-280	117-424	392+	260	20-822
Age IV							
IN Muscatatuck R.: M317	33	198	–	142-236	33	99	31-167
IL Big C.: L108	3	230	–	–	–	–	–
IN White R.: B416	46	245	–	–	46	178	–
OH: R42	+	279	–	–	+	269	–
KY, WV rivers: C175-6, H218	20	291	–	282-302	–	–	–
TX reservoirs: C256	5	315	–	264-401	5	648	284-1,514
OK Illinois R.: J25, L106	7	315	–	290-335	–	–	–
VA Claytor L.: R117	57	332	–	254-447	–	–	–
OK Wister L.: H127	3	345	–	–	–	–	–
AR Bull Shoals L.: V94	16	356	–	304-410	15	696	449-932
TN Norris L.: E22, E26, S97	26	403	–	348-459	–	–	–
Combined	216+	302	261-377	142-459	99+	390	31-1,514
Age V							
IN Muscatatuck R.: M317	12	226	–	185-267	12	142	79-230
IN White R.: B416	18	263	–	–	18	231	–
IL Big C.: L108	5	287	–	–	–	–	–
OH: R42	+	330	–	–	+	425	–
KY Floyd's Fork R.: C175-6	5	330	–	–	–	–	–
OK Illinois R.: J25, L106	5	368	–	358-381	–	–	–
VA Claytor L.: R117	31	370	–	326-423	–	–	–
WV Elk R.: H218	3	373	–	–	–	–	–
TN Norris L.: S97	1	411	–	–	–	–	–
AR Bull Shoals L.: V94	1	414	–	–	1	1,135	–
TX reservoirs: C256	5	448	–	345-505	5	1,984	677-3,065
Combined	26+	332	263-270	185-505	36+	673	79-3,065
Age VI							
IN Muscatatuck R.: M317	4	231	–	208-249	4	162	116-207
IL Big C.: L108	3	260	–	–	–	–	–
IN White R.: B416	8	270	–	–	8	351	–
KY Floyd's Fork R.: C175-6	1	345	–	–	–	–	–
WV Elk R.: H218	2	421	–	–	–	–	–
AR Bull Shoals L.: V94	2	432	–	420-444	2	1,134	1,043-1,225
VA Claytor L.: R117	2	441	–	411-472	–	–	–
TX reservoirs: C256	4	472	–	434-498	4	1,922	1,684-2,161
Combined	26	356	260-441	208-498	18	832	116-2,161
Age VII							
KY Floyd's Fork R.: C175-6	1	323	–	–	–	–	–
TN Norris L.: S97	1	435	–	–	–	–	–
TX Park L.: C256	1	486	–	–	1	1,928	–

	No.	Average calculated TL at each annulus							
		1	2	3	4	5	6	7	8
R292 TN Dale Hollow L. (from Nichols & Turner, 1966)	+	140	170	185	206				
C175 KY Floyd's Fork R.	162	74	145	208	249	290	300	323	
L110 IL creeks	8	79							

	No.	Calculated TL at each annulus							
		1	2	3	4	5	6	7	8
A87 AR Bull Shoals L.	190	99	183	211					
R116 WV Elkhorn C.	5	122	152						
H442 AR Catherine L.	13	122	178						
J41 OK Ft. Gibson L.	28	109	183	224					
F64 OK Little R.	68	84	165	224	277	320			
F87 OK Little R.	34	89	175	236					
tributaries	16	89	175	236					
cutoff lakes	18	81	160	224	279	320			
S391 OK Rock C.	16	114	198						
H150 OK Illinois R. 1953	23	109	173						
J25, L106 OK Illinois R. 1954	173	114	168	251	307	345			
E61 OK Other rivers	276	109	183	246	305	340	338		
E61 OK Salt Cr.	17	117	198	264	312	353			
P113 MO statewide	743	86	183	254	292	323	353		
middle river stretches	638	84	175	239	290	328	353		
lower river	97	94	185	272	292	307			
poorest station	+	74	152	221	246	259	345		
best station	+	102	218	287	338	368	368		
S100 NC Hiwassee L.	8	79	206	251					
R117 VA Claytor L.	359	79	178	272	333	384	409	455	
with Fraser modification	359	104	198	282	340	386	411	455	
S100 TN Douglas L.	23	66	198	284					
S100 TN Cherokee L.	53	94	218	284					
H163, P98 South Africa	3	127	249	264					
S266 KY streams	66	142	239						
H45 OK Verdigris R.	6	127	226						
tributaries	23	109	211	290					
H442 AR Ouachita L.	14	124	203						
H442 AR Hamilton L.	9	127	231						
T54 KY streams	52	152	239	290	340	409	508		
T44 OK Grand L.	25	107	198	295					
J31 OK Grand L.	65	104	213	300	356	396	419		
J101 OK Spavinaw L.	114	127	239	300	353				
T54 KY lakes	52	135	234	300					
C169 TN Norris L.	+	114	259	333	386	417	424		
P73 MO Wappapello L.	68	132	259	315	345	368	363		
E26 TN Norris L.	179	135	257	333	378				
E22 TN Norris L.	51	155	290	343					
R292 TN Center Hill L. (from Hargis, 1965)	+	168	264	351	406				
R116 WV Harrington L.	1	130	292						
R116 WV Dale Hollow L.	1	216	323	406	457				

		1	2	3	4	5	6	7	8
	Average TL and annual increment at each annulus								
L108 IL Big C.		86	152	196	224	231	249		
	Incr.	86	65	45	33	28	29		
	No.	29	21	16	11	8	3		
C256 TX Crockett L.		104	168	203	257				
	Incr.	104	86	56	64				
	No.	20	18	8	3				
R292, R308 TN Dale Hollow L.									
1963-65		111	188	224					
	Incr.	111	77	55					
	No.	25	17	3					
1970		108	192	251					
	Incr.	108	86	71					
	No.	119	97	32					
C256 TX Park L.		130	196	251	318	371	429	455	
	Incr.	130	64	56	61	20	56	53	
	No.	49	29	15	10	9	5	1	
C256 TX Timberlake L.		49	198	312	388	480			
	Incr.	94	91	84	66	86			
	No.	22	18	8	2	1			
C373 KY Cumberland L.		135	224						
	Incr.	135	81						
	No.	261	221						
S97 TN Norris L.		124	262	335	378	409	417	424	
	Incr.	124	144	91	42	22	19	7	
	No.	304	288	97	18	2	1	1	

Difficulty in recognizing annuli beyond age III was reported from Cumberland L. (C373). Most growth was calculated on a direct proportion basis, but Fraser modifications were used by C373, P73, and R117. In Norris L., TN, the annulus of age II fish was formed in June and of age III fish, in mid-June to mid-July (S97). Most growth of age II and older fish occurred in July to September but age I fish made most of their growth in June to August (S97).

In general, the tabulations indicate more rapid growth in reservoirs than in streams. Growth also seems to be faster in rivers than in tributary or smaller streams (F87, J45, P113). Little change in growth in various years after impoundment was noticed in Norris Lake (S97). Uniformity of growth of young reduces the amount of cannibalism compared to smallmouth and largemouth bass (H452, B393). A tendency for few spotted bass to get beyond age III was noted (C373). In the Illinois R., OK, (H150) spotted bass grew 1.4 mm/day from May 23 to October 3. Growth of spotted bass was not affected by introducing threadfin shad in Dale Hollow L. (R292). The Alabama spotted bass, *M. p. henshalli*, is reported to grow more slowly but reach a greater size (G197).

The ability to grow at different lengths is suggested by the following data:

Annual increment as percentage of TL at beginning of year
Number of specimens in parentheses

	OK, Grand L.	TN, Norris L.	
TL	T44	E26	S97
76-101	103 (10)	148 (76)	-
102-126	58 (1)	118 (45)	108 (124)

	Annual increment as percentage of TL at beginning of year		
	OK, Grand L.	TN, Norris L.	
TL	T44	E26	S97
127-151	77 (4)	118 (152)	95 (30)
152-177	60 (1)	67 (4)	-
178-202	-	52 (2)	-
203-228	-	44 (52)	-
229-253	26 (4)	33 (25)	32 (29)
254-278	-	33 (14)	28 (23)
279-304	-	29 (4)	-
305-329	-	18 (3)	-
330-355	-	11 (14)	-
356-380	-	10 (4)	-
381-405	-	5 (2)	-
406-431	-	2 (1)	-

Spawning females were usually larger than the males (R298). Breeding spotted bass were 216-356 mm (H452). The number of eggs per female increased with the size of the female (V94):

Age	No.	Length		Weight		Number of ova	
		Mean	Range	Mean	Range	Mean	Range
III	10	315	270-356	487	298-659	8,891	3,249-14,901
IV	13	357	327-391	690	449-932	11,680	5,205-19,140
V	1	414	-	1,135	-	19,109	-
VI	2	432	420-444	1,134	1,043-1,225	25,517	20,448-30,586
Combined	27	351	-	677	-	11,911	-

Females from 190-390 mm had 1,150 to 26,500 eggs (R312).

Spawning occurred when water temperatures reached 17.8°C (H453), with spawning usually later than that of the largemouth (V88) or about the same time (B393, R312). Nests in Bull Shoals L., AR, were cleared as early as April 10 (V94) or April 15 (V89) at water temperatures of 12.8°C (V89) or 14°-15°C (V94), and spawning occurred between then and May 14 at depths of 1-3 m. Fresh nests were found when temperatures were as high as 22.8°C (V94). Most active spawning was in late afternoon, and courtship and spawning sequences required up to 3.5 hours. Some males renested when the first nest was unsuccessful (V94). Spotted bass migrate to smaller streams prior to spawning when water temperatures are about 10°C in Ohio (T113). They spawned at 15.6°-18°C (H453, B393), at 15.6° to 21°C (R312), and at 20°-21°C (R298). In Louisiana spawning was in May and June (R298). Nests are built on mud bottoms (H453) or commonly on gravel (V88, V89). Nests were at depths of 1.5 to 6.7 m in Bull Shoals Res. (V94). There were 2000-2500 fry per nest in 3 nests (H452) and an average of 2000 fry in 16 nests (H453). Numbers of eggs in 21 nests ranged from 763 to 18,995 (mean 5,016), and the numbers of larvae in 15 nests ranged from 87 to 3,820 (mean 1,476) (V94). Egg diameters ranged from 1.60 to 2.30 mm in 21 nests, but the greatest range in a single nest was 0.50 mm (V94). Eggs hatched in about 5 days at 14.4° to 15.6°C, and larvae remained in the nest about 8 days at 15.6° to 17.2°C. Males defended the eggs and the compact schools of fry until the fry were about 30 mm (V94). The fry scattered from the nest earlier than smallmouth of largemouth fry (H452).

Schools of smallmouth and spotted bass fry were sometimes mixed and guarded by either or both male parents (V89). Male spotted bass do not school the young as closely as do smallmouths (H452, H273).

Fry first feed on entomostraca but soon take chironomids and other aquatic insects (H453, M523, A87, and W246).

At 51-99 mm, fish comprised 56% of the diet in Bull Shoals L. (A87, M524). In an older lake, Beaver L., 100-200 mm spotted bass utilized crayfish, but largemouth bass of the same size utilized terrestrial insects to make up for a shortage of forage fish (M524).

Adult spotted bass feed on crayfish and minnows (H273, M523, A87, D163, L140, F91, and R117) or on crayfish and young shad (A99).

Spotted bass consumed only about one-half as much fish as largemouth bass of similar size (M523). Immature insects were the major food of spotted bass in an Illinois river (S555); entomostraca were fairly important in bass under 75 mm; and terrestrial insects, crayfish, and fish, in bass 150-320 mm. Most feeding is in early morning and late evening (H453).

The width of spotted bass mouths (M) are described by the following formulas (L237):

TL of bass	
under 100 mm	M = -1.615 mm + 0.117 TL
100-199	M = -1.923 mm + 0.121 TL
200-299	M = 3.875 mm + 0.085 TL
300-399	M = -28.495 mm + 0.198 TL

The largest sizes (TL of various forage fish which could be swallowed by spotted bass) were tabulated for several sizes but are here given only for 190 mm and 267 mm spotted bass (L237):

	190 mm	267 mm
bluegill	69	83
green sunfish	68	83
golden shiner	100	121
gizzard shad	87	106
threadfin shad	86	106

The average TL of spotted bass taken in gill nets of various bar measure mesh sizes were (L216):

mesh (mm)	No.	TL
25	1	201
38	2	302 (299-303)
51	2	361 (356-366)
64	1	406

Preferred temperatures of spotted bass acclimated at various temperatures were (C396):

Acclimated (C°)	6	9	12	15	18	21	24	27	30
Preferred	16.9	17.9	20.1	24.8	26.7	29.5	30.2	31.4	32.1
95% CI	14.7-	17.2-	19.8-	22.2-	24.4-	26.5-	28.4-	30.2-	31.9-
	19.5	21.2	23.1	24.9	27.0	29.3	31.7	34.2	36.8

Spotted bass preferred 23.5°-24.4°C in Norris Res., TN (D207). Spotted bass were mostly at depths of 0.9-1.2 m in Altoona L., GA (K148), but about 10% were at depths of over 6 m, about the same depth distribution as largemouth bass. Almost 40% of the tagged spotted bass, 267-325 mm, were caught by anglers, but the exploitation rate for those over 216 mm was 31% (K148), compared to 17% for largemouth bass.

Standing crops of spotted bass in White R., IN, were 39.1 and 82.2 kg/ha, compared to 18.5 and 47.9 kg/ha in areas of heated effluent (B417). L. Martin, AL, had 1,300 young, 30 yearling, and 12 older spotted bass/ha (B268).

LARGEMOUTH BASS, *Micropterus salmoides* (Lacèpède)

The largemouth bass is one of the most important freshwater game fish in the United States. It has been widely used as the major predator and sport fish in warmwater ponds and reservoirs. The species originally ranged east of the Rocky Mountains from southern Quebec and Ontario through the Great Lakes and Mississippi Valley to the Gulf of Mexico and from northeastern Mexico to Florida and north to the Carolinas (H382). Largemouth bass have been introduced in the eastern states to New England and in suitable waters in the west. They were introduced into California in 1874 (E145).

The largemouth bass also has been introduced into Austria, Finland, France, Hungary, Czechoslovakia, Germany, Italy, Holland, Spain, Sweden, England, Russia, Philippines, Japan, Cuba, South Africa, Natal, West Cameroons, and Madagascar (M553) and Belgium (V92). In the British Isles, introductions have usually failed (W245). An extensive summary of introductions is given in R312. It was decided in 1968 not to introduce it in New Zealand (M553) because of possible danger to native species. Introductions into Saskatchewan prior to 1950 failed, but some more recent ones may be successful (M547).

Largemouth bass prefer nonflowing waters with aquatic vegetation and clear water (T113). In the Sierra Nevada foothill streams, CA, the abundance of largemouth bass showed significant positive correlation with turbidity, rooted vegetation, floating vegetation, % pools, % riffles, modification by man, species numbers, and abundance of green sunfish, bluegill, mosquitofish, golden shiner, and hitch, and a negative correlation with temperature, elevation, shade, % native fish, and abundance of California roach and rainbow trout (M566).

Young largemouth bass (16-30 mm) can be distinguished from spotted bass fry by the 20-33 pyloric caeca compared to 10-13 in spotted bass (A92). Bass over 350 mm were sexed with 92% accuracy on the basis that the scaleless area around the urigenital openings is nearly circular in males but pear-shaped in females (P200).

In Nebraska (M552), largemouth bass were classified as suitable for slightly alkaline waters (total alkalinity less than 900 ppm, carbonate alkalinity less than 250 ppm, and K + Na less than 200 ppm). Largemouth bass were frequently taken in brackish water, to 24.4 ppt salinity, whereas spotted bass were not (B399). They were taken in tidal marshes with up to 11.8 ppt (K151), or to 10 ppt, but rarely beyond 3.5 ppt (T141).

	SL	No.	FL/SL	No.	TL/FL	No.	TL/SL
C37 MN	under 150	-	-	-	-	-	1.196
	under 249	191	1.132	-	-	170	1.170
	250-349	65	1.120	-	-	65	1.159
	over 349	9	1.111	-	-	9	1.150
M302 IA	53-457	-	-	-	-	55	1.18
B30 MI	under 212	-	-	-	-	527	1.220
	212-381	-	-	-	-	843	1.212
	over 381	-	-	-	-	86	1.199
S97 TN Norris L.	50-450	-	-	-	-	789	1.205
C33 IA ponds	40-259	33	1.168	-	-	33	1.206

	SL	No.	FL/SL	No.	TL/FL	No.	TL/SL
C30 IA	-	16	1.15	-	-	22	1.21
L29 IA East and Red Haw L.	60-359	126	1.165	-	-	-	-
	10-439	-	-	-	-	185	1.217
	25-127	-	-	-	-	37	1.236
	128-203	-	-	-	-	49	1.226
	204-280	—	-	-	-	24	1.211
	281-357	-	-	-	-	25	1.205
	358-432	-	-	-	-	9	1.192
C256 PA	under 101	-	-	-	-	25	1.256
	over 101	-	-	-	-	25	1.140
S289 MA	50-469	-	-	-	-	305	1.219
R125 IA Ike L.	-	-	-	108	1.050	108	1.226
S350 UT	-	-	1.164	-	1.041	-	1.211
T139 IA Clear L.	46-521	86	1.17	82	1.04	85	1.22
R116 VA Claytor L.	TL = 11.34 mm + 1.15 SL			-		-	-
R178 VA Back Bay	TL = 7.94 mm + 1.17 SL			-		-	-
S575 CA ponds	SL = -1.7 mm + 0.861 FL			-		-	-
	TL = 1.6 mm + 1.034 FL			-		-	-

Ratios for TL/SL in Norris L., TN, (S97) were given by 20 mm intervals, SL, and declined from 1.236 at 50-69 mm to 1.174 at over 450 mm.

In making conversions to total lengths where conversion factors are not given, it is assumed that TL = 1.22 SL up to a TL of 200 mm; TL = 1.215 SL up to 380 mm; and TL = 1.21 SL over 380 mm; and that TL = 1.08 FL.

	Weight in grams				
	Mean of	Central			
TL	means	50%	Range	No.	Citations
23	0.2	-	-	3	C148: FL
25-50	0.5	-	0.2-0.9	12	C148: FL
	1.0	-	0.6-1.5	14	C256: PA
	1.5	-	-	139	L189, L226: AL
51-75	2.3	2.0-2.3	1.4-4.8	410	C148, L189, L226, S472: FL, AL
	3.2	3-3	1.7-6	284+	B30, C33, C60, C256, D5, F31, G48, M302, M536: IA, IN, MI, OH, PA
	3.6	-	1-5	25	B233, K80, S350: OR, UT
	4.6	-	4-5	22	J46, S97: OK, TN
	3.2	2.3-4	1-6	744+	Combined
76-101	5	-	-	1	W126: France
	6	5-7	2-14	1,195	A39, A40, L189, L226, S472: AL
	7	-	5-10	6	C148: FL
	7	6-9	5-9	42	C30, C33, C107, L29, M246: IA
	8	6-8	5-15	21+	C256, G48, M536: IN, OH, PA
	9	7-9	3-20	72	C202, H156, J46, T92, S97: AR, OK, TN
	14	9-18	9-18	96	K80, L126, S350: CO, OR, UT

TL	Weight in grams Mean of means	Central 50%	Range	No.	Citations
76-101	20	-	-	74	B30: MI
(cont.)	9	6-9	2-20	1,507+	Combined
102-126	13	10-15	3-30	1,612	A39, A40, L189, L226, S472: AL
	14	-	13-14	10+	K71, S97: TN
	16	-	9-30	11+	G48, M499: IL, IN
	18	17-19	17-23	120	C33, C107, F31, L29, M246, M536: IA
	20	18-23	14-31	140	J26, C202, H156, T92: AR, OK
	21	-	16-26	3	C148: FL
	26	23-34	14-40	75	B233, L126, S350, W71: CA, CO, OR, UT
	27	-	20-43	61	B30, C20, C256, J15, W39: MI, MN, PA, WI
	27	-	-	1	W126: France
	19	14-20	3-43	2,033+	Combined
127-151	23	22-25	14-59	1,330	A39, L189, L226, S472: AL
	25	22-30	22-30	30+	C373, K71, P113, S97: KY, MO, TN
	32	-	25-45	18	C256: PA
	33	31-34	21-82	98	B150, C30, C33, C107, F31, I10, M246, M536: IA
	37	32-44	28-59	214	C202, H156, J46, T92: AR, OK
	38	-	24-47	10	C148, H202: FL
	39	30-40	23-75	124+	G48, K43, M499: IL, IN
	40	38-40	25-51	100	B233, K80, L126, M140, S350: CA, CO, OR, UT
	45	-	31-75	72	B30, C20, E32, J15, M20: MI, MN, WI
	35	30-40	14-82	1,996+	Combined
152-177	42	38-46	16-91	968	A39-40, L189, L226, S472: AL
	43	-	38-47	2	W126: France
	46	-	40-48	43+	C373, K71, S97: KY, TN
	53	48-59	36-91	216+	J46, P113, C202, T92, H156: AR, MO, OK
	55	51-59	40-68	77	B150, C105, C107, I10, L29, M246, M536: IA
	55	50-60	36-65	170+	E32, M499, R80, K43: IL, IN, OH
	57	45-71	28-73	93+	C20, J15, W39, S62, B30, M20, C256, M359: MI, MN, PA, WI
	62	-	-	9	H61: MD
	63	-	55-73	4	C148: FL
	68	60-71	28-96	97	B233, K80, L126, M140, S350, W71: CA, CO, OR, UT
	55	48-60	16-96	1,679+	Combined

TL	Weight in grams Mean of means	Central 50%	Range	No.	Citations
178-202	71	68-77	23-141	502	A39-40, S472, L189, L226: AL
	75	71-79	65-79	47+	S97, K71, C373: KY, TN
	80	71-86	45-104	238	J46, P113, C202, T92, H156: AR, MO, OK
	87	65-94	54-113	104	B233, L126, S350, W71: CA, CO, OR, UT
	88	85-99	65-227	63	C30, C33, C105, C107, F31, L29, M246, M536: IA
	92	86-93	59-127	171+	J15, W39, B30, E32, M20, M359, N94, N115: MI, MN, WI, SD
	94	-	85-113	12	C148, H202: FL
	100	82-91	50-200	74+	R80, R47, G48, M499, H61, C256: IL, IN, OH, PA, MD
	86	77-93	23-227	1,211+	Combined
203-228	102	98-109	86-181	354	A39, L189, L226, S472: AL
	110	-	-	1	W126: France
	112	109-116	108-119	69+	H61, C373, S97, K71: KY, MD, TN
	119	108-127	91-141	283+	J46, M115, P113, H156, C202, T92: AR, MO, OK
	121	99-136	96-170	105	B233, K80, L126, M140, S350, S366, W71: CA, CO, OR, UT
	125	-	110-167	12	H202: FL
	139	130-142	122-195	386+	C20, B30, E32, J15, W39, M20, S62, M359, N94, N115: MI, MN, SD, WI
	145	125-149	105-399	83	C107, I10, L29, M246, M536, M302: IA
	150	122-187	91-250	147	R80, R47, G48, C256, M499: IL, IN, OH, PA
	129	109-136	86-399	1,440+	Combined
229-253	150	-	91-227	329	L189, L226, S472: AL
	160	153-159	153-198	106+	S11, C373, S97, K71: KY, TN
	161	142-177	91-181	290+	J46, M115, P113, C202, T92: AR, MO, OK
	167	-	145-191	2	W126: France
	172	153-192	74-225	83	C366, B233, L126, K80, M140, S350, S363, W71: CO, CA, OR, UT
	178	-	145-213	24	C148, H202, M242: FL
	189	181-198	176-268	300+	C20, B30, M20, J15, W39, M359, N94, N115: MI, MN, SD, WI
	191	177-204	113-354	88	C30, C107, B150, M302, M246, M536, I10, L29: IA

| | Weight in grams | | | | |
| | Mean of | Central | | | |
TL	means	50%	Range	No.	Citations
229-253 (cont.)	202	184-204	132-250	83+	R80, E32, R47, G48, M499: IL, IN, OH
	178	156-188	74-354	1,305+	Combined
254-278	202	204-209	136-340	349	A39-40, L189, L226, S472: AL
	230	218-290	218-250	177+	S97, K71, C373: KY, TN
	239	204-250	159-261	32	B233, L126, S350, W71: CA, CO, OR, UT
	240	227-256	170-397	364+	C20, J15, W39, S62, B30, M20, M359, N94, N115, R237: MI, MN, NY, WI, SD
	244	-	195-272	5	W126: France
	247	210-275	119-454	320+	M115, J46, P113, H156, C202, T92: AR, MO, OK
	249	232-272	179-281	46	C148, H202, M242: FL
	269	252-275	200-425	73+	R80, E32, R47, G48, C256, M499: IL, IN, OH, PA
	270	255-275	184-454	67	B150, C30, C107, F31, I10, L29, M302, M246, M536: IA
	246	218-258	119-454	1,433+	Combined
279-304	288	272-308	227-635	316	L189, L226, S472: AL
	298	295-300	295-300	151+	C373, S97, K71: KY, TN
	312	280-345	119-454	358+	M115, J46, P113, H156, C202, T92: AR, MO, OK
	319	297-339	255-346	44	B233, K80, M140, S350, S366, W71: CA, CO, OR, UT
	329	312-349	252-351	52+	F74, M242, C148, C384, M572, H202: FL
	343	326-358	170-680	468+	C20, B30, J15, W39, M20, M359, N94, N115: MI, MN, SD, WI
	348	333-360	227-449	127+	R80, E32, R47, G48, C256, M536, M399: IL, IN, OH, PA
	349	320-375	241-482	97	B150, C30, C33, C107, F31, I10, L29, M246, M302: IA
	350	-	-	1	W126: France
	330	308-349	119-680	1,614+	Combined
305-329	393	390-395	113-570	201	L189, L226, S472: AL
	409	363-435	318-510	96+	F74, H202, M242, M572: FL
	411	369-440	272-454	239	M115, J46, P113, C202, T92, H156, AR, MO, OK
	414	403-425	198-851	475+	C20, S62, B30, J15, W39, M20, M359, N94, N115: MI, MN, WI

	Weight in grams				
	Mean of	Central			
TL	means	50%	Range	No.	Citations
305-329 (cont.)	417	377-442	374-480	31	K80, M140, S350, S366, W71: CA, OR, UT
	427	-	400-455	2	W126: France
	440	409-524	394-524	182+	C373, S97, K71: KY, TN
	470	445-500	340-700	89+	R80, R47, C35, E32, H61, R24, G48, C256, M499, M536: IL, IN, OH, PA, MD
	481	436-535	397-652	57	B150, C30, C107, F31, I10, L29, M246, M302: IA
	432	405-445	113-851	1,372+	Combined
330-355	500	-	-	1	W126: France
	504	490-518	363-626	104	L189, L226, S472: AL
	529	510-558	436-665	31	B233, L126, M140, S350, S366, W71: CA, CO, OR, UT
	531	460-592	318-694	207+	M115, J46, P113, C202, T92, H156: AR, MO, OK
	539	454-601	381-783	128+	F74, H202, M242, M572: FL
	555	527-570	468-712	365+	C20, B30, J15, W39, M20, M359, N94, N115, S330: MI, MN, SD, WI
	567	-	-	1	P39: South Africa
	569	531-624	531-624	61+	W132, C373, S97, K71: KY, TN
	580	553-567	340-850	51+	R47, R80, G48, C256, M499, M536: IL, IN, OH, PA
	633	499-765	499-1,021	24	B150, C30, C107, L29, I10, M246: IA
	555	518-570	318-1,021	973+	Combined
356-380	622	613-635	408-907	165	A39, L189, L226, S472: AL
	645	620-647	587-743	34	B233, K80, L126, M140, S350, S366, W71: CA, CO, OR, UT
	669	635-737	544-819	150+	F74, H202, M242, C384, M572: FL
	683	606-732	590-746	72+	W132, S97, K71, C373: KY, TN
	696	691-726	364-1,304	293+	C20, S62, B30, J15, W39, M20, M359, N94, N115, S330: MI, MN, SD, WI
	698	599-731	454-1,009	196+	M115, J46, P113, J26, H156, C202, T92: AR, MD, OK
	752	707-794	578-907	37+	R47, R80, E32, R24, C256, M499, M526: IL, IN, OH, PA

TL	Weight in grams Mean of means	Central 50%	Range	No.	Citations
356-380 (cont.)	770	730-894	482-907	10	M302, C30, C107, I10, L29, M246: IA
	690	634-730	369-1,304	954+	Combined
381-405	769	758-803	544-1,021	126	A39, L189, L226, S472: AL
	782	-	-	5	P39, P41: South Africa
	821	738-934	689-1,037	26	B233, K80, L126, M140, S350, S366: CA, CO, OR, UT
	831	680-900	227-1,297	159+	M115, J46, P113, H156, C202, T92: AR, MO, OK
	850	-	-	1	W126: France
	834	802-896	604-1,049	112+	F74, H202, C242, C384, M572: FL
	893	848-949	848-969	58+	C373, K71, S97: KY, TN
	912	860-950	397-1,701	270+	C20, J15, W39, B30, M20, C13, G25, M359, N94, N115, S330: MI, MN, NY, SD, WI
	930	902-1,000	647-1,250	32+	C107, M246, M302, I10, M536, R47, E32, R80, G48, M499: IA, IL, IN, OH
	867	822-930	227-1,701	789+	Combined
406-431	975	934-1,016	794-1,135	81	L189, L226, S472: AL
	1,065	893-1,200	851-1,433	99+	M115, J46, P113, H156, C202, T92: AR, MO, OK
	1,065	998-1,162	758-1,559	99+	F74, H202, M242, C284, M572: FL
	1,081	1,038-1,185	1,038-1,185	55+	C373, K71, S97: KY, TN
	1,093	1,021-1,120	916-1,406	31+	R47, R80, E32, C256, M499, L29, M536: IA, IL, IN, OH, PA
	1,107	1,021-1,179	482-2,098	179+	S62, J15, W39, B30, M20, M359, N94, N115, S330: MI, MN, SD, WI
	1,116	1,009-1,124	983-1,497	15	B232, B233, M140, S350, S366: CA, MT, OR, UT
	1,418	-	-	1	H164: South Africa
	1,078	1,016-1,134	482-2,098	560+	Combined
432-456	1,176	1,193-1,215	907-1,401	68	A39, L189, L226, S472: AL
	1,213	-	907-1,418	4	H163-4: South Africa
	1,220	1,125-1,310	1,085-1,568	26	K80, M140, S350, S366: CA, OR, UT
	1,292	1,167-1,401	1,009-1,814	67+	F74, H202, M242, C384, M572: FL
	1,331	1,191-1,438	857-1,556	101+	C20, B30, M20, M359, N115, M1: MA, MI, MN, SD, WI
	1,333	1,234-1,412	1,234-1,412	27+	C373, K71, S97: KY, TN
	1,342	1,219-1,455	1,044-1,556	113+	M115, J26, H156, C202, T92: AR, MO, OK

| | Weight in grams | | | | |
TL	Mean of means	Central 50%	Range	No.	Citations
432-456 (cont.)	1,384	1,219-1,602	1,121-1,786	19+	R47, R80, E32, M302, C256, M499, C107, M246, I10: IA, IL, OH, PA
	1,301	1,193-1,412	857-1,814	425+	Combined
457-482	1,399	904-1,529	904-1,956	9	B233, L126, M140, S366: CA, CO, OR
	1,499	1,400-1,593	1,352-1,764	11+	R47, R80, M499, C89, E32, L29: IA, IL, IN, OH
	1,510	1,451-1,570	907-1,724	29	L189, L226, S472: AL
	1,487	1,315-1,588	1,275-2,013	23+	F74, H202, M242, C384, M572: FL
	1,563	1,389-1,667	1,270-1,808	98+	C20, S62, J15, W39, B30, M20, C13, M359, N115: MI, MN, SD, WI
	1,565	1,290-1,651	1,225-2,110	90+	M115, J46, C225, T92, H156: AR, MO, OK
	1,743	-	1,460-1,937	13+	C373, K71, S97: KY, TN
	1,542	1,406-1,638	904-2,110	273+	Combined
483-507	1,716	1,515-1,918	1,375-1,918	67+	M115, J46, C202, T92: AR, MO, OK
	1,765	-	1,471-1,982	11	B233, S350, M140: CA, OR, UT
	1,806	1,693-1,950	1,247-2,177	58+	L189, L226, S472, F74, M242, C384, M572, H202, V27: AL, FL, LA
	1,815	-	1,619-1,950	5+	C373, K71, S97: KY, TN
	1,828	-	1,361-2,665	6	H163, P95: South Africa
	1,893	-	1,588-2,134	5+	R47, R80, M499, C89, E32, IA, IL, IN, OH
	1,894	1,687-2,132	879-2,722	71+	C20, J15, W39, M20, M359, S330: MI, MN, SD, WI
	1,824	1,678-1,950	879-2,722	223+	Combined
508-532	1,970	1,843-2,214	1,191-2,722	24+	S62, J15, W39, M20, C13: MI, MN, WI
	2,024	-	1,389-2,438	9	H163, P99: South Africa
	2,034	1,772-2,223	883-2,835	47+	C373, J26, J46, K71, P27, M115: KY, MO, OK, TN
	2,038	-	1,701-2,722	19	C20, L29, M310, M499, S97: IA, IL, PA, TN
	2,199	2,091-2,197	1,701-2,948	64+	F74, H202, L189, L226, M242, M572, S472: AL, FL
	2,664	-	2,381-2,948	2	B233, S350: OR, UT
	2,102	1,937-2,214	883-2,948	165+	Combined
533-558	1,861	-	1,588-2,367	4	P92, P99, H163: South Africa
	2,152	-	1,616-3,005	21	J15, M20, W39: WI
	2,423	-	2,155-2,540	45	C202, T92, M115, J46: AR, MO, OK
	2,521	-	2,120-2,722	3	H62, M499: IL, IN

TL	Weight in grams Mean of means	Central 50%	Range	No.	Citations
533-558 (cont.)	2,548	2,522-2,626	1,834-3,175	83	F74, H202, L189, L226, M242, M572, S472: AL, FL
	2,656	-	2,495-2,763	3	B233, S350: OR, UT
	2,443	2,185-2,583	1,616-3,005	159	Combined
559-583	2,159	-	1,588-3,458	28+	C20, J15, M20, S62, W39: MN, WI
	2,792	-	2,410-3,175	2	H163, P42: South Africa
	2,843	-	2,552-2,999	17	M140, J46, M499: IL, OK, CA
	2,906	2,803-3,130	2,325-3,443	34	F74, H202, M242, L189, L226, M572: AL, FL
	2,660	2,240-2,999	1,588-3,458	81+	Combined
584-609	2,144	-	1,559-2,948	14	C20, J15, W39, S97: MN, WI
	3,120	-	2,005-3,884	6	H440, J46, M140, W13: IL, CA, CT, OK
	3,386	-	2,835-3,946	26	L189, L226, H202, M242, M572: AL, FL
	2,947	2,122-3,538	1,559-3,946	46	Combined
610-634	2,563	-	2,041-3,288	8	C20, J15, N52, W39: MN, WI
	3,147	-	-3,628	5	P34, P42, P54, P53: South Africa
	3,217	-	2,693-3,742	2	J46, M310: OK, PA
	4,232	-	3,719-4,905	12	H202, M242, L122, M572: FL, LA
	3,430	2,590-4,082	2,041-4,905	27	Combined
635-659	3,346	-	-3,458	3	P42: South Africa
	2,700	-	-	1	B372: IL "very emaciated"
	4,289	-	3,765-4,581	9	F74, M242: FL
	2,892	-	2,268-3,402	3	J15, W13, W39: CT, WI
	3,634	-	2,268-4,581	16	Combined
660-685	3,458	-	-	1	J15, W39: WI
686-700	2,807	-	-	1	E105: MD
	4,658	-	-	1	A42: FL, a xanthic fish
711-736	1,644	-	-	1	B60: IN
	3,380	-	-	1	P53: South Africa
940	10,500	-	-	1	H52

For lengths up to about 500 mm, there is a tendency for the recorded weights to be higher in the northern than in the southern parts of the range, but the variation in all regions is so high that regional differences are probably not significant.

	SL	No.	
C148 FL Silver Springs	18-242	33	$\log W = -4.5 + 2.95 \log SL$
B30 MI	-	-	$\log W = -4.6252 + 2.993 \log SL$
L29 IA East L.	60-439	64	$\log W = -4.777 + 3.058 \log SL$
S350 UT	41-459	466	$\log W = -4.75 + 3.06 \log SL$

	SL	No.	
L29 IA Red Haw L.	96-291	62	log W = -4.789 + 3.075 log SL
S97 TN Norris L.	-	-	log W = -4.8776 + 3.115 log SL
R178 VA Back Bay	-	378	log W = -5.089 + 3.187 log SL

	FL	No.	
K62 CA Lower Susan R.	-	-	log W = -4.142 + 2.727 log FL
S575 CA Porter P.	-	93	log W = -4.723 + 2.932 log FL
S575 CA Applegate P.	-	357	log W = -4.792 + 2.983 log FL
K63 CA Big Sage L.	-	-	log W = -4.846 + 3.037 log FL
C399 FL L. Wier	15-296	482	log W = -4.947 + 3.041 log FL
S575 CA Reuter P.	-	359	log W = -4.967 + 3.062 log FL
W71 CA Salt Springs L.	-	-	log W = -5.0046 + 3.076 log FL
T132 CA Folsom L.	110-430	-	log W = -5.224 + 3.163 log FL
L227 CA Sutherland L. 1960	-	-	log W = -5.194 + 3.174 log FL
L227 CA Sutherland L. 1958	-	-	log W = -5.521 + 3.308 log FL
L227 CA Sutherland L. 1959	-	-	log W = -5.669 + 3.358 log FL

	TL	No.	
T146 AL pond, young of year	-	-	log W = -3.372 + 2.223 log TL
AL pond, young of year	-	-	log W = -3.457 + 2.287 log TL
AL pond, young of year	-	-	log W = -4.419 + 2.773 log TL
M567 IA Green Valley L. 1973	-	-	log W = -4.223 + 2.781 log TL
J37 OK Chickasaw L.	125-280	29	log W = -4.517 + 2.809 log TL
M567 IA Red Haw L. 1973	-	-	log W = -4.619 + 2.912 log TL
S472 AL	51-254	5,984	log W = -4.80 + 2.96 log TL
	275-533	490	log W = -5.26 + 3.16 log TL
M567 IA Bobwhite L. 1972	-	-	log W = -4.843 + 3.010 log TL
T92 AR Ft. Smith L.	84-546	185	log W = -5.099 + 3.069 log TL
P203 IA Big Creek Res.			
spring	170-290	-	log W = -4.58 + 2.90 log TL
fall	240-400	-	log W = -5.58 + 3.32 log TL
C305 FL pond 1	-	-	log W = -4.99 + 3.07 log TL
pond 2	-	-	log W = -5.14 + 3.19 log TL
pond 3	-	-	log W = -4.98 + 3.05 log TL
pond 4	-	-	log W = -4.93 + 3.07 log TL
pond 5	-	-	log W = -4.90 + 3.03 log TL
M567 IA Bobwhite L. 1971	-	-	log W = -4.866 + 3.019 log TL
M554 UT Powell L. 1968	10-129	1,019	log W = -5.029 + 3.039 log TL
1969	10-129	1,691	log W = -5.490 + 3.292 log TL
T139 IA Clear L.	51-530	172	log W = -4.999 + 3.091 log TL
P113 MO Clearwater L.	125-410	+	log W = -5.1212 + 3.094 log TL
H156 OK Canton L.	86-475	104	log W = -5.0625 + 3.1076 log TL
F31 IA ponds	51-305	34	log W = -5.138 + 3.113 log TL
S188 RI	-	50	log W = -5.1915 + 3.1312 log TL
M246, C107 IA ponds	97-445	257	log W = -5.199 + 3.136 log TL
C355 PA Alanconnie L.	56-551	688	log W = -5.287 + 3.163 log TL ✓
M567 IA Bob White L. 1973	-	-	log W = -5.237 + 3.165 log TL
J27 OK Lower Spavinaw L.	-	61	log W = (-7.879?) + 3.186 log TL
J27 OK Upper Spavinaw L.	-	174	log W = (-7.029?) + 3.210 log TL
M567 IA Red Haw L. 1971	-	-	log W = -5.374 + 3.216 log TL
P122 WI Flora L. 1952	-	119	log W = -12.600 + 3.222 log TL
1956	-	119	log W = -11.765 + 3.084 log TL

	TL	No.	
M567 IA Green Valley L.			
1971	-	-	log W = -5.417 + 3.239 log TL
M567 IA Red Haw L. 1972	-	-	log W = -5.448 + 3.244 log TL
H419 OK	-	-	log W = -5.933 + 3.249 log TL
J46 OK	51-610	3,174	log W = -5.469 + 3.248 log TL
H202 FL Canals	150-559	93	log W = -5.469 + 3.250 log TL
C373 KY Cumberland L.	127-518	577	log W = -5.562 + 3.274 log TL
H202 FL Blue Cypress L.	216-627	64	log W = -5.587 + 3.285 log TL
R125 IA Ike L.	117-447	72	log W = -5.774 + 3.330 log TL
J37 OK Rod and Gun Club L.	125-280	89	log W = -5.391 + 3.333 log TL
J108 NM Elephant Butte L.	125-530	50	log W = -5.765 + 3.378 log TL
B398 AR Beaver L. 1969			log W = -5.773 + 3.378 log TL
1970			log W = -5.731 + 3.344 log TL
Bull Shoals L. 1968			log W = -6.104 + 3.483 log TL
1969			log W = -5.629 + 3.309 log TL
1970			log W = -6.016 + 3.462 log TL
K78 MN George L.	3-6.4	-	log W = -3.798 + 1.343 log TL
	6.45-11.95	-	log W = -5.801 + 3.896 log TL
	12-80	-	log W = -4.798 + 3.962 log TL
Z7 OK L. Carl Blackwell			
1968	Age I	100	log W = -3.32 + 2.35 log TL
	Age II	65	log W = -5.55 + 3.28 log TL
	Age III	55	log W = -4.41 + 2.84 log TL
	Age IV	50	log W = -5.06 + 3.09 log TL
	Age V	35	log W = -4.05 + 2.72 log TL
	Age VI + VII	12	log W = -3.28 + 2.45 log TL
1969	Age I	90	log W = -4.81 + 2.99 log TL
	Age II	100	log W = -5.85 + 3.41 log TL
	Age III	35	log W = -5.06 + 3.10 log TL
	Age IV	70	log W = -5.44 + 3.24 log TL
	Age V	75	log W = -5.7 + 3.15 log TL
	Age VI - VII	40	log W = -4.10 + 2.75 log TL

The regressions for largemouth bass for individual months (or seasons) and age groups had the following slopes (V95):

	No.		No.		No.
1.759	1	2.96-3.05	2	3.607	1
2.112	1	3.06-3.15	3	3.783	1
2.56-2.65	2	3.16-3.25	2	4.057	1
2.66-2.75	3	3.26-3.35	4		
2.76-2.85	2	3.36-3.45	2		
2.86-2.95	1	3.46-3.55	5		

When age groups were combined within a season, the regression slopes were 3.014, 3.035, 3.021, 3.078, and 3.091.

The inclusion of a wider range of lengths reduces the variance in the estimated slopes.

The 1.759, 2.112, and 3.783 slopes were based on 3 or 4 fish each.

	SL	No.	K(SL)	Range
B409 SC Heated pond	226-426	77	-	1.3-2.9
E27 TN Norris L.	-	108	1.90	-

	SL	No.	K(SL)	Range
C256 TX Pond 52	34-287	102	1.92	0.64-2.54
E20 MI Howe L.	70-100	36	1.93	-
C256 TX Dealey's L.	55-255	50	1.96	1.51-2.37
M536 OH ponds unbalanced	41-394	84	1.99	1.64-2.59
balanced	41-419	296	2.12	1.63-2.84
C202 AR Ft. Smith L.	58-448	185	2.04	-
B114 MO ponds	-	118	2.09	1.19-3.23
C399 FL L. Wier	3 months	257	2.14	1.78-2.34
C256 TX Greenbriar L.	101-440	51	2.16	1.45-3.06
C256 TX Windhaven L.	91-346	28	2.18	1.75-2.74
F97 MI Wintergreen L.				
netting	-	44	2.18	-
angling	-	184	2.33	-
E24 TN Clinch R.	200-299	118	2.19	-
C257 PA Pymatuning L.	-	48	2.24	-
C33 IA	40-259	33	2.25	-
S100 TN	-	33	2.26	-
E60 IL ponds	-	11	2.26	-
S100 TN Douglass L.	-	12	2.28	-
E20 MI Howe L.	101-322	156	2.29	-
C399 FL L. Wier	15 months	80	2.29	-
L29 IA East L.	60-439	64	2.29	-
B30 MI	46-392	+	-	2.27-2.31
S100 TN Cherokee L.	-	139	2.30	-
S372 NY pond (probably errors				
in weighing)	51-108	56	5.9	3.9-10.6
	140-178	17	2.3	2.1-2.5
S350 UT	41-459	466	2.31	2.04-2.89
C256 TX Ferndale L.	66-548	109	2.33	0.66-3.29
L29 IA Red Haw L.	96-291	62	2.35	-
C23 MN	-	156	2.37	-
S366 CA San Vincente L.	180-397	54	2.37	1.97-3.20
D179 TX before shad-thinning	145-440	106	2.37	1.86-3.55
after shad-thinning	145-480	153	2.70	1.07-4.92
E28 TN	-	344	2.4	-
C34 MN Leech L.	-	3	2.40	-
C36 IA	58-274	25	2.40	1.54-3.21
E3 MN	76-532	430	2.44	-
C148 FL Silver Springs	18-242	33	2.45	-
C399 FL L. Wier	Age II	-	2.46	-
E24 TN Chickamauga L.	Age II	26	2.47	-
S97, S100 TN Norris L.	-	774	2.48	-
K80 OR	56-372	191	2.52	2.12-2.98
C399 FL L. Hollingsworth	Age 0-I	459	2.55	2.11-2.98
M302 IA	43-370	55	2.59	1.81-3.49
C257 PA Sanctuary L.	-	23	2.60	-
E24 TN Chickamauga L.	yearlings	3	2.63	-
W105 IL Crab Orchard L.	-	11	3.06	-
L238 IL ponds at start	152-470	+	-	2.03-2.79
after 47 days, no extra feed	-	+	-	1.89-2.07

	SL	No.	K(SL)	Range
after 47 days, extra feed, tadpoles	-	+	-	2.95-3.07
afte⸱ 68 days, no extra feed	-	+	-	1.82-1.94
after 68 days, extra feed, Purina	-	+	-	2.71-3.01
C23 MN standards - poor		under	2.00	
average			2.25-2.50	
excellent		over	2.65	

	FL	No.	K(FL)	Range
S575 CA Potter P.	-	93	1.45	-
S575 CA Applegate P.	-	357	1.47	-
S575 CA Reuter P.	-	359	1.47	-

	TL	No.	K(TL)	Range
B430 SC Par P.	20-120	+	-	0.9-1.1
C105 IA Jensma P.	165-201	21	1.08	0.94-1.22
J37 OK Rod and Gun Club L.	127-279	89	-	0.94-1.25
J37 OK Chickasaw L.	127-279	29	-	1.02-1.16
J96 OK Franklin P. before	-	+	1.14	-
2 years after 90% removal	-	+	1.30	—
Rod and Gun Club L. before	-	+	1.16	-
2 years after 90% removal	-	+	1.30	-
M539 LA Idlewilde L.	-	+	-	0.94-1.36
M246 IA ponds	97-445	257	1.19	0.75-2.30
M306 MD Snowden P., May	246-351	43	1.19	-
H424 OK Rod and Gun Club L.	152-279	+	1.19	1.16-1.22
T92 AR Ft. Smith L.	84-546	185	1.20	1.02-1.40
C302 PA poor growth year	-	-	-	1.14-1.32
good growth year	-	-	-	1.05-1.55
V93 IA McFarland L.	145-295	583	1.20	1.11-1.69
tagged bass	140-220	153	1.13	0.95-1.87
C202 AR Ft. Smith L.	71-544	185	1.21	0.61-1.55
B373 IL ponds	Age 0	2,080	1.24	1.05-1.58
	Age I	621	1.18	1.05-1.50
	Age II	190	1.27	1.11-1.44
	Age III+	41	1.25	1.00-1.44
B408 IL Ridge L. 1963-9, monthly means	-		-	1.13-1.37
B30 MI	-	+	-	1.27-1.30
B219 MI Wintergreen L.	178-462	225	1.33	-
M499 IL	102-445	700	1.36	0.86-1.61
S472 AL	51	190	1.56	-
	76-254	5,794	1.27	1.19-1.39
	279-406	436	1.39	1.36-1.41
	432-533	52	1.51	1.48-1.66
S481 ON Heart L.	180-370	588	-	1.30-1.43
J27 OK Lower Spavinaw L.	-	161	1.38	-
J27 OK Upper Spavinaw L.	-	174	1.38	-
T19 IL	-	+	1.39	-
M567 IA Red Haw L. 1971	111-511	-	1.39	1.21-1.54
M567 IA Red Haw L. 1972	136-561	-	1.39	1.21-1.54
1973	111-561	-	1.48	1.27-1.67

	TL	No.	K(TL)	Range
T146 AL monthly averages	-	-	-	1.22-1.8
M539 LA ponds	Age 0	+	-	1.16-1.22
	Age I	+	-	1.27-1.55
	Age II	+	-	1.61-1.66
B41 IL Fork L.		+	-	1.00-1.80
Z7 OK L. Carl Blackwell	Age 0	189	1.24	1.11-1.67
	Age I	624	1.41	1.28-1.53
	Age III-IV	247	1.55	1.46-1.63
	Age V-VII	188	1.59	1.48-1.72
M667 IA Green Valley L.				
1971	236-511	-	1.58	1.41-1.81
1973	111-561	-	1.75	1.54-1.95
M567 IA Bob White L. 1971	111-536	-	1.52	1.27-1.74
1972	161-361	-	1.54	1.43-1.70
1973	211-561	-	1.56	1.41-1.81
G164 NM Navajo L.	101-200	4	1.44	
	301-400	6	1.85	
B150 IA Little Wall L.	140-335	15	1.51	1.33-2.02
C305 FL ponds	-	-	-	1.21-1.82
C89 IA	457-483	3	1.53	1.39-1.80
P203 IA Big Creek Res.				
spring	170-290	-	1.51	-
fall	240-400	-	1.61	-
S393, N94, N115, F105, S255-9,				
S276 SD reservoirs	102-457	143	1.54	1.16-1.86
R125 IA Ike L.	117-447	72	1.41	1.11-1.86
D68 NC	-	195	1.58	0.80-3.30
T139 IA Clear L.	51-530	162	1.62	1.02-2.33
H58, H163-5, P34, P39, P41-2,				
P53-4, P92, P95, P99:				
S. Africa		386	1.66	1.19-2.08
R58 OH		+	1.66	-
C302 FL	-	-	-	1.63-1.83
H156 OK Canton L.	86-475	104	1.83	0.97-1.99
B41 IL standards	poor			0.97-1.25
	average			1.26-1.52
	good			1.53-1.80
W126 France standards	poor			1.25-1.6
	average			1.65-1.95
	good			2.0-2.3

Increase of K with increase in length was noted in M246, M536, S225, T139, B408, C399, V92, V93, S97, G164, and S472, a condition that was also evident in the fact that the length-weight regression slopes were often above 3.0. In C202, the K values were lowest in mid range, 190-300 mm.

Mean K increased with age in Z7 (Age 0 and I differed significantly and differed from III-IV and V-VII, but the latter two groups did not show a significant difference). Mean K also increased with age in C399 and M539 but showed a decrease in D68. Ages II-V had higher average K values than younger or older bass in Norris L. (S97). In ponds that were cropped, K did not decrease with age as it did in ponds that were not cropped (B373).

Several studies suggest no seasonal cycle in plumpness of largemouth bass (B382, S575, Z7). K values were higher in summer than in winter in Oklahoma (H424) and in Florida (C399), but the highest K was in January in another Florida study (C302). K improved from spring to late July and then dropped but increased slightly in early September to drop again in Illinois ponds (B373). In an Iowa lake, K of yearling and age II bass increased from April to November (V93). Condition factors of young largemouth bass were higher in May, June, and July and declined in August to September (B430). Condition factors usually were higher in heated water than in unheated water in Par P., SC (B430). In Pennsylvania, K was highest in late July (1.05-1.55) in a year of good growth but was highest in May (1.14-1.32) in a year of poor growth (C302). Mean K was highest during the spawning season for Age III and IV bass but was not significantly high for Age 0-II or V-VII bass (Z7). Seasonal changes in condition factors were related to weights of the stomach contents (K78). When an extensive dieoff of aquatic plants in August permitted bass to feed more readily on small fish, the average K increased from 1.39 in August to 1.68 in October (B41, B382).

K was positively related to growth rates in four comparisons in Pennsylvania (C302) and in Florida (C305) but not in a Wisconsin study (C355). Mean K was not related to abundance of zooplankton in Ohio ponds (M536). Bass in balanced ponds had higher mean K values than those where the population was unbalanced, except for one balanced pond that was turbid and had a low mean K (M536). In the unbalanced ponds, the largest bass, however, tended to be in good condition (M536). Young bass in cropped ponds had higher average K values in fall than those from uncropped ponds (B373). The higher the population density the lower was the average K in a series of ponds (B373). In Florida, however, bass seemed to be more crowded in canals than they were in Blue Cypress L., but the adjusted mean weights were significantly higher in the canals (H202). The slopes of the length-weight regressions in the canals and the lake did not differ significantly but were significantly greater than 3.0. The poor condition of bass in Bull Shoals L. in 1967 was related to low predation on and high survival of bass, and the good condition in Beaver L. in 1969 was related to the availability of abundant young threadfin shad (B398).

Reduction in standing crop of other species increased the standing crop and average K of bass in Franklin Pond and in Rod and Gun Club L. (H96). Plumpness was negatively related to population density in New York ponds (R239), and average condition factors of bass decreased with time after stocking, probably because of increased population density. Mean condition factors were inversely related to bass abundance (C302, K78).

The mean weight of 254 mm bass decreased from about 227 to 280 g the 2nd year after stocking to about 204 to 254 g in the 5th year in New York ponds (R237). Tagging had no effect on K (Z7), but jaw-tagged bass showed significantly lower K values in V93. However, no difference in condition factors or growth rate was shown after injecting latex dyes (B93). Bass heavily parasitized with roundworms, *Contracaecum spiculigerum*, were in good condition (W71).

S561 recommends using Kn, adjusted for changes in length, with the following standard weights:

TL in inches	6	7	8	9	10	11	12	13	14	15
Weight in pounds	0.10	0.17	0.24	0.33	0.46	0.68	0.87	1.08	1.35	1.67
TL in mm		152	178	203	229	254	279	305	330	356	381
Weight in g		45	77	109	150	209	308	395	490	613	758

However, it is obvious that the length-weight regressions differ in various populations, and the assumption that there is a standard length-weight regression is hard to justify.

The mean depth at various TL is given by L189:

TL	25	51	76	102	127	152	178	203	229	254	279	305
Depth	7.0	10.2	14.5	19.4	26.2	31.1	38.0	43.9	50.7	55.3	64.0	70.7
No.	139	212	106	53	46	123	99	119	127	139	197	80
TL	330	356	381	406	432	457	483	508	533	559	583	
Depth	79.0	84.5	91.6	103	109	119	130	131	147	149	152	
No.	34	98	92	54	42	16	9	7	12	8	5	

In Minnesota (K78), the mean TL of various experimental groups were:

2 weeks	8-17	10 weeks	43-60
4 weeks	16-36	12 weeks	51-68
6 weeks	26-45	14 weeks	54-75
8 weeks	34-54	16 weeks	55-81

Young largemouths were 3 mm at hatching, 5 mm at 3 days, 13 mm at 3 weeks, 18 mm at 4 weeks, 25 mm at 6 weeks, and 46 mm at 2 months in French ponds (W126). In L. George, MN, (K78) young largemouth were 3 mm at hatching, 6.16 mm (5.92-6.31) at rising from the nest, and 32.5 mm at dispersal from the brood.

		TL Mean of means	Central 50%	Range	No.	Weight Mean of means	Range
	No.						
Age 0-April							
SC: B430 unheated	20	19	-	s.d. 0.28	-	-	-
heated	19	23	-	s.d. 0.29	-	-	-
FL: C305, C399	975	32	19-52	12-52	123	1.6	-
Age 0-May							
UT: M554	+	-	-	6.5-8.3	-	-	-
AR: A93	+	-	-	18-24	-	-	-
AL: S117, T146	+	-	-	25-90	+	0.1	-
SC: B430 unheated	100	31	-	-	-	-	-
heated	76	41	-	-	-	-	-
FL: M32, C399	23+	39	34-48	22-61	-	-	-
TX: C256	36	42	-	-	36	5	-
LA: V27	+	-	-	-89	+	-	-0.6
FL: C305	103	109	-	107-112	103	17	16-18
Age 0-June							
UT: M554	+	-	-	12-31	-	-	-
KS: R309	+	-	-	20-50	-	-	-
MI, MN: C60, L132	994	32	-	18-75	957	0.02	-
OH: C284, M535, R42, R82, R83	31+	35	25-41	19-91	-	-	-
IN: M75, R40	311	37	-	30-67	-	-	-
OK: B183, C134, C271, H150, J33, Q5	277	38	38-48	13-79	-	-	-
IA: D5, L29, M246	36+	40	-	13-58	+	-	0.03-0.06
TN, KY: C92, D43, E28	202+	40	-	20-81	-	-	-
SC: B430 unheated	115	41	-	-	-	-	-

	No.	TL Mean of means	Central 50%	Range	No.	Weight Mean of means	Range
Age 0-June (cont.)							
SC: B430 heated	76	50	-	-	-	-	-
FL: M32, C399, T142	859+	51	40-58	20-93	-	-	-
IL: B41, B86, R26, S166, W105	127+	53	30-79	13-102	-	-	-
AR: A93, M523	759+	59	-	30-106	-	-	-
AL: S117, T146	+	-	-	70-140	+	3	-
NY: F160	1	86	-	-	-	-	-
FL: C305	110	147	-	141-156	110	46	42-52
Combined	3,897+	47	30-48	13-156	1,067+	-	0.02-52
Age 0-July							
NY: S372	25	41	-	24-47	25	9	7-14
UT: M554	+	-	-	30-55	-	-	-
KS: R309	+	-	-	40-55	-	-	-
MI, MN: C60, K78, L132	2,524	50	46-60	18-107	2,242	0.14	0.06-0.23
SC: B430 unheated	119	45	-	-	-	-	-
heated	75	62	-	-	-	-	-
IA: B2, B3, C91, C101, C102, C109, D5, L29, M246, T89, V93	309+	54	36-62	23-183	+	-	0.03-0.2
MD; M306	7	56	-	48-61	-	-	-
OH: C284, G32, M535, R42, R71, R82, R83	22+	58	46-58	43-152	-	-	-
KY, TN: C92, D43, E28	307+	65	-	28-127	-	-	-
IN: B44, R39, R40, M74	790+	65	43-61	43-203	-	-	-
IL: B373, R26, W105	1,437+	65	66-71	25-86	-	-	-
OK: B183, C134, C271, H150, J42, O37	612	78	64-91	31-196	2	1.5	-
SD: S225	+	81	-	-	-	-	-
TX: C256	4	81	-	-	4	4	-
AR: M523	206	-	-	35-155	-	-	-
AL, FL: C305, C399, M32, S117, T142, T146	211+	102	49-176	36-186	78+	91	76-102
LA: V44	18	170	-	135-203	18	85	37-147
Combined	6,682+	69	46-74	23-203	2,369+	30	0.03-147
Age 0-Aug.							
L. Erie: F5	45	9	-	-	-	-	-
KS ponds: R309	+	-	-	41-60	-	-	-
IN: M74, R40	712	60	-	51-79	-	-	-
CA: K63-4, M148	561+	60	48-64	18-137	-	-	-
IL: W105	2	61	-	57-66	-	-	-
NJ: H40	+	-	-	51-76	-	-	-
UT: M554	+	-	-	60-68	-	-	-
FL, GA, ponds: C42, M32	+	65	-	51-64	-	-	-
OH: C284, G32, M535, R71, R82-3	8+	67	60-81	51-135	-	-	-
SC: B430 unheated	17	50	-	-	-	-	-
heated	8	78	-	-	-	-	-

		TL				Weight	
	No.	Mean of means	Central 50%	Range	No.	Mean of means	Range
Age 0-Aug. (cont.)							
MI, MN, SD: C60, K78, L56, L132, S225	513+	77	58-109	28-146	250+	10	3-28
IA: B2, B3, C91, C101-3, C109, D5, L29, T89, V93	513+	79	56-95	33-196	162+	9	1.7-11
TN: E28	112	81	-	41-152	-	-	-
OK: J26, M107	38+	85	-	66-109	38	11	6-20
AR: M523	116	-	-	35-185	-	-	-
FL lakes: C399	201	99	90-111	50-159	-	-	-
KS: S207	106	150	-	79-201	106	48	-
MO: W73	+	178	-	-	-	-	-
FL, LA, TX lakes: C305, C256, V44	125	213	-	124-251	125	153	26-284
AL ponds: S117, S325, T146	+	-	-	110-160	+	190	136-227
Combined	3,007+	84	56-97	18-251	781+	62	1.7-284
Age 0-Sept.							
KS ponds: R309	+	-	-	50-70	-	-	-
TN, WV: E28, S124	488+	69	-	36-173	-	-	-
MI, MN, SD, WI: E20, L132, M265, S225-6	38+	73	54-84	48-107	-	-	-
CA, UT: K64, M554	32+	75	-	53-99	-	-	-
SC: B430 unheated	11	61	-	-	-	-	-
heated	11	91	-	-	-	-	-
OH: B102, G32, G82, R42, R71, R82	8+	91	69-109	58-137	-	-	-
MD: M306	1	94	-	-	-	-	-
FL: C399	41	102	92-113	57-167	-	-	-
AR, MO, OK: C271, C353, B183, M523, W73	300+	116	41-174	40-220	-	-	-
AL, GA: C42, S117, S558, S559	+	127	-	76-229	+	164	80-300
IA: C101, C103, F43, I10, D5, L29, T89, V93	57+	131	102-145	53-300	24	45	24-82
TX: C256	1	135	-	-	1	26	-
NC, SC: P198, B80	+	200	-	185-229	+	136	90-227
LA: V44	21	300	-	282-315	21	568	363-553
Combined	1,009+	110	71-137	36-315	46+	175	24-553
Age 0-Oct.-Dec.							
SC: B430 unheated	4	64	-	-	-	-	-
Ontario: S481	105	70	-	-	-	-	-
FL ponds: M32	+	71	-	66-84	-	-	-
OH: C284, R82, T140	32+	80	-	70-112	-	-	-
IN: K23, R38-40	634+	93	71-107	71-173	+		6-9
AZ, CA, OR, UT: C76, K64, M399, M554, S29, W137	107+	94	66-130	38-152	+	227	-
AR: B393, C202, M523	430	104	-	40-220	+	340	-
MD: M306	18	97	-	53-150	-	-	-
NY, PA: F160, C302, S372	36+	106	-	47-150	29	17	14-28

	No.	TL Mean of means	Central 50%	Range	No.	Weight Mean of means	Range
Age 0-Oct.-Dec. (cont.)							
MI, MN, WI: C13, C60, L72, L132, W232	690+	109	78-155	41-208	1,185	20	3-51
NJ: S258	1,753	113	-	61-178	-	-	-
SC: B430 heated	14	119	-	-	-	-	-
IL: B41, B373, B407, B480, H440, H466	818+	134	109-160	99-206	127+	45	4-109
IA: B3, C157, D5, H398, M197, S86, T89, V93	38+	129	76-152	57-279	3+	51	6-255
TN, WV: E24, E28, S124	684+	130	76-246	43-295	-	-	-
TX: C256, L22	19+	141	112-177	76-208	19	35	14-102
FL lakes: C299	86	143	114-173	72-244	-	-	-
KS, MO: B115, L63, T79, W73, B73	+	152	-	64-229	-	-	-
OK: A8, A10, B181-2, C353, D31, H77, H150, J34, L63, R201	154+	174	140-221	74-259	+	173	6-340
AL: E146, S117, S121	214	183	-	157-208	+	225	156-292
South Africa, H163	+	-	-	218-274	-	-	-
FL, C305	132	266	-	250-295	132	318	242-441
LA, MS: F45, V26-7, V44	82+	280	-	160-351	73+	446	65-907
Combined	6,047+	134	79-165	38-351	1,565+	124	4-255
Age 0							
MA, NY, PA: E9, M310, S289	96+	71	43-97	23-107	77	3	-
MI, WI: B31, E20, C213, H250	18,795+	72	53-85	36-102	18,777	5	1-9
FL, TX: C256, M33	8+	76	-	64-137	-	-	-
KS, MO: D201, T133	329+	79	64-81	71-168	+	25	5-75
AR: H442, T92	99	102	77-126	66-150	-	-	-
CA, CO: D13, L126	18	109	-	38-150	-	-	-
IA, IL, IN, OH: B41, G80, H62, H278, M323, M503, S199, T20, T35	274+	110	76-145	11-216	-	-	-
KY, TN, WV: C373, E26, R292, S18, S124, S266	380±	137	140-157	28-243	28	36	27-54
OK: H156, T44, W7, W53	68	143	137-157	86-193	39	36	6-79
South Africa: H21	+	-	-	64-318	-	-	-
Cuba: H441	+	260	-	-	-	-	-
Combined	20,067+	107	76-145	23-318	18,921+	18	1-79
Age I							
Italy: D204	+	122	-	88-168	+	31	10-77
PA: C302, M310	1,090	135	101-173	101-213	1,090	41	11-122
OR: B233, B236, G76, K80, G77, L117, O24, O34, O36	252+	136	104-180	51-279	100	15	3-45

	No.	TL Mean of means	Central 50%	Range	No.	Weight Mean of means	Range
Age I (cont.)							
MA: M1, M87, M146, S289	170	138	117-157	117-160	72	37	28-48
WI: B39, C115, C117, H250, J15, M20, M97, W232	152+	139	91-182	76-229	3	57	-
MO: B73, D201, M569, R306	182	138	109-185	89-201	24+	255	23-720
IN: B60, H62, K23, K41, R37, R38, R40, W220	51+	141	116-169	67-224	+	42	6-113
MD: E76, M306	60	150	119-203	107-254	-	-	-
Cuba: H441	+	156	-	136-176	+	102	65-140
NY: E9, E59, E112, F160, G24-5, R237, R239, S184, S372	80+	156	130-186	81-241	22+	75	14-191
France: W126	6	157	-	81-221	6	62	6-145
CA: D13, K63-4, K61, M140, S29, S350, T132, W137	604+	169	145-193	53-277	34	77	-
MI: B31, B131, B219, C13, C213, C260, B77, E20, L72, M531	758+	170	150-212	53-254	555+	70	3-227
OH: H52, L14, M536, P8, R42, R82, S117, T140, W48	175+	173	127-224	28-330	+	57	-
IA ponds, lakes: B150, L29, M246, M323, R125, R218, M197, P203, V93	793+	177	132-190	113-338	462	36	22-82
MN, SD: C270, E3, S229	61	177	151-193	89-254	-	-	-
Ontario: J107, S481	439+	183	145-195	145-229	+	195	-
WV: S124	+	183	-	51-358	-	-	-
IL: B41-2, B86, B373, G80, H278, H440, S199, T18, T20, B408, H466	897+	186	168-215	89-356	121	339	150-217
NC: D68, D167, H97, M368, P198, R217, S395, T103, T64, S57	364+	189	178-216	107-279	+	114	15-350
KS: C226, C349, H332, S207, S340, S374, T79, T133	124+	193	160-203	109-287	46+	146	20-267
AR: C202, H442, S504, T92	270+	202	188-241	124-262	+	191	-
NJ: S258	950	202	-	76-320	-	-	-
OK: B181-3, C271, H150, H156, J26, J33, J41-2, K65, L106, M107, T21, T44, W7, W73	562+	204	163-244	64-391	369+	137	37-954

		TL				Weight	
	No.	Mean of means	Central 50%	Range	No.	Mean of means	Range
Age I (cont.)							
KY: C175-6, C373, S266	433	212	170-244	145-264	183	163	77-195
TX: B61, B109, C256, V27	85+	221	178-257	114-300	85+	207	28-482
AZ, CO, NM, NV: L126, M67, M81, M399	48+	223	170-262	109-277	23	293	290-295
LA: M539, V23, V27	76+	224	190-226	178-330	11+	335	113-992
FL: C305, D37, C399	830+	234	163-293	78-327	339+	354	142-1,043
IA river, lakes: H177, I10, U8	64+	248	231-262	190-376	37	321	113-989
TN: B380, E26, E22, R292, S18, S97, S389	839	262	221-292	86-388	22	213	-
South Africa: H18, H163, T13	5+	290	-	229-406	5	443	198-1,361
AL: B350, E146, P27, S117, A100	241+	312	285-335	245-356	112+	272	23-680
Combined	10,604+	189	150-226	28-406	3,721+	169	3-1,361
Age II							
MA: M1, M87, M146, S289	142	194	-	178-221	69	99	94-108
UT: M340, S350, W137	126+	203	178-269	127-302	-	-	-
ID, MT: B108, I7	22	205	-	109-277	-	-	-
WI: C115, C117, B39, D200, H250, J15, M20, W232	651+	210	152-287	109-356	74	136	-
OR, WA: B233, B236, C279, G76, L117, O24, O33-4, O36, W11	156+	213	173-249	102-384	8+	272	14-595
IN: B60, G38, H62, K41, R40, M317, W220, B410	83+	215	182-279	117-312	12	114	17-250
OH: L14, M536, O8, R42, T140, T145	329+	215	178-254	114-229	2	213	-
PA: M310	5	218	-	183-249	5	190	-
NY: E9, E59, E112, F160, G22, G24-5, R237, S184, S372	100+	219	218-241	71-315	45+	203	51-485
Cuba: H441	+	221	-	196-246	+	270	209-331
MD: D103, E76, M306	109	234	-	145-366	3	176	34-454
MI: B31, B77, B92, B131, B219, C13, C213, E20, L72, M531	815+	236	210-269	150-353	559	156	51-587
KS: MO: C226, D201, H332, S340, S374, S207, T79, T133, M569, R306	281+	239	194-295	142-442	32+	442	40-836
France: W126	6	244	-	221-244	6	235	190-272
IA ponds, lakes: B150, L29, M246, M323, M372, M503, R125, R218	60+	251	227-282	157-432	7	162	59-284

| | TL | | | | Weight | | |
	No.	Mean of means	Central 50%	Range	No.	Mean of means	Range
Age II (cont.)							
Ontario: J107, S481	581+	252	–	218-282	+	381	–
TX: C256	120	252	246-251	185-345	120	234	94-595
IL: B41, B86, B373, G80, H278, H440, S199, T18, T20, B408, H466	599+	261	218-300	165-394	84	673	327-1,297
OK: B183, C271, H150, H156, J26, J33, J41, K65, L106, M107, O37, T44, W7, W53	151+	271	231-319	185-376	147+	249	139-680
KY: C175-6, C373, S266	416	275	252-305	208-366	249	376	249-626
NC: D68, D167, H97, M368, R217, S297, T64, T103	255+	275	236-328	168-381	–	–	–
MN: C270, C356, E3	80	279	261-297	155-339	–	–	–
IA river-lakes: H177, I10, U8	63+	282	246-305	203-368	9	524	254-1,021
WV: S124	+	282	–	213-356	–	–	–
AR: C202, H442, S504, T92	128+	285	257-302	254-318	+	372	–
CA: D13, K63-4, M140, T132	272+	290	264-325	137-414	23	344	235-454
NJ: S258	337	296	–	170-419	–	–	–
TN: B380, C111, E28, E22, E26, J10, R292, S11, S18, S389, S97	1,301+	314	294-331	207-414	44	310	198-338
AL, FL, GA: D37, E146, P27, W213, C399, A100	312+	321	302-368	226-517	241+	499	99-3,420
AZ, CO, NM, NV: L126, M67, M81, M399	88+	322	305-340	221-386	79	529	454-604
LA: M539, V27	142	338	272-361	241-490	1	2,155	–
South Africa: H163, T13	12	369	–	305-432	12	896	340-1,474
Guatemala: L247	–	–	–	–	+	3,000	–
Combined	7,742+	259	218-300	71-517	1,831+	359	14-3,420
Age III							
UT: M340, S9, S350	109+	205	–	178-216	+	–	454-1,361
OR: B233, B236, C279, L117, O24, O34, O36	855+	228	168-290	119-358	744	26	–
OH: L14, M536, O8, R42, T145	103+	251	221-279	200-343	+	340	–
WI: B39, C85, C115, C117, D200, H250, J15, M20, M97, W232	780+	259	196-312	160-467	230+	511	250-1,814
MO: B115, D201, M569, R306	266	265	229-300	220-328	38	411	287-907
IN: B60, G38, H62, M317, R40, B410, W220	92+	266	207-295	180-388	10	243	102-312

		TL				Weight	
Age III (cont.)	No.	Mean of means	Central 50%	Range	No.	Mean of means	Range
MA, ME: F70, K94, M1, M87, S289	118	267	237-287	237-300	34	238	133-397
MI: B31, B77, B219, C13, C213, C215, E20, B92, L72, M531	1,066+	275	254-294	178-424	706	283	122-1,072
MD: D103, E76, M306	48	284	257-323	226-338	12	321	218-454
WV: S26	+	290	-	-	-	-	-
LA, TX: C256, M539	74	293	277-295	193-414	55	347	91-655
NJ, NY, PA: E58, E112, F160, G22, G24-5, M310, R237, S258, S372	221+	293	272-315	188-401	77+	295	68-510
Quebec: C78	+	295	-	-	+	406	-
France: W126	4	295	-	269-310	4	425	351-499
ID, MT: I7, B108	19	297	-	269-335	-	-	-
KS: C226, G74, H332, S340, S374, T79, T133	36+	301	279-310	200-414	2+	406	100-840
MN, SD: C270, C356, E3, M359, S330	320+	315	-	185-463	+	136	-
Ontario: S481	17	315	-	295-330	-	-	-
IA, ponds, and lakes: L29, M197, M246, M323, M372, M503, R125, R218	86+	317	295-343	236-483	2	412	277-548
IL: B41, B86, B373, G80, H278, H440, S199, T18, T20, B408, H466	436+	319	274-361	216-470	59	947	680-1,361
Cuba: H441	+	325	-	276-374	+	871	533-1,209
IA, river-lakes: H177, I10, U8	72+	326	305-340	254-437	4	1,036	349-1,724
NC: H97, D68, M368, R217, S395, T103	143+	329	290-368	208-437	-	-	-
TN: B380, C111, E22, J10, R292, S11, S18, S97, S389	422+	344	307-380	231-462	73	440	284-510
OK: B183, H150, H156, J26, J33, J41, K65, M107, O37, R201, T44, W7, W53	148+	345	318-391	241-455	135	594	275-1,247
AR: C202, H442, T92	84	346	-	315-388	-	-	-
KY: C175-6, C353, S266	71	361	312-409	279-475	19	1,021	749-1,511
CA: D13, K63-4, M140, T132	283+	368	351-388	221-511	23	499	-
South Africa: H163, H99, T13	5	376	-	305-427	6	1,002	510-1,418
CO, NM, NV: L126, M67, M81	95	380	340-411	340-457	92	898	581-1,216
AL: E146, P27	97	418	353-464	309-506	45	381	142-536
GA: V27	-	-	-	-	1	2,495	-
Combined	6,070+	307	270-350	119-511	1,371+	492	26-2,495

	TL				Weight		
	No.	Mean of means	Central 50%	Range	No.	Mean of means	Range
Age IV							
MO: M569, R306	102	285	253-304	234-343	8	397	–
OR: B233, B236, H447, L117, O24, O34, O36	84+	297	279-328	178-421	14	178	31-329
KY: C175-6, C353, S266	415	312	282-353	282-419	-	-	-
WI: B39, C77, C115, C117, D200, H215, J15, M20, W232	806+	313	277-328	211-559	263	387	357-539
IN: B60, G38, H62, M317, R40, W220, B410	56	319	272-345	213-488	35	355	133-468
MI: B31, B77, B219, C13, C215, E20, L72, M531, C213	1,009+	321	292-356	165-447	153+	432	198-1,622
OH: L14, M536, O8, R42, T145	223+	326	288-343	210-406	+	468	-
NV, UT: M67, S350	137	322	-	272-371	99	723	-
MA, ME: K94, M1, M87, M146, S289	40	328	311-347	307-353	27	489	434-737
NJ, NY, PA: E112, F160, G22, G24-5, M310, R237, S258	131+	329	297-353	190-457	33+	415	71-862
ID, MT: I7, B108	9	329	-	277-406	-	-	-
MD: D103, E76	23	334	323-343	305-376	5	590	454-907
IL: B86, G80, H278, H440, S199, T18, T20, B408	97+	348	315-375	201-483	-	-	-
Quebec: C78	3+	352	-	343-368	3+	652	624-680
France: W126	1	356	-	-	1	851	-
IA, ponds, lakes: C77, L29, M197, M246, M323, M372, M503, R218	75+	362	315-394	241-533	15	1,044	513-1,270
MN, SD: E3, M359, S330	325+	364	339-394	217-555	+	431	-
KS: C226, G74, S340, T133	29+	366	328-421	380-472	+	778	280-1,191
IA, river-lakes: I10, U8	22	369	345-396	318-470	1	862	-
OK: H150, H156, J26, J41, K65, L106, M107, O37, R201, W7, W53, T44	40+	372	340-421	218-500	33	922	479-1,588
NC: D68, H97, M368, R217, S395, T103	100+	385	343-439	221-475	1	-	-
TN: B380, J10, R292, S11, S18, S97, S389	395	394	345-419	292-469	95	659	454-992
CA: D13, K63-4, M140, T132	112+	426	399-442	318-544	-	-	-
LA, TX, AL: C256, M539, P27	51	432	380-487	295-526	49	844	312-1,588
AR: C202, H442, T92	15	439	-	366-531	-	-	-
Cuba: H441	+	457	-	447-468	+	2,570	2,217-2,923

		TL Mean of	Central			Weight Mean of	
	No.	means	50%	Range	No.	means	Range
Age IV (cont.)							
CO, NM: L126, M81	55	467	-	460-495	54	1,964	-
South Africa: H163, P41, P99, T13	4	467	-	384-533	4	1,460	680-2,296
Combined	4,359+	354	310-396	165-559	892+	679	31-2,923
Age V							
MO: M569, R306	54	328	279-388	262-388	21	871	-
WI: B39, C115, C117, D200, H250, J15, M20, W232	557+	338	312-363	254-536	163	505	-
MI: B31, B77, B219, C13, C213, C77, C215, E20, L72, M531	701+	351	333-363	229-439	350	611	227-1,264
NJ, NY, PA: E112, F160, G22, G24-5, M310, S258	42+	352	309-395	226-480	24+	495	159-819
MA, ME: K94, M1, M87, M146, S289	20	366	360-381	340-391	14	736	601-964
IN: B60, G38, H62, K41, M317, W220	31	371	340-382	257-486	18	616	275-907
NV, UT: M67, S29, S350	88	381	-	376-386	67+	1,214	800-2,041
OR: B233, B236, L117, O24, O34, O36	30+	382	358-434	287-470	-	-	-
OH: L14, O8, R42	18+	395	-	381-406	+	851	-
MD: E76	14	396	-	-	-	-	-
Quebec: C78	4+	397	-	330-432	4	1,049	567-1,247
MN, SD: E3, M359, S330	219+	398	353-434	217-617	+	726	-
KS: C226, G74, H332	13+	400	-	295-521	+	2,523	-
IA: L29, M197, M218, M246, M323, M372, M503, R125, R218, U8	48+	404	386-432	305-546	22	1,743	1,420-2,041
IL: B86, G80, H278, H440, S199, T18, T20, B408	93+	414	358-488	254-559	-	-	-
KY, TN: B380, C175-6, S11, S18, S97, S266, S289	163	429	381-445	351-521	72	910	680-1,134
OK: H150, J26, J41-2, K65, L106, M107, R201, T44, W7, W53	26+	439	388-472	353-572	27	1,591	501-2,580
NC: D68, H97, M368, R217, S395, T103	62+	440	414-486	348-503	-	-	-
AL, LA, TX: C256, M539, P27	37	463	351-528	315-554	36	1,056	369-3,070
AR: C202, H442, T92	10	501	-	434-546	-	-	-
CA: D13, K64, M140, T132	41	506	445-526	434-561	-	-	-
South Africa: P99, H163	5	519	-	493-549	5	1,973	1,616-2,296
NM: M81	20	523	-	-	20	2,455	-
Combined	2,296+	400	348-445	217-617	843+	1,016	159-3,070

		TL				Weight	
	No.	Mean of means	Central 50%	Range	No.	Mean of means	Range
Age VI							
WI: B39, C115, C117, D200, H250, J15, M20, T77	332+	375	363-384	254-572	99	811	-
NJ, NY, PA: F160, G22, G24-5, M310, R237, S258	45	376	340-450	221-493	30	666	128-1,633
MI: B31, B219, C13, C213, C215, B77, L72, M531	326	380	366-388	300-462	145	752	397-1,531
MO: R306	13	382	310-446	269-488	-	-	-
OR: B236, C279, L117, O24	7+	390	-	351-434	-	-	-
IN: B60, W220	4	394	-	242-459	-	-	-
MN: E3, M359	78+	415	376-435	309-593	+	885	-
MA, ME: F70, K94, M1, M87, M146	13	415	-	381-442	5	1,100	851-1,361
Quebec: C78	6+	420	-	406-457	6+	1,296	995-1,724
NV, UT: M67, S29, S350	22	430	-	417-457	19+	1,565	1,044-2,608
OH: L14, O8, T145	12+	420	-	376-445	-	-	-
IL: H440, G80, S199, T18, T20, B408	91+	443	388-498	269-549	-	-	-
IA: L29, M323, M372, U8	16+	444	450-472	307-572	7	2,410	-
KY, TN: B380, C175-6, S11, S18, S97, S266, S389	67	455	417-488	399-535	45	1,202	992-1,531
MD, NC: D68, E76, R217, S395, T103	37	474	437-522	406-546	-	-	-
OK: H150, L106, P26, R201, T44, W7, W53	6+	484	452-508	427-610	6	1,893	1,764-2,098
KS: C226, H332	6	488	-	450-513	-	-	-
CA: K64, M140, T132	20	528	470-554	470-589	-	-	-
South Africa: H163, P54, P99	4	535	-	508-549	4	2,165	1,928-2,367
NM: M81	7	541	-	-	7	2,880	-
AR: C202, H442, T92	4	542	-	526-559	-	-	-
AL, LA, TX: C256, M539, P27	25	545	498-582	455-613	24	1,587	893-2,611
Combined	1,141+	434	381-488	221-613	397+	1,454	128-2,880
Age VII							
OR: B233, B236, C279, L117, O24	7+	412	381-439	361-457	-	-	-
NY, PA: F160, G22, M310	20	413	373-470	330-508	12	635	-
MI, MN, WI: B31, B39, B77, B219, C13, C117, C213, C215, D200, E3, H250, J15, L72, M20, M359, M531, T77	446+	417	396-427	292-585	151+	1,040	811-1,398
MA, ME: K94, M87, S289	8	438	-	427-457	7	1,330	1,247-1,474

		TL				Weight	
	No.	Mean of means	Central 50%	Range	No.	Mean of means	Range
Age VII (cont.)							
Quebec: C78	2+	444	-	437-452	2+	1,495	1,361-1,721
OH: L14, O8, T145	2+	451	-	430-483	-	-	-
IL, IN: G80, H62, H440, B408, S199, T18, T20, W220	50+	461	454-493	307-533	2	1,211	661-1,764
MO: R306	3	467	-	404-498	-	-	-
IA: L29, M323	5+	473	459-515	352-584	-	-	-
TN: S11, S18, S97, S389	40	479	461-486	452-583	25	1,573	1,361-1,814
NV, UT: M67, S29, S350	3	486	-	437-511	3	2,435	2,178-2,948
OK: B115, K65	+	490	-	432-549	-	-	-
KS: C226	2	508	-	483-536	-	-	-
MD, NC: D68, E76, M368, R217, T103	13	510	450-451	437-589	-	-	-
CA: K64	1	561	-	-	-	-	-
AL, TX: C256, P27	11	573	-	486-642	11	1,944	1,332-3,147
NM: M80	2	584	-	-	2	3,288	-
South Africa: P41, P42, P99	4	586	-	516-643	4	3,093	2,410-4,004
Combined	619+	456	405-493	292-643	219+	1,447	635-4,004
Age VIII							
MI, MN, WI: B31, B39, C13, C117, C213, C215, D200, E3, H54, H250, J15, M20, M359	243+	441	424-445	308-594	77+	1,251	851-1,375
OR: B233, O24	4	445	-	439-457	-	-	-
Quebec: C78	2+	451	-	437-457	2+	1,531	1,474-1,588
NY, PA, ME: F70, F106, K94, M310	9	473	439-520	411-523	2	1,914	1,843-1,985
IA, IL, IN, OH: B60, C77, H62, H440, G80, L14, L29, T18, T20, W220	32+	486	467-516	307-590	6	1,291	1,049-1,644
KY, TN: S11, S18, S289, S266	29	497	478-513	460-572	22	1,824	1,701-2,268
UT: S350	5	508	-	-	-	-	-
NC: R217, M368, T103	7	533	-	488-589	-	-	-
MO: R306	2	546	-	-	-	-	-
South Africa: P53, P95, H163	3	546	-	493-605	3	2,476	2,155-2,665
KS: C226	1	572	-	-	-	-	-
NM: M80	1	597	-	-	1	3,629	-
AL: P27	2	617	-	-	2	1,673	-
Combined	340+	478	439-508	307-605	116+	1,613	851-3,629
Age IX							
MA, NY, PA, ME: G25, K94, M146, M310	8	448	427-481	343-503	8	1,317	825-2,098
IL: T18, T20	4	460	-	424-495	-	-	-
MI, MN, WI: B31, B39, C13, C117, C213, C215, H250, J15, M20, M359	127+	465	455-467	292-577	45	1,599	1,313-2,240

	No.	TL Mean of means	Central 50%	Range	No.	Weight Mean of means	Range
Age IX (cont.)							
OR: B233, O24	5	467	-	434-503	-	-	-
TN: S11, S18, S389, T67	15	522	-	500-559	9	2,107	1,956-2,211
MO: R306, P30	2	525	-	503-546	-	-	-
NC: M368, R217	4	535	-	505-572	-	-	-
AL: P27	2	633	-	-	2	1,673	-
Combined	167+	480	455-504	292-633	64	1,683	825-2,240
Age X							
IL, IA: L29, T20	4	405	-	345-526	-	-	-
OR: B233	1	437	-	-	-	-	-
MI, MN, WI: B39, C213, C215, E3, J15, H250, M20	72+	487	472-495	312-587	11	1,551	-
MO: R306	1	498	-	-	-	-	-
TN: S11, S289	10	537	-	521-531	10	2,427	1,871-2,671
NC: M368	1	610	-	-	-	-	-
South Africa: P42	1	630	-	-	1	3,124	-
Combined	90+	495	472-523	312-630	22	2,172	1,871-3,124
Age XI							
OR: B233	2	470	-	-	1	2,948	-
PA: M310	1	483	-	-	1	2,296	-
MI, MN, WI: B39, C215, E3, H250, J15, M20	62+	505	490-503	343-648	5	2,192	-
Combined	65+	500	478-503	343-648	7	2,364	2,192-2,948
Age XII							
IL: T20	3	498	-	-	-	-	-
TN: S18, S389	2	556	-	546-566	1	3,175	-
CT, MA, PA: M146, M310, S289, W13	4	554	-	511-610	2	3,222	3,042-3,402
MI, WI: B39, C215, J15, M20	42	516	513-528	406-681	2	2,129	-
Combined	51	526	511-533	406-681	5	2,775	2,129-3,402
Age XIII							
MI, WI: B39, C215, J15	19	520	-	394-559	1	2,438	-
South Africa: H163	1	521	-	-	-	-	-
Age XIV							
MI, WI: B39, C215, J15	4	492	-	445-508	-	-	-
CT: W13	1	584	-	-	1	2,807	-
Age XV							
MI, WI: B39, C215, J15	6	517	-	495-546	-	-	-
IL: T20	1	521	-	-	-	-	-
OR: B233	1	503	-	-	-	-	-
Age XVI							
IN: H62	1	541	-	-	1	2,121	-

Mean calculated TL at each annulus

	No.	1	2	3	4	5	6	7	8	9	10	11	12	13	14	15
M551 WI Punch L. (a stunted pop.)	+	56	137	193	236	254										
K47 WI south	743	53	135	206	259	302	330	356	384	424	450					
G152 WI George L.																
before muskies	42	83	160	213	246	269	296	332	348	378						
after muskies stocked	24	71	182	231	267	293	312	331	343	352						
Corrine L.																
before muskies	27	80	152	195	239	279	337	362	354							
after muskies stocked	10			–	222	267	309	344	380	412	424					
M267 WI Brown's L.	1,770	91	170	229	272	305	345	411	452	467	478	498				
M427, T77 WI Brown's L.	+	91	168	231	277	302	335	404	457				521	511		
P122 WI Flora L.																
after harvesting	139	107	175	231	284	315	345	378	404	437	470					
K33, M258 MN	457	91	173	221	287	335	394									
C118 WI Day L.	577	89	170	237	292	333	384	414	447	460						
after thinning	+	79	142													
B39 WI north	+	119	193													
D89 WI Murphy Flowage	+	71	165	246	297	335	353	386	424	450	470	490	486	490		
B39, M551 WI state average	+	130	208	259	340	373	391	427								
E3, E2 MN	618	84	188	267	318	356	384	414	442	460	475	495	505	513	523	533
S62 MN	2,128	99	190	270	325	392	434	437	467	516	560	604				
C270 MN Miller L.	+															
(backwater)	65	99	155	292	351	388	414	429	457	503	551	592				
G180 SD Francis Case L.	96	120	262	340	343	417										

Mean calculated TL and increment at each annulus

	1	2	3	4	5	6	7
L72 MI Loch Alpine Ponds	81	165	244	315	343	358	378
Incr.	81	84	79	71	28	15	20
No.	316	248	172	90	41	14	2
S225 SD Ft. Randall L. before impoundment	112	165	251	279			
Incr.	79	91	84				
No.	6	4	3				

	No.	Mean calculated TL and increment at each annulus (cont.)														
		1	2	3	4	5	6	7	8	9	10	11	12	13	14	15
after impoundment		124	71	117	48											
Incr.		23	2	1	3											
No.																
B39 WI		84	185	267	318	356	384	414	442	460	475	495	505	513	523	533
Incr.		84	107	79	51	38	28	25	23	18	15	15	13	10	18	23
No.		618	610	580	510	375	249	174	138	104	73	51	32	11	3	1
N94, N115, S229, S276 SD Gavins Point L.		140	209	277	337	383	424									
Incr.		140	83	70	59	57	64									
No.		58	43	33	16	9	1									
F100 SD Oahe L.		51	152	290												
Incr.		51	100	137												
No.		2	2	1												
B219 MI Wintergreen L.		112	229	302	343	381	396	409								
Incr.		112	117	76	46	46	23	23								
No.		225	219	124	91	50	23	7								
C270, C356 MN Mississippi R. backwaters		102	244	325												
Incr.		102	142	107												
No.		32	13	3												
S226, S228 SD. Ft. Randall L.		117	261	342	374	417										
Incr.		117	145	90	73	36										
No.		78	68	27	5	1										
Unweighted mean, MI, MN, SD, WI		94	184	255	294	336	364	390	414	443	448	529	504	507	523	533
Mean calculated TL at each annulus																
M344 PA Cowan's Gap L.	26	33	41	99	122	150	185	221	259	290						
N50 NJ north	158	65	152	224	282	330	366	396	419	439	472	505	526	554		

	No.	Mean calculated TL at each annulus (cont.)														
		1	2	3	4	5	6	7	8	9	10	11	12	13	14	15
S187, S188 RI 14 ponds	137	76	163	236	300	363	396	424	460	470						
M344 PA Upper Wood L.	22	74	163	264	348	401										
M344 PA Brody's L.	14	61	193	269	338	366	399	427	490							
M344 PA Reinings L.	38	84	196	274	328	366	434	486	508							
T23 W13 CT	142	130	211	272	328	373	411	445								
N50 NJ south	168	86	185	277	348	399	434	480	513	531	546	549				
N50 NJ	263	89	193	279	330	363	427	439	478	513	533	528	538	554		
G83 MA Micjah P., after reclaiming	9	122	226	295	351											
large year class	17	76	114	137												
G83 MA 3 reclaimed ponds	59	124	257	320												
M292 MA Quabbin L. 1946–52	299	102	234	325	381	417	445	467	478	439						
after smelt added 1953–57	573	97	234	310	351	378										

	No.	Mean calculated TL and increments at each annulus														
		1	2	3												
S372 NY ponds		56	91	249												
Incr.		56	28	69												
No.		20	6	1												
Unweighted mean, CT, MA, NJ, PA, RI, NY		85	177	255	316	355	389	421	451	447	517	527	532	554		

	No.	Mean calculated TL at each annulus														
		1	2	3	4	5	6	7	8	9	10	11	12	13	14	15
B232 MT Three Fork ponds	158	48	97	145	196	251	292	328	351	373	351					
H455 MT Nixon Rapids L.	6	48	130	168												
P159 MT lakes	449	56	130	190	236	272	320	358	378	384	396	455				
B108 MT ponds	22	53	163	262	274											
I7, ID Strike L.	25	152	269	338	406											

Mean calculated TL and increments at each annulus

	No.	1	2	3	4	5	6	7	8	9	10	11	12	13	14	15
S350 UT ponds		76	142	193	264	340	394	429	472							
Incr.		76	69	58	64	64	64	53	43							
No.		396	298	177	68	30	9	6	5							
Unweighted mean, ID, MT, UT	72	72	155	216	229	288	335	372	400	378	373	455				

Mean calculated TL at each annulus

	No.	1	2	3	4	5	6	7	8	9	10	11	12	13	14	15
K80 OR ponds, central	+	35	80	179	249	262	296									
K80 OR South (from Becker)	+	41	124	183	279											
O36 OR (from Becker)	57	66	140	224	284	338	391	409								
K80 OR ponds, west	111	71	157	224	282	336										
K80 OR (from Herrman)	+	99	150	235	269											
O36 OR (from Herrman)	66	97	173	237	282	302										
O36 OR (from Locke)	+	160	234	246	307	378	366	439								
K80 OR (from Oakley)	+	92	185	252	284											
G77 OR 43 waters	+	66	175	254	312	373	420	457	472							
G161 OR 96 waters	934	76	175	259	318	361	401	439	470	498	523	531				
G99 OR	17	79	190	290	335	358	455	488								
K80 OR ponds, south	48	82	185	307	368											
G161 OR Fern Ridge L.	+	81	237	312	361	396										
K80 OR ponds, northeast	12	166	251													

Mean calculated TL and increments at each annulus

	No.	1	2	3	4	5	6	7	8	9	10	11	12	13	14	15
O36 OR Willamette Valley	89	89	180	246	279	297										
Incr.		89	74	43	38	30										
No.		53	35	22	8	1										
O36 OR ponds, west	81	81	183	264	310	353										
Incr.		81	97	61	41	33										
No.		65	47	33	14	4										

Mean calculated TL and increments at each annulus (cont.)

	No.	1	2	3	4	5	6	7	8	9	10	11	12	13	14	15
O36 OR Coast L.		76	185	279	338	371										
Incr.		76	109	89	53	36										
No.		12	12	11	6	3										
Unweighted mean, OR		86	177	249	304	344	388	446	471	498	523	531				

Mean calculated TL at each annulus

	No.	1	2	3	4	5	6	7	8	9	10	11	12	13	14	15
K63 CA Big Sage R.	43	58	117	183	292											
E145 CA Millerton L.	+	112	213	307	373	467	480	488								
S575 CA Potter P.	65	126	254	340												
S575 CA Reuter P.	106	132	279	352	397											
S575 CA Applegate P.	111	152	278	365	450	512										
L227 CA Havasu L. (from Beland)	72	127	264	368	445											
B134 CA Loveland L.	3	178	272	361												
K63 CA Millerton L.	+	160	279													
L227 CA Sutherland L.	6,808	178	312	391	447	495										

Mean calculated TL and increment at each annulus

	No.	1	2	3	4	5	6	7	8	9	10	11	12	13	14	15
J108 NM Elephant Butte		77	139	200	269	322	364	407	446	489						
Incr.		77	62	61	71	58	44	46	71	35						
No.		264	255	253	241	185	86	25	7	1						
T132 CA Folsom L.		152	284	351	399	434	465									
Incr.		152	130	60	33	33	18									
No.		523	181	84	33	6	2									
K64 CA Colorado R.		124	246	353	427	475	526									
Incr.		124	132	119	81	58	76									
No.		279														
Unweighted mean, CA, CO, NM		131	245	325	389	451	459	447	446	489						

	No.	Mean calculated TL at each annulus														
		1	2	3	4	5	6	7	8	9	10	11	12	13	14	15
E30 OH (slow)		58	132	208	254	325	356	406								
L29 IA Red Haw L.	57	95	173	234	288	312	335	433	564							
C284 OH St. Mary's L.	+	89	170	236	279	330	373	411	447	486	500	483				
R40 IN Foot's P.	25	102	183	239	371											
M567 IA Red Haw L.	291	98	170	246	316	391	439	472	484							
R83 OH L. Vesuvius	+	89	178	249	297	356	391	417	457							
R79 OH L. Meander	+	91	190	251	358	394										
B102 OH	+	89	178	254												
M197 IA McBride L.	113	124	175	-	330	432	457	503								
M201 IA Thayer L.	177	127	190	254	330	394	450	508	546							
M320 IA East Osceola L.	40	117	203	257	318	381	424	483								
B150 IA Little Wall L.	15	114	224													
R82 OH L. Alma	+	102	198	257	297	363	419	450	480	503						
E30 OH state average	71	89	178	257	318	368	409	450	480							
R125 IA Ike L.	37	76	178	259	335	417										
G80 IL Horseshoe L.	64	99	179	261	326	375	428	458	496	504	526					
L29 IA East L.	+	110	202	266	309	367	425	452	506							
B373 IL Red Hills L.	274	119	196	267	302											
M567 IA Bob White L.	+	103	193	268	324	374	420	450	478	505						
B373 IL Johnson Sawk Trail L.	25	102	183	277	330											
B39 NB	119	91	193	277	343	401	447	480	503	518	561					
U4 IL Mississippi R.	+	112	218	282	328	361										
B373 IL Ramsey L.	144	109	224	287	348											
T19 IL Sportsman's L.	30	97	206	290	338	378	421	498	495							
M567 IA Green Valley L.	+	118	216	291	346	401	442	464	456	533						
R48 OH		102	203	292	353	406	445	483	508							
L129, L228 IL length in midsummer	5,280	160	229	295	343	401	442	480	503	516	526					
C107 IA ponds	137	119	211	312	338	363										
B143 IL Ridge L.	4,373	206	262	320	348	363	424	442	478							

Mean calculated TL at each annulus (cont.)

	No.	1	2	3	4	5	6	7	8	9
E30 OH fast	+	170	267	333	361	406				
M198 IA Keomah L.	120	173	272	338	391	437	483	504	559	
B86 IL Onized L.	81	86	269	356	419	472				
T16 IL Horseshoe L.	+	216	318	361	399	424	432			
R40 IN Grassy P.	1	107	254	363						

Mean calculated TL and increments at each annulus

		1	2	3	4	5	6	7	8	9
M536 OH ponds		124	196	246	267					
	Incr.	124	55	45	30					
	No.	320	229	180	58					
P204 IA L. Rathbun, 1973		130	204	258	290	329	358	319		
	Incr.	130	72	52	38	39	27	21		
	No.	20	19	15	11	8	7	1		
M246 IA 18 ponds		132	231	310	335	363				
	Incr.	132	112	76	46	41				
	No.	93	31	24	11	2				
M246 IA pond 25		86	168	318	343					
	Incr.	86	81	109	69					
	No.	44	15	5	1					
I10 IA Missouri R. Cutoff lakes		132	241	320	340					
	Incr.	132	104	66	25					
	No.	51	14	5	1					
T139 IA Clear L.		104	241	328	376	405	431	456	480	470
	Incr.	104	134	84	66	25	19	18	15	10
	No.	206	165	121	84	66	34	17	8	3
E60 IL 2 ponds		119								
	Incr.	119	101	76	52	39				
	No.	22	6	3	1	1				
S199 IL Crab Orchard L.		114	241	330	394	439	470	488		
	Incr.	114	133	87	64	52	36	28		
	No.	166	157	138	87	58	25	6		

Mean calculated TL and increments at each annulus (cont.)

	No.	1	2	3	4	5	6	7	8	9	10	11	12	13	14	15
Unweighted mean, IA, IL, IN, NB, OH		114	208	283	334	385	421	457	496	504	528	483				

Mean calculated TL at each annulus

	No.	1	2	3	4	5	6	7	8	9	10	11	12	13	14	15
W27 MO L. of Ozarks	16	58	104	147												
H467 MO Vandalia Res.	195	119	160	226	266	287	361	417								
H467 MO Sterling Price Res.	196	107	196	226	287	310	-	384								
H467 MO Henry Sever Res.	155	102	175	230	277	312										
H467 MO Jo Shelby Res.	230	109	193	234	310	325	-	404								
F163 MO Deer Ridge Res.	+	112	196	234	277	-	417									
H467 MO Ella Ewing Res.	325	102	203	235	310	378										
F163 MO Wakado Res.	+	151	216	251	292	352	523									
W27 MO L. of Ozarks	55	122	196	267												
A98 MO Clearwater Res.	129	89	196	277	348											
A98 MO Wappapello Res.	533	112	206	282	361	434	490	533								
M115 MO L. of Ozarks	+	135	208	284	335	363	421	470	500	508						
P113 MO Statewide	303	109	218	292	335	353										
headwaters	149	102	234	282	323	340										
middle	98	114	221	300	353	386										
lower streams	56	104	196	272	312											
poorest station	+	97	175	272	312	335	378									
best station	+	119	269	325	384	386										
B114 MO ponds	574	94	224	302	376	447										
P73 MO Wappapello L.	100	137	277	338	409	460	498									
A87 AR Bull Shoals L.	116	152	211													
B398 AR Beaver L.	861	152	277	333	396	462	474									
B398 AR Bull Shoals L.	1,075	176	297	377	427	457	492	519	524							
S207 KS Fall River L.	20	193	292													

Mean calculated TL and increments at each annulus

		No.	1	2	3	4	5	6	7	8	9	10	11	12	13	14	15
H456 MO Little Dixie L.																	
1959	Incr.		130	130													
1960–63	Incr.		120	85	66	55											
1964 year of drawdown	Incr.		109	109	89	66											
P71 MO Clearwater L.																	
	Incr.		157	114	64	48											
	No.		1,445	1,263	731	130											
H454 MO Bull Shoals L.																	
1957	Incr.		213	168	102	64	38	25									
	No.		171	44	13	26	14	4									
1958	Incr.		152	76	23	13	18	13									
	No.		45	74	21	20	8	16									
1959	Incr.		165	117	79	28	20	18	13	10							
	No.		15	34	53	19	16	21	9	1							
C202 T92 AR Ft. Smith L.			94	190	269	325	391	498									
	Incr.		94	104	76	64	64	81									
	No.		172	103	63	13	4	2									
Unweighted mean, AR, KS, MO			127	211	268	334	381	463	455	512	508						

Mean calculated TL at each annulus

	No.	1	2	3	4	5	6
C175-6 KY North Fork and Floyd's R.	182	104	188	259	320	394	404
T127 KY Kentucky L.	33	109	213	300	371	437	
S100 TN Hiwassee L.	29	142	259	328	363	386	
E22 TN Norris L.	136	178	262				
E28 TN Chickamauga L	360	175	267				
K71 TN White Oak L.	92	102	236	330	406	429	
slowest		33	102	203	259	384	
fastest		180	384	442	457	432	

	No.	Mean calculated TL at each annulus (cont.)														
		1	2	3	4	5	6	7	8	9	10	11	12	13	14	15
T54 KY streams	6	114	231	338												
T127, F151 TN eastern reservoirs 1954–62	345	119	254	343	411	452	498	569	635							
S266 KY farm ponds	248	147	254	356	399	442	480	526	556							
C373 KY Cumberland L.		170	264	358	470											
T54 KY lakes	49	150	292	368	396	483										
S100 TN Cherokee L.	97	198	297													
S100 TN Douglas L.	8	190	343	371												
S97, C169 TN Norris L.	1,589	175	315	373	409	445	490									
C110 KY Herrington L.	31	163	315	386	419	483	503	528								
T67 TN Center Hill L.	1	127	254	432	457	483	521	533	546	559						

Mean calculated TL and increments at each annulus

	1	2	3	4	5	6	7	8	9
R292 TN Dale Hollow L.	124	227	302	369					
Incr.	124	105	80	60					
No.	77	45	24	10					
B380 TN Woods L.									
before impoundment	147	257	340						
Incr.	147	127	114						
No.	111	20	3						
after impoundment	254	330	419	462					
Incr.	254	173	94	46					
No.	249	146	124	72					
E26 TN Norris L.	178	315	371						
Incr.	178	140	64						
No.	722	370	19						
Unweighted mean: KY, TN	153	268	352	404	446	484	539	579	559

Mean calculated TL at each annulus

	No.	1	2	3	4	5	6	7	8
L157 NC Singletary L.	2	53	124	229	343	381	445	465	480
S192 MD Potomac R.	+	124	193	251	300	351	386		

Mean calculated TL at each annulus (cont.)

	No.	1	2	3	4	5	6	7	8	9	10	11
L157 NC Maccamaw L.	17	117	231	318	386							
R95 NC lakes	142	203	274	323	381	429	483	526	561	584	610	635
M306 MD Snowden P.	90	180	287	343								
H158 DA	+	198	290	345	363	388	411					
R117 VA Claytor L. Direct proportion	48	114	257	345	399	427	452					
R117 VA Claytor L. Fraser correction	48	142	274	356	404	429	452					
R178 VA Back Bay	378	130	274	358	404	452	503	541	554			

Mean calculated TL and increments at each annulus

		No.	1	2	3	4	5	6	7	8	9
M368 NC Glenville L. F = 23			89	127	173	206					
	Incr.		89	38	53	38					
	No.		16	8	5	2					
D68 NC			89	152	211	259	312	366	401		
	Incr.		89	63	59	48	53	43	38		
	No.		195								
M368 NC Nantahala L. F = 20			114	206	269	366					
	Incr.		114	104	69	117					
	No.		27	15	9	1					
R217 NC Rhodhiss L. F = 25			124	206	290	376	450	488	536	559	
	Incr.		124	86	74	79	61	38	36	25	
	No.		79	35	12	7	6	5	4	1	
R217 NC L. Lure F = 23			117	196	292	368	421	486			
	Incr.		117	69	36	68	48	48			
	No.		25	17	9	5	2	1			
R217 NC Hickory L. F = 46			127	231	300	373	399	429	493	516	561
	Incr.		127	104	84	64	53	53	46	23	33
	No.		128	68	28	17	9	5	2	2	1

Mean calculated TL and increments at each annulus (cont.)

	No.	1	2	3	4	5	6	7	8	9	10	11	12	13	14	15
R217 NC Lookout Shoals L.																
F = 43		132	244	312	366	406	439	465	472	490						
Incr.		132	107	74	48	43	28	26	18	8						
No.		70	33	24	16	10	4	4	3	2						
M368 NC Santeetlah L.																
F = 13		127	241	325	394	434	467	495	513	554	592					
Incr.		127	97	69	56	40	33	28	20	23	18					
No.		26	17	5	4	4	4	4	3	2	1					
M368 NC Fontana L.		142	251	325	406	478	498	536	564							
Incr.		142	89	69	69	41	18	38	13							
No.		19	12	5	4	3	2	2	1							
T103 NC Catawba L. F = 13		104	239	328	381	455	516									
Incr.		104	119	84	51	61	30									
No.		53	20	13	9	5	1									
T103 NC Mt. Island L.																
F = 51		147	251	328	388	427	465	472	498							
Incr.		147	102	71	58	46	38	23	26							
No.		30	11	5	3	2	2	1	1							
T103 NC Tillery L. F = 61		124	244	333	411	462	490	511	533							
Incr.		124	89	66	38	36	25	20	23							
T103 NC Blewett Falls L.																
F = 20		119	241	355	401	457	498									
Incr.		119	122	91	56	30	41									
No.		26	15	8	3	1	1									
S395 NC Kerr L.		178	279	348	401	442	455									
Incr.		178	94	81	61	36	30									
No.		65	57	39	26	10	3									
T10e NC Falls L. F = 23		122	241	351	414	455	472	503								
Incr.		122	119	119	58	41	43	31								
No.		11	11	7	5	4	1	1								

Mean calculated TL and increments at each annulus (cont.)

	No.	1	2	3	4	5	6	7	8	9	10	11	12	13	14	15
T103 NC Badin L. F = 38		130	246	361	429	465	500	493								
Incr.		130	122	102	69	36	28	20								
No.		75	24	9	7	7	4	2								
R217 NC James L. F = 64		145	241	396	452											
Incr.		145	89	74	58											
No.		20	11	4	2											
D167 NC Kitty Hawk L.		175	335													
Incr.		175	135													
No.		12	2													
Unweighted mean; DA, MD, NC, VA		133	235	312	374	424	461	489	525	547	601	635				

Mean calculated TL at each annulus

	No.	1	2	3	4	5	6	7	8
J25 OK Clarmore L.	73	94	198	264	335	378	472	526	
J25, L106 OK Illinois R.	60	117	198	264	333	401	503		
J31 OK Ft. Gibson L.	54	117	216	267	333	330			
H150 OK Illinois R.	8	102	190	269	345	391	432		
J33 OK Franklin P.	277	135	206	272					
J38 OK Chickasaw L.	29	157	244	279					
J25 OK Shawnee L.	64	127	203	282	315	356	381	406	424
F64 OK Subprison L.	9	124	193	282	351	404			
restocked	87	117	279	356					
F63 OK Little R.	151	107	216	287	361	424	470	498	
J45 OK Verdigris R.	2	185							
tributaries	59	119	224	297					
J38 OK Rod and Gun Club L.	64	140	229	302	376	417	460		
S391 OK Rock C.	21	137	198	312					
F87 OK Illinois R.	3	117	241	312	388	452	490	513	
tributaries	13	94	206	269	338				
cutoff lakes	135	109	216	292	361	419	465	493	

	No.	Mean calculated TL at each annulus (cont.)														
		1	2	3	4	5	6	7	8	9	10	11	12	13	14	15
J46, H419 OK 152 lakes	4,060	140	246	318	378	434	472	505	531	574						
slowest	+	64	124	170	264	328	356	406	424	531						
fastest	+	284	391	511	554	579	556	579	605	597						
E61 OK Salt C.	16	140	231													
J27 OK Spavinaw L.	62	137	213	323	381	439	518	549								
J28 OK Lower Spavinaw L. (old)	151	140	269	323	394											
Upper Spavinaw L. (new)	144	185	320	386	450											
J37 OK Ardemore City L.	221	137	241	325	396	455	500	551								
K66 OK Hiwassee L.	123	124	226	325	371	457	490	531								
H220 OK Ft. Gibson L. 1948-55	592	145	259	335	411	478	518	559								
J25 OK Grand L.	174	-	251	351	391	455	498									
T44 OK Grand L.	174	160	269	358	411	467	498									
J101 OK Spavinaw L.	94	150	277	356	429	486	523	546	561							
T44 OK Great Salt L.	+	241	307	358	388											
O37 OK Wister L.	+	185	300													
Z9 OK Carl Blackwell L.	1,166	140	279	369	425	462	485	504	531							
E126 OK Rocket Plant L.	35	163	363	384												
J101 OK Eucha L.	457	165	300	386	447	505	531	561	584	612						
J31 OK Grand L.	235	183	320	391	437	478										

Mean calculated TL and increments at each annulus

	No.	1	2	3	4	5	6	7
J26 OK Claremore L.	114	114	216	323	378	442	467	505
Incr.		114	71	61	38	23	25	15
No.		73	35	11	8	4	3	1
H424 OK Rod and Gun Club L.	160	160	229	323				
Incr.		160	66	51				
No.		27	11	2				

Mean calculated TL and increments at each annulus (cont.)

	No.	1	2	3	4	5	6	7	8	9	10	11	12	13	14	15
O37 OK Hegburn L.		130	213	343	421											
Incr.		130	83	81	61											
No.		91	91	72	5											
R201 OK		165	277	361	427	493	498	490								
Incr.		165	122	97	56	51	36	15								
No.		195														
B183, C271 OK Canton L.		122	295	363												
Incr.		122	175	66												
No.		509	172	24												
H303 OK Lawtonka L.		178	287	366	424	470	508	538								
Incr.		178	109	71	56	48	46	33								
No.		45	44	39	29	24	20	13								
Z7 OK L. Carl Blackwell		141	282	371	426	462	485	503								
Incr.		141	134	86	55	38	21	22								
No.		1,124	755	468	352	197	68	12								
H156 OK Canton L. 1951		168	307	376	396											
Incr.		168	142	64	25											
No.		31	24	19	5											
Unweighted mean: OK		142	249	325	386	437	487	516	526	588						

Mean calculated TL at each annulus

	No.	1	2	3	4	5	6	7	8	9	10	11
B39 LA	30	193	287	368	478	531	597	630	658	688	706	719
M539 LA ponds	170	218	318									
M187 GA	17	251	320	381	421							

Mean calculated TL and increments at each annulus

	No.	1	2	3	4	5
C256 TX Windhaven L.		79	155	226	297	318
Incr.		79	86	94	84	91
No.		28	20	15	4	2
C256 TX Dealey's L.		117	188	226		
Incr.		117	79	58		
N.		80	26	5		

Mean calculated TL and increments at each annulus (cont.)

	No.	1	2	3	4	5	6	7	8	9	10	11	12	13	14	15
C256 TX Greenbriar L.		114	180	241	335	394	439	470								
Incr.		114	61	56	79	58	64	43								
No.		51	24	11	6	5	4	2								
P27 AL Silver L., males		93	217	259	308	353	484									
Incr.		93	124	53	90	41	44									
No.		52	52	21	8	3	1									
M539 LA Idlewild L.		218	267	287	351	388	498									
Incr.		218	43	18	46	30	43									
No.		80	61	22	3	2	1									
C256 TX pond 52		178	241	310												
Incr.		178	79	61												
No.		102	53	7												
C256 TX Ferndale L.		155	234	318	391	450	483	528								
Incr.		155	81	69	58	46	38	41								
No.		109	93	58	31	13	7	1								
P27 AL Silver L., females		124	268	345	405	493	559	599								
Incr.		124	134	69	41	57	47	53								
No.		49	49	18	11	8	4	2								
P27 AL Auburn L., males		202	322	408	460	499	543									
Incr.		202	120	69	53	41	30									
No.		89	68	52	32	18	5									
P27 AL Auburn L., females		213	325	408	460	503	541	559	587	602						
Incr.		213	114	85	62	47	37	24	28	16						
No.		97	63	51	36	26	18	10	4	2						
Unweighted mean: AL, GA, LA, TX		166	256	315	391	437	418	557	622	645	706	719				

Mean calculated TL at each annulus

	No.	1	2	3	4	5	6	7	8	9	10	11	12	13	14	15
H163 South Africa	1	81	297	335												
T13 South Africa	20	262	356	386	421											
Mean of unweighted means, North America		118	215	287	341	389	434	463	495	510	528	554	518	530	523	523

Calculated weight at each annulus

	No.	1	2	3	4	5	6	7	8	9	10
R239 NY ponds observed at end of growing season	+	50	122	113							
	+	54	136	200	213						
	+	64	109	181							
	+	100	195	168							
	+	113	318	386							
	+	229									
R125 IA Ike L.	71	4	57	200	463	1,000					
M567 IA Red How L.	291	15	74	219	457	833	1,175	1,411	1,585		
M567 IA Bob White L.	274	14	103	275	515	813	1,152	1,450	1,770	2,125	
M201 IA Thayer L.	177	57	113	284	454	1,276	1,503	1,956	2,495		
M320 IA East Osceola L.	40	43	128	312	499	1,045	1,542	2,449			
L228 IL	+	59	200	340	513	907	1,134	1,361	1,814	2,268	2,495
M567 IA Green Valley L.	30	35	189	447	679	1,069	1,355	1,592	1,475		
H419 OK statewide	+	32	200	454	807	1,261	1,646	2,059	2,418	3,116	
slowest	+	-	23	59	249	504	653	1,016	712	2,413	
fastest	+	304	893	2,127	2,772	3,202	2,812	3,202	3,683	3,538	
M198 IA Keomah L.	120	65	264	539	1,304	1,531	1,956	2,778	3,233		
T139 IA Clear L.	206	18	236	600	916	1,152	1,397	1,656	1,950	1,823	
R178 VA Back Bay	378	20	268	626	934	1,370	1,923	2,427	2,602		
Z8 OK Carl Blackwell L.	+	43	309	787	1,205	1,603					

Largemouth bass stocked at 250-350 mm grew 450-900 g per year for 3 years when first stocked in Glendale Lake, IL (H440). Bass finclipped at 75-280 mm in the fall of 1946 were 250-410 mm in 1948, 280-435 mm in 1949, and 340-460 mm in 1950 (H440). On artificial feeds in Alabama ponds, bass averaging 203 mm and 109 g increased to 240 mm and 195 g in 118 days (S557).

	No. fish	Size of fish	Average growth per day	
W111 MI aquaria	4	117-133 mm	0.2-0.6 mm	feeding study
C399 FL L. Wier 1968	-	18-97 mm	0.36 mm	April - Aug.
	-	-	0.46 mm	July - Jan.
C399 FL L. Wier 1969	-	27-132 mm	0.47 mm	June - Aug.
	-	51-138 mm	0.595 mm	Aug. - Sept.
L. Wier 1970	-	20-147 mm	0.620 mm	June - Aug.
FL L. Hollingsworth	-	33-159 mm	1.03 mm	June - Aug.
	-	65-244 mm	0.659 mm	Aug. - Nov.
	-	96-257 mm	0.062 mm	Nov. - Jan.
K78 MN L. George	-	fry	0.47 mm (0.4-0.6)	from hatching to rising from nest
L73 MI aquaria	-	105-132 mm	0.4, 0.5, 0.6 mm	a feeding study
A93 AR Bull Shoals L.	-	young	0.52 mm	May - June
H86 MI Portage L.	-	young	0.57 mm	July
A93 AR Beaver L.	-	young	1.17 mm	May - June
H150 OK	-	young	1.75 mm	May 26-Oct. 3
Z7 OK Carl Blackwell L.	-	Age I	0.3 mm, 0.7 g	tagged fish, 39 days
C399 FL L. Wier 1969	-	Age I	0.17 mm	Jan. - June
	-	Age I	0.42 mm	June - Nov.
C399 FL L. Wier 1970	-	Age I	0.507 mm	June - Aug.
FL L. Hollingsworth	-	Age I	0.625 mm	June - Sept.
S229 SD Lewis and Clark L.	-	Age I	0.50 mm	summer
S225 SD Ft. Randall L.	-	Age I	1.27 mm	summer
Z7 OK Carl Blackwell L.	1	Age I	1.5 mm, 4.4 g	best growth, 39 days
W111 MI aquaria	1	312 mm	0.0 mm	62 days feeding
C399 FL L. Wier 1970	-	Age II	0.20 mm	Jan. - June
	-	Age II	0.26 mm	June - Aug.
Z7 OK Carl Blackwell L.	-	Age II	0.3 mm, 1.7 g	tagged fish
S563 AL	-	2-6 g	1.2-5.9%	
C305 FL	-	young	up to 10% weight	
W111 MI aquaria	4	19-27 g	0.17-0.56 g	feeding study
W111 MI aquaria	1	464 g.	1.0 g	62 days feeding

The first summer growth of largemouth bass in Louisiana was described by the following formulas, where A = number of days after April 1st (L74):

	No. of fish		Length at end of year
L. Bistineau	163	TL = 3.07 mm + 0.53 A	102-114
Clear L.	613	TL = 19.5 mm + 0.43 A	102-114
L. Providence	298	TL = 30.5 mm + 2.08 A	86
Bayou De Siaid	66	TL = 11.7 mm + 0.71 A	

First summer growth in Par P., SC was described with A = number of days after April 1st (B430):

unheated,	1969	TL = 0.362 mm + 0.0019 A
	1970	TL = 0.228 mm + 0.0026 A
heated,	1969	TL = 0.371 mm + 0.0025 A
	1970	TL = 0.251 mm + 0.0030 A

Daily instantaneous growth rates were reported for young largemouth bass in L. Powell, UT (M554):

0–June	0.064 - 0.183
0–July	0.044 - 0.126
0–Aug.	0.023 - 0.044
0–Sept.	0.12 - 0.020
0–Oct.	0.012 - 0.014

Daily instantaneous growth rates in a feeding study (W111) were:

for 4 bass, 19-27 g	0.0054 - 0.0133
and 1 bass, 464 g	0.0018

Annual instantaneous growth rates for Carl Blackwell L., OK were (Z8):

Age	I	II	III	IV	V	VI-VIII
g	1.97	0.93	0.42	0.28	0.22	0.16

Annual increments expressed as percentages of lengths at beginning of the year. The numbers of fish are given in parentheses.

TL	17 ID Snake R.	B39 WI	R40 IN Foots P.	H156 OK T44	J26 OK Clare-more L.	E26, S97 TN Norris L.	P27 AL lakes
51-75	-	145 (9)	116 (8)	-	-	319 (2)	-
76-101	-	128 (601)	-	-	-	-	134 (51)
102-126	-	-	94 (7)	-	-	144 (51)	129 (31)
127-151	86 (9)	58 (2)	64 (9)	128 (15)	48 (24)	104 (132)	126 (1)
152-177	73 (19)	40 (37)	-	63 (114)	58 (6)	92 (260)	60 (26)
178-202	-	43 (471)	-	61 (31)	43 (3)	75 (393)	64 (24)
203-228	-	38 (72)	-	-	-	-	52 (85)
229-253	38 (3)	24 (38)	30 (1)	49 (5)	27 (4)	-	52 (32)
254-278	27 (9)	19 (452)	-	-	21 (16)	34 (10)	23 (5)
279-304	-	15 (78)	-	17 (40)	21 (1)	30 (70)	26 (26)
305-329	-	12 (276)	18 (1)	19 (14)	3 (4)	24 (138)	20 (15)
330-355	19 (3)	9 (175)	-	13 (31)	19 (3)	13 (22)	21 (62)
356-380	-	8 (171)	-	7 (5)	4 (1)	13 (38)	12 (1)
381-405	-	6 (16)	-	9 (5)	11 (3)	11 (75)	15 (42)
406-431	-	5 (141)	-	10 (6)	7 (2)	8 (42)	13 (27)
432-456	-	4 (124)	-	5 (10)	-	5 (9)	15 (28)
457-482	-	3 (84)	-	4 (6)	4 (2)	7 (9)	9 (19)
483-507	-	3 (64)	-	-	3 (1)	5 (8)	8 (25)
508-532	-	4 (1)	-	-	-	3 (4)	6 (24)

Photographs of bass scales with true and false annuli indicated are given in B42 and B382. Known-age yearling bass, which had grown rapidly the first year but slowly the second, formed false annuli that had all the visible characteristics of true annuli in August or September. The true annulus was formed on these bass in an Illinois lake from mid-April to mid-July with a few not showing enough scale growth to form an annulus by September. Some of the scales showed lateral erosion prior to annulus formation.

Annulus formation has been reported at March and April in CA (S575), in April and May in TN (B380), most in May in OK (Z7, Z9), June to August in OK (C271), late May to late July in MD (M306), first week of May in OH (M536), and late May to June in SD (S225, S226). In Norris Lake, TN, the first annulus formed in May to mid-June, but annuli did not form on 2-year-olds until mid-May to late June, or on older fish until after mid-June (E26, S97). Age II bass in OK did not have the second annulus in late June, but did in mid-July (O37). Annulus formation occurred in winter in a constant-temperature spring in Florida and was probably due to photoperiod changes (C148).

In Norris L., TN, young bass grew most in May and June; older bass, in September; and many did not show any growth until June (S97). In Chickamauga Res., TN, most growth occurred in spring and early summer (E28). Bass showed some growth in the winter in southern Florida (C305) but not in Oklahoma, where shrunken stomachs in the spring indicated lack of feeding (C353). No growth was observed after October in Louisiana (L74).

In one study of known-age bass (age 0-II) in Alabama (P194), 16.2% of 272 bass did not have the right number of annuli even when the scale reader knew what the age was, but 80.1% were correctly aged without knowledge of the age. Good correspondence between the number of annuli and known age (up to age II) was reported in Louisiana (M539). A bass tagged with a disk dangler at age II in Clear L., CA, had the proper number of annuli when recaptured 9 years later (L214). Tag returns on 6 bass in Norris L., TN, verified the scale readings (E26). Separate readings by two persons agreed (E26). In an Iowa study, however, second readings by the same individual agreed on only 64% of 239 bass (T139). Difficulty with the first annulus was reported.

The body-scale relationship of largemouth bass in Fork L., IL, was shown to be a straight line through the zero intercept (B42), and direct proportion calculations have been used in most studies. In many, however, a Fraser-type correction has been used (B114, B183, C270, C271, C373, E60, G83, H456, J28, J108, K63, K80, L227, M306, M368, O36, P73, P122, R117, R217, R292, S188, S372, S395, S575, T103, T132, T139, and Z7). Calculated lengths at the first year agreed closely with observed lengths, indicating that a Fraser-type correction was not needed in Norris L., TN (S97). Scales were not present on 4 bass, 18-22 mm, but were on 2 bass 20-23 mm (S97). A single row of scales was found on a 24 mm bass but scalation was not complete until bass were 36 mm (C373). Scales formed on bass at 26 mm in Oregon (K80) and this was used as the Fraser correction factor. Scales from above the lateral line gave lower calculated lengths than scales from below the lateral line (K80). Scales were formed at 20 mm in California (K62). A direct proportion relationship between body-scale measurements was observed in Indiana for largemouth bass from 23-323 mm, but above that length the line showed a pronounced deflection (R40).

One danger in using corrections from the simple direct proportion formula is illustrated in data from North Carolina lakes (M368, R217, T103). The number of fish in each collection varied from 11-128 with a mean of 45. The intercept values varied from 13 to 64 mm, and these were used in the computation of lengths for the proper lake. The accuracy with which the intercept can be computed is affected by the range of sizes in the sample as well as the number of specimens, and in some of the lakes the bass were mostly of one size. I believe that the differences in calculated lengths at the first year, and to a lesser extent the next year or two, are more a function of the intercept value used than of the actual differences in growth.

Intercept value	Number of lakes	Mean lengths at first year Range	Mean
13	2	104-127	116
20-25	6	89-124	114
38	1	-	130
45-51	3	127-147	135
61-64	2	124-145	135

I would not recommend using a Fraser-type correction for a given population unless considerable confidence can be given to the computed intercept value. Another paper (H442) demonstrates the dangers of computing lengths from scale measurements using only the body-scale relationship as is done in some computer programs, for in this paper some of the average computed lengths are greater than any reported for the fish at the time the scales were collected.

Parabolic body-scale relationships were used in B398, M267, and S350, and it was noted in B398 that there were considerable differences in these body-scale relationships from year to year. In L. Carl Blackwell, OK (Z7), no significant improvement in calculation was evident with a curvilinear rather than rectilinear body-scale relationship.

Lee's phenomenon was reported in only P122 and was reported as not evident by B39. A reverse Lee's phenomenon, with calculated growth increasing as older fish were used, was reported by Z9.

Growth compensation was reported by C373, but B39 reported that bass which grew rapidly the first year retained this advantage and continued good growth, not demonstrating the growth compensation phenomenon.

Cannibals, young bass that begin feeding on fish, including other young bass earlier than others of their brood, grow more rapidly than the others during the first year (B39, C60, H77, S86) and retain the advantage in later years, (H440, B39). Cannibals at age II were 250-400 mm compared to the rest of the bass at 203-230 mm, and at age III they were 290-460 mm compared to 180-255 mm (H440).

Rapid growth in Louisiana, intermediate in Nebraska, and slower average growth in Wisconsin was believed to be related to the length of the growing season (B39). In general growth is slower in the north and more rapid in the south, as shown in the tabulations, but variations in each region are great. Largemouth bass also are longer-lived in the north than in the south. The short life span in Cuba, 5 years, was noted by H441. Bass have been kept 11 years in captivity (F6). Growth was faster in south than north New Jersey (N50) and south than north Wisconsin (B39). However, growth was faster in the Wisconsin-Iowa section of the Mississippi R. than in the Illinois-Iowa section just to the south (U8).

Difference between the growth of offspring of bass from Florida and from Iowa when grown in Florida were believed to be due to temperature relationships rather than genetic differences in growth potential (C305). The bass fingerlings may have reached carrying capacity and stopped growing prior to the end of the growing season. No differences in growth of southern and northern stocks of bass were noted when introduced into California (E145). There was no difference in the growth of M. s. salmoides and M. s. floridanus when grown in Alabama ponds, but the hybrids of these two races grew more rapidly (A100).

Most largemouth bass fry would not feed below 15.9°C, and the growth rate increased with temperature from 17.5°C to 27.5°C (S383). Growth at 30.0°C did not differ from that at 27.5°C. When fry were started on April 8, they averaged on April 29 as follows at the various temperatures (S383):

Temperature (°C)	17.5	20	22.5	25	27.5	30
TL	9.5	12.8	15.2	22.0	25.5	25.0

Crowding was thought to have slowed the growth at 22.5°C somewhat. Optimum growth was reported to occur at temperatures above 26.7°C (F70). Largemouth bass fed at libitum at 20° to 35.5°C grew fastest at 26°-28°C (N140).

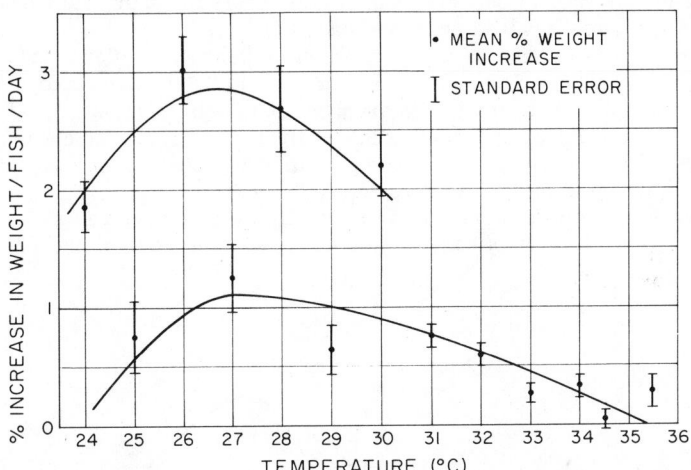

Fig. 6. Growth and food consumption of largemouth bass being fed pellets dropped in mid-September when water temperatures dropped below 22°C (B412). Growth was faster in heated areas (33°-38°C of Par P. than in unheated waters (27°-32°C) (B430). (Redrawn from N140.)

The average lengths of bass at the end of the first year in different years in L. George, MN, were directly related to the seasonal temperatures (K78). Growth of sac fry was related to water temperature (r = 0.885[xx]). Temperature was also associated with growth to August 1st, but thereafter other factors masked any temperature effect (K78). No correlation between annual increments and summer temperatures was evident in Clear L., IA (T139). Growth ceases at below 10°C (M551). Growth began in the spring at 10° to 15.6°C in Ridge L., IL (B142). Bass did not increase in length from November to March in an Alabama pond (E146).

Growth was slower in two coldwater reservoirs than in other North Carolina reservoirs (T104).

Growth was faster in the Illinois R., OK, than in its tributaries (F87).

The dystrophic condition of a lake in Italy was believed to be the cause of slow growth of largemouth bass (D204).

In Iowa, growth of bass was better in artificial lakes with deep basins and protected shorelines than it was in broader unstratified reservoirs, and it was slowest in water-supply reservoirs treated for algae and fluctuating in water levels (M530).

Raising water levels in a Michigan marl lake increased the growth of bass (M531). The increment in the 4th year of life was 76 mm compared to 64 mm before the water rise. Annual increments were lower in Clear L., IA, during years of low water level (T139). Growth was more rapid in Norris L., TN, when water levels raised, flooding new areas (S97). Growth during the 1st year of life and water levels were positively correlated in L. Carl Blackwell; but increments during the 2nd year were negatively correlated with water levels, and those of the 3d and 4th years showed negative but not statistically significant correlations (Z9). It was thought that reduced water levels favored Age I and older bass by concentrating the f rage fish but hindered young because of loss of littoral benthos. High water levels in spring and early summer in Norris Res. favored growth of bass (S97).

Growth rates of bass were higher when oxygen was held near saturation than when held at lower or higher concentrations (S560). Growth dropped rapidly as dissolved oxygen levels dropped, and the bass grew only 60%-75% as rapidly at 4 ppm O_2 as at 8 ppm. Diurnal fluctuation of oxygen concentration also markedly impaired growth. The growth at various oxygen concentrations at $26°C$ is indicated in the following figure from S560.

Hardness of water seemed to be correlated with growth rates of bass in WI (B39). Growth rates of bass were correlated with the total phosphorus

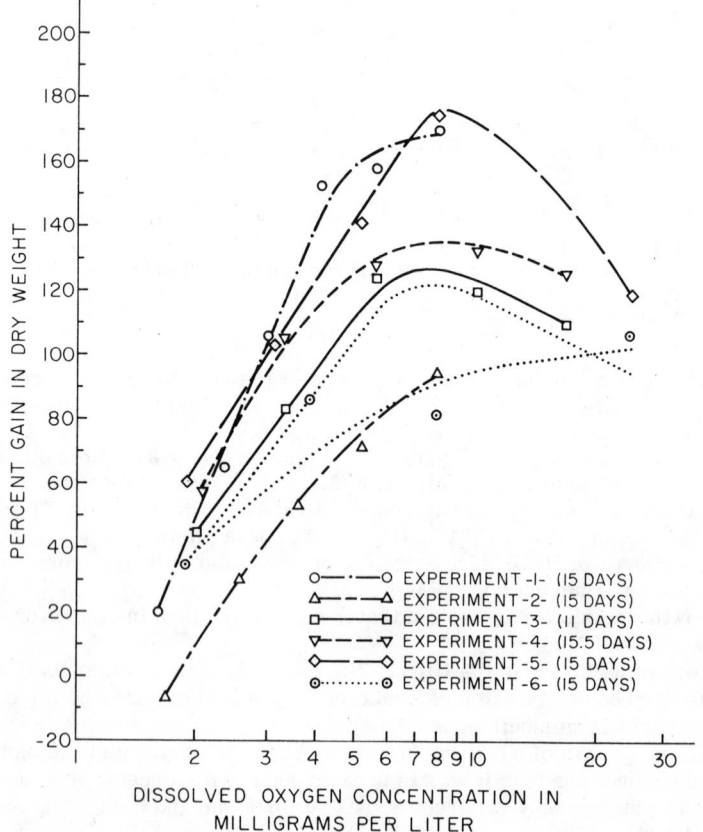

Fig. 7. Growth of largemouth bass in relation to dissolved oxygen concentration. (Redrawn from S560.)

contents of the water in one set of ponds (J107). Growth rates were slower in ponds of low fertility in New York, even though population densities were lower than in fertile ponds (R239). Stunted populations were usually in small infertile lakes in Wisconsin (M551). No relationship between water chemistry or illumination and growth of fingerling bass was detected (K78). Fertilization of Obed R., TN, increased the population and average growth of young bass (P196).

Differences in turbidity did not show a relationship with annual growth of bass in Clear L., IA (T139), nor with growth in North Carolina (T104). In Clear L. the turbidity was more algae than silt or clay. Clay turbidity was associated with slower bass growth in Oklahoma (J46).

Growth was faster in clear than turbid ponds in OK (B181-2).

	Age 0-fall Weight		Age I-fall Weight	
	Mean	Range	Mean	Range
Turbid	74	48-119	122	57-198
Intermediate	173	82-233	218	91-516
Clear	238	150-332	499	207-954

F42 used these data from B181-2 to show that 75% of the good fish faunas listed by E147 were in turbidities less than that reducing the growth of bass. In Kansas ponds growth was more rapid in clear than turbid ponds despite greater population densities in the clear ponds (H332). Turbidity slowed growth rates of bass in experimental ponds in Missouri (G190).

Later year classes rarely grow as rapidly as the 1st bass stocked in ponds or reservoirs (H440, R237, R536, C349, and M306). The 2nd generation did not go beyond 200 mm at age II, while the stocked fish reached 200-279 mm at age I in Kansas ponds (C349). The first large year class, 2 years after stocking, had slow growth (L227, W232).

Growth of young bass was slower the 1st year than the 2nd year of impoundment when more predators kept the numbers of young bass down (C373, E26, O37). In new reservoirs there may be an abundance of vulnerable prey, but the prey are rapidly reduced by the bass (L240). In general, growth was fast after impoundment (B380, C271, H150, H220, K63, S225, D200, W1, J31, J46, L227, and T132). In Francis Case L., SD, growth improved the first 4 years of impoundment, but the number of bass dropped later (G180). In Cox Hollow L., WI, the bass population showed severe stunting by the 8th year (D200). In Norris L., TN, growth declined after the 1st years of impoundment but improved later when water levels were raised, flooding new areas (S97). The slowed growth of largemouth bass in Grand L., OK, was neither thought to be the result of overpopulation of largemouth bass nor of white bass, but of rough fish (J31). Rapid growth in Badin Res., NC, was thought to be related to the high level of intake, which raised the thermocline, thereby limiting the living space and concentrating the forage (T104). An abundant year class produced in the 1st year of impoundment of a California reservoir grew slowly and limited recruitment of all centrarchids for three years (H465).

Growth of young bass in May and June was faster in a new than in an old reservoir, despite the shortage of forage fish in the new reservoir (A93). Entomostraca were abundant, however, and heavily utilized. Faster growth in new than in old reservoirs was also reported by J28.

Growth is usually better at low population densities (R239, C302, B232, H141, and B39). Low population density was associated with disruption of reproduction by dredging (B232). Growth of bass appeared to be related inversely to number stocked (R237, J107), but the differences at age I were not statistically significant in one series of New York ponds (R237). In Pennsylvania ponds growth was faster at 123 kg/ha than at 149 kg/ha (C302). At the latter density there was practically no growth. Growth and condition were negatively related to population density in three Iowa reservoirs (M567).

Growth was slow in ponds with no fish harvest, better at 20% harvest, and best at 40% (G190). At 60% harvest, growth of the few surviving large fish was rapid, but that of the later year classes was poor. Biennial culling resulted in more rapid growth than just supplemental feeding and biennial drawdown (B408). Growth rates of at least the younger age groups of largemouth bass increased following winterkill in Batteese L., but not in Goose L. (B96).

Thinning the population increased the average growth rate (C85, W1, C118, and P122), and growth was more rapid after reclamation (H177, S340, G83, M292, and R201). The growth increase after population thinning was short-lived (P122). The 1st year growth averaged 231 mm compared to 41 mm prior to reclamation (R201). Growth of bass when introduced into new areas has often been very rapid (see South Africa, California data in the tabulations), probably as the result of small population density and minimal competition. Bennett (B382) noted that although the growth rates reported from South Africa are near the maxima reported for largemouth bass, there have as yet been no reports of 3.5-4.5 kg bass, which are not uncommon in United States. The generally slow growth of largemouth bass in Oregon indicates that other factors may mask the increased growth expected after introduction. Even here growth immediately after introduction is probably better than after population densities build up. A 306 mm minimum size limit resulted in slow growth of bass 180-300 mm (R306, F163). A 381 mm minimum size limit also resulted in slow growth (H467). Bluegill growth benefitted, however.

Abundant year classes showed slower growth (G83, E146, J112). Abundance of fingerlings, yearlings, and other fishes did not noticeably affect the growth of young bass in George L., MN (K78).

Largemouth bass in Bull Shoals L., AR, showed slow growth and scale resorption in 1958 when the population of predators (bass and white bass) was highest (H454). Tag returns suggested high mortality of bass that year. Young gizzard shad were scarce that year, also.

Variation in growth the 1st year was greatest when the population density was greatest and when the average growth was slowest (P198).

Bass can grow to a maximum of 136 g in ponds on an insect-plankton diet, but they need larger food for more growth (S561). Slow growth the 2nd year in Fork L., IL, was believed to be the result of scarcity of forage fish (B42).

Average growth of bass is often related to the abundance of vulnerable prey (L240). Rapid growth was associated with abundant young gizzard shad (T104). The growth of young bass was related to the abundance of bluegill fry, and the instantaneous growth rate increased with appearance of bluegill fry in September (E146).

Addition of smelt to Quabbin Res., MA, increased the population but not the average growth rate of largemouth bass (M292). Threadfin shad introduction in Dale Hollow Res., TN, did not noticeably affect bass growth (R292, R308), but in L. Havasu, CA, bass growth increased (K64). In Millerton L., CA, however, introduction of threadfin shad was followed by slower growth of age 0 and age I bass but faster growth of age II (E145).

In reservoirs where gizzard shad are the major forage, the age 0-II bass grow slowly because the shad fry are in deeper water and are too big for the bass by the time the shad enter the water occupied by these bass (S561). The larger bass are rapid growing since they are able to feed on shad throughout the growing season. An inverse relationship was found between the abundance of adult threadfin shad and the 1st year growth of bass, probably because of competition by the shad in the shallow water (V90).

In Oklahoma bass grew faster when with bluegills and with coarse fish than they did alone or with fathead minnows and most slowly with golden shiners (J107). Growth of bass alone was slower than in mixed populations (B373). Growth of bass was faster with golden shiners than with bluegills in New York ponds (E112). The bass also gave better reproduction and higher standing crops with golden shiners, but the golden shiner populations decreased after 2 years (R239). All bass stocked with golden shiners were overcrowded and stunted by 4 years after stocking. Bass fingerlings stocked with bluegills, channel catfish, or java tilapia grew faster than those stocked alone or with golden shiners (B400). Bass alone in water conditioned by bluegills grew better than bass in unconditioned water or in water conditioned by channel catfish, tilapia, or golden shiners. In another experiment tilapia-conditioned water provided better growth of bass than bluegill-conditioned water. Addition of tilapia to a bass-bluegill-catfish population increased the growth of bass, even though the total population density was increased (B400). Young bass stocked with young bullheads grew more slowly than when bass were alone in ponds (R309).

Bass stocked at 50 mm with 30 breeder bluegills averaged 224 mm and 231 mm at 1 year, whereas in none of the ponds where they were stocked with fingerling bluegills did they average over 160 mm (M536).

Growth of bass fry was directly related to the amount of food in their stomachs during the first 2 weeks after rising from the nest. Near the end of the 1st growing season growth was directly related to the ratio of large to small organisms in the stomachs (K78). Bass, at 160 g grew faster to 241 g on live minnows than to 206 g on Oregon moist pellets (C380), but both groups reproduced successfully (C380). Bass grew faster with biennial culling in Ridge L., IL, than with supplemental feeding of bluegills and drawdown (B408).

The growth of bass fingerlings in tanks and fed ground fish flesh was slower than in nature and gave poor survival (S376):

Date	May 8	June 19	July 17	Aug. 20	Sept. 3
Mean weight, g	0.417	1.59	2.414	3.70	4.2

Dieoff of the heavy growth of submerged aquatic plants in August in a small Illinois lake was followed by rapid growth of the bass, which could then readily feed on the smaller fish that had hidden in the plant growth (B41, B382). Bass of 265-280 mm in August increased to 330 mm in October, those 175-205 mm in August averaged 267 mm, and the fingerling bass were nearly 165 mm in October.

Growth of age I, II, and III but not age 0 bass was increased significantly during mid-July drawdown (H456).

Circulation of the water to destroy thermal stratification in Cox Hollow L., WI (W232), failed to increase the average growth of bass, but the harvest increased.

Growth of largemouth bass in Reelfoot L., TN, was not noticeably affected by abolition of commercial fishing for it in 1939 and for all game fish in 1955, but the number of small bass seemed to be up in 1961 (S538, S389).

Average weight of tagged bass decreased over a period of 3 months (W230).

Jawtagged yearling bass grew more slowly than those without tags, but injection of latex dye did not affect growth (V93).

No sex differences in growth were evident in most cases (K78, M306, M551, S97, and S575), but females at most ages were longer than males (P27, H447); and males were larger than females at age I (P198) and perhaps in Norris L. (E22). Females may live longer than males (P27, females IX, males VI; S97). Males may reach sexual maturity at a lower weight and length than females (P198).

In male bass, the scaleless area surrounding the urigenital opening is nearly circular, but in females it is elliptical (M565). At 40 mm ovaries were typically distended, while testes were thin strands (C399).

In Cuba, bass reproduced at 8 months (H441); and in Florida, at 9 months (C305). Both males and females mature at age I in Alabama (S542, S578), Arkansas (B393), Florida (C305, C399), California (L227), North Carolina (P198), and Kentucky (M536); at age II in Florida (C399), Arkansas (B398), Oklahoma (C271, a few at age I), Kansas (T133, occasionally at age I), Illinois (J115), Iowa (M247), South Dakota (S228, S276), New York (E112, R237), New Hampshire (N95, most at age III), and Oregon (G77, O33); at age III in Ontario (S481); and at age IV in South Dakota (W192). In general these figures are for the earliest maturity, and slower-growing fish do not mature until 1 or more years older. Females less than 250 mm did not spawn at age II in L. Weir, Florida (C399).

Although the bass stocked in ponds often spawned at age II, the later year classes with slower growth did not spawn until age III to V in New York (R237) and in Iowa (M247). All males were mature at age II in two Arkansas reservoirs (B398), but only 38% of the females, those over 300 mm, were mature at this age.

Citation		Size of females	No.	Number of eggs per female	
				Mean	Range
S57 NC		-	-	-	2,000-10,000
E4 MN		-	-	-	2,000-26,000
B73		-	-	-	2,000-10,000
T7 TN		-	-	-	4,000-8,000
V21 MN		370 g	1	7,000	-
V21 MN		425 g	1	15,000	-
		1,135 g	2	30,500	29,000-32,000
		1,360-1,420 g	2	33,100	26,200-41,000
O8 OH		-	-	-	11,000-15,000
S577 AL Pellet fed					
Natural Spawn	Age I	160 g	-	9,551	-
	Age II	608 g	-	21,744	-
	Age III	800 g	-	15,223	-
Ova over 1 mm	Age I	160 g	-	45,600	-
	Age II	600	-	39,337	-
	Age III	800	-	32,398	-
Fed forage fish					
natural spawn	Age II	1,040 g	-	19,410	-
ova over 1 mm	Age II	1,040 g	-	36,068	-
B73		1,050 g	1	17,000	-
C77		450-800 g	3	27,930	22,300-31,008

Citation		Size of females	No.	Number of eggs per female	
				Mean	Range
C77		1,130-1,250 g	2	94,157	79,000-109,314
K94 ME	Age III, 295-300 mm,	370-400 g	3	14,080	5,549-22,857
	Age IV, 330-355 mm,	595-740 g	3	17,601	10,420-24,891
	Age V, 390 mm,	965 g	1	13,419	–
	Age VI, 395-440 mm,	850-1,360 g	5	21,751	11,071-31,588
	Age VII, 435-460 mm,	1,245-1,475 g	4	59,164	42,640-81,582
	Age VIII, 465-472 mm,	1,840-1,985 g	2	56,041	47,649-64,434
	Age IX, 462-503 mm,	1,530-2,100 g	2	32,822	26,261-39,383

There are about 4,400 to 15,400 eggs/kg of female (N25) or 2,025 eggs/kg (M553). One age II female, 1,040 g, produced 18,845 fry and still had 38,271 ova when killed (S577).

K94 commented that fecundity declines after age VII and that fecundity was better correlated with age than with length or weight. Ova larger than 0.75 mm in diameter were considered mature (K94). Egg diameters, after water hardening, were 1.4-1.8 mm (M525), 1.49-1.67 mm, with a mean of 1.59 mm (C399), and 1.74 mm ± 0.058 (M529). Better production was secured from age V bass than age I to IV bass even though age I bass were mature in Alabama (S578).

The egg is significantly larger than eggs of *Pomoxis* or *Lepomis*, and this may be a factor in the poor hybrid success with these species (M529). The mean size of the egg increased with the weight of the female (M529):

$$\text{Egg diameter} = 1.51 \text{ mm} + 0.000311 \text{ W}$$

However, this regression was based upon only 3 females.

Spawning has been reported as:

Southern CA, early April to late May (S366)
UT, mid-April to mid-June (M554)
FL, March-April, L. Panasoffkee (M242), April to June (H460), January to May (C399)
SC, March-April (B425)
TX, mid-April to May (S571)
OK, late April, early May (Z7)
IL, May-June (B143), late April-May (J115)
MI, late April-June (D178)
MN, April 19-June 2 (K78), July 22-August 5 (in laboratory in Duluth, C380)
WI, early May to mid-June (M427) late April to early June (J105), late April to early July (M551)
SD, completed by July 1 (W192), June 12-30 (S228)
NH, May-June (N95)
NY, May (B395)
Ontario, late May-early June (S481)

In California, largemouth bass from Florida reached the peak of spawning two weeks before other largemouth bass in the same lake (H341). In southern Florida, spawning started when the water cooled to about 16°C in mid-December to mid-January, peaked in February, and stopped in April or May when water temperatures reached 26.7°C (C361). Bass spawned at 15°C in February in Florida (C399), but most spawned at 15.6°-18.3°C. In marshes with cooler water, spawning may continue into July. Bass were induced to spawn in

December in a Minnesota laboratory by raising the water temperature gradually from 4.5°C to 20°C in 7 days and maintaining it at 20°C (C381). A spawn
15 days later gave 96% hatch. Reproductive cycles in a heated area of a 1,120
ha reservoir in South Carolina receiving thermal effluent did not differ from
those in unheated areas (B425). It was suggested that bass would not spawn in
Florida unless water temperatures dropped to about 16°C (C361).

The first spawning was 2-5 days after the mean water temperature
reached and remained above 15.6°C, and a sharp drop in water temperature
followed by a rise stimulated repeated spawning (K78). Spawning occurred
at 11.5°-29°C (B395), at above 12.8°C (B143), at 12.8°-21.1°C (N95), at 14.4°-
23.9°C (M554), at 15.6°-17.2°C (E4), at 16.7°-17.8°C (M427), at 16.7°-18.3°C
(J105), at 16.7°-18.4°C (M551), at 17.2°-20°C (H342), at 17.8°-21.1°C (S366), at
20°-24°C (S565), at 20.5°-22.5°C (C380), and at 21.1°C (B370). Nest building
occurred at 15.6°C (M427), at 11.7°-13.9°C (M554), and at 15.6°C (M551).
Males prepared nests 1-2 days prior to egg laying (K78).

Males build nests by cleaning areas on substrates such as sand, gravel,
roots, or aquatic vegetation (K95, E145, B370, C40, and M551). Bass will not
nest on silt bottoms (R300). Nests are often near boulders or pilings (H341).
Most nests in Powell L., UT, were on or under sandstone ledges at 0.46-8.2
m, but water levels rose during the spawning season so that the deeper nests
may not have been as deep when built (M554). Nests are usually at depths of
30-76 cm (C361), 1.2-1.8 m (J109), or about 60 cm but range from 15 cm-2 m
(B370). Nests were more often at 1.2 m than 45 cm in a California reservoir
(H341). Nests at 0.9-1.8 m were more successful than those at 25-90 cm in a
Minnesota lake (K95). Wave action destroyed some nests on sandstone rubble
at 0.45-1.07 m (M554). Floating nesting boxes have been successful (S366,
R300). Nylon mats were used for 56%-79% of the bass nests in one study
(C367). Absence of spawning sites in the lower part of Tuckahoe C., MD, was
believed to be the cause of the low population of bass (F148). Bass spawned
successfully in laboratory tanks with better results on Spanish moss than on
nylon mats or gravel (C380).

In George L., MN, 0%-100% of the nests at different periods were successful, and the egg survival to hatch varied from 0%-94% (K95). Wind was a
major factor in nest failure, and water temperatures were related to survival
and to nest success (K95).

When aquatic plants were so abundant that photosynthesis changed the pH
to over 10.2 in the daytime, the large bass were listless and did not spawn.
Weeds were killed on June 4 and bass nested by June 15 (B373). When the
phyllopods, *Apus* spp., are so abundant as to make the water turbid, bass
would not spawn, and stocked fingerling bass were not big enough to eat *Apus*
(H457). Abundant fairy shrimp eliminated the plankton, so that bass fry survival was almost nothing (H457). Reproduction occurred in ponds as turbid as
83 ppm but not at 348-612 ppm (T133). There was no successful reproduction
at 0.5% salinity, or at pH below 5.0, or above 10.0 (S550). Anglers removed
male bass from 34 nests, all of which were then unsuccessful (M554).

Relatively stable water levels at spawning time gave the most abundant
year classes, rising water levels had only a little negative effect, and dropping
water levels resulted in poor year classes (J108, V90). When the water level
dropped 66 mm per day during the spawning season, survival was poor (V90).
Drawdown in the spawning season reduced bass reproduction in Elephant Battle
Res., NM, but stabilized water levels during spawning did not increase the
subsequent catch (J109). Late summer drawdown was demonstrated to be an
effective method of promoting largemouth bass reproduction the next year

(B98, B142, B372, and B382). Severe and prolonged winter drawdown of a NC reservoir was followed by peak reproduction of bass (T104). A large year class of largemouth bass produced in the 1st year of impoundment dominated the population 3 years, limiting survival and recruitment of all centrarchids (H465).

In a California reservoir, strong year classes came when water levels were high because there was less cover in the shallows for small bass when water levels were low (V90). In L. George, MN, year class strength was set before brood dispersal but not at the time of egg deposition (K95). Growth rate, food, predation, cannibalism, and brood stock strength did seem related to year class strength (K95). Introduction of muskies in a lake practically eliminated young bass survival (G152). A negative relationship or no relationship was shown between numbers of spawners and number of bass fry (B372). More young largemouth bass were taken in shoreline seining in a year when bass were stocked in Rathbun L., IA, than in years of no stocking (P204).

Factors limiting bass reproduction were listed as unstable water temperature at spawning time, predation, and depletion of plankton when feeding began (M427).

Low numbers of age I bass follow abnormally high numbers of age I in ponds, and it was suggested that fingerlings be stocked in midsummer when there are large numbers of age I bass (S561).

A dominant bluegill population may control bass reproduction through predation on eggs and fry (B98). When there were 2,800 or more small bluegills per hectare, bass fry survival was low (B372). With high bluegill populations, bass may fail to spawn and then resorb the eggs (S390). An abundant adult threadfin shad population was detrimental to young bass survival (V90). However, the highest numbers of shad were produced the same years as the highest numbers of bass (T104). Reproduction of bass is retarded by excretions from bluegills or from bass when either are abundant (S390). When bass from such habitats were transferred to uncrowded ponds in February and March (in Florida), some spawned (C384). Other bass in the crowded habitat reproduced when injected with human gonadotropin (C384). Some females can be ovulated 2 or 3 times with multiple gonadotropin injections, but the success rate is low (W251). Most females ovulated within 48 hours of injection. Ovulated eggs, if not released and fertilized, became inviable within 12-16 hours of ovulation (W251).

Golden shiners often deposited their eggs in bass nests 1 or 2 days after bass spawned, and nests with over 20% shiner eggs were less successful than the others (K95). Golden shiner eggs were found in 20 of 31 bass nests in L. Wier, FL; and Notropis maculatus eggs, in 5 of 75 nests (C399). Hydra caused some mortality of fry in nests (M554). Embryonic development was described and photographed (C399). At 22°C eggs hatched in 45 hours (C399). Three broods spawned in the laboratory gave hatches of 70%, 80%, and 90% (C380).

Largemouth bass eggs (12-15 hours old) taken from a bass nest in water of 14.4°-17.2°C and chilled to 11.7°, 8.9°, 6.1°, or 3.3°C within 3 hours and then returned to 17°-23°C water hatched normally (J113). Eggs kept at 3.3° to 5.6°C for 3.5 or 14.75 hours also hatched normally—though later than usual—when returned to 17°-23°C water.

In another study, the mortality of embryos was least at 20°C and increased as eggs were transferred to lower or higher temperatures (10°-30°C). The most critical time for temperature shock was just after fertilization, and acclimation reduced the mortality increases (B395). Eggs spawned at

$17.2°-21.1°C$ and acclimated at temperatures from $10°-26.7°C$ gave above 80%, usually 92%-100%, hatch (K150). This was also true at $29.4°C$ in 3 cases, but in the 4th, the hatch was 40.6%. Without acclimation, results were comparable at $12.8°-23.9°C$ but were lower at $10°$, $26.7°$, and $29.4°C$. In lakes in Maine, however, almost complete loss of bass nests was noted when temperatures fluctuated between $12.2°$ and $21.1°C$ (K150). Observations indicated that the male bass all deserted, and the mortality may have resulted from lack of aeration rather than temperature shock (K150). The more crowded central eggs died and fungused, and then the fungus spread. When the temperature dropped below $10°C$, there was complete loss of eggs in nests (K95). Pond snails, *Viviparus georgianus*, ate large numbers of bass embryos in laboratory experiments and probably in natural nests (E151).

Bass eggs deposited on nylon netting were transferred to Heath Vertical Incubators and gave 57.6% survival to swim-up, despite 22% infertile (S578).

Some embryos hatched at 1.0, 1.1, and 1.3 mg O_2/l, but survival rates dropped sharply at oxygen levels below 2.0, 2.0, and 2.8 mg/l at $15°$, $20°$, and $25°C$, respectively (D205). The time of hatching was the most critical for low dissolved oxygen. At $20°C$, a 90% hatch occurred at 64 hours at 70% and 100% saturated O_2 (5.8-9.0 mg/l), but it occurred at 59, 58, and 57 hours, respectively at 50%, 35%, and 20% saturation (C383). At $23°C$ hatching occurred at 31 to 47 hours. Initial feeding was delayed at both temperatures as O_2 saturation decreased. Survival to 20 days was 87%-90% at O_2 saturations of 100%, 70%, and 50% at $20°C$ but dropped to 67%-84% at 50% saturation and 0%-0.9% at 20% saturation (C383). Mean length at 20 days also decreased with O_2 decrease. Field studies showed no correlation between survival and oxygen concentrations ranging from 5.8 to 12.8 mg/l (K95).

Embryos developed and hatched at oxygen concentrations as low as 1.0, 1.1, and 1.3 mg/liter at $15°$, $20°$, and $25°C$, respectively; but concentrations below 2.0, 2.1, and 2.8 mg/liter significantly lowered survival. Most mortality occurred at hatching. If containers moved up and down to stir the eggs, embryos survived better at oxygen levels below 1 mg/liter than unmoved eggs, but there was nearly complete mortality at hatching. Bass eggs hatched in 20% but not in 30% sea water, and fry did not survive in more than 10% sea water (T141). The 96-hour median tolerance limits for fingerlings 12-16 mm and 23-27 mm were 31% and 35% seawater, respectively (T141).

The first hatch was when the water was at $21°C$, 15 cm from the surface (S334), or at $21.1°-23.3°C$ (L189).

The incubation period was 4 days, at $14.4°-23.9°C$, with 80.4%-92.2% survival. The daily instantaneous embryo mortality rates were 0.037-0.014 (M554) at $18.3°C$. The first eggs hatched at 51.2 hours and the last at 62.5 hours with a median of 55 hours (M544). At $23°C$ hatching occurred in 4 days (C380).

The incubation time in NY experiments (B395) was:

Temp. °C	10	12.5	15	17.5	20	22.5	27.5	30
Days	13.2	9.8	6.8	4.8	2.87	2.87	2.04	1.5
	13.2	-	6.8	4.0	-	-	-	-

A direct correlation (r = 0.91) was found between water temperature and the number of fry the third day after spawning (K95).

Incubation took 3-4 days, and sac fry remained in the nest 5-8 days after hatching (K95, K78). Hatching was reported as requiring 16-21 days in one Wisconsin study (M427) and averaging 7 days at the Burlington, WI, hatchery (M551).

At 19°C hatching occurred at about 80 hours, the mouth formed at 192 hours, the fry were free swimming at 240 hours, and the yolk sac was absorbed at 312 hours (L241). A slight energy deficit occurred about 288 hours after fertilization, but by this time over 50% were feeding and thus showed no deficit. Some individuals never started feeding. At 21°-22°C, hatching was completed in 48 hours, and all fry were free swimming with yolk sac absorbed by 168 hours (T141). Hatching occurred in 2-3 days at 23°-25°C (C40), with free swimming fry about 5 days later.

The fry at hatching were 5.4 mm (M525) or 3 mm (K78). They became free swimming larvae at 6.1-6.3 mm (M525) or at an average of 6.16 mm (5.92-6.31 mm) (K78). All fin rays were developed by 9.7-9.9 mm, and the dark lateral band developed at 9.7 mm (M525). At dispersal from the brood, the fingerlings averaged 32.5 mm (K78). In a North Carolina study, 5%-50% of the fry had caudal curvature (M544).

There were about 5,000 fingerlings per brood in 1956 and 1957 but only 3,600 in 1958 in George L., MN (K95). Brood size ranged from 500 to 12,715 (K95) or 715-11,457 with a mean of 4,375 (C6), or 3,000 (C115).

Survival rates as percentages for periods other than annual were reported as follows:

	Period	Mean	Range
S558 AL 3 ponds	90 days	91	74-98
L238 IL ponds	47 days	-	93-95
	68 days	-	81-95
S559 AL ponds	114 days	81.5	69-98
R239 NY	2-3 years	83	-
	next 2.7 years	94	-
R237 NY in ponds with bluegills	summer	80.3	60-95
S244 AL new ponds, fingerlings	to fall	75	-
new ponds, advanced fry	to fall	-	82-90
B372 IL Ridge L., 0.07-0.34 kg bass	2-3 years	67	-
0.35-1.13 kg bass	2-3 years	86.6	-
1.14-1.81 kg bass	2-3 years	85.4	-
1.82-2.27 kg bass	2-3 years	61.0	-
2.28-2.72 kg bass	2-3 years	27.1	-
S176 AL 67 ponds with bluegills	6 months after 1" fingerlings	66.5	19-100
29 ponds with bluegills	18 months	54.3	24-92
7 ponds with bluegills		76	60-92
2 ponds with only bass		76	-
B109 TX stocked as fingerlings, 14 ponds	18 months	65.4	47-85
B110 TX stocked as fingerlings, 6 ponds	18 months	64.7	47-83
B111 TX stocked of fry, 3 ponds	14 months	77.6	72-85
stocked of fry, 3 ponds	15 months	64.9	64-66
S176 AL stocked as fingerlings, 9 ponds	5 months	76.0	60-92
stocked as fingerlings, 67 ponds	6 months	66.5	19-100
stocked as fingerlings, 29 ponds	18 months	54.3	24-92
B373 IL finclipped bass	to fall	60	56-67
S557 AL 10 ponds	90 days to Oct.	58	39-87
same another year	90 days to Oct.	-	69-93
E146 AL ponds stocked with 150-330 mm bass with 45% caught by anglers	42 months	49	-
ponds	to Age I	-	31-55

	Period	Mean	Range
E 146 AL ponds, fish age I and up	Oct. - April	62	-
J 107 ON ponds	to Age I	-	55-65
	to Age II	42	-
when stocked with coarse fish	to Age I	25	-
	to Age II	18	-
M306 MD Snowden P., stocking	to Age II	40	-
P198 NC ponds	to Age I	-	1-34.5

Little correlation was found between survival rates of bass and bluegills in New York ponds (R237). Overwinter mortality was relatively high, so that the standing crop was lower in spring than fall (B373). In Ridge L., IL, 59% of the marked fish were caught by anglers, 18% were recovered when the lake was drained, and 23% disappeared (B372).

Annual mortality rates (percentages which die per year), a, expectation of death due to fishing or rate of exploitation, u, and expectation of death due to natural causes, v, have been reported as follows:

	a	u	v
R237 NY ponds, no fishing	0.195	-	-
M267 WI Brown's L.	0.24	0.12	-
S175 AL ponds after 1st year	0.30		
G110 MN Francis L.	0.296	-	-
M567 IA Bobwhite L.	0.32		
B143 IL Ridge L.	0.35-0.40	0.25-0.30	0.05-0.11
M567 IA Red Haw L.	0.39		
C302 PA Breon's P.	-	-	0.003-0.117
K148 GA Altoona L., tagged	-	0.17	-
B92 MI ponds	0.40		
Z7 OK Carl Blackwell L.	0.40; 0.47	-	-
Z9 OK Carl Blackwell L. (increased with age)	0.54		
K149, R280 CA Clear L.	0.56	0.20	0.36
M359, R280 MN Gladstone L.	0.60	0.15	0.45
B410 IN Bischoff Res., Age II and older	0.68	0.39	0.29
C214, R280 MI Sugarloaf L.	0.70	0.26	0.44
B410 IN Mollenkramer Res., Age II and older	0.69	0.47	0.22
C374 AR Fort Smith L. (virtual population)	-	0.32	0.38
L227, R280 CA Sutherland Res.	0.70	0.36	0.34
G85 IN Oliver L.	0.74	0.17	0.57
IN Gordy L.	0.60	0.36	0.24
R32 IN Shoe L.	-	0.20	-
H441 Cuba	0.76; 0.89	-	-
R280 CA Folsom L.	0.89	0.40	0.49
C168, R280 TVA reservoirs	-	0.41, 0.42	-
E23, M105 TN Norris Res.	-	0.184, 0.185	-
M306 MD Snowden P., tagged	-	0.57	-
R304 CA Merle Collins Res. 1965	0.92	0.36	0.56
1966	0.71	0.45	0.26
1967	0.86	0.62	0.24
1968	0.76	0.65	0.11
1969	0.86	0.65	0.21

The rate of exploitation decreased with age (Z7). Only 2 of 692 jaw-tagged bass were later caught by anglers (F160), although 41.7% of the year class taken by shocking were marked. Of bass caught and released after weighing in at tournaments, 30% showed immediate or delayed mortality (M570), or 22.3% to 43.8% (M572), or 32% (S596). Injection of bass with 25 mg Terramycin per pound of weight did not significantly reduce delayed mortality (S596).

In San Vincente L., CA, 52.9% of the tagged fish were caught in 6 months, and 51.7% of another tagging were caught in 3 months with an additional 6% in the next 8 months (K149).

Home ranges were evident in a 3.4 ha pond (L239), a 6.1 ha lake and a 3.2 ha pond (H433), and a 16 ha lake (F97). Most bass seemed to home and have territories in Carl Blackwell L., but some roamed quite a bit (Z7). One travelled 217 m/day for 5 days between capture. In the St. John's R., FL, tagging indicated that there was a resident and a transient population (M304). Large bass moved more than small ones, and one traveled 198 km (M304). Average movements of tagged bass in California lakes were 1.1 km, 1.9 km, and 7.2 km (E145). When 2 bass were placed in aquaria, the heavier fish usually became dominant (M374).

Turbidity at 4-6 and 14-16 Jackson Turbidity Units decreased the activity of bass compared to controls, but resulted in more scraping and "coughing" action (H434). Turbidity was not considered a major limiting factor of bass in Louisiana waters, but largemouths were not found in water of more than 4.1 ppt salinity (C368).

The LC50 values for 24-hour tests of largemouth bass (13-102 mm) were at pH 3.9 for hydrochloric acid, 4.2 with acetic acid, 10.3 with sodium hydroxide, and 10.5 with calcium hydroxide. Larger bass (178-279 mm) took a pH of 11.0 with calcium hydroxide. The small bass were less tolerant of pH change than were bluegills (C363). Bass (90-152 mm) tolerated rapid change of pH from 8.1 to 6.0, from 7.2 to 9.3, from 9.2 to 6.1, and from 6.1 to 9.5 (W237), even though in some cases the dissolved oxygen was less than 4 ppm. Bass (38-82 mm) tolerated fairly rapid changes from pH 8.0 to 9.35 and from 8.6 to 6.0 when the dissolved oxygen was 7 ppm (W237). Fry were less resistant to high pH values than smallmouth bass fry (W237).

Differences in temperature tolerances of largemouth bass from Florida, Tennessee, and Put-in-Bay, Ohio, were found (H141).

	Acclimated temperatures (°C)	Lethal temperatures (°C)	
		Upper	Lower
H255	10	28	-
H141 Florida stock	20	31.8	5.2
Ohio stock	20	32.5	5.5
B200	20-21.8	28.9	-
H141 Florida stock	25	32.7	7.0
Ohio stock	25	34.5	-
H255	30	35	-
H141 Florida stock	30	33.7	10.5
Tennessee stock	30	36.4	-
Ohio stock	30	36.4	11.8

The LD50 temperature of bass was reported as 36.7°-37.2°, 37.2°, and 37.8°-38.9°C (T102). Bass in 35.6°C water had body temperatures of 35.6°C.

Fingerling bass, 47.8 ± 5 mm and 1.13 ± 0.4 g, when transferred from 30°C and from 24°C to 17°C water, showed higher mortality to adult bass predation than did bass transferred from 20°C or bass kept at 17°C (C382).

Adult largemouth bass with ultrasonic temperature-sensing tags generally selected the warmest temperatures when temperatures were below 25°C but remained at 26° to 28°C when higher temperatures were available (N140). In the winter, however, most bass from an area with heated effluent had body temperatures of 16° to 21°C, and only a few were measured at 26° to 27°C, suggesting that current and temperature may both be important (B409). With higher metabolism in the winter, a few bass became emaciated.

Largemouth bass preferred water temperatures of 26°-32°C in L. Monona and 28°-31°C in laboratory tests (N141).

Bass, when acclimated at 25°C, avoided water at temperatures of 30.6°-32.8°C (M545). The final preferred temperature of largemouth bass was 30°-32°C (F161) or 28°-32°C (N141). However, in L. Opinican, ON, largemouth bass remained quiet in the shade at temperatures of 28°-30°C in August, not feeding even though their stomachs were empty (J83, D206). Respiration seemed to be the problem, since a 5-fold increase in opercular pulse rate was required at 32°-35°C compared to 5°-10°C. At 5°-10°C, an average pulse rate of 16 per minute was adequate (D206). In Norris Res., TN, largemouth bass preferred 26.6°-27.7°C (D207). In the winter in Par P., SC, more largemouth bass were seen and caught per given effort in 27°C water than in 16.5°C or 11.1°C (G188).

Bass (13-155 mm) were maintained under atmospheres of pure oxygen, with greater than normal atmospheric pressure and dissolved oxygen values up to 40 ppm with no distress (W237).

The minimal oxygen concentrations which bass could stand in a sudden change were 0.92, 1.19, and 1.40 ppm at 25°, 30°, and 35°C, but when acclimated, these values were 0.78-0.87, 0.79-0.87 and 1.20-1.32 ppm (M351). Young bass showed stress at 5 ppm O_2 at 25°C, with average opercular movements of 130 per minute compared to 88 per minute at 6 ppm O_2 (P202) general activity rate was down. At 2 ppm O_2 the bass did not eat, although they did at 3 ppm. After 3 hours at 2 ppm the bass became bleached, and after 10 hours the fish continuously respired with open mouths. None of the fish died in 24 hours, and all resumed normal color and behavior when returned to 6 ppm O_2. At 1.5 ppm the bass bleached in one hour, but none died in 24 hours. At 1 ppm O_2 and 25.2°-25.5°C all bass died in less than 11 hours.

At temperatures of 17°-22.5°C in summer bass showed no avoidance of water at 6 ppm O_2, some avoidance at 3-4.5 ppm, and strong avoidance to water with only 1.5 ppm O_2 (W226).

Standard and active oxygen consumption increased with temperature from 5°-29°C (J83). Scope for activity, the difference between standard and active oxygen consumption, was low at 5°C, and the bass at this temperature were sluggish and had difficulty maintaining equalibrium. The scope for activity increased with temperature to 17°C but differed little at 17°, 27°, and 29°C (J83). Active oxygen consumption increased linearly with weight on a logarithmic grid, and at each temperature, 10°-34°C, the regression slope was 0.8-0.99 (B396). The logarithm of oxygen consumption for a 150 g bass increased linearly with log temperature from 10°-30°C but decreased from 30°-34°C. Over the linear phase the Q_{10} was 1.65. The Q_{10} for standard oxygen consumption was 1.32. Log oxygen consumption increased linearly with sustained swimming speeds. At 5°C, bass did not swim long enough to determine oxygen consumption at any speed (B396).

Metabolic rates decreased with increase in weights to 15 g but did not change with increasing size beyond this point (M351).

Respiration rates were used to test effects of copper, cadmium, phenol, ammonia, and cyanide upon largemouth bass (M568).

The following came from a study in Ontario (J83) on 83-97 mm largemouth bass:

Temperature (°C)	Oxygen uptake, cc/kg/hr., standard	active	Cruising speed, m/minute	Food, minnow/bass/week
5	38	48	5.8	0
12	130	198	13.4	1.6
17	134	250	14.6	2.3
22	185	303	25.0	4.1
29	320	450	24.4	4.3

Oxygen consumed by 11.8 g bass averaged 0.7 mg/g/hr, both when isolated and when in groups (P201).

Maximum sustained swimming speeds increased linearly with length or log weight of the bass (B396). At 150 mm TL, maximum swimming speeds were about 24 cm/sec at 10°C and 48 cm/sec at 30°C. At 250 mm they were about 46 cm and 63 cm respectively (B396). Final swimming speeds of 75-85 mm bass were reduced at oxygen concentrations below 5-6 ppm at 25°C, but no effect of oxygen concentration was noted above that level (D203). Concentrations of carbon dioxide up to 48 ppm had no noticeable effect on swimming speeds. Mean swimming speeds were 37-41 cm/sec or 4.6-5.0 lengths/sec at favorable oxygen concentrations but only 20-30 cm/sec when O_2 was below 2 ppm (D203).

Largemouth bass, 52-64 mm TL, collected from 27°-31°C lake water and immediately transferred to 15°, 20°, 25°, 30°, and 35°C water, showed reduced swimming speeds at temperatures other than 30°C when tested 14°-20° hours later (H464).

Temperature (°C)	Critical swimming speed m/min.	lengths/sec.
15	18.29	5.18
20	21.40	5.59
25	23.96	7.06
30	29.81	8.08
35	26.15	7.62

Where X = temperature (°C):

$$m/min = -2.0227 + 1.6717X - 0.0236 X^2$$
$$lengths/sec = 0.8638 + 0.3307X = 0.0037X^2$$

Larval largemouth bass fed during the first week of swimming improved their swimming speed to a sustained velocity of 4.0 cm/sec, while starved larvae attained a velocity of only 1.5 cm/sec (L246).

In Altoona L., GA, most of the largemouth bass were at 0.9-1.2 m, but about 10% were taken at depths over 6 m (K148). Bass moved close to shore at night and rested on the bottom in shallow water (W238).

Fingerling bass start feeding on microcrustaceans (M523, R299, M148, R212, C402, S372, K95, A87, and E145). Insects and cladocera were the food of bass from 30-50 mm (K129) or from 22-75 mm (M310). Insects become the major food by the time the bass reach 40 mm (M523, R212, M554, M148, and K63) or 50 mm (K129, K95, A87) or 70 mm (M558). Young fathead minnows were eaten by 10-22 mm bass (T146). In Florida, cladocera were the major

major food of bass 13-30 mm; amphipods, insects, and decopods, of bass 31-75 mm; and fish, and decapods, of bass 76-296 mm (C399).

At 50 mm, when bass discontinue feeding on entomostraca, 10-12 gill-rakers are on the first arch, but by 110-160 mm, there are only 8-10 gill-rakers (K153).

In a new reservoir, the average volume of food in young largemouth bass was greater than in an older reservoir. In the new reservoir, they fed primarily on entomostracans, but in the older reservoir they shifted to midge larvae at 36 mm and to gizzard shad at 40 mm (A93). In hatchery ponds in Alabama, 5-10 mm bass fed on copepods, cladocerans, and a few rotifers, and shifted to midge larvae thereafter. Since there were no fish or larger insects available, the food items did not increase in size after the bass reached 30 mm (R299). Rotifers were eaten by bass 6-20 mm (K78). Bass did not begin feeding until 8 mm, and small fish were eaten when bass were only 20-29 mm (M554). Bluegill fry became the major food of bass at 66 mm in 1956 in Clear L., CA, but not until 81 mm in years when the bluegills were not as abundant (M148). Although a few insects were eaten by 6-9 mm bass, insects were the most important food items for bass 40-100 mm (K78). Fish started to enter the diet when bass were about 20 mm (K95), 25 mm (K78), or 44 mm (L86). Chironomids were the major insects in some studies (N117, D161), and scuds (*Hyalella*) were fairly common (E121, K78, R212, and M336). In South Dakota reservoirs, ants, corixids, and grasshoppers were the main insects taken (N93, N94). Dragonfly, damselfly, and mayfly immatures are the major insects eaten by the larger bass (K129, S481).

Bass over 80 mm (K129, K125) and over 100 mm (M523), feed mainly on fish (K63) or fish and large insects (S481, S357, B233, M422, K74, and M524). The principal foods of adult bass were small centrarchids (B368, F91, D103, and S372), centrarchids and crayfish (H456, P205, S350), perch ("growth was good on this forage" S464), crayfish and fish (S564, S556, S366, and L140), gizzard shad (L245), gizzard shad and crayfish (A99, D209, Z7), gizzard shad and crappies (M310), gizzard shad and yellow bass (K73), and fish (B406, G97, D161, D163, R212, and M340).

Young bass in ponds with crayfish 5.1-5.8 g ate more crayfish and fewer small bass than in ponds with crayfish 7.2-8.3 g (R309).

Bass in ponds without other fish relied on crayfish, frogs, large insects, and young bass (J107). They greatly reduced the crayfish population when stocked in a pond (T140). Crayfish of the year were eliminated before July in Fork L., IL, and when forage fish were scarce, bass and bluegills competed for insects, with bass taking more dragonfly and damselfly immatures and more flying insects (B42, B41). Small rainbow trout were eaten in numbers by largemouth bass shortly after the trout were stocked in an Ontario lake (S481) and in Oregon (B233).

Two pound bass ate 180 mm channel catfish stocked in ponds but did not reduce survival of catfish over 200 mm (K154). When rainbow trout, mostly over 230 mm TL were stocked in L. Owachita, AK, 48.6% of 35 bass over 400 mm had trout in their stomachs when collected in January and February and 10 of 11 over 456 mm had eaten trout, but none of 51 under 400 mm had (K155). In Cub L., MI, where forage fishes were not abundant and growth of bass was slow, bass did not eat bass of their own age group, but yearlings did eat some young of the year (C402). Adult bass took small perch and practically no small bass.

In January to May, rising water levels in Beaver Res., AR, brought in many terrestrial foods on which bass fed (M523, M524).

In brackish water, bass 245-525 mm ate blue crabs, white shrimp, and a few fish and insects (L137, L140, D56). Less usual items in bass stomachs include a mouse (D163) and a shrew (C271).

Largemouth bass would not voluntarily eat minnows at temperatures below 10°C (M557), and smaller individuals fed in colder water than the larger ones. They did not readily take food if the stomach was not empty (M557). In Ontario, largemouth bass are sometimes caught by angling through the ice, but the stomachs are usually shrunken with little evidence of winter feeding (K137). In Wisconsin, largemouth bass feed little between October and May (M427). In Arizona, bass ate sparingly in the winter, mostly on aquatic insects in contrast to small centrarchids in the rest of the year (B368). In Beaver Res., AR, crayfish, frogs, and salamanders were eaten in the winter (M523). In Bull Shoals Res., AR, crayfish and yearling shad were the major foods. In Canton Res., OK, only 13% of the bass taken from September to April had food in the stomach compared to 38-50% in the summer; and the winter bass had taken only shad, whereas the summer food was varied (C271). About 50% of the stomachs were empty each month from April through October, although average water temperatures varied from 16° to 30°C (L245). Bass in South Carolina fed in the winter both in areas receiving heated effluent (12.2°-34.4°C) and in other areas of the reservoir (9°-14°C) (B406).

Feeding was reduced during the spawning season (Z7). Even in experimental tanks bass quit feeding about two weeks in the spring as gonads matured (L154).

Bass fed from mid-morning through the afternoon but little at night (Z7). Most empty stomachs were found from 2-8 A.M. (D209, M427). Adult largemouth bass were seen resting inactively on or just above the bottom at night (N141).

A marked increase in the use of forage fish when these are drugged (559), as with rotenone, indicates that low vulnerability keeps bass from eating all they would (L240). Gizzard and threadfin shad seem to be quite vulnerable, but they are not eliminated by bass predation (L240).

Drawdown of water levels in reservoirs increases vulnerability of prey (B411, L240, H456, and S564). Daily food intake of 5.6 g per 454 g of bass in May decreased to 4.0 g in early July but increased to 10.7 g in mid-July when vegetative cover was reduced by drawdown (H456).

Bass consumed their body weight in food in 15 days compared to 36 days for gar (H244). Bass ate 2.2% of their body weight per day when fed forage fish (L154).

The percentage of body weights eaten per day by bass was:

4.5 g bass	at 10°C	2.4-5.8	at 20°C	17.4	H288
9 g bass	at 10°C	3.4-4.6	at 20°C	11.1-14.0	H288

Bass usually ate a single gizzard shad per meal, which comprised the following average percentage of the bass weight (L245):

Weight of bass (g)	No. of bass	% food	Weight of bass (g)	No. of bass	% food
90-450	116	9.2	1,351-1,800	15	3.2
451-900	83	7.6	1,800+	10	2.1
901-1,350	42	3.7			

The average size of the shad eaten increased with the size of the bass but not proportionately (L245).

Food consumption and food conversion were better when dissolved oxygen was near saturation than either below or above, and drops in dissolved oxygen markedly affected both (S560). Dry feeds were accepted better as water temperatures increased and were better above 27°C (N139).

The curves showing the effects of dissolved oxygen on food consumption and food conversion at 26°C are given from S560:

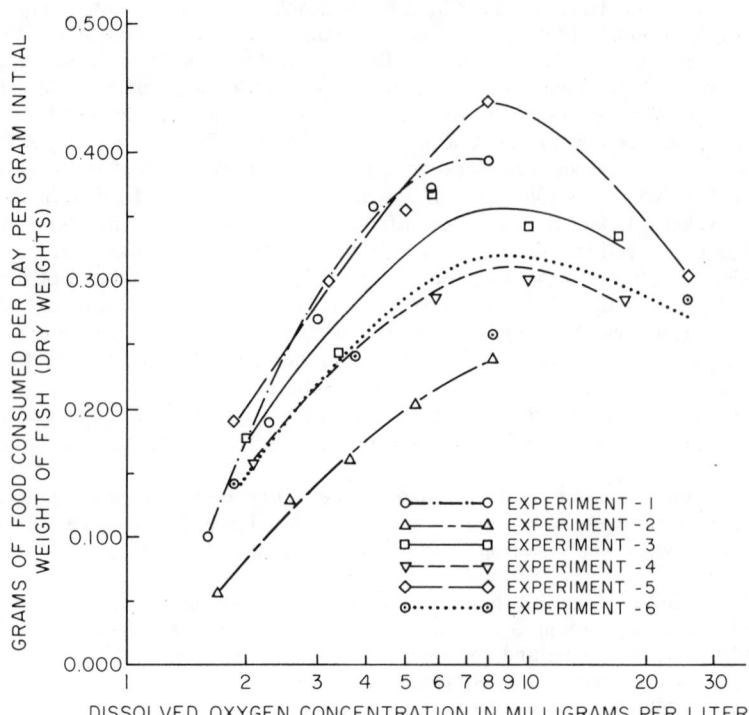

Fig. 8. Food consumption rate in relation to dissolved oxygen concentration. (Redrawn from S560.)

Bass stopped feeding when dissolved O_2 approached 1 ppm (S557). Food conversion rates (S562) with fathead minnows were 3.2, 4.4, 4.9

with goldfish,	4.3, 5.6
with bluegill,	3.7, 4.9, 5.2
with pellets, in troughs,	2.6
with pellets, in concrete tanks,	3.3

Survival in the tanks was 39% compared to 69% in the troughs, but the tanks would care for more fish (S562).

Food conversion (L73, W111) of 105-132 mm bass was 2.87, 3.92, 5.57

19-27 g bass,	2.1-5.7
464 g bass,	6.6

Digestion rates (when fed bluntnose minnows) were as follows (M557):

Temperature (°C)	Weight of bass (g)	Hrs. for digestion	g digested/hr./100 g bass
4	40-54	360	0.021-0.028
4	170-200	360	0.007-0.008

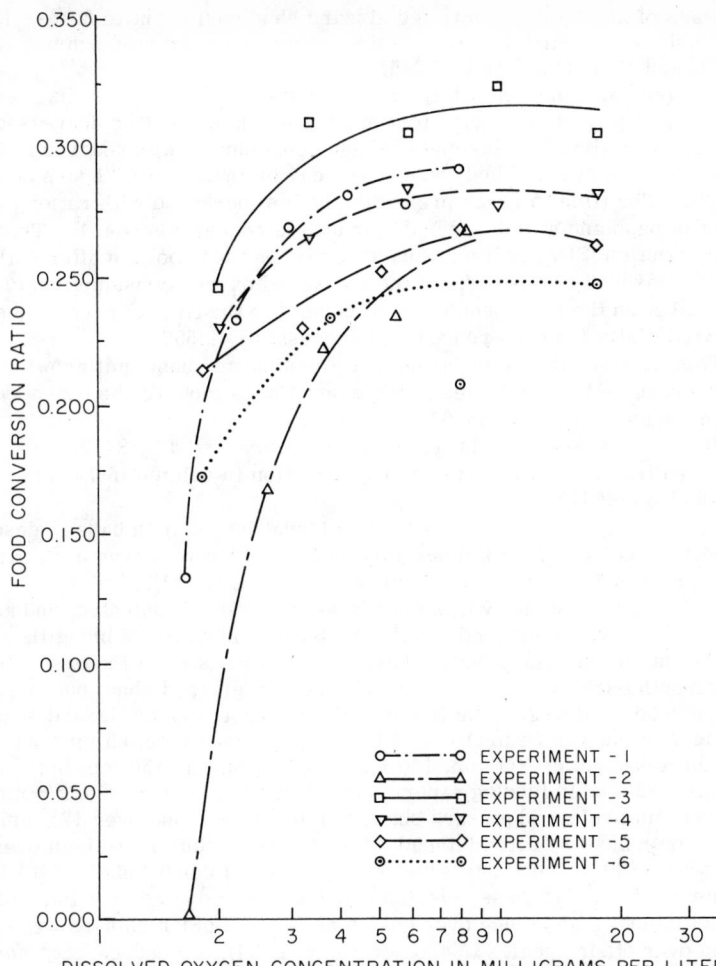

Fig. 9. Food conversion ratio in relation to dissolved oxygen concentration. The food conversion ratio of those fish that lost weight was considered to be 0, regardless of the weight loss. (Redrawn from S560.)

Temperature (°C)	Weight of bass (g)	Hrs. for digestion	g digested/hr./100 g bass
10	33–39	168	0.06-0.07
10	121–158	92	0.03-0.04
16	30–39	84	0.12-0.16
16	123–160	32	0.10-0.13
22	42–47	22	0.38-0.43
22	175–177	14–16	0.18-0.20
28	27–40	16	0.63-0.93
28	130–140	12	0.30-0.32
34	46–54	12	0.62-0.72
34	140–153	10	0.33-0.36

Bass of 200 to 400 g were fed gizzard shad equivalent to 3.0%-4.1% of their body weight, and the average time to evacuate the stomach was 20 hours at 27°C and 30 hours at 18°C (L245).

Oxygen consumption of largemouth bass forced to swim at fixed speeds increased abruptly after feeding to a maximum and thereafter decreased to pre-feeding levels (B414). Maximum oxygen consumption was reached in 6.0 ± 3.3 hours when fed at 2% of body weight per day or in 13.3 ± 3.3 hours at 8% feeding rate. The time to reach prefeeding levels increased with ration size and weight of bass and was described by multiple regression (B414). The specific dynamic action (SDA) is a measure of increased metabolism after eating and can be measured on the basis of the increased oxygen consumption. The apparent SDA on the bass seemed to be related to the time of protein absorption.

Artificially fed bass reproduced successfully (S557).

When 5 bass were kept in aquaria, the smallest bass quit growing at 10 weeks because the territories of the other 4 bass took all the space even though food was left over (S562).

Digestive enzymes of largemouth bass were listed by S450.

When fed perch, there was a 36% reduction in volume in 7 hours and 70% in 20 hours (S464).

Evacuation time of actively feeding larval largemouth bass is described as 354.6 minutes - 10.41 temperature in C, and that of larvae not actively feeding as 638.5 minutes - 18.1 temperature in C (L242).

In Murphy Flowage, WI, bullheads were positively selected, and accessibility and abundance decided the choice (S564). The size of bluegills and crayfish, but not of bullheads, eaten increased with the size of the bass (S564). Largemouth bass less than 1 kg ate 75-150 mm gizzard shad, but larger bass took gizzard shad to 250 mm (D209). The average size of gizzard shad in stomachs of bass in Crab Orchard L., IL, increased from 65 mm in bass at 175 mm to 92 mm at 200 mm, 130 mm at 225 mm, and 150 mm in bass over 275 mm (W236). In feeding experiments, however, bass at 225-275 mm and at 315-365 mm preferred gizzard shad 75-100 mm to those over 125 mm (W236). Largemouth bass selected fathead minnows over goldfish and both over bluegills when offered the species under experimental conditions (S562). In another study (L217), bass selected crayfish and tadpoles over bluegills and fathead minnows and these over golden shiners. Golden shiners were preferred over other species in another study (L154). Spinedace, *Lepidomeda mollispinis*, and larval salamanders, *Ambystoma tigrinum*, were preferred over golden shiners and goldfish in an experiment in Nevada (E152). When offered equal numbers of 7-16.5 g green sunfish averaging 11 g and equal numbers of 1.9-4.0 g green sunfish averaging 3 g (T90), 88% of the sunfish eaten by an 840 g bass were the larger sunfish, 45% of the sunfish eaten by a 714 g bass were the larger sunfish, 25% of the sunfish eaten by a 699 g bass were the larger sunfish, and 14% of the sunfish eaten by an 564 g bass were the larger sunfish. Sacramento blackfish, *Orthodon microlepidotus*, seemed to be the selected food of fingerling bass in Clear L., CA, in 1948, but not in 1956, 1957, and 1958 when bluegills were taken most commonly (M558). The width of the mouth of largemouth bass at various lengths and the average sizes of forage fish, which can be swallowed, were given by L97.

Crayfish were swallowed tail first in 96.8% of the cases, and gizzard shad headfirst in 66% (Z7).

Largemouth bass introduced into L. Atitlan, Guatemala, are believed to be responsible for elimination of several native species of fish, reduction in total biomass of fish, predation on young flightless giant grebes, and competition for the insects and crustaceans eaten by the grebes (L247).

Of 1,800 bass caught by angler, 31.8% had food in the stomach (S556).

Artificial lures took 75% of the bass, although 50% of the fishing was with natural baits (S564).

The following population data are not comprehensive; and a review of population data has been planned for a later volume, but these data are presented at this time for use until a better summary is available.

The mean standing crop of bass in North American lakes and reservoirs containing bass has been reported as 16.8 kg/ha (B382).

Populations of bass have been reported as (kg/ha):

Natural lakes

M564 IA Clear L.	0.55	0.64		or only in littoral area		2.4
L131 LA Clear L.	3.4	10.1	14.6	22.4	33.6	
M84 MN 68 gamefish lakes	mean	7.7	range	0-34.2		
44 rough fish lakes	mean	7.1	range	0-74.5		
C154 FL, WI 30 lakes	mean	8.0	range	0.2-23.2		
T73 WI Brown's L., 13 seinehauls	mean	10.2	range	4.0-20.0		
T77 WI Brown's L. (bass over 150 mm)	33.6					
M267 WI Brown's L. (bass over 150 mm)	37.3					
T40 WV Cacapon L.	10.3					

B13 MI 14 lakes (0.3-8.7 ha) 13.0 16.9   5.4   0.2 9.1 4.7 0.2 3.7 14.2
3.6   6.2 21.3   8.1 1.1

B77 MI Third Sister L.	14.2				
M96 FL 5 lakes, 1-10 ha	14.6	18.0	7.4	8.5	3.3
only legal sizes	6.8	11.4	3.3	5.0	0.03
L76 LA backwater lakes (bass over 250 mm)	26.6				
L72 MI Lower Loch Alpine	26.9				
G94 Lower Loch Alpine	33.8				
F97 MI Wintergreen L	53.9				
H441 Cuba lakes, cove sampling	56.2	128.8			

Reservoirs

C355 PA   0.52 0.24 0.15* 0.32 0.62 0.46 0.18 0.40

M567 IA   Green Valley L. 0.26; Bobwhite L. 8.15;   Red Haw L. 26.7

J37 OK Chickasaw L., Rod & Gun Club L.	1.1	17.9	
Z7 OK L. Carl Blackwell	1.4		
S131 IA Beed's L.	2.9		
P66 TN reservoir	3.8		

H456 GA L. Lanier   9.7 4.9 11.2   5.8   6.7 12.6 9.4

P73 MO L. Wappapello, catchable bass only	7.4		
W71 CA Salt Springs Valley Res.	7.5		
C154 TX, MA 34 reservoirs	mean	21.5	range 0.1-66.1
C99 IA	81	242	97 82
gravel pit	18.9		
M320 IA East Osceola Res.	33		
S597 TX Bastrop L.	37		
B372 IL Ridge L.	54	56.6	56 52.3

Ponds

F1 IA ponds balanced	0.3	1.0	17.1	18.4	19.3	29.5
overpopulated with bluegill		4.0	7.8	21.0	27.9	
overpopulated with bass		14.7	37.5			
overpopulated with bullheads		2.2				
L72 MI pond	13.7	33.9				

H278 IL 3 fertilized ponds　　57　82　93　(only large bass,　48　75　78)
　　　　　3 unfertilized ponds　　48　77　109　(only large bass, 24　50　69)
E101 MD Urieville P.　　　　12
T94 KY　9 balanced ponds　　54
　　　　13 unbalanced ponds　41
O35 OR pond with bass only　60.7
B407 IL end of 1 year　　　30　31　35　38　39　48
B407 IL end of 2 years　　　56　65　83　97　107
B382 IL Arrowhead P.　　　84
S372 NY ponds　　　　　　　　73　100
MI ponds　　　　　　　　73　169　179　240
D105 AL pond　　　　　　　255
I16 OR ponds, bass over 101 mm　　　　64　69　37　17　28　216　58　410
C154 TX, MI 166 ponds　　mean　　62.9　range　0.3-173.7
G190 MO ponds, no harvest　　　　　69　68　78
　　　　20% annual harvest　　　　55　63　75
　　　　40% annual harvest　　　　65　82　95
　　　　60% annual harvest　　　　28　46　59
W247 OK ponds　　　5.9　13.3　(net production 7.2, 11.0 respectively)
W252 FL L. Tohopekaligo, littoral zone　　52,　37,　40,　56,　72,　28,　54
　　　　　　　　　　　limnetic zone　　8,　9,　10,　4,　3,　4,　3,　7
S176 AL bass-bluegill ponds 39　44　54　55　58　61　61 67 69 80 90 94
　　　　　　　　　　　98　99　100　106　107　115　144　173
　　pond with cover　　　　31.7
　　ponds without cover　　78　83
S124 WV pond　　　　　　88.1
B373 IL 4 control ponds　　80 mean; range 24-142
　　　　3 cropped ponds　　57 mean; range 24-94
　　　　4 ponds with added bass　109 mean; range 42-179
S558 AL bass intensively fed　　2,018　2,869　3,676
S557 AL ponds bass fed　　　2,360

*The low figure of 0.15 kg/ha was related to heavy exploitation that year (C355). A 31% harvest of largemouth bass did not decrease later catches in L. Tanycomo, MO (H454). The decline in largemouth bass in Bull Shoals Res. after 6 years of impoundment was believed to be the result of competition from white bass and not angling (H454). Only 11% of the bass in Bull Shoals Res. were caught in the best year, 1958 (H454). In Illinois, carrying capacity for bass in ponds was estimated at 11-28 kg/ha in forest land, 28-56 kg/ha on light colored soils, and 56-140 kg/ha on black soils (L228). In ponds where bass were added to established populations (B373), the bass lost weight over time. Growth and production rates were best in the cropped ponds. Carrying capacity was not reached until the second year in Illinois ponds (B407).

Largemouth bass standing crops in midwest reservoirs increased with abundance of crappies, sunfish, and other forage fish but decreased with bullheads, carp, gizzard shad, and other rough fish (C154).

Populations have also been reported as numbers of bass/ha:

Natural lakes
T144 IA Clear L. 1964　　　0.29
M564 IA Clear L. 1958-59　　0.47, 0.55, or in littoral zone 2.0
B410 IN Bischoff Res., age I and over　　154 in 1973
　　　　　　　　　　　　　　　　109 in 1974
B410 IN Mollenkramer Res., age I and over　49 in 1973, 74 in 1974

W252 FL L. Tohopekaliga, littoral zone   216, 168, 265, 1,267, 723, 111, 277
                          limnetic zone   80, 17, 18, 13, 73, 46, 25, 24
          over 254 mm, littoral zone   54, 43, 39, 69, 102, 44, 60
                          limnetic zone   6, 11, 15, 8, 5, 3, 8, 3
S595 MI Mill L., after 5 years of no fishing   35
P122 WI Flora L.   3.1   2.7   5.0   6.4   10.5
M96 FL 5 lakes, legal size 12.8   0.5   5.4   4.0   6.2
          all sizes   64   47   47   30   67
B13 MI 14 lakes (0.3-8.7 ha) 277 3,473 117 199 0.5 268 59 14 29 116 13
                    52   281   131
M359 MN Gladstone L. (bass over 150 mm)   57   30
M551 WI Murphy Flowage (bass over 200 mm)   19.8
T73 WI Brown's L., 13 hauls of seine, mean 40 range 24-63
M367 WI Brown's L., bass over 150 mm   137
L72 MI Lower Loch Alpine (56% young bass)   313, about 42 legal bass
H441 Cuba lakes, cove sampling   267   2,744

Reservoirs
P73 MO L. Wappapello, Cove 17 catchable + 34 intermediate + 210 fingerlings
H454 AR Bull Shoals Res., bass over 230 mm, 115 in 1958, 27.7 in 1959,
                                        20.7 in 1960
G189 OK Eufoula, bass over 254 mm, 1.3-10.1, but bass over 254 mm, probably
                                        less than 5% of total.
M569 MO Poney Express Res., bass over 220 mm   63   35   26   44
D103 MD Snowden P.   61.5   19.8   24.9   49.9
W71 CA Salt Springs Valley Res.   67
L227 CA Sutherland Res., (bass over 2 years) 203   143   138   111   279

Ponds
H278 IL 3 fertilized ponds   385   432   687
          only "large" bass   121   143   158
          3 unfertilized ponds 511   766   771
          only "large" bass   59   136   146
B411 IL Ridge L. 1943-70   655 225 343 279 220 285 324 341 961 374
          over 254 mm   3   97   54   50   90   21   35   17   27 142
C302 PA Breon's P. (all sizes)   17,680
S63 MN rearing ponds 25-140,000; mean 18,315
R77 OH rearing ponds 14,820
S171 WV rearing ponds 18,992
D5 USA rearing pond   21,062
F43 IA rearing pond   24,206

The annual production of bass in a 3 hectare reservoir was estimated at
7.7 to 28.6 kg/ha/year of age I bass and 10.5 to 50.9 kg/ha/year of age II bass
(V93). In a large reservoir, the annual production was 0.62 kg/ha/year (Z8),
but only a small proportion of the reservoir produced bass. The production/
biomass ratio was 0.41 (Z8) and 0.44 to 2.11 (V93).

It was estimated that about 126 legal-sized bass per ha would be needed to
provide an average catch of 1 bass per hour (L72). The harvest of bass has
been reported as (kg/ha):

L127 MN Linwood L.   0.7   0.7   1.1
S597 TX Tournaments   0.01(3), 0.02(7), 0.03(9), 0.04, 0.06(2), 0.07,
                                            0.08(2), 0.09, 0.10(3), 0.11(2), 0.12(2), 0.16,
                                            0.18, 0.19, 0.21(2), 0.22, 0.25, 0.28, 0.31, 0.35,
                                            0.37, 0.39, 0.43, 0.56, 0.83, 2.02, 2.09
M359 MN Gladstone L.   3.6
B189 AL Public lakes   33.1
M551 WI Murphy Flowage   6.9-35.6
        Yellowstone L.   10.0-31.4
H441 Cuba 2 lakes   43   87
M296 MA Quabbin Res.   0.13   0.11   0.09   0.11   0.13
K64 CA Havasu Res.   0.8   1.5
S597 TX L. Bastrop   11
E145, L227 CA 3 reservoirs   4.9   50.7   31.0
R304 CA Merle Collins Res. 1965-70   1.36   1.58   3.30   2.18   2.18   2.21
B410 IN Bischoff Res.   89
        Mollenkramer Res.   104
R306 MO Limpp Community Res.   32.6   38.1
B408 IL Ridge L., before drawdown   13.6   12.6
B408 IL Ridge L. with supplemental feeding   19.4   18.6   17.8   20.3   14.1
H278 IL 3 fertilized ponds, 5 year averages   7.8   13.5   19.0
        3 unfertilized ponds, 5 year averages   9.0   15.7   16.8
M306 MD Snowden P.   8.1   24.3
H278 IL 36 ponds   1-37, mean 13.5
H270 AL 32 ponds   0.8-39, mean 18.5
E146 AL 3 years of 1 pond   30.3   18.8   25.7
                                            Catch as numbers of bass/ha.
M551 WI Murphy Flowage   3.7-9.9
        Yellowstone L.   64-180
M569 MO Poney Express Res.   26   11   12   23
M267 WI Brown's L.   67
E145 CA 3 reservoirs   67.2   89.9   20.5
H278 IL 3 fertilized ponds   126 37 62 91 32 89 15 47 47 74 44 35 5
                                            32 20 35 32 30
        3 unfertilized ponds   37 32 49 35 49 62 15 37 35 30 35 99 12
                                            17 17 62 30 30

When Mill L., MI, was opened after 5 years of no fishing, there were 35
bass per hectare, and 35% were taken in 3 days with 96 hours of fishing per
hectare (S595). Fishing success dropped over the 3 days more than the bass
population, indicating a decreased catchability.

In the following summary, S refers to a statewide average or one covering
several waters or several seasons:

Catch per hour	No. of lakes or years	Citations
0.01-0.09	18 + 7 S	CA, E145, P91(6); FL, M355(2); IA, C251; IL, H278(4); CT, T22(2); MI, E37, E38; MN, H72 (7S); WI, K47
00.1-0.19	32 + S	CA, E145, P91(4); R304(2); FL, M355; IL, H278(14), B99, H129(3); IN, T32, R36(2); MN, H72 (S); MO, M569(3); IA, L131
0.2-0.29	14	CA, R304; IL, H129(2), H278(10); VA, R118

Catch per hour	No. of lakes or years	Citations
0.3-0.39	13	IL, H278(8); IN, R36; MI, T31; IA, L131(3)
0.4-0.49	2	IL, H129; CA, R304
0.5-0.59	7	IL, H278(3); CA, R304; MO, M569; IA, L131, FL, M572
0.6-0.69	3	IL, H278(2); WI, K47
0.7-0.79	3S + 1	IA, C123(S), C124(2S); CA, R304
1.21	1	IL, H278
1.4-1.49	1	MI, T31

Catch per angler day

E145, K64, L227, C94: CA 2.0 0.07 0.04 0.5 0.51 0.72 0.82 1.76 0.37
0.30 0.34 0.55 0.5 0.35 0.4 0.87 0.45 0.56 1.6

Experiments (A94) showed that vulnerability of bass to angling decreased with previous fishing experience. The catch per hour in Murphy Flowage, WI, was better in May and June and before 10 A.M. and after 6 P.M. (S556). The largest bass were also caught before 10 A.M. or after 6 P.M.

An inverse relationship between fishing pressure and largemouth bass catch per unit effort was found, suggesting a "harvestable" surplus, which is easily caught and after which increased pressure reduces the catch per unit effort (B143, M178).

Longer angling seasons increased the annual catch without depleting the bass population (M555, M296). Overharvest was thought to be the cause of declines in California reservoirs where 47% and 48% of tagged bass were harvested in the first year (R303). In Ridge L., IL, however, numbers of large bass remaining after heavy fishing suggested that overfishing is unlikely. Numbers and weight of bass could be replaced within one season with average removals of 56% of the numbers or 63% of the weight (B411). Only a small percentage of the bass over 1.3 kg were ever caught. In a small Missouri reservoir, only 8.6 bass over 230 mm/ha survived a fishing season with 77 hours/ha of fishing (M569). A 40% harvest was estimated as the maximum allowable, and this could be taken in a few days (M569). The first 4 days of fishing on 6 Missouri impoundments resulted in harvest of 40% to 69% of the adult bass (R305).

Hatchery stocking was shown to have little effect on the population or later catch in Grand L., OK, (J31) and in New Mexico (J109). Sustained stocking costs per pound of yield in reservoirs were estimated as less for largemouth bass than for northern pike or walleyes when using yearling stock, but costs with fingerlings were higher for bass than the other two species (G181).

Largemouth bass growth and harvest was better with biennial culling in Ridge L. than in a management program including supplemental feeding and drawdown (B408).

On the basis of analysis with models of 40% and 80% annual mortality, the following warning signals for management biologists were suggested (A98): populations with few bass over 680 g; high mortality rates as indicated by catch curves; inadequate recruitment; and forage or panfish populations showing signs of overpopulation. Slow growth of age I-III bass suggests excess recruitment.

The desirability of controlling parasites of largemouth bass stocked in reservoirs was discussed in an extensive survey of bass parasites in Arkansas (B394).

In stocking of ponds, stocking rates and the times of stocking bluegills had no detectable effect on the bass survival (R237). In Nebraska, fingerling bass stocked in August had better survival, 54%-100%, than those stocked in October-November, 10%-48% (B397). Stocking of bass fingerlings in early summer and bluegills and channel catfish in fall gave better results than when all three were stocked in the fall (D201). Stocking did not improve fishing in Lower Loch Alpine (L72). Stocking additional bass in a pond with excessive cover and stunted bluegills eliminated bluegill reproduction, but the stocked fingerling bass grew very slowly with little forage left (S176). Stocking suggestions are reviewed (E145, S176). Propagation instructions are given in M565. Largemouth bass and channel catfish were raised successfully together on pelleted feeds (B412).

By stocking fathead minnows and *Gambusia* in early spring and following this with 124 bass, 494 bluegills, 494 redears, and 247 channel catfish/ha and a few adult bass which spawned the first year, an excellent population was developed with many sizes of bass (R305, H468). A minimum size limit of 356 mm permitted a harvest of 90 kg/ha in 14 months and continued to provide good fishing for at least 5 years (R305). Reduced creel limit, opening the lake immediately after stocking to avoid the opening season intensity, midwinter opening of the season, refuge area, and a 40% harvest quota were not as successful in maintaining bass fishing.

Young largemouth bass were taken in greater numbers in shoreline seining in Rathbun Res., IA, in the year stocked than in years when not stocked (P204).

Stable water levels until bass were through nesting followed by rapid drawdown were recommended (V90). A pelagic predator, the striped bass was suggested to control threadfin shad (V90). Drawdown decreased the catch of bass at first, but later more and larger bass were caught (H456).

The average size of the bass caught in gillnets increased with the mesh size (L216).

Bar measure of mesh	TL of bass		No.
	Mean	Range	
25	295	190-470	3
38	325	241-508	15
51	381	343-559	14
64	455	406-533	7
76	472	432-521	3

The following hematological data were given by D206:

red blood cells	1,720,000-3,481,000 per ml
hemoglobin	5.52-12.3 g/100 ml
oxygen capacity	42-86.4 cc/l
oxygen in blood	1.9-48.8 cc/l, most 5.7-28.8 cc/l
lactic acid	28-270 mg per cl

The number of chromosomes has been reported as 48 or as 46 in tissue culture (R286).

Largemouth bass have per 100 grams, dry weight: 11.46 g nitrogen, 3.81 g phosphorus, 0.95 g sulfur, 6.51 g calcium, 0.17 g magnesium, 1.37 g potassium, 0.44 sodium, 3.42 mg iron, 0.59 mg manganese, 4.85 mg zinc, and 0.62 mg copper (G186). Nitrogen, sulfur, and potassium levels were slightly higher in bass less than 300 g than in larger fish.

An extensive bibliography on largemouth bass compiled by Roy Heidinger can be purchased from the Bass Research Foundation, PO Box 3385, Montgomery, Al. 36109. (H479).

GUADALUPE BASS, *Micropterus treculi* (Vaillant and Bocourt)
This species is limited to the Edwards Plateau region of Texas, it is most abundant in turbid downstream waters (R312), and it has been established in some reservoirs in central Texas. The maximum size is about 2,000 g (R312).

WHITE CRAPPIE, *Pomoxis annularis* Rafinesque
White crappie originally were found in ponds, lakes, bayous, and slow-moving streams and rivers from eastern South Dakota to western New York and south in the Mississippi R. and Gulf of Mexico drainage to Alabama and Texas. They have been introduced into other suitable waters in United States. The California populations probably are the progeny of 16 fish planted near Morena Res. in 1917 (G185). The white crappie is quite tolerant of turbidity and siltation (W192) but does not become abundant in clear waters in competition with bluegills, black crappies, or largemouth bass (T113, G185, N138).

	No.	SL	FL/SL	TL/SL	TL/FL
L29 IA East L.	60-85	30-229	1.209	1.285	
C30 IA	46-60	-	1.21	1.29	
L29 IA Red Haw L.	15-29	88-251	1.228	1.288	
H10 IL Horseshoe	96	40-250	-	1.25-1.50	
S100 TN	8	140-179	-	1.279	
	46	180-219	-	1.259	
	7	220-259	-	1.246	
E48 IA Clear L.	44	-	1.222	1.284	1.053
W138 MO	61	60-99	-	1.319	1.079
	189	100-179	-	1.304	1.061
	95	180-219	-	1.288	-
	128	180-299	-	-	1.049
	141	220-369	-	1.270	-
	13	300-369	-	-	1.042
	345	60-219	1.227	-	-
	141	220-309	1.213	-	-
C257 PA	25	104	-	1.218	
	15	105-195	-	1.297	
	8	over 196	-	1.285	
M320 IA	10	40-117	-	1.311	
	52	118-179	-	1.288	
	13	180-280	-	1.267	
H153 OK	1,176	76-277	-	1.268	
W105 IL Crab Orchard L.	1,285	95-169	-	1.301	
	46	170-316	-	1.260	
R166 VA Clayton L.	200	60-339	TL = 8.08 mm + 1.222 SL		
H153 OK	1,176	76-277	TL = 9 mm + 1.212 SL		
H200 IA Ahquabi L.	238	46-280	TL = -4.5 mm + 1.321 SL		

Because of the evidence that the relationship is not constant with growth, a single ratio is not adequate. In the tabulations which follow, I have used the following equivalents, interpolating in between:

SL	22	40	60	80	100	120	140	160	180	200	220	240	260	280	300	320	340
FL	24	49	74	99	123	147	172	196	220	244	266	290	314	339	363	388	412
TL	25	50	76	102	128	154	180	205	230	255	280	305	330	355	380	405	428

W138   112 crappie 35-120 mm   fresh TL = 1.014 preserved W
                                 fresh W = 1.17 preserved W
B220   preservation in formalin   fresh TL = 1.007 - 1.023 preserved TL
                                 fresh W = 0.920 - 0.952 preserved TL

Stunted fish changed less on preservation than did fast-growing fish (B220).

TL	Weight in grams Mean of means	Central 50%	Range	No.	Citations
25-50	0.2	-	0.1-0.9	173	S472 AL
	1	-	0.6-1.3	11	C256 PA
51-75	1.7	-	0.5-11	1,796	S472 AL
	3.6	-	3-6	37	M248, M302: OH, IA
	3.8	-	1.1-5	22+	C257, F105, S225: PA, SD
	4.3	-	3-9	9	B220, E97, C202, T92: OK, TX, AR
	17	-	-	3	L126 CO
	4.5	2.7-5.0	0.5-17	1,867+	Combined
76-101	5	-	1.6-11	2,325	S472 AL
	7	5-9	5-10	61+	C106, E84, M246, M428, W138: IA, OH, MO
	9	-	5-18	23	B220, E97, H156, C202, T92: OK, AR, TX
	10	-	9-14	9+	C257, F105, S225: PA, SD
	20	-	-	6	L126 CO
	9	6-9	1.6-20	2,424+	Combined
102-126	12	-	5-34	778	S472 AL
	14	-	-	+	K71 TN
	15	-	9-18	6	M499 IL
	16	-	-	346	W138 MO
	18	-	11-23	4+	R81, M248, E84: OH
	18	-	14-24	32	C257 PA
	19	17-22	9-36	99+	B220, H156, W68, C202, T92: AR, OK
	20	-	14-27	+	F100, F105, F124, N94, N115, S225, S227, S229, S276, S330, S393: SD
	20	-	-	14	L126 CO
	21	18-23	4-44	83	C30, L29, E48, C106, I10, M302, M246, R150, A88: IA
	27	-	-	8	E97 TX
	19	16-23	4-44	1,370+	Combined
127-151	26	-	11-50	339	S472 AL
	29	-	-	+	K71 TN
	30	27-32	18-64	117+	B220, H156, W68, J92: AR, OK
	30	-	-	339+	P113, W138: MO
	30	26-33	22-34	48+	O7, R81, E84, M248: OH

TL	Weight in grams Mean of means	Central 50%	Range	No.	Citation
127-151 (cont.)	30	26-33	14-41	189	C30, L29, E48, C106, I10, M302, M246, A88: IA
	32	-	24-57	13	C257 PA
	32	-	27-36	9	M499 IL
	34	-	-	7	E97 TX
	35	-	27-41	+	F100, F105, F124, N94, N115, S225, S227, S228, S229, S276, S330, S393: SD
	40	-	37-42	30	E51, L126: CO
	45	-	-	+	U8 Mississippi R.
	31	29-34	11-64	1,082+	Combined
152-177	37	-	-	6	A39, A40 AL
	45	-	34-113	173	S472 AL
	46	-	40-57	856+	W138, P113: MO
	48	-	20-65	6	C257 PA
	51	-	-	+	K71 TN
	51	-	45-91	34	T92 AR
	53	48-59	29-99	41+	C30, L29, C89, E48, C106, I10, M302, M246, A88: IA
	53	41-54	28-147	195+	R78, R81, O7, H96, M248, E84: OH
	56	-	45-68	+	F100, F105, F124, F125, N94, N115, S225, S227-9, S276, S330, S393: SD
	57	50-57	36-100	157+	B220, H156, W68: OK
	59	-	-	9	M499 IL
	68	-	-	+	U8 Mississippi R.
	71	-	-	43	L126 CO
	53	47-57	20-147	1,520+	Combined
178-202	68	-	45-127	210	S472 AL
	74	70-82	43-91	155+	R78, R81, O7, M248, E84: OH
	75	-	65-79	355+	P113, W138: MO
	81	-	-	+	K71 TN
	85	-	28-113	4	C257 PA
	86	85-88	57-142	244+	C30, L29, C89, E48, C106, I10, M302, M246: IA
	86	-	73-95	+	F100, F105, F124, S225, S227-9, S276, S330, S393, N94, N115: SD
	90	-	71-105	72	L126, E51: CO
	91	-	77-100	11	M499 IL
	91	-	-	1	A39 AL
	91	-	-	5	E97 TX
	93	76-111	45-150	129+	B220, H156, W68, T92: OK, AR
	100	-	-	+	U8 Mississippi R.
	85	79-91	28-150	1,186+	Combined
203-228	77	-	-	4	A40 AL
	94	-	45-136	14	T92 AR
	112	-	104-127	12	M499 IL

TL	Weight in grams Mean of means	Central 50%	Range	No.	Citation
203-228	113	-	64-318	162	S472 AL
(cont.)	113	100-121	99-121	85+	O7, R78, R81, M248, E48: OH
	114	-	91-125	1+	K71, W132: TN
	118	-	113-124	48	E51, L126: CO
	119	-	105-125	399+	F113, W138: MO
	135	-	109-172	+	F100, F105, F124, N94, N115, S225, S227-9, S276, S330, S393: SD
	136	128-142	85-181	131	C30, L29, E48, C106, I10, M302, M246, B150: IA
	139	-	105-170	12	C257 PA
	145	-	-	+	U8 Mississippi R.
	145	142-154	95-209	247+	B220, H156, W68: OK
	147	-	-	6	E97 TX
	123	113-142	45-318	1,121+	Combined
229-253	159	-	86-281	157	S472 AL
	164	150-174	147-193	50+	O7, R78, R81, M248, E84: OH
	170	-	-	3	L126 CO
	170	-	136-204	2	M499 IL
	177	-	166-181	257+	P113, W138: MO
	179	-	173-187	6+	K71, W132: TN
	184	-	-	74	E97 TX
	187	172-204	130-260	45	C30, L29, E48, C106, I10, M302, M246, B150: IA
	193	-	163-227	+	F100, F124, N94, N115, S225, S227-9, S276, S330, S393: SD
	193	-	142-255	29	C257 PA
	194	185-204	163-263	173+	B220, H156, T92, W68: AR, OK
	204	-	-	+	U8 Mississippi R.
	181	169-200	86-281	796+	Combined
254-278	227	-	-	3	E51, L126: CO
	230	-	-	+	K71 TN
	236	-	136-286	67	S472 AL
	240	213-258	213-298	45+	O7, R78, R81, M248, E84: OH
	241	-	-	68	E97 TX
	245	-	221-255	328+	P113, W138: MO
	250	-	236-286	+	N94, N115, S225, S228-9, S276, S330, S393: SD
	269	-	212-340	27	C257 PA
	271	241-295	227-358	52+	B220, H156, T48, T92, W68: AR, OK
	286	-	-	+	U8 Mississippi R.
	288	295-315	181-331	14	C30, L29, C106, I10, M302, M246: IA
	257	230-284	136-358	604+	Combined
279-304	290	-	-	1	M499 IL
	313	-	100-395	51	S472 AL
	321	-	293-340	7	C257 PA
	323	-	-	36	E97 TX

TL	Weight in grams Mean of means	Central 50%	Range	No.	Citations
279-304	328	298-332	298-397	19+	O7, R78, R81, M248, E84: OH
(cont.)	340	-	-	1	N52 WI
	344	-	313-367	+	N94, N115, S225, S228-9, S330, S393: SD
	349	-	-	3	L126 CO
	358	-	340-397	155+	P113, W138: MO
	364	-	272-468	21	C30, L29, C106, I10, M302, M246: IA
	372	-	-	+	U8 Mississippi R.
	372	334-408	318-486	54+	B220, H156, T92, W68: AR, OK
	346	323-372	100-486	348+	Combined
305-329	378	-	-	19	E97 TX
	390	-	200-590	41	S472 AL
	428	-	346-539	+	O7, R78, R81, E84: OH
	463	-	408-622	19+	B220, H156, T92, W68: AR, OK
	468	-	448-518	128+	P113, W138: MO
	468	-	-	+	U8 Mississippi R.
	482	-	454-567	4	C257 PA
	497	-	366-612	7	C30, L29, I10, M302: IA
	501	-	431-636	+	F100, N94, N115, S330, S393: SD
	550	-	-	10	E51: CO
	449	399-486	200-636	228+	Combined
330-355	472	-	340-608	19	S472 AL
	536	-	-	16	E97 TX
	542	-	539-545	2	E105, S327: MD
	556	-	545-567	2	L29, M302: IA
	574	-	561-601	79+	P113, W138: MO
	584	-	-	+	U8 Mississippi R.
	611	-	543-709	2+	R78, R81, M248: OH
	629	-	499-728	8+	B220, H156, T44, T92, W68: AR, OK
	709	-	-	1	L126 CO
	579	536-601	340-728	129+	Combined
356-380	663	-	-	40	W138 MO
	665	-	504-879	4	B220, E97, T48: OK, TX
	712	-	635-794	8	S472 AL
	767	-	765-771	3	E51, U8: CO, Miss. R.
	701	630-712	504-879	55	Combined
381-405	650	-	408-794	4	S472 AL
	794	-	595-906	5	C257, E51, W7, W239: PA, CO, OK
	831	-	-	8	W138 MO
406-431	680	-	-	1	S398 MD
	993	-	-	2	S472 AL
493-507	851	-	-	1	M310 PA

There do not seem to be any consistent regional differences in the average weights at various lengths.

	No.	SL	
R313 TX Meridian L. 1968	1,000	100-300	log W = -4.381 + 2.882 log SL
1969	3,633	-	log W = -4.556 + 2.957 log SL
1970	213	150-320	log W = -4.290 + 2.842 log SL
W105 IL			log W = -4.698 + 3.007 log SL
E60 IL Upper P.	26	60-209	log W = -4.646 + 3.014 log SL
Lower P.	46	60-209	log W = -4.702 + 3.057 log SL
R116 VA Claytor L.	200	60-339	log W = -5.1395 + 3.23 log SL
E48 IA Clear L.	42		log W = -5.308 + 3.376 log SL

	No.	FL	
G185 CA Pine Flat L.	-	-	log W = -5.621 + 3.556 log FL

	No.	TL	
P208 IN White R.,			
Sec. A 1970	-	-	log W = -4.098 + 2.673 log TL
Sec. A 1969	-	-	log W = -4.341 + 2.783 log TL
Sec. B 1970	-	-	log W = -4.551 + 2.877 log TL
Sec. B 1969	-	-	log W = -4.619 + 2.904 log TL
Sec. C 1970	-	-	log W = -4.795 + 2.999 log TL
Sec. C 1969	-	-	log W = -6.851 + 3.847 log TL
T92, C202 AR Ft. Smith L.	163	48-340	log W = -4.453 + 2.776 log TL
M556 IA Coralville L.			
1964	-	135-350	log W = -4.678 + 2.910 log TL
1966	-	135-350	log W = -4.890 + 2.868 log TL
1968	-	135-350	log W = -5.763 + 3.3697 log TL
1969	-	135-350	log W = -5.549 + 3.298 log TL
B220 OK Boomer L.	-	-	log W = -5.589 + 3.003 log TL
P204 IA Rathbun L. 1973	27	164-315	log W = -4.92 + 3.02 log TL
J37 OK Rod & Gun Club L.	54	125-290	log W = -4.908 + 3.028 log TL
S472 AL	6,305	50-410	log W = -4.659 + 3.05 log TL
H200 IA Ahquabi L.	359	56-366	log W = -4.394 + 3.048 log TL
N144 SD Oahe L.	259	111-341	log W = -4.792 + 3.065 log TL
S569 SD Lewis and Clark L.,			
Spring and Summer	392	-	log W = -5.091 + 3.087 log TL
Fall	301	-	log W = -5.432 + 3.213 log TL
J37 OK Rod & Gun Club L.	174	100-240	log W = -5.237 + 3.114 log TL
W68 OK	-	-	log W = -5.153 + 3.118 log TL
M246 IA pond 2	52	97-284	log W = -5.308 + 3.208 log TL
pond 6	105	147-206	log W = -5.335 + 3.177 log TL
B377 OK Boomer L.	150	-	log W = a + 3.179 log TL
C106 IA ponds	156	89-284	log W = -5.331 + 3.205 log TL
H156 OK Canton L.	799	75-350	log W = -5.434 + 3.260 log TL
W240 IA Decatur L.	150	125-320	log W = -5.478 + 3.267 log TL
J28 OK Lower Spavinaw L.	478	-	log W = a + 3.469 log TL
Upper Spavinaw L.	140	-	log W = a + 3.285 log TL
W138 MO	3,319	84-386	log W = -5.595 + 3.299 log TL
S225 SD Ft. Randall L.	-	60-240	log W = -5.561 + 3.310 log TL
C373 KY Cumberland L.	540	110-360	log W = -5.772 + 3.346 log TL
P204 IA Rathbun L. 1972	33	160-312	log W = -6.16 + 3.52 log TL
H153, H419 OK	1,176	76-275	log W = -5.732 + 3.369 log TL
B220 OK Blackwell L.	-	-	log W = -5.872 + 3.413 log TL
G42 Upper Mississippi R.	259	102-343	log W = -5.912 + 3.43 log TL

	No.	TL
B220 OK Canton L.	-	-	$\log W = -4.870 + 3.431 \log TL$
C160 KY Kentucky L.	-	-	$\log W = -5.966 + 3.457 \log TL$

The slopes are mostly above 3.0, but few tests of significance were reported. The b value of 3.267 for Decatur L., Iowa was not significantly different from 3.0 (W240). The b value for the crappie in Coralville in 1968 was significantly different from the 1964 and 1966 values (M556).

	No.	SL	K(SL)	Range
C23 MN	18	125-210	1.48	-
T48 OK Clinton L.	54	-	1.85	0.28-2.79
C36 IA	58	82-266	2.08	-
W105 IL Spring	152	-	2.02	-
May-June	832	-	2.20	-
Summer	158	-	2.16	-
H10 IL Senachwine L.	16	36-51	2.20	2.01-2.46
M248 OH	75	51-100	1.98	1.89-2.21
	450	101-177	2.35	2.01-2.99
	196	178-320	2.52	1.91-3.39
R313 TX Meridian L. 1968	1,000	100-300	2.276	s. d. = 0.202
1969	3,633	-	2.240	s. d. = 0.206
1970	213	150-320	2.286	s. d. = 0.173
S100 NC Hiwassee L.	25	-	2.37	-
T48 OK Fairfax L.	51	-	2.40	-
T48 OK Cleveland L.	52	-	2.41	-
C202 AR Ft. Smith L.	163	71-343	2.41	-
E60 IL Upper P.	26	60-209	2.42	-
Lower P.	46	60-209	2.66	-
R116 VA Claytor L.	200	60-339	2.43	1.90-2.60
T48 OK Pawnee L.	44	-	2.48	-
L29 IA Red Haw L.	29	88-251	2.50	-
L29 IA East L.	90	30-259	2.52	-
E28 TN Chickamauga L.	434	-	2.6	-
C257 PA Sanctuary L.	31	-	2.60	-
Upper Pymatuning L.	92	-	2.69	-
Lower Pymatuning L.	55	-	2.31	-
S100 TN Douglas L.	8	-	2.61	-
D179 TX	663	74-298	2.72	1.14-4.39
G198 TX Benbrook L.	51	94-117	2.41	2.14-2.70
	368	118-141	2.56	2.24-2.68
	392	142-164	2.70	2.41-2.86
	251	165-188	2.87	2.65-3.06
	149	189-212	3.01	2.80-3.55
	67	213-235	3.13	2.85-3.59
	54	286-263	3.16	2.92-3.45
M304 IA Lakes	72	71-333	2.75	1.67-4.28
H10 IL Horseshoe L.	94	127-305	2.77	2.42-3.32
H10 IL Decatur L.	3,327	76-279	2.80	2.0-3.7
H10 IL Senachwine L.	266	102-203	2.84	2.66-3.12
S100 TN Cherokee L.	25	-	2.93	-
E48 IA Clear L.	42	80-184	2.95	-
M235 IA Backbone L.	-	-	3.13	-

	No.	SL	K(SL)	Range
S419 IA Backbone L., before	-	-	3.13	-
after thinning	-	-	4.75	4.41-5.03
H10 IL Crab Orchard L.	13	203-279	3.88	3.66-4.18
H128 IL Decatur L.	9	76-101	2.55	2.23-2.81
	307	102-126	2.38	2.12-3.20
	747	127-151	2.48	2.08-3.00
	638	152-177	2.76	2.42-3.50
	852	178-202	2.97	2.20-3.70
	772	203-228	3.08	2.30-3.70
	105	229-253	3.15	2.00-3.70
	23	254-278	3.01	2.69-3.40
	5	279-304	3.02	2.57-3.38
other lakes	39	102-126	2.80	2.65-2.85
	167	127-151	2.79	2.66-2.84
	82	152-177	2.74	2.48-2.93
	38	178-202	2.94	2.42-3.10
	11	203-228	3.15	2.76-3.70
	9	229-253	3.36	3.12-3.66
	10	254-278	3.49	3.03-4.18
	12	279-304	3.44	3.32-4.03
	5	305-329	3.17	-

	No.	SL	K(TL)	Range
W240 IA Decatur L.	150	125-320	-	0.55-1.02
A88 IA Little Wall L. 1967	95	114-157	0.82	0.66-1.05
B373 IL pond-population	21	100-280	0.97	-
added to				
Control pond	80	150-260	1.08	-
Cropped pond	25	150-310	1.19	-
B377 OK Boomer L.	150	-	0.98	-
B220 OK Canton L.	17	66-150	0.99	-
	131	152-302	1.40	1.20-1.69
C89 IA	5	152-190	1.05	1.00-1.11
B220 OK Boomer &	492	114-202	1.06	1.02-1.08
Blackwell L. (stunted)	90	203-361	1.28	1.19-1.50
S566 IA Backbone L. 1956-66	-	-	-	0.83-1.39
H200 IA Ahquabi L.	359	56-366	1.14	-
J96 OK Franklin P.,				
before	-	-	1.16	-
after thinning	-	-	1.99	-
B220 OK pond	35	109-363	1.19	0.93-1.84
S225, 228, 330, 393: SD				
Ft. Randall L.	1,203+	50-330	1.23	1.19-1.40
S227, S229, S276, N94, N115:				
SD Gavins	1,837	102-330	1.24	1.08-1.40
H156 OK Canton L.	364	75-175	1.25	0.69-1.86
	345	179-250	1.52	1.14-2.16
	90	254-350	1.61	1.39-1.94
B220 OK Ft. Supply Basin	5	157	1.27	-
J96 OK Rod & Gun Club L.,				
before	-	-	1.27	-
after thinning	-	-	1.36	-

	No.	TL	K(TL)	Range
T92 AR Ft. Smith L.	163	48-340	1.27	0.82-1.80
N138 IA Clear L.	1,028	102-260	1.29	1.11-1.42
M556 IA Coralville L. 1964	-	-	1.30	-
1966	-	-	1.61	-
1968	-	-	1.34	-
1969	-	-	1.43	-
E97 TX Koon Kreek Klub L.	266	-	1.30	0.96-1.56
M499 IL Northeast lakes	50	100-280	1.30	0.86-1.41
S472 AL	1,969	25-60	1.30	1.28-1.40
	3,442	75-125	1.10	1.05-1.26
	894	150-510	1.34	1.17-1.59
J28 OK Lower Spavinaw L.	478	-	1.31	-
Upper Spavinaw L.	140	-	1.25	-
P204 IA Rathbun L. 1972	83	160-312	1.33	1.04-2.31
B150 IA Little Wall L.	36	115-250	1.39	-
R49 OH	-	-	1.39	-
P204 IA Rathbun L. 1973	27	164-315	1.40	1.23-2.00
M246 IA Pond 2	52	97-294	1.41	-
Pond 6	105	147-206	1.22	-
M575 IA Williamson L.	45	-	1.41	1.23-1.60
H153 OK	4,243	90-410	1.41	-
F100, 124, 125: SD Oahe L.	339	50-310	1.48	1.38-1.49
B220 OK Ft. Supply L.	91	168-325	1.50	1.46-1.54
G164 NM Navajo L.	11	171-209	-	1.71-1.75
P204, B416, B417 IN White R.				
1969	73+	100-280	1.44	0.9-2.1
1970	91+	100-280	1.53	1.3-1.7

The average K value increased with length (B220, G198, H128, H156, and M248) but showed no such trend in other populations (N138, S472). In Clear L., IA, an increase in average K from 1960 to 1961 accompanied an increase in water level (N138), and in Coralville L., IA, average K was highest in 1966, the year following the highest water levels (M556). The length-weight relationships of male and female crappies in Lewis and Clark L., SD, differed only slightly from each other in spring or in fall (S569). K values were higher in May-June than in spring or summer in Illinois (W105), but they were lower in late June through early July than early June or later in the summer in Clear Lake, IA (N138). In Texas, K values were highest in the winter when metabolic rates were low and were lowest in the spring, when food was scarce, building up in summer and fall (G198).

Reduction in numbers of crappies resulted in increased average K (B373, J96, R313, S419, and S566).

		TL				Weight	
		Mean				Mean	
		of	Central			of	
	No.	means	50%	Range	No.	means	Range
Age 0-June							
AR: A96, B413	50+	12	-	5-21	-	-	-
IA Rathbun L.: M561	20	16	-	8-23	-	-	-
IN Foot's P.: R40	+	30	-	18-48	-	-	-
OH St. Mary's L.: C284	+	38	-	-	-	-	-
TN Chickamauga L.: E28	392	38	-	18-53	-	-	-

		TL				Weight	
	No.	Mean of means	Central 50%	Range	No.	Mean of means	Range
Age 0-June (cont.)							
OK Canton L.: Q5	20+	58	-	-	-	-	-
Combined	464+	28	16-38	5-58	-	-	-
Age 0-July							
IA: H200, N99	143	34	-	14-56	-	-	-
IN Foot's P.: R40	+	36	-	23-51	-	-	-
OH: G32, C284	+	40	-	25-47	-	-	-
PA Pymatuning L.: M560	+	41	-	36-52	-	-	-
CA: H146	18	41	-	40-48	-	-	-
AR: B413	+	44	-	-	-	-	-
OK: B183, C134, G183, O37	84+	48	41-57	36-57	1	0.03	-
SD: N136, S569, W242	2,153	-	-	12-87	-	--	-
TN Chickamauga L.: E28	58	53	-	36-71	-	--	-
MO: W138	266	56	-	36-91	-	--	-
Combined	2,722+	43	38-53	12-91	1	0.03	-
Age 0-August							
SD: N136, S225, S569, W242	1,621+	47	35-66	13-108	-	-	-
IA, IN, KS, OH, PA: C284, G32, H200, M560, N99, R40, S207, M561	184+	65	48-71	16-150	11	23	-
AR, OK, MO, TN: E28, G183, H150, J26, R292, W138	264+	70	58-81	20-124	25	6	3-9
CA: H146	16	74	-	61-94	-	-	-
Combined	2,085+	65	48-74	13-150	36	14	3-23
Age 0-September							
AR, OK, MO: G183, W138, B413	209+	74	-	46-112	-	-	-
SD: F105, S225, W242	1+	85	-	26-117	-	-	-
IA: H200, I10	87	87	-	79-120	1	27	-
PA: M560	+	97	-	80-102	-	-	-
Combined	287	80	74-84	26-120	1	27	-
Age 0-October							
SD: W242	1	40	-	-	-	-	-
IA: H265	49	53	-	37-61	-	-	-
OH: G32, C284	+	63	-	56-70	-	-	-
TN: E28	62	71	-	48-94	-	-	-
IL, IN: H10, R40, H128, B373	164+	84	74-107	33-114	-	-	-
KS: G74	12	94	-	-	12	14	-
PA: M560	+	107	-	99-117	-	-	-
OK: A10, T44, T41, L63, J25, J41, H150	27+	116	86-137	79-165	8+	6	2-9
AL: S118	-	-	-	-	+	43	+-227
AR: C202, T92	54	127	-	-	-	-	-
OK: T45 (introduced)	39	251	-	-	9	227	-
Combined	408+	91	71-107	33-251	29+	61	2-227
Age 0-November							
SD; W242	1	22	-	-	-	-	-
PA: M560	+	118	-	102-123	-	-	-
Age 0-December							
MO: W138	167	65	-	30-165	8	22	-

		TL				Weight	
	No.	Mean of means	Central 50%	Range	No.	Mean of means	Range
Age 0-December (cont.)							
TN: R292	1	88	-	-	-	-	-
PA: M560	+	121	-	114-124	-	-	-
OK: J95	4,569	140	-	112-168	-	-	-
Age 0							
SC: S343	3	51	-	-	-	-	-
PA: M174, M310	96	56	-	30-76	96	3	1-3
KY: C159, W68	13	73	-	43-137	-	-	-
IA: M323, M448, W240	39+	78	58-97	41-124	15	3	-
IN: B416	1	86	-	-	1	22	-
AR, OK: H156, H442, W68	274	102	61-142	43-183	-	-	-
CO: L126	18	109	-	64-142	-	-	-
LA: C368	+	-	-	71-132	-	-	-
WI: H250	+	-	-	78-108	-	-	-
IL: L129	+	135	-	-	-	-	-
Combined	444+	82	58-97	30-142	112	3	1-22
Age I							
MN, SD, WI: W192, H250, S348	18+	107	69-162	64-175	-	-	-
KY rivers: C175	42	124	-	119-137	-	-	-
PA: M174, M310, M560	146+	124	-	91-175	62	27	26-28
OH: L14, R58, O9, K78, R49, M248	4,312+	125	102-147	58-165	-	-	-
TX: E97	9	130	-	94-196	9	28	-
IA, IL, IN, KS rivers: M215, M448, U4, U8, B416	129+	132	91-173	91-208	50	31	28-34
MO, KS: S207, G74, C226, S340, H332, W119, W138	1,077+	136	107-140	58-251	818	72	11-156
CA: H146	50	142	-	112-185	50	31	-
NC: R217, S395, T103	302	143	130-155	122-168	-	-	-
CO: E51, L127, P207, P214	104+	147	131-150	80-193	-	-	-
OR: O24, O34, K80	110	152	140-175	124-183	-	-	..
IA: L29, E48, M198, M302, H200, B150, M323, I10, N99, W240, W241, M503	477+	153	137-168	102-251	397+	52	11-261
AR: T92, C202, H442	83	153	-	142-165	-	-	-
IL: T18, H10, B97, S166, H128, G80, E60, L129, B373	6,266+	155	155-183	71-241	16	79	
MN, WI: Mississippi R., C270, C356, U8	93	157	-	117-208	-	-	-
OK: W7, T44, J26, W68, T48, H127, J25, J26, W53, T45, J25, J41, H150, B183, H156, O37, C272, B220, A95, A97	2,278+	160	135-188	89-309	430	44	14-142
TN: E28, S329, S538, S540, R292	462	167	132-183	132-186	280	76	74-79
IN: R40, R34, J9, W220	159+	167	141-186	104-210	-	-	-

| | TL | | | | Weight | | |
	No.	Mean of means	Central 50%	Range	No.	Mean of means	Range
Age I (cont.)							
KY lakes: C160, C162	121	188	-	145-259	-	-	-
SC: S342	2	234	-	-	-	-	-
OK introduced: T45	3	310	-	-	-	-	-
Combined	16,184+	151	131-173	58-310	2,112+	49	14-261
Age II							
PA: M174, M310	94	154	140-160	124-213	94	35	28-40
KS, G74, C226, S340, H332	495+	155	145-165	122-236	122	43	-
MN, SD, WI: H250, S348, W192	8+	161	-	102-225	-	-	-
OH: M248, H96, L14, R58, O9, R78, R48	7,050+	170	152-193	119-216	-	-	-
IA, IL, IN, KS rivers: M215, M448, U4, U8, B416	259+	172	127-213	127-264	85	61	53-71
KY rivers: C175	56	173	-	157-308	-	-	-
NC: R217, S395, T103	334	175	157-192	155-213	-	-	-
OR: C197, G75, K80, O24, O34	64+	177	170-183	130-251	-	-	-
CA: H146	12	178	-	155-229	12	65	-
CO: E51, G158, L126, P207, P214	228+	178	171-196	114-229	3	50	10-80
OK: W7, J26, T48, W53, W68, T44, T45, H127, J41, H150, B183, L248, H156, O37, C272, B220, A95, A97	2,696+	179	142-200	102-323	809	52	14-386
IL: E60, L129, B373, T18, H10, T16, H128, S166, W105, G80	5,831+	196	169-221	136-255	39	136	-
IA: M323, H265, I10, M99, W240, W241, L29, E48, M198, M201, H200	177+	198	173-203	119-325	116+	126	31-608
IN: W220, R34, J9, R40	184+	200	195-208	169-234	-	-	-
TX: E97	42	203	-	122-274	38	133	31-215
TN: E28, S136, S138, T99, S329, M368, S540, R292	950+	205	188-218	178-279	364	130	77-198
AR: C202, T92, H442	60	208	-	201-213	-	-	-
MN, WI Mississippi R.: C270, C356, U8	115	208	-	157-310	-	-	-
MO: W119, W138, P30	788+	227	180-277	117-318	786	78	37-184
KY: C160, C162	267	266	-	196-366	-	-	-
SC: S343	63	307	-	257-358	-	-	-
Combined	19,773+	189	162-213	102-366	2,468+	79	10-608
Age III							
MN, SD, WI: H250, S330, S384, W192	78+	187	165-205	147-234	-	-	-
PA: M174, M310	96	200	193-206	122-257	96	96	68-124
CO: E51, G158, L126, P207, P214	144+	205	175-246	160-292	47	68	20-120

	TL				Weight	
No.	Mean of means	Central 50%	Range	No.	Mean of means	Range
Age III (cont.)						
OK: W7, J26, L248, T48, W53, C272, A95, A97, W68, T45, H127, T44, H156, O37, B220 — 755+	206	178-211	124-320	438	110	18-544
OR: C197, G75, O24, O34 — 47+	210	-	183-297	-	-	-
IA: H96, L29, E48, M198, M201, H200, M323, H265, I10, N99, W240, W241 — 484+	214	196-229	127-310	412+	147	57-608
OH: T18, O8, O9, R58, L14, R49, R78, M248 — 2,421	214	185-224	132-295	-	-	-
NC: R217, S395, T103 — 59	215	175-251	170-297	-	-	-
KY, river: C175 — 44	216	-	196-246	-	-	-
KS: S207, G74, C226, S340, H332 — 235+	216	185-229	168-297	13	204	65-343
IL, IN, IA, KS rivers: M215, M448, U4, U8, B416 — 158+	220	201-269	180-310	76	111	103-127
MO: W119, W138 — 201+	227	-	173-305	201	169	127-321
IL: H10, H128, S166, W105, G80, E60, L129 — 2,883+	230	218-259	152-328	1	187	-
AR: C202, T92, H442 — 28	230	-	211-246	-	-	-
TN: S13, S136, S138, T99, S329, S540, M368, R292 — 491+	240	205-239	196-330	472	157	130-320
IN: R40, J9, R34, W220 — 88+	242	211-246	190-313	-	-	-
MN, WI, Mississippi R.: C270, C356, U8 — 86	257	-	190-310	-	-	-
KY: C160, C162 — 482	306	-	208-386	-	-	-
TX: E97 — 117	262	-	145-356	115	230	147-323
SC: S343 — 55	320	-	292-358	-	-	-
Combined — 8,952+	221	190-244	122-387	1,871+	131	18-608
Age IV						
AL: M111 — +	218	-	-	-	-	-
IA: H96, L29, M198, M201, H200, M235, M323, H265, N99, W240, M503 — 235+	219	213-236	137-363	192+	132	68-425
MN, WI, SD: H250, S330, S384, W192 — 66+	221	183-235	180-290	-	-	-
OR: C197, C198, O24 — 23+	222	-	183-236	-	-	-
CO: E51, G158, L126, P207, P214 — 49+	223	-	167-343	34	77	40-120
PA: M174, M310 — 164	234	-	165-282	164	159	136-181
KS; MO: H332, G74, C226, F162, W138 — 62+	245	211-279	175-343	45	240	102-567
OK: W7, J26, T48, T45, W68, W53, L248, H127, T44, C272, A95, A97, B220 — 362+	246	224-260	130-338	124	235	23-644
OH: T18, L14, R58, O9, R78, R49, M248 — 636+	256	226-287	155-328	-	-	-

		TL				Weight	
	No.	Mean of means	Central 50%	Range	No.	Mean of means	Range
Age IV (cont.)							
IL: H10, T16, H128,							
W105, G80, E60, L129	722+	259	249-286	175-371	-	-	-
KY river: C175	19	259	-	213-295	-	-	-
TN: S13, S136, S138, T99,							
S329, S540, M368	910+	261	254-267	175-381	309	235	187-397
IL, IN, IA, KS rivers:							
M215, U4, U8, B416	44+	259	228-312	211-361	8	165	-
TX: E97	86	277	-	165-376	99	307	221-439
NC: R217, S395, T103	10	289	-	201-388	-	-	-
MN Mississippi R.:							
C270, C356, U8	33	291	-	254-348	-	-	-
AR: H442	9	295	-	-	-	-	-
IN: R40, R34, W220	38+	300	-	278-325	-	-	-
KY lakes: C160, C162	250	329	-	284-361	-	-	-
SC: S343	20	336	-	315-363	-	-	-
Combined	2,229+	253	224-287	130-388	951+	190	23-644
Age V							
CO: E51, G158, P207,							
P214	9+	235	-	165-343	8	60	40-100
MN, WI: H250, S384,							
W192	26+	246	-	193-326	-	-	-
IA: H96, L29, M201,							
M323, N99	11+	248	217-287	173-323	5+	204	136-284
PA: M174, M310	232	251	-	203-310	232	318	190-241
OR: C198, K80, O24	27+	263	-	229-339	-	-	-
KS, MO: C226, W138,							
H332	9	268	-	175-356	4	510	460-666
OH: T18, L14, R49, R78,							
M248, O9	207+	277	254-325	168-368	-	-	-
IA, IN, IL, KS rivers:							
M215, M448, U4, U8,							
B416	10+	276	-	250-353	5	264	262-269
TN: S13, S136, S138,							
S329, S540	149	280	264-300	264-338	149	323	255-680
OK: W7, W68, T48, W53,							
H127, T44, C272,							
B220, L248	71	292	262-309	175-378	60	393	50-1,197
IL: H10, T16, H128,							
W105, G80	164+	293	292-307	173-358	-	-	-
KY river: C175	4	305	-	297-330	-	-	-
TX: E97	15	307	-	224-356	14	349	-
MN, WI Mississippi R.:							
C270, C356, U8	16	314	-	292-340	-	-	-
KY lakes: C160, C162	35	352	-	299-404	-	-	-
SC: S343	13	375	-	356-394	-	-	-
Combined	1,011+	282	254-308	168-404	477+	299	40-1,197
Age VI							
SD: W192	37	246	-	-	-	-	-
PA: M174, M310	148	269	-	221-318	148	275	204-334
IA: L29, M201, U4, M323,							
N99	73+	281	231-335	231-376	1+	456	179-595
OR: C198, K80, O24	10+	296	-	268-361	-	-	-
OH: R58, L14, R49, M248	54+	303	274-333	193-356	-	-	-
OK: W68, C272, B220,							
L248	16	310	279-319	180-364	8	437	59-879

		TL				Weight	
	No.	Mean of means	Central 50%	Range	No.	Mean of means	Range
Age VI (cont.)							
TN: S13, S136, S138, S229, S540	104	310	296-324	279-351	104	437	351-680
IL: H10, H128, W105, G80	30	318	305-331	226-391	-	-	-
WI, Mississippi R.: U8	9	320	-	272-348	-	-	-
CO: E51, P207, P214	1+	338	-	330-353	-	-	-
KS, MO: C226, W138	7	347	-	274-366	3	590	516-757
TX: E97	1	356	-	-	1	635	-
SC: S343	6	380	-	358-402	-	-	-
Combined	496+	302	269-333	193-402	265+	391	59-879
Age VII							
SD: W192	14	269	-	-	-	-	-
PA: M174, M310	38	286	-	244-388	38	363	241-421
OR: C198, O24	3+	316	-	305-328	-	-	-
IA: L29, M198, N99	2+	334	-	271-368	1+	535	471-567
OH: R48, M248	3+	340	-	292-361	-	-	-
TN: S13, S136, S138, S329, S540	53	352	348-357	335-381	53	640	595-830
KS, MO: C226, W138	3	356	-	330-376	1	587	-
WI, Mississippi R.: U8	1	361	-	-	-	-	-
IL: H10, S166, W105, H128	8	363	-	201-402	-	-	-
SC, NC: S343, T103	6	383	-	323-505	-	-	-
OK: W7, W53, C272	2	390	-	386-394	2	910	740-1,080
Combined	133+	336	303-361	201-505	95+	553	241-1,080
Age VIII							
SD: W192	2	310	-	-	-	-	-
IL, OH: H128, R49, W105	3+	349	-	326-394	-	-	-
TN: S13	8	361	-	351-371	8	854	624-865
MO: W138	1	366	-	-	1	718	-
OR: O24	1	368	-	-	-	-	-
Combined	15+	354	-	310-394	9	820	624-865
Age IX							
OR: O24	1	384	-	-	-	-	-

		Mean calculated TL at each annulus								
	No.	1	2	3	4	5	6	7	8	9
Western waters										
G136 OR, Black L.	+	36	84	122	178	188	196			
G136 OR, Miller L.	+	30	84	122	165	183				
G161 OR, 51 waters	1,117	48	135	180	206	226	236	279		
G185 CA, Isabella L.	+	53	112	183	208	246				
G185 CA, Anderson L.	+	74	155	190						
G77 OR, 24 waters	+	58	140	190	218	236	241	272		
G185 CA, East Park L.	+	94	175	198	287					
G185 CA, Stony Gorge L.	+	76	165	218	241					
G185 CA, Bullards Bar L.	+	69	163	221	244					
G185 CA, Englebright L.	+	114	193							
G99 OR	7	53	150	239	264					
B134 CO, Loveland L.	2	97	157	249	282					
G161 OR, Fern Ridge L.	+	36	185	279	312	343	368	373		

	No.	Mean calculated TL and increment								
		1	2	3	4	5	6	7	8	9
Western waters (cont.)										
K80 OR, West		76	170	244	287	320	342			
	Incr.	76	80	58	43	33	22			
	No.	44	14	3	3	3	2			
Mean for Western waters		65	148	202	241	249	277	308		

	No.	Mean calculated TL at each annulus								
		1	2	3	4	5	6	7	8	9
Northern waters										
W192 SD, Lewis & Clark L.	+	69	102	147	180	208	244	269	310	
E3 MN	16	66	118	173						
C270 MN, Miller L.	30	48	119	203	231					
K33 MN	273	64	140	221	251	274				
G180 SD, Francis Case L.	830	65	164	223	260	289	308	322	336	

	No.	Mean calculated TL and increments								
		1	2	3	4	5	6	7	8	9
S227, S229, S276, N94, N115, SD, Gavins P.		58	109	150	188	221	275	323		
	Incr.	58	61	48	36	28	43	66		
	No.	748	458	260	172	60	10	1		
F105 SD, Oahe		53	104	168						
	Incr.	53	48	71						
	No.	177	58	14						
F105 SD, Oahe tailwaters		53	145	168	249					
	Incr.	53	92	53	104					
	No.	4	4	3	1					
S225, S226 SD, Ft. Randall tailwaters		66	140	178	229	280				
	Incr.	66	86	53	53	38				
	No.	79	30	16	11	2				
S569 SD, Lewis & Clark L.		78	144	189	214	242	268	287	308	363
	Incr.	78	68	50	34	34	29	22	18	11
	No.	1,225	1,062	846	668	560	391	167	35	1
F100 SD, Oahe L.		81	157	190						
	Incr.	81	84	66						
	No.	65	36	5						
S225, 226, 228, 393 SD, Ft. Randall L.		66	165	224	259	295				
	Incr.	66	107	66	43	36				
	No.	630	328	84	19	3				
C270 MN, Mississippi R. backwater		66	152	231	277	297				
	Incr.	66	99	76	41	25				
	No.	165	94	42	7	1				
N144 SD, Oahe		93	176	239	275	302	327	362		
unweighted	Incr.	93	84	62	33	22	15	19		
	No.	475								
Mean for Northern waters		66	138	193	238	268	286	313	318	363

	No.	Mean calculated TL at each annulus								
		1	2	3	4	5	6	7	8	9
Central waters										
E30 OH, poorest lake	-	48	86	122	147	224				
O37 NB, Harlan Co. L.	+	58	97							
H195 NB, Harlan Co. L.	318	69	109	145						
M248 OH, Buckeye L.	276	55	107	150	193	231	259	302		

	No.	Mean calculated TL at each annulus								
		1	2	3	4	5	6	7	8	9
Central waters (cont.)										
M245 IA, Pond 6	45	58	104	145	190	231				
M245 IA, Pond 2	50	51	130	168	198	244				
L29 IA, East L.	85	53	135	170	196	213	231	272		
H265 IA, Fairport	17	65	117	173	226					
M575 IA, Williamson P.	45	66	135	176	201	218				
B102 OH	+	76	127	178						
M197 IA, Macbride L.	+	99	165	180						
E48 IA, Clear L.	44	51	130	188						
M195 IA, Macbride L.	531	99	165	190						
C284 OH, St. Mary L.										
1946-47	+	66	142	196	234	264	300	343	371	
1953-54	186	56	104	155	201	234	258			
E30 OH, average	1,200	66	142	198	282	307	351	391		
L129 IL	669	135	183	211	269	310				
G80 IL, Horseshoe L.	168	67	155	216	257	297	331			
L29 IA, Red Haw L.	28	79	193	213	251	284	307			
R40 IN, Grassy P.	9	64	160	218						
P204 IA, Rathbun L. 1972	33	136	176	222	270					
R40 IN, Foot's Pond	230	69	145	224	297					
H194 NB, Minature L.	35	71	157	236	279	310				
P204 IA, Rathbun L. 1973	27	90	186	248	315					
E30 OH, best lake	-	94	211	300	345	368				

	No.	Mean calculated TL and increment								
		1	2	3	4	5	6	7	8	9
E84 OH, St. Mary's L.	-	58	130	160	206	251	262	262	315	
	Incr.	58	74	43	56	48	41	51	53	
	No.	220	138	89	44	19	4	1	1	
C193 IA, Backbone L.										
1957		53	99	160	201	218				
year after removal of										
19 kg/ha		56	127	142	185	224				
year after removal of										
29 kg/ha		69	160	203						
before removal	Incr.	56	46	58	43	18				
	No.	94	63	62	61	3				
after removal	Incr.	58	102	38	51	38				
	No.	10	41	10	1	26				
H200 IA, Ahquabi L.		91	140	165	196					
	Incr.	91	48	28	33					
N138 IA, Clear L.		74	152	178	216	236	254	323		
	Incr.	74	80	25	23	20	17	10		
	No.	457	364	249	86	23	5	1		
W105 IL, Crab Orchard L.		84	135	183	221	272	315	343	325	
	Incr.	84	51	46	37	39	32	29	33	
N99 IA, Clear L.		71	145	185	208	234	272	323		
	Incr.	71	76	41	23	15	13	10		
	No.	156	127	105	19	6	2	1		
S419 IA, Backbone L.										
1959-61		84	145	190	218	246	274	287	300	312
	Incr.	84	61	45	36	33	28	30	25	12
	No.	140	140	139	93	45	23	5	3	3

	No.	Mean calculated TL and increment								
		1	2	3	4	5	6	7	8	9
Central waters (cont.)										
M556 IA, Coralville L.		84	160	211	251	277	297	310		
	Incr.	84	79	48	38	33	18	23		
	No.	754	739	730	634	345	224	72		
J9 IN, Greenwood L.		122	188	213						
	Incr.	122	54	74						
	No.	105	30	2						
W240 IA, Decatur L.		58	147	225	234					
	Incr.	58	71	66	30					
	No.	150	72	11	4					
E60 IL, Upper P.		79	183	230	268					
	Incr.	79	98	64	29					
E60 IL, Lower P.		91	193							
	Incr.	91	102							
I10 IA, 3 Missouri R. Cutoffs		79	180	245						
	Incr.	79	94	96						
	No.	46	18	8						
Mean for Central waters		74	146	192	232	259	285	316	328	312

	No.	Mean calculated TL at each annulus								
		1	2	3	4	5	6	7	8	9
South central waters										
G74 KS, El Dorado L.	153	99	140	155	185	211				
W27 MO, L. of the Ozarks	65	56	122	163						
K71 TN, White Oak L.	108	86	137	175	163					
P112 MO, Salt R. Middle	72	64	122	175	211					
Lower	20	84	165	201						
W27 MO, L. of the Ozarks	35	53	127	183						
C175-6 KY, 2 rivers	165	58	124	183	236	287				
P113 MO, statewide	549	69	145	206	249	279	292			
Middle regions of streams	217	69	142	208	257	312	325			
Lower streams	332	71	147	203	244	264	277			
Poorest station	+	61	124	175	226	257	269			
Best station	+	84	165	234	279	312	325			
S207 KS, Fall River L.	49	112	130	216						
S100 TN, Douglas L.	11	74	185							
C373 KY, Cumberland L.	531	79	157	231						
F151 TN, Eastern reservoirs	4,462+	64	173	239	284					
C110 KY, Herrington L.	+	76	190	251	279					
C169 TN, 3 reservoirs	+	53	196	254						
P73 MO, Wappapello L.	468	76	178	259	297	320	338	353		
C160, C162 KY, Kentucky L.	925	117	201	264	302	325				
tailwaters	237	86	165	224	264	302				
T54 KY, streams	22	152	226	272						
T54 KY, lakes	31	91	208	274						
S100 TN, Cherokee L.	25	38	221	295						
B380 TN, Woods L.	9	203								

	No.	Mean calculated TL and increment								
		1	2	3	4	5	6	7	8	9
R292, R308 TN, Dale Hollow L.										
1963-64		80	152	184						
	Incr.	80	83	41						
	No.	29	9	2						
after shad stocking, 1970		75	172	235						
	Incr.	75	97	63						
	No.	84	50	10						

		Mean calculated TL and increment								
	No.	1	2	3	4	5	6	7	8	9
South central waters (cont.)										
W138 MO, L. of the Ozarks		84	152	216	279	330	343	358	366 .	
	Incr.	84	69	66	61	28	15	15	15	
	No.	1,805	1,038	248	50	9	5	2	1	
W138 MO, Wappapello L.		76	178	259	297	320	338	353		
	Incr.	76	102	81	41	25	20	15		
	No.	468	372	157	64	43	23	7		
W138 MO, Taneycomo L.		94	180	259	302	340	363	384		
	Incr.	94	89	71	38	25	15	13		
	No.	307	199	110	76	29	12	5		
W138 MO, Norfolk L.		102	211	284	335	376				
	Incr.	102	112	76	43	33				
	No.	200	99	44	11	2				
P71 MO, Clearwater L.	Incr.	94	114	79	43					
	No.	827	480	154	18					
W138 MO, Clearwater L.		107	226	315	376					
	Incr.	107	122	81	71					
	No.	326	147	38	3					
Mean for south central waters		84	166	226	263	303	319	362	366	

		Mean calculated TL as each annulus								
	No.	1	2	3	4	5	6	7	8	9
AR, OK Waters										
B377 OK, Boomer L.										
1966	145	99	122	141	182	229	268	343		
1953	+	61	104	142	193	256	274	391		
K65-6 OK, Hiwassee L.	145	71	109	145	178	221	216			
T48 OK, Cleveland L.	52	75	112	146	178	211				
B377 OK, Boomer L.										
1950	+	98	128	156	179	222	254			
J95 OK, Ardmore L.	139	53	109	160	216	259	300	330	363	
F62 OK, SubPrison L.	10	61	109	163						
after poison & restocking	49	58	203	246	320					
T48 OK, Pawnee L.	44	70	117	163	191					
J26 OK, Claremore L.	251	104	137	165	236					
L195 OK, Cimmarron R.	12	56	122	168	241	257				
T48 OK, Clinton L.	54	113	158	179	193	280	299			
F64 OK, Little R.	272	43	117	180	244	300	351			
J25 OK, Grand L., 1948	234	79	135	183	267	338				
F87 OK, Little R.	18	51	137	183	244					
tributaries	2	38	112							
cutoff lakes	252	43	114	178	244	300	351			
W68 OK, Texoma L.	862	97	142	185	226	254	259			
J26 OK, Henryetta L.	+	119	175	193	211	234 .				
T48 OK, Fairfax L.	51	94	163	198	300	345	352			
B220 OK, Ft. Supply	95	94	157							
J25 OK, above Tenkiller L.	77	56	183							
H303 OK, Lawtonka L.	143	58	132	201	264	333	363	386		
E61 OK, Salt C.	4	76								
J27 OK, Lower Spavinaw L.	27	51	178	203						
W239 OK, Texoma L.	653	86	152	208	257	295	328	353	391	
H153 OK	10,560	74	150	198	249	302	335	363	381	
slowest	+	30	69	102	132	165	221	320	358	
fastest	+	251	328	361	353	386	401	406	421	
8 new reservoirs	2,651	112	229	259	312	373				
17 old reservoirs	4,649	71	150	190	234	295	333	335	358	
17 lakes, 45-202 ha	1,996	61	124	175	239	290	315	361	363	

	No.	Mean calculated TL at each annulus								
		1	2	3	4	5	6	7	8	9
AR, OK Waters (cont.)										
H153 OK (cont.)										
43 lakes, 2-44 ha	709	69	145	203	254	312	340	358		
36 ponds	397	71	145	201	249	305	363	406	421	
4 streams	116	66	142	201	257	282	310			
45 clearwaters	+	71	152	206	257	305	338	366		
31 turbid waters	+	64	127	178	231	297	325	353		
J25 OK, Redwine L.	47	64	147	213	284	310				
W244 OK, Texoma L., 1942-62	2,610	66	147	216	249	284	323	353	366	
J31 OK, Grand L.	1,408	89	147	218	262	312	361			
J45 OK, Verdigris R.	11	66	185	221						
tributaries	82	53	130	203						
J37 OK, Rod & Gun Club L.	62	91	183	236						
J25 OK, Texoma L., 1948	967	91	150	236	257	318				
J41 OK, Ft. Gibson L., 1952	101	56	152	244						
J28 OK, Upper Spavinaw L., (new)	140	114	236							
Lower Spavinaw L., (old)	460	117	208	239	287	335				
J25 OK, Illinois R.	14	61	150	251	318					
H150 OK, Illinois R.	1	43	178							
J37 OK, Chickasaw L.	58	132	203	262	305					
O37 OK, Wister L. (from Latta)	+	104	201	269	330					
J25 OK, Wister L.	141	119	198	269	330					
J25 OK, Poteau R.	62	107	221	279	305					
H220 OK, Ft. Gibson L.	1,862	61	180	284	343	409				
J101 OK, Spavinaw L.	2,291	127	218	287	325	358	378	411		
J101 OK, Eucha L.	889	119	241	305	335	361	401			
J25 OK, Tenkiller L.	41	140	254							
B220 OK, Canton L.	102	216	283							
E126 OK, Rocket Plant L.	9	150	290	345						

		Mean calculated TL and increment								
		1	2	3	4	5	6	7	8	9
C272 OK, Boomer L.		61	104	142	193	257	274	391		
	Incr.	61	43	33	43	69	25	41		
	No.	330	315	123	51	32	5	1		
B220 OK, Blackwell L.		97	127	155	185	218	224			
	Incr.	97	32	30	33	41	33			
	No.	254	223	126	62	19	1			
B220 OK, Boomer L.		99	124	160	188	231	254			
	Incr.	99	25	24	25	51	51			
	No.	315	284	65	22	7	2			
C202, T92 AR, Ft. Smith L.		53	119	163						
	Incr.	53	69	58						
	No.	111	42	14						
A95, A97 OK, Keystone L.		111	144	189	244					
	Incr.	111	33	34	35					
	No.	1,759	1,286	16	4					
T44 OK, Grand L.		89	137	201	272	358	368			
	Incr.	89	46	64	41	23	8			
	No.	234	226	62	23	3	2			
L240 OK, Grand L.		75	150	209	255	302	348			
	Incr.	75	75	57	44	41	24			
	No.	234	234	232	169	12	4			

		Mean calculated TL and increment								
	No.	1	2	3	4	5	6	7	8	9
AR, OK Waters (cont.)										
J31 OK, Grand L		89	147	218	262	312	361			
	Incr.	89	58	71	43	48	36			
B220 OK, pond		69	127	224	300	325	358			
	Incr.	69	60	76	74	57	23			
	No.	36	23	8	7	2	1			
H156 OK, Canton L.		97	190	264						
	Incr.	97	94	64						
	No.	1,193	860	100						
B183 OK, Canton L.		102	201	269	279	283	310			
	Incr.	102	99	66	38	84	56			
	No.	834								
O37 OK, Heyburn Res.										
1949 year classes		46	264	305						
	Incr.	46	218	41						
	No.	4	4	4						
1950-52 classes		81	142	173						
	Incr.	81	61	25						
	No.	230	220	132						
Mean for AR, OK waters		84	162	210	254	291	316	369	375	

		Mean calculated TL at each annulus								
	No.	1	2	3	4	5	6	7	8	9
Southeast waters										
R119 VA, Drummond L.	+	74	137	188	234	269	297			
R117 VA, Claytor L.	200	66	135	203	267	305	348	409		
R116 VA, Claytor L.	200	64	119	216	272	318	353	409		
S100 NC, Hiwassee L.	25	61	173	241	241					

		Mean calculated TL and increment								
	No.	1	2	3	4	5	6	7	8	9
M368 NC, Cedar Cliff L.		66	109	142	160					
	Incr.	66	43	38	25					
	No.	40	40	8	1					
R217 NC, James L.		58	124	152	183					
	Incr.	58	41	38	23					
	No.	25	15	4	1					
S395 NC, Roanoke Rapids L.		76	130	157	231					
	Incr.	76	56	46	36					
	No.	37	31	14	3					
T103 NC, Falls L.		104	140	163						
	Incr.	104	33	20						
	No.	25	9	2						
T103 NC, Badin L.		112	145							
	Incr.	112	33							
	No.	73	46							
T103 NC, Tillery L.		74	130	178	340	399	447	478		
	Incr.	74	56	48	51	59	48	31		
	No.	126	119	8	1	1	1	1		
T103 NC, Blewett Falls L.		94	147	180	201	231	262	300		
	Incr.	94	53	38	33	30	31	38		
	No.	53	3	1	1	1	1	1		
R217 NC, Hickory L.		66	140	196	343					
	Incr.	66	86	91	71					
	No.	61	18	3	1					

		Mean calculated TL and increment								
	No.	1	2	3	4	5	6	7	8	9
Southeast waters (cont.)										
S395 NC, Kerr L.		76	150	216	366					
	Incr.	76	74	48	30					
	No.	28	23	10	2					
T103 NC, Catawba L.		102	163	218	224					
	Incr.	102	64	56	33					
	No.	40	26	13	2					
R217 NC, Rhodhiss L.		76	160							
	Incr.	76	89							
	No.	54	4							
T103 NC, Mt. Island L.		84	160	224	257					
	Incr.	84	64	48	36					
	No.	53	22	6	1					
S343 SC, Marion L.		48	175	251	284	312	320	333		
	Incr.	48	127	97	61	46	36	48		
	No.	129	127	75	28	13	6	2		
R217 NC, Lookout Shoals L.		56	137							
	Incr.	56	89							
	No.	76	30							
R217 NC, L. Lure		61	140	274						
	Incr.	61	94	114						
	No.	8	3	1						
S343 SC, Moultrie L.		56	208	287	340	371	381	378		
	Incr.	56	152	112	56	25	18	18		
	No.	34	34	23	15	10	4	2		
Mean for southeast waters		74	146	205	263	315	344	385		

		Mean calculated TL at each annulus								
	No.	1	2	3	4	5	6	7	8	9
Gulf States										
S157, MS, Sardis L.	+	114	221	292	338	363				

		Mean calculated TL and increment								
	No.	1	2	3	4	5	6	7	8	9
E97 TX, Koon Kreek Klub L.		86	178	234	282	325				
	Incr.	86	109	58	36	38				
	No.	278	226	108	15	1				
Mean for Gulf States		100	200	263	310	344				
Mean of regions		78	158	213	257	390	304	342	347	338

		Calculated weight at each annulus							
	No.	1	2	3	4	5	6	7	8
W105 IL, Crab Orchard L.	1,331	3	19	44	85	144	175	206	307
E48 IA, Clear L.	44	1	28	99					
H153, H419, OK	10,560	-	39	102	220	421	594	771	908
slowest	+	-	-	11	26	54	146	508	744
fastest	+	241	550	762	707	958	1,093	1,140	1,290
R119 VA, Drummand L.	+	17	48	108	187	295	436		
W139 MO, Niangua L.	1,805	6	40	130	301	510	592	674	725
M556 IA, Coralville L.									
1966	+	9	68	150	245	322	404	463	
1968	+	5	45	118	209	290	372	413	
W139 MO, Taneycomo L.	307	9	71	236	386	570	706	842	
M139 MO, Wappapello L.	468	3	68	236	369	462	555	650	
M138 MO, Norfolk L.	200	11	116	315	548	794			
M138 MO, Clearwater L.	326	11	148	442	609				

		No.	1	2	3	4	5	6	7	8	9
				Calculated weights and increments							
H200 IA, Ahquabi L.		389	8	31	51	86					
	Incr.		8	22	22	37					
E60 IL, Upper P.	Incr.	6		59	99	68					
Lower P.	Incr.	9		80							
Mean of calculated weights			8	60	169	295	423	479	574	647	

Young crappies of the year grew 1.3 mm per day in June and July in Tenkiller L., OK (H150), and in Larto L., IA, the total lengths of young crappies equaled 29.7 mm + 0.22 A, where A = number of days after April 1 (L74). In South Dakota, growth of age I white crappies was reported as 25 mm in 18 days (S228), 19 days (S225, F105), 20 days (S227), 23 days (S228), and 28 days (S229); and of age II, as 25 mm in 33 days (S228) and 34 days (S276).

Descriptions and photographs of scale features are given in A97 and H458. Annuli are distinguished by incomplete and fragmented circuli in a wide space and are followed by widely spaced circuli (A97). These features are most pronounced in the anterior field and on the antero-lateral ridge. The first and last annuli were easily identified on fish over age III, but the intermediate annuli were not distinct (A97). Two types of annuli were found (H458): the first did not go completely around the scale and the second was visible around the scale and usually showed evidence of scale resorption before the final complete circulus of the annulus, thus probably representing a spawning period.

False annuli were usually restricted to the anterior part of the scale and were not followed by widely spaced circuli as was true of true annuli (A97). False annuli were often on some but not all scales of a fish. Commonly, an annuluslike feature was present between the focus and the first annulus, however, this was not followed by widely spaced circuli (A97).

Scales from about 30 of 683 crappies were discarded as of questionable interpretation (W239). First and second readings of scales gave 83% agreement in one study (N138). The proper number of annuli were on the scales of known-age crappies (up to age II) in a pond (W138), and in other populations there appeared to be good agreement of calculated growth of different year classes. It was stated that Lee's phenomenon was not evident in N138. In the L. of the Ozarks study (W138), there was some evidence of Lee's phenomenon and of growth compensation. Lee's phenomenon was evident in direct proportion calculated data on Grand L., OK, but was not when corrected body-scale relationships by age groups were used (J31, J35).

Annulus formation in Oklahoma was reported as in late March to early June (W244), in mid-April to late May (H153), as early as mid-April (B220), and in early April to late May with the age I-III mostly in April and older crappies in May (W239). In Missouri, age I fish may form annuli in late April, Age I-II usually in May to early July, and age III-IV in July and August (W138). In Ohio, annulus formation was from early March to early fall (M248) or in July and August (R78); in Indiana, late April to mid-June (J9); in Illinois, late May to late August (H128) or May to early June for young crappies and June to late July for older age groups (H458); in Iowa, late May and early June (W240) or June and July (N138); and in South Dakota, late May (N94, S227, S229, S276) or early June (S228, S225, S393), in mid-May for age I and late May for age II (S226) and in early May for ages I-III and mid-May to early August for mature crappies (S569).

In L. of the Ozarks, MO, (W138) the growing season is May to October, when water is over 15.6°C. The younger fish complete more of the annual growth early in the season than do the older age groups, which are spawning. In Illinois, the growing season was reported as mid-May to late September or November, with mature crappies not starting growth until mid-June to mid-August (H389).

Direct proportion calculations were used in most studies, but a Lee-type correction was used in the following studies: OK (A95, A97, H156, W68, T48), NC (M368, T103), VA (R116), KS (G74), IN (R40), IL (E60, W105), IA (H200, N99, L29), MN (C270), SD (N144), and OR (K80). Empirical body-scale relationships were used in calculating growth in MO (P73, W138) and in OK (B200, J31). In stunted populations the body-scale regression was curvilinear because the scale grows faster than the length after growth slows (B220). The sigmoid body-scale relationships appeared to be so different in various lakes that 2 different relationships were used for separate series of lakes (W138). Faster growing populations had significantly greater numbers of scales in the lateral line (W138). An effect of the correction factor on the calculated length at the first annulus is illustrated by the following data from M368 and T103:

Correction factor	20	23	46	51	58	64	76
Calculated TL	74	66	94	84	102	104	112

With the sample sizes used, mostly under 50, the correction factors probably could not be accurately estimated. It is not expected that the true factors would vary as much as from 20 to 76. The calculated lengths at the first annulus are thus probably unduly affected by the correction. Crappies at 30 mm had a single row of scales, and at 43 mm they were completely scaled (C373). In this population the body-scale regression intercept was calculated at 37 mm.

The growth tabulations indicate that growth may be rapid or slow in all parts of the range but that in general growth is more rapid in the south than the north. Growth was better in streams than in lakes in Kentucky (T54), but this difference was not generally evident in data from elsewhere (e.g. H153). Growth in the Mississippi R. and its backwaters (U8, C270, C356) was better than in lakes at the same latitude. Growth was generally better in oxbow lakes open to the river than in those cutoff, perhaps due to higher population densities in the latter (W241). Crappies in lakes of 2-44 ha showed better growth than in those of 45-200 ha (H153).

Growth was more rapid in new than old reservoirs (H153, W192, F100, F105, H220, B220, J35), or it showed an increase shortly after impoundment (G180, S225, W138). In Arkansas (H442), growth was better through age II in new than in old reservoirs but not in older groups. In older lakes where crappie growth was slow, the food was mostly small centrarchids, but in newer lakes with more rapid growth, the food was mostly threadfin shad (B220). In Ft. Randall Res., SD, (S225) increments before impoundment were 41 (13 fish) for the 1st year and 94 (5 fish) for the 2nd year of life, and during the 1st year of impoundments the corresponding increments were 81 (74) and 130 (8). No consistent differences in average growth of white crappies were noted in three types of reservoirs in Iowa, although other species showed differences (M323, M530).

In South Dakota, growth was better in tailwaters than in the reservoir (F105, S225, S226), but the reverse was true in Kentucky (C160, C162).

White crappies are known to be more tolerant of turbid water than black crappies. Growth of both species improved with reduction of turbidity and increased volume of L. Ahquabi (H200). In Oklahoma, growth of white crappies was more rapid in clear than turbid ponds (H153). In Kansas, white crappie from a turbid pond grew better than those in clear ponds (H332). An inverse relationship between turbidity and growth was reported in North Carolina, where the 4 turbid reservoirs were among the 6 with slowest crappie growth (T104). White crappies had high mortality, and many had popeye conditions in clear-water ponds but had good success and reproduction in roily ponds (L244). Dr. James Henshall, however, reported successful spawning of white crappies in clear-spring ponds (L22).

No correlation between annual growth of white crappies with air temperatures or water levels was found in Clear L., IA (N138), but years of good growth for white crappie were years of poor growth for black crappie. Water level was not detectable as a factor affecting growth in OK (A97); but temperatures above 24°C caused slow growth, and crappies beyond age II did not grow at 27°C. Growth was better in years of high water in Coralville Res., IA (M556). Water drawdown appeared to be related to slow crappie growth in some California lakes (G185). Lowered water level in July slowed crappie growth (J9). Water level changes and introduction of white crappie were not reflected in the growth of white crappie in Grand L., OK (J35). Little annual difference in growth was noted in L. Texoma from 1942 to 1962 (W244, W239), suggesting little effect of water level or of angling pressure. White crappies averaged 338 mm at age IV in L. Taneycomo, MO (F162), in 1959 but only 211 mm after the lake became a coldwater reservoir.

In L. Oahe, SD, 52% of the variation in annual growth was related to percent change in surface area, average depth, and water level fluctuation (N144), with the first two being negatively correlated with growth.

Lack of forage fish was suggested as a cause of slow growth of crappies in reservoirs (M556, K64, C272, G185).

Stocking of threadfin shad was followed by increased growth of white crappies (R292, R308, G185). Stocking pike as predators resulted in increased growth but not in greater average size (P214).

Growth was fastest in ponds with lowest population density in Illinois ponds (B373).

Growth of the dominant year class was slower than others (A95, A97, L74, M530). Growth was improved after the population was thinned (S566, R313, S340), but there was no clear-cut improvement with removal of carp (S384). Population density was believed to be the most important factor affecting growth in Illinois (H389).

Growth rates decreased with elimination of commercial fishing in Reelfoot L., TN (S540). Growth increments in Coralville Res. were best in 1965, following a partial fish kill and a flood (M556).

Competition with buffalo, *Ictiobus*, may have been a cause of slow growth of white crappie (M556).

Growth was slower in ponds where white crappies were alone than in ponds with mixed populations (B373).

Characteristics of a stunted population were listed as (B220): scale grows faster than length, low condition factor, less shrinkage on preservation, and smaller average volume of food in digestive tract. After an extensive literature review of factors affecting crappie growth and dominant year classes, R313 suggested the following methods of correcting stunting and indicated further references to each:

1. Increasing food supply
2. Stocking of predators: e.g. largemouth bass, northern pike, largenose gar (little evidence of success in control of stunting)
3. Use of sterilized fish to control year class abundance
4. Encouraging crappie harvest
   a. encouragement of selective fishing
   b. removal of restrictive laws (bag and season limits)
   c. markers at locations where crappie congregate
   d. brush shelters
   e. encouragement of night fishing
5. Density thinning or population reduction
   a. rotenone or other chemical treatment
   b. trapping or seining
   c. water level fluctuation
6. Complete eradication and restocking

Growth and average K were not necessarily correlated (B373).

There is little evidence of sex differences in growth of white crappie. Males seemed a little faster growing in the 2nd year of life (B183) and in the 2nd to 5th year (W138), but females grew slightly faster (M248) and no difference was reported (S393, N144, H389, M248). The oldest female was age VI, and the oldest male was age VIII in one study (W138); but in another study (M248) more females than males were in age VI or VII, the oldest groups, and the females were faster growing at these ages.

White crappies matured at age II to III in Ohio (M248), Illinois (H128), and Tennessee (E28), but some matured at age I in Texas (H126). In L. Texoma white crappies matured at age II if they were 178 mm TL (W239). In Lewis and Clark L., SD, 47% matured at age I in 1966 when growth was fast, but only 23% matured at age II in 1963 and 1964 when growth was slow (S569). In the same lake, some males and females matured at age I (S226, S227, S393), but in another year only males matured at age I (S228). In L. Oahe, SD, most males matured at age II, but only 5.7% of the females (N144) matured. The smallest mature male in an Illinois study was 140 mm TL (H389), and the smallest mature female in Buckeye L., OH, was 150 mm (M248). In stunted populations they may spawn when only 109 mm (T113).

Ovaries prior to spawning contain small oogonia to mature ova at the same time, often with little sharp segregation of sizes (C377). Testes are attached only at the anterior end by mesorchia from the swim bladder, but are otherwise grossly similar to the ovaries (C377).

		No. of eggs per female				
		Mature eggs		Total eggs		
	No. of		Range		Range	
TL	females	Mean	(thousands)	Mean	(thousands)	Citation
150	1	970	-	13,591	-	M248 OH
160-170	18	12,683	8-14	22,723	13-24	M248 OH
180-219	24	21,457	16-32	31,977	22-52	M248 OH
220-280	13	76,509	40-123	91,505	57-152	M248 OH
290-330	4	164,126	125-213	245,408	206-326	M248 OH
-	42	7,120	2.9-14.8	-	-	H96 OH
190	1	14,750	-	-	-	L14 OH
-	+	53,000	25-92	-	-	W239 OK
211-316	24	-	23-194	-	-	S569 SD

		No. of eggs per female				
		Mature eggs		Total eggs		
	No. of		Range		Range	
TL	females	Mean	(thousands)	Mean	(thousands)	Citation
150	1	22	-	44	-	M248 OH
158	3	158	-	264	-	M248 OH
165-216	89	260	208-337	401	254-477	M248 OH
220-330	17	368	204-496	480	286-642	M248 OH
		No. of eggs per g of female				
211-250	+	277	-	-	-	S569 SD
251-290	+	260	-	-	-	S569 SD
291-330	+	338	-	-	-	S569 SD

The average diameters of mature eggs were 0.89 mm (0.82-0.92) (W239); 0.82-0.90 mm (M248); 0.89 mm (H128). Eggs in nests averaged 0.89 mm (H459). Embryonic development is described (M248).

Immature eggs were resorbed at the end of the spawning season, and the ovaries were smallest in July but increased slowly in size until March or April in Oklahoma (W239). The maximum ovary size comes in mid-March to mid-April for ages IV-VIII, mid-May for age III, and late May for age II (W239). Testes increased a little earlier than the ovaries.

During the spawning season, males develop dark pigmentation on the sides of the head, lower jaw, and breast. This coloration is lost soon after spawning (W239).

In Texas (S571) and Oklahoma (W239), spawning occurred from late March to early May; in Ohio, late March to July (M541) or late April to July with most in May and early June (M248); May-June in MO (W238, L243), IL (H128, H389, H459), and IA (M561); and from mid-May to mid-July in SD (S569, S226, S228, S567, S229, S393, N136). The gonads develop in the fall, and thus white crappie are prepared to spawn soon after spring temperatures are favorable (M541).

Nesting started when water temperatures were 15.6°C (H389). Since they usually spawn under overhanging banks or tree roots or at depths to 6 m and often in turbid water, observation of spawning is difficult (M248).

In SD ponds, males defended territories of about 1 m², with nests at 20-97 cm depth (S567). The depths were shallower in a cool year than when waters were warmer. Nesting usually occurred within a day of territory establishment, and polygamy and multiple spawning by an individual female were observed. Spawning occurred at 14°-23°C but mostly at 16°-20°C, and photoperiodism was believed to play a factor (S567).

In IL, males guarding nests, May 26, under an overhanging bank were 0.6-1.2 m apart and at 10-20 cm depth. Eggs were clinging to dead grass and root fibers (H459). Nesting may be in the open on hard or soft bottoms, flat or sloping, 10 cm to 1.8 m deep, usually in colonies (H389). Pituitary injections induced spawning (R302). In a reservoir, spawning was mostly in protected bays and shallow island areas (N136), and suitable spawning areas were limited as the bays were sealed off and as the reservoir aged (S569).

Incubation required 93 hours at 14.4°C, 43-51 hours at 18.3°-19.4°C, and 42 hours at 22.8°C, with about 20 hours between the first and last hatch of a single brood (S567). Egg mortality (0%-51.2%) was highest when incubation was longest (S567). The first hatch is expected when water reaches 20°C at water depth of 15 cm (S334), but hatches were reported at 18.3°C (S550). Embryonic development is described (M248). Length at hatching was reported as

1.21 to 1.98 mm (M248) or as a minimum of 2.56 mm compared to 2.32 mm
for black crappies (S568) or as 4.1 to 4.6 mm in nature (N136). The latter is
probably the swim-up stage. The last fry leave the nest about 95 hours after
the first hatch, at 4.1-4.6 mm, and the yolk sac is resorbed at 4.5-4.6 mm
(S567). No schooling of fry was noted, and fry leave the shallows at 50-60 mm
(N136).

Postlarval and fingerlings of white and black crappie could be distinguished
with the following criteria (S568).

	White crappie	Black crappie
5-6.5 mm	19 or fewer postanal myomeres	21 or more
6-16 mm	30 (rarely 31) myomeres	32
over 16 mm	5-6 dorsal spines	7

Strong year classes of crappies do not occur when there is a strong age I
bass population because of predation, but age 0 bass do not control young
crappies because the crappies spawn earlier than the bass and the fry go to the
limnetic regions before young bass begin feeding (S561). Age 0 crappies were
most abundant where older crappies were least abundant, where crayfish were
absent, and where chora was most abundant (B373). Age 0 crappies showed
only 10% mortality from mid-July to October 1 in an Illinois pond, whereas the
8 adults had 37% mortality in the same period (B373).

In L. of the Ozarks, MO (B404), white crappies were found at the surface
and to a depth of 8.5 m, with most at 2.4-4.8 m in June and early July, but in
late July and August they were seldom below 3.7 m because of low oxygen in
the hypolimnion. In September they moved down as oxygen concentrations improved. In L. Texoma, OK (G182), white crappies were concentrated near the
bottom, at a mean depth of 7.9 m, from December to February. The mean
depth of capture was 5.15 m in April, which was the closest mean to the surface. From March to May, the white crappies were nearer the surface than
from June to August, but none were found in water with less than 3.3 ppm O_2.
Crappies were deeper when the water was clear than when turbid. Crappies
over 250 mm TL were at greater depths than the smaller crappies, but the
smaller ones were more often taken in open water than along the shoreline,
while the reverse was true of the larger white crappies (G182).

The catch (kg/gillnet/day) of white crappie in L. Carl Blackwell, OK
(S570), was negatively correlated with water depth and positively correlated
with the biomass of mayflies and gizzard shad, both major forage items.
Multiple regression indicated that up to 47.6% of the variability in catch rate
was accounted for by depth of water, organic content of sediments, and particle
diameter and that 59% accounted for by biomass of gizzard shad, chironomids
and mayflies, water depth, and organic content. Catches were highest in shallow littoral zones, with an abundance of mayflies from August through
September.

The annual mortality rate was estimated at 0.36 or an i of 0.4447 in
Coralville Res., IA (M556).

In L. of the Ozarks (B404), 10.7% of the white crappies taken from July 31
to September 4 were fungused, with those around 100 mm showing 31% infesta-
tion, around 130 mm, 16%; around 150 mm, 8%; and those 180-310 mm, not
fungused. Perhaps crowding of small sizes was responsible. A heavy mortality of white crappie, but not other species, resulted from a columnaris infection in Oregon (D208). Winterkill in 1965 reduced the biomass 60%-75% in
Coralville Res., IA (M556).

	kg/ha Standing crop	Annual production
M556 IA Coralville Res.	8.5, 10.6	-
B417 IN White R.		-
	29.8, 38.5	-
thermally affected areas	8.3, 8.8	-
R313 TX Meridian L.	36.5	-
B373 IL ponds, 1 year*	39.7, 41.5, 48.6	38.3 41.1 42.6
2 years	72.2, 78.0, 100.4	
	Mean 80.7 Max. 225	
W247 pond	117	50.5

*In the second year the annual production was 39.2 kg/ha, where additions were made to the population, 92.0 kg/ha in a pond left alone, and 98.0 kg/ha where cropped. The highest production occurred with the lowest population density (B373).

The biomass of white crappies was lower than that of bluegills in the same ponds in other years (J40).

	yield in kg/ha
S572 IL Chautauqua L., commercial	3.0
R313 (from others) TN Reelfoot L., commercial	4.0, 6.0, 5.7, 5.0
R313 TX Meridian L., netting	10.7
C193 IA Backbone L., netting	14.6, 29.2
S213 MN Clear L., angling	28, 31

In Illinois (S572), it was noted that large year classes had heavy mortality of age II white crappies and that there was only about a 9% harvest but over 50% annual mortality. Thus, angling and commercial fishing appeared to have little effect on the population.

Crappies tagged in Wheeler Res., AL (W243), moved 129 km upstream and 13 km downstream, which were greater distances than for 3 species of catfish released at the same time. A 5.7% return of white crappie tags resulted in one year.

Young crappies feed mostly on entomostraca (G185, N117, E121, N115, N130, S357, N136, S567, M560, S569, A96, and M561). Rotifers, copepod nauplii, and *Bosmina* were the food of 8-14 mm fry (A96). Nauplii were the principle food to 6 mm; *Cyclops,* to 10 mm; but *Daphnia* were taken at 8-14 mm (S567). *Cyclops* were selected from 11-20 mm but were unimportant in crappies 30-100 mm, when *Daphnia, Diaptomus,* and *Leptodora* were the major foods (N136). Zooplankton were almost the sole food the 1st year, with *Daphnia* dominant June-October and *Cyclops,* September-April; but a few amphipods and insects were taken in late fall and spring (M560). Chironomids were usually the first insects taken, at 75 mm (N117), before 60 mm (E121), or 66 mm (M130).

Entomostraca continue to be a significant food throughout the life of white crappies in most water (W238, S569, M562, S357, N99, F91, and H250), but insects and forage fish are usually the major foods of larger crappies (C272, H243, G185, M562, R288, and W221). Small shad and other fish were dominant June to January, but chironomids, mayflies, and entomostraca were dominant from February to May in L. Texoma (W238). Threadfin shad comprised over 97% of the volume of food of white crappies in Benbrook L., TX, except in March to May when plankton and insect larvae contributed up to 45% (G198).

The daily ration was greater in the summer than at other seasons and was
2%-4% of the body weight. Smaller crappies consumed more food for their
weight than the large ones (G198). White crappie fed on shad and other fish
more than did black crappie in Beaver Res., AR (N413). In Arkansas reservoirs, threadfin shad were the major food July to September or December
(B402), particularly in years when young shad were abundant, but *Chaoborus*
was also an abundant food. *Hexagenia* mayflies were important after crappies
reached 150 mm (S569). *Hexagenia* were more commonly taken by large than
intermediate sized crappies (R288). Perch, small crappies, and gizzard shad
were the major goods of crappies, 185-360 mm, in Pymatuning L., PA (M310).
The lengths of shad, *Dorosoma,* eaten by white crappies at different lengths
were given as /B220):

| TL crappies | TL *Dorosoma* | | | | | |
| | Control | | Ft. Supply L. | | Boomer L. | |
	Mean	Range	Mean	Range	Mean	Range
163-177	69	-	46	33-56	-	-
178-202	64	56-69	53	51-58	102	-
203-228	69	41-89	53	38-64	-	51-102
229-253	69	33-89	-	-	-	51-89
254-278	76	33-89	56	-	-	38-114
279-304	89	-	-	-	-	-
305-329	-	-	114	-	-	64-127

A 259 mm crappie ate a 140 mm crappie, a 342 mm crappie had eaten a
58 mm carp, an 89 mm carp, and a 114 mm shad (B220). Minnows were eaten
by 80 mm crappies.
    Other species of fish eaten by white crappies included: *Notropis hudson-
ius, N. spilopterus, Notemigonus crysoleucas, Ictalurus punctatus, Lepomis*
spp, *Etheostoma olmstedi,* and *Percina caprodes* (M562).
    Argulus in stomachs of four white crappies suggested that they may show
a cleaning symbiosis with other fishes (S545).
    Young crappies fed mostly in the daytime (morning and early afternoon
chiefly), and the food passed through the stomach in 14-17 hours (M560).
Adult crappies feed day or night but mostly at dusk (H389). Feeding experiments showed the greatest feeding activity at dawn and dusk, and white crappies fed more in the daytime than did black crappies (C376). Night anglers
caught crappies at a rate 16 times faster than in the daytime at L. Havasu, CA
(G185). During the summer white crappies fed all night, but in other seasons
most of the feeding was at dawn and dusk (G198). Activity peaks in the summer in the Mississippi R. were at 8-10 P.M. and 4-6 A.M. (R288). Crappies
feed throughout the year (H389 and others) but less when temperatures are
below 10°C (M562).

    The lengths of white crappies taken in gillnets of various mesh size are
as follows:

| | | MO (W138) | | | LA (L216) | |
Bar measure	No.	Mean TL	Range	No.	Mean TL	Range
19	182	119	104-147	-	-	-
25	482	155	114-196	1	165	-
38	10	196	114-236	1	378	-
51	1	211	-	5	353	305-406
76	-	-	-	3	386	381-394

The red blood cells per mm were reported as 1,795,000 ± 687,222 on 12 Illinois crappies (S151), and the haemoglobin was 4.87 ± 5.4.

Additional growth data are apparently available in C379 and J111, which were not seen and food data in M563. The number of chromosomes is reported as 40 to 50 (R286).

BLACK CRAPPIE, *Pomoxis nigromaculatus* (Lesueur)

The range of black crappies extends from southern Manitoba to Quebec south to Florida and Texas (S574, E119). It prefers clearer, deeper, and cooler water than the white crappie, and its range extends further north. The black crappie was widely distributed in fish rescue work (S574) and probably did not occur in Kansas prior to about 1900 (C349).

	No.	SL	FL/SL	TL/SL	TL/FL
C22 MN L. of Woods	199	20-286	1.176	-	-
	7	30-50	-	1.263	1.058
C35 MN L. Vermillion			1.182	1.204	1.018
C37 MN	40	90-130	1.247	1.308	-
	70	131-170	1.242	1.299	-
	40	171-210	1.229	1.289	-
C37 MN SL to end					
of flesh	60	x-100	1.200	1.270	-
	91	100-149	1.186	1.241	-
	283	150-249	1.172	1.230	-
	75	250-299	1.154	1.216	-
J9 IN Greenwood L.	-	-	1.228	-	1.04
C30 IA	76-123	-	1.21	1.28	-
C33 IA ponds	10	80-163	1.223	1.285	-
L29 IA East L.	19	70-189	1.203	-	-
	30	40-189	-	1.274	-
E48 IA Clear L.	205	70-250	1.221	1.284	1.060
M302 IA lakes	76	53-360	-	1.388	-
S289 MA	186	30-309	-	1.298	-
H226	1,088	40-259	-	1.303	-
	862	260-371	-	1.300	-
L29 IA Red Haw L.	96	70-189	1.218	-	-
	132	260-371	-	1.284	-
H200 IA Ahquabi L.	195 crappies, 55-240 mm,				
		SL = 0.138 mm + 0.7878 TL			

Where standard lengths were given without conversion factors, they were converted on the basis of TL = 1.27 SL and fork lengths on the basis of TL = 1.05 FL. Black crappies preserved in 10% formlin shrank 1%-3% of their length and increased in weight 6%-5% with most changes taking place in 6 days (J116). When transferred to alcohol the weight dropped to 95% of the original weight.

			Weight in grams		
	Mean of	Central			
TL	means	50%	Range	No.	Citations
15-24	0.06	-	0.02-0.08	22	B220 OK
25-50	0.4	-	0.3-0.9	295	S472 AL
	0.55	-	0.1-1.4	93	B220 OK

	Mean of	Weight in grams Central			
TL	means	50%	Range	No.	Citations
51-75	1.4	-	-	2	B220 OK
	1.8	-	0.5-3.2	3,772	S472 AL
	3	-	-	+	K71 TN
	5	-	-	+	F105, S225: SD
	6	-	-	1	M302 IA
	3	1.8-5	0.5-6	3,775+	Combined
76-101	4	-	2-6	2,839	S472 AL
	9	-	-	+	F105, S225: SD
	11	9-13	8-14	102+	B220, E97, K71, S97: South
	17	-	3-30	55	L29, C30: IA
	76	-	-	3	C20 MN
	15	9-13	2-76	2,999+	Combined
102-126	14	-	9-27	7,592	S472 AL
	15	14-18	9-18	35+	A88, E48, E84, C33, M499, R81: Central
	20	-	17-24	57+	B220, E97, K71: South
	21	-	19-23	+	F100, F105, F124, N94, N115, S225-9, S276, S393: SD
	30	-	28-34	11	E97, M20: MN, WI
	19	14-23	9-34	7,695+	Combined
127-151	23	-	9-45	2,163	A39, S472: AL
	34	31-37	26-50	105+	B220, E97, K71, S97, V44, South
	37	-	-	5	E51 CO
	38	28-40	15-70	170+	A88, C30, C33, E48, E84, I10, L29, M302, M499, O7, R81: Central
	40	-	36-44	+	F100, F105, F124-5, N94, N115, S225-9, S276, S393: SD
	47	42-51	42-51	109+	B365, E97, M20, M136: MN, WI
	50	-	-	+	U8 Mississippi R. MN, WI
	59	-	43-71	9	B317, OR
	37	30-42	9-71	2,561+	Combined
152-177	36	-	28-50	470	A39, S472: AL
	61	54-65	36-100	81+	C89, E38, E84, I10, M302, M499, O7, R81: Central
	62	57-66	31-79	92+	B220, E97, K71, V44: South
	63	-	54-77	+	F100, F105, F124-5, S225-9, S276, S393, N94, N115: SD
	73	-	-	1	E150 AZ

TL	Mean of means	Central 50%	Range	No.	Citations
152-177 (cont.)	75	60-81	60-122	472+	B365, C20, E97, M20, M136: MN, WI
	78	-	57-100	33	B317 OR
	82	-	-	+	U8 Mississippi R. MN
	62	56-71	28-122	1,149+	Combined
178-202	73	-	45-113	50	S472 AL
	91	82-100	56-142	134+	C30, C33, C89, E48, E84, I10, M302, M499, O7, R81: Central
	95	85-105	68-169	76+	B220, E97, F74, H156, H226, H228, K71, P189, S97, V44: South
	96	-	82-118	+	F100, F105, F124-5, N94, N115, S225-9, S276, S393: SD
	109	82-136	82-138	603+	B365, C20, E97, M20, M136: MN, WI
	122	-	-	+	U8 Mississippi R.
	128	-	100-156	24	B317 OR
	97	85-106	45-169	887+	Combined
203-228	104	-	59-181	42	S472 AL
	105	-	99-125	4	M140 CA
	125	-	122-127	3	B232, E51: CO, MT
	139	122-159	79-268	110+	C30, C33, E48, E84, I10, L29, M302, M499, O7, R81: Central
	141	130-154	105-224	670+	B220, F74, H156, H226, H228, H242, E97, K71, P189, S97, V44: South
	152	-	139-181	693+	B365, C20, M20, M136, MN, WI
	154	-	136-172	+	F105, F124-5, N94, N115, S225-8, S330, S393: SD
	176	-	115-260	245	B317 OR
	177	-	-	+	U8 Mississippi R. MN
	186	-	159-209	3	E150 AZ
	143	123-159	59-268	1,770+	Combined
229-253	141	-	77-227	52	S472 AL
	170	-	125-198	6	M140 CA
	189	175-211	110-293	82+	C30, E48, E84, I10, L29, M302, M499, O7, R81: Central
	211	179-232	168-340	1,763+	B220, F74, H156, H226, H228, K71, E97, S97, V44: South

Weight in grams

BLACK CRAPPIE

Weight in grams

TL	Mean of means	Central 50%	Range	No.	Citations
229-253 (cont.)	225	-	216-244	533+	B365, E97, M20, M136: MN, WI
	227	-	195-263	+	F100, F105, F124-5, N115, S225-8, S330, S393: SD
	233	-	200-290	49	B317 OR
	238	-	186-277	10	E149, E150, AZ
	245	-	-	+	U8 Mississippi R. MN
	262	-	227-317	20	C208, E51: CO
	210	178-233	77-340	2,515+	Combined
254-278	191	-	104-272	51	S472 AL
	235	-	198-252	2	M140 CA
	286	201-340	201-363	868+	B365, C20, E97, M20, M136: MN, WI
	289	256-306	181-454	888+	B220, E97, F74, H156, H226, H228, K71, M242, S97: South
	296	256-343	189-406	22+	C30, E48, E84, I10, L79, M302, R81: Central
	297	-	241-358	13	E149, E150: AZ
	305	-	270-340	5	B317 OR
	338	307-363	277-372	+	F124, N115, S226-8, S330, S393: SD
	345	-	-	+	U8 Mississippi R.
	289	268-336	104-454	1,849+	Combined
279-304	301	-	275-325	3	M140 CA
	308	-	272-345	13	S472 AL
	345	301-375	227-422	799+	B365, C20, E97, M20, M136, N52: MN, WI
	354	320-374	251-482	18+	C30, E48, E84, I10, L29, O7, M302, R81: Central
	361	348-372	272-504	672+	B220, E97, F24, H156, H226, H228, M242, S97: South
	380	-	372-390	+	S228, S330, S393: SD
	418	-	341-518	9	E149, E150: AZ
	421	-	354-488	21	C208, E51: CO
	449	-	-	+	U8 Mississippi R.
	363	330-394	227-518	1,535+	Combined
305-329	408	-	363-454	2	S472 AL
	419	-	350-502	10	M140 CA
	452	420-454	318-544	489+	E97, F74, H226, H228, M242, S97, South
	462	417-488	417-530	142+	B365, C20, E97, M20, M136: MN, WI
	490	406-587	343-609	17+	C30, E48, L29, M302, R81: Central
	506	-	486-526	+	S228, S330, S393: SD

TL	Mean of means	Weight in grams Central 50%	Range	No.	Citations
305-329	511	-	454-636	7	E149 AZ
(cont.)	587	-	-	+	U8 Mississippi R.
	468	419-511	318-636	667+	Combined
330-355	525	-	425-652	14	M140 CA
	560	482-593	454-699	9	C30, E48, L29, M302: Central
	582	517-596	454-998	259	F74, H226, H228, M242, S97, S472: South
	593	-	563-663	111+	B365, C20, M20, M136: MN, WI
	609	-	-	9	E51 CO
	726	-	-	+	S228 SD
	731	-	-	+	U8 Mississippi R.
	591	520-644	425-998	402+	Combined
356-380	663	-	627-750	9	M140 CA
	667	-	569-765	2	C30 IA
	694	-	680-718	2+	B365, M20, M136: MN, WI
	705	655-703	544-907	66	F74, H226, H228, M242, S97, S472: South
	746	-	-	2	E51 CO
	862	-	-	+	U8 Mississippi R.
	708	-	569-862	81+	Combined
381-405	712	-	-	1	E150 AZ
	717	-	675-915	10+	B365, C20, M20: MN, WI
	828	-	680-998	15	F74, H226, H228, M242, S97: South
	964	-	-	1	E51 CO
	785	680-860	675-998	27+	Combined
406-431	844	-	550-1,225	8+	B365, C20, M20: MN, WI
	942	-	-	1	S97 TN
	1,350	-	-	1	M140 CA
	920	-	550-1,350	10+	Combined
432-456	1,206	-	1,097-1,315	3+	B365, C20: MN
	1,220	-	1,097-1,315	1	S327 MD
	1,209	-	-	4+	Combined
457-482	794	-	-	1	S398 MD
	1,352	-	-	3	C20 MN
483-507	2,268	-	-	1	S343 SC
559	2,098	-	-	1	C349 KS

	No.	SL	
L29 IA East L.	-	-	log W = -4.231 + 2.870 log SL
E48 IA Clear L.	169	70-250	log W = -4.459 + 2.985 log SL
S97 TN Norris L.	-	-	log W = -4.632 + 3.053 log SL
L29 IA Red Haw L.	-	-	log W = -4.710 + 3.081 log SL

	No.	TL	
T146 AL	-	-	log W = -4.695 + 2.908 log TL
	-	-	log W = -4.710 + 2.914 log TL
	-	-	log W = -4.928 + 3.028 log TL
V91 SD Lewis & Clark L.	30	males, May	log W = -5.310 + 3.192 log TL
	35	females, May	log W = -4.804 + 2.993 log TL
	1,229	-	log W = -5.019 + 3.075 log TL
S472 AL	17,328	25-360	log W = -4.846 + 2.97 log TL
J37 OK Chickasaw L.	104	100-280	log W = -4.862 + 3.001 log TL
J110 IA Spirit L.	303	115-300	log W = -4.928 + 3.066 log TL
S225 SD Ft. Randall L.	-	-	log W = -4.995 + 3.074 log TL
P204 IA Red Rock L.	36	134-343	log W = -4.990 + 3.080 log TL
H200 IA Ahquabi L.	82	56-235	log W = -5.190 + 3.143 log TL
J95 OK Ardmore City L.	64	-	log W = -5.294 + 3.177 log TL
N144 SD Oahe L.	318	95-394	log W = -5.252 + 3.198 log TL
C355 PA Alanconnie L.	629	30-221	log W = -5.213 + 3.262 log TL
B403 SC Par P. heated	88	-	log W = -5.220 + 3.264 log TL
control	13	-	log W = -4.906 + 3.044 log TL
H153 OK	442	-	log W = -5.646 + 3.341 log TL
G42 Upper Mississippi R.	728	125-330	log W = -5.630 + 3.35 log TL
B220 OK hatcheries	117	15-56	log W = -5.659 + 3.351 log TL
B220 OK Rod & Gun Club L.	125	100-230	log W = -5.780 + 3.370 log TL
H156 OK Canton L.	40	200-290	log W = -5.832 + 3.475 log TL

The intercept values in B403 were changed from those reported in the paper, even though it was stated that lengths were in mm and weights in g because it appeared that lengths in cm had been used.

	No.	SL(mm)	K(SL)	Range
W105 IL Crab Orchard L.	8	age III	2.31	-
	1	age IV	3.48	-
C34 MN Leech L.	2	-	2.40	-
H62 IN	140	-	2.47	-
C23 MN 1941-43	661	-	2.57	-
C33 IA ponds	10	81-163	2.75	-
S100 TN Hiwassee L.	9	-	2.79	-
E28 TN Chickamauga L.	131	-	2.9	-
L29 IA Red Haw L.	139	30-279	2.94	-
S100 TN Douglass L.	27	-	3.04	-
M235, S419: IA Backbone L.	+	-	3.05	-
S419 IA Backbone L., after thinning population	+	-	5.26	4.84-5.79

	No.	SL(mm)	K(SL)	Range
S97 TN Norris L.	643	55-317	3.07	-
E3 MN	1,917	76-380	3.08	-
L29 IA East L.	39	70-189	3.09	-
M302 IA	55	47-260	3.20	2.50-4.18
E48 IA Clear L.	169	70-250	3.25	-
S100 TN Cherokee L.	85	-	3.38	-
C36 IA	56	86-312	3.39	2.30-4.04
B315 AZ Roosevelt L.	1,008	160-300	4.05	3.08-4.79
G184 AZ Roosevelt L.	770	-	3.43	3.09-3.90
E150 AZ Roosevelt L.	219	160-380	3.62	3.47-3.78
K152 AZ Roosevelt L.	203	100-399	3.72	2.66-4.72

	No.	TL	K(TL)	Range
A88 IA Little Wall L.	36	124-147	0.97	0.80-1.08
B220 OK pond	53	94-160	1.04	0.78-1.10
	4	179-275	1.36	1.30-1.44
hatcheries	120	15-56	0.79	0.58-0.89
Ft. Supply L.	32	147-285	1.63	1.55-1.69
Canton L.	18	163-236	1.57	1.41-1.72
B403 SC Par P.	11	80-110	ca 1.07	0.90-1.35
	+	310-410	ca 1.55	1.07-1.93
J37 OK Rod & Gun Club L.	125	100-230	-	0.91-1.25
S566 IA Backbone L. 1956-66	+	-	-	0.86-1.61
C89 IA	7	173-198	1.25	1.11-1.39
J37 OK Chickasaw L.	104	100-280	-	1.08-1.55
S472 AL	295	25	2.38	-
	16,818	50-155	1.21	0.89-1.35
	215	175-370	1.31	1.15-1.76
T146 AL	+	-	-	1.00-1.45
M499 IL, Northeastern	117	102-230	1.35	0.86-1.47
J95 OK Ardmore City L.	64	-	-	1.19-1.52
H200 Ahquabi L.	82	56-235	1.36	-
N138 IA Clear L.	125	127-270	1.41	1.19-1.52
V91 SD Lewis & Clark L.	1,229	based on mean lengths	1.42	1.23-1.72
H226 FL	263	175-385	1.44	1.33-1.88
N94, 115; S227, 229, 276; SD Gavins Pt. L.	883	102-260	1.45	1.41-1.69
V44 LA Springhill L.	57	125-255	1.47	1.16-2.35
S225, 226, 228, 330, 393: SD Ft. Randall L.	1,304+	102-310	1.51	1.40-1.63
H153 OH	442	-	1.52	-
E94 TX Koon Kreek Klub L.	413	89-315	1.52	1.09-1.85
F100, 105, 124, 125: SD Oahe L.	248	75-260	1.59	1.49-1.62
J110 IA Spirit L.	303	115-300	1.66	1.41-1.86
H156 OK Canton L.	40	200-290	1.68	1.30-1.97
E97 WI Chetek L.	82	109-325	1.68	1.42-1.88
P204 IA Red Rock L.	36	134-343	1.71	1.15-2.10

	No.	TL	K(TL)	Range
C23 MN Standards poor			less than 1.05	-
average			1.22-1.50	
excellent			over 1.88	-

While the length-weight relationships vary considerably, it is difficult to distinguish any regional differences. In general, weights at most lengths seem less in the south than the north. The weight seems to increase more than the cube of the length—as indicated by the slopes of most regression lines and the increase in K with increased length (S226, S227, C23, S472, H156, H226, B220, and B403). The very high K for 25 mm crappies in S472 is probably an error in weighing these small fish. No trend in K with increased length was noted in S229, S276, S226, and B315 but a slight decrease with increased length was noted in S225. K increased with age through age III (E97).

The slopes of regression lines calculated for monthly samples for several years from Lewis and Clark Lake, SD (V91), were distributed as follows:

	2.75-2.84	2.85-2.94	2.95-3.04	3.05-3.14	3.15-3.24	3.25-3.34
No.	10	4	3	10	5	5
No.*	1	0	0	1	2	3

	3.35-3.54	3.55-3.74	4.31
No.	4	3	1
No.*	3	1	0

Those marked (*) are those which differed at the 95% confidence level from 3.0. Many of the samples were small with large variances. The 4.31 slope was based upon 3 fish. The slopes were less than 3.0 in June each year, probably because the longer fish lost weight in spawning. In most years the slopes increased in July and August. K values and growth increments were not correlated (V91).

In Roosevelt L., AZ, K averaged lowest in June and July and highest from October to February (E150). In Koon Kreek Klub L., TX, mean K was lowest in April and May and highest in February (E97). In Alabama, condition was better in August-September than in May-June (T146). No trends in mean K values during the summer were evident (S227, S228, S276), but S393 reported a slight increase during the summer and S229 a slight decrease.

No sex differences in K were evident in V91, F100, S393, and S226-229.

Drawdown of water levels to control populations did not improve the average K (P189). Thinning the population did improve the average K in Backbone L., IA (S419, S556).

Introduction of threadfin shad as forage increased the average K of black crappies in Roosevelt L., AZ (B315).

| | TL | | | | Weight | | |
	No.	Mean of means	Central 50%	Range	No.	Mean of means	Range
Age 0-April							
TX: S150	58	10	-	-	-	-	-
Age 0-May							
TX: S150	101	20	-	-	-	-	-
AL: T146	+	-	-	18-70	-	-	-
AZ: E148	+	50	-	-	+	-	x-15

	TL				Weight		
	No.	Mean of means	Central 50%	Range	No.	Mean of means	Range
Age 0-June							
ON: A102	+	-	-	5-13	-	-	-
TN: S97	479	18	-	5-33	-	-	-
IA, OH: C284, H200,							
M561	42+	25	16-34	8-38	-	-	-
OK: B220	41	25	-	15-38	-	-	-
TX: S150	67	30	-	-	-	-	-
AL: T146	+	-	-	55-90	-	-	-
Combined	629+	26	18-30	5-90	-	-	-
Age 0-July							
TN: S97	625	23	-	10-33	-	-	-
IA, IN, OH: B2, B3,							
C30, C91, C102,							
C109, C284, E48,							
H200, L29, M561,							
N99, R40	727+	35	29-37	14-61	-	-	-
WI: P7, N141	6+	37	-	20-42	-	-	-
CA: K59, M140	1+	49	-	31-76	-	-	-
AL: T146	+	-	-	80-110	-	-	-
OK: B220, J34, J42	304	87	-	30-152	-	-	-
Combined	1,663+	45	30-41	10-152	-	-	-
Age 0-Aug.							
OK: B220	30	47	-	41-56	-	-	-
CA: H146	8	48	-	28-66	-	-	-
WI: P7, N141	10+	50	-	37-65	-	-	-
TN: E28, S97	416	54	-	25-81	-	-	-
Central B2, B3, C30,							
C91, C102-3, C109,							
E48, H200, L29,							
M561, N99, R40	792+	54	45-61	11-91	-	-	-
SD: S225	+	61	-	-	-	-	-
France: D28	1	84	-	-	-	-	-
AL: T146	+	-	-	105-180	-	-	-
Combined	1,257+	55	46-61	11-140	-	-	-
Age 0-Sept.							
WI: M265, M308	+	64	-	48-81	-	-	-
TN: E28, S97	207	67	-	36-81	-	-	-
IA: C102, C103, E48,							
H200	158	74	68-80	46-97	-	-	-
SD: S225	+	97	-	64-122	-	-	-
AL: T146	+	-	-	115-170	-	-	-
LA: V44	19	142	-	127-239	19	48	28-261
Combined	384+	90	58-81	36-239	19	48	28-261
Age 0-Oct.							
IA: B4, E48, J104	94+	72	69-71	57-86	-	-	-
TN: E28	10	81	-	-	-	-	-
OK: J42	3	135	-	x-160	-	-	-
AL: T146	+	-	-	130-200	-	-	-
LA: V44	38	190	-	147-249	38	116	40-340
Combined	145+	106	69-135	57-249	38	116	40-340
Age 0-Nov.-Dec.							
OR: B317	+	52	-	25-80	-	-	-
WI: P7	22	61	-	-	-	-	-
TN: S97	403	76	-	56-117	-	-	-
OK: J34	11,934+	132	-	102-160	-	-	-
Combined	12,359+	92	61-132	25-160	-	-	-

| | | TL | | | | Weight | |
	No.	Mean of means	Central 50%	Range	No.	Mean of means	Range
Age 0							
CA: H146	1	41	-	-	-	-	-
MA, NY: E9, S289	1+	57	-	51-66	-	-	-
IA, IL: G80, M323, M503, M530	12+	74	53-90	46-119	-	-	-
MN, WI: E3, H250	66+	74	-	66-132	-	-	-
AR: H442	21	82	-	76-84	-	-	-
FL: H226, H228	70	103	-	51-147	-	-	-
TN: E26, K71	48+	112	-	76-140	-	-	-
AZ: E148	+	201	-	194-230	+	160	115-206
Combined	219+	89	71-102	41-230	+	160	115-206
Age I							
MD: B218, E76	100+	57	-	48-86	-	-	-
OR: B236, B317	195+	78	-	56-132	-	-	-
MA, NY: E9, M87, M146, S289	45+	118	114-122	79-140	12	27	23-34
MI, MN, WI, SD: C35, C270, E3, E97, H250, M20, M308, M531, P7, T76, V91, S384, W192	641+	123	95-148	58-206	118	46	23-64
AR, OK: B220, C272, H127, H442, J42, M107	310+	140	117-150	94-216	165	29	11-65
NC, SC: D47, D68, H97, R217, T103, S343, S395	214+	141	127-150	112-211	-	-	-
MN Mississippi R.: C356, U8	74	145	-	107-196	-	-	-
IL, IN, OH: J9, L14, L129, B86, R34, S166, W220, B416	1,273+	146	130-170	70-206	6	20	-
IA: E48, I10, J110, J104, L29, H200, M323, M372, M503, M530, N99	274+	147	124-163	82-229	134	49	14-181
CA: H146, K59, M140	24+	149	84-249	71-249	-	-	-
KS: C226, H332, S207, S340	88+	157	89-201	66-351	3	124	-
IA Mississippi R.: U8	69	160	-	114-201	-	-	-
CO: E51	+	165	-	-	-	-	-
FL: F74, H226, H228	299	176	135-234	124-234	-	-	-
TX: B109, E97	198+	185	165-234	104-246	134+	120	26-238
TN: E22, E26, K71, S97	316+	203	117-259	65-259	-	-	-
AZ: E148	+	258	-	226-301	+	331	202-547
Combined	4,120+	143	122-160	48-301	572+	59	11-547
Age II							
MD: B218, E76	100+	102	-	97-104	-	-	-
MT: B108	1	132	-	-	-	-	-
OR: B236, B317, C279, K80	150+	138	135-160	66-170	-	-	-
SD: V91, W192	410	141	-	114-213	-	-	-
MI: B31, B75, C213, C215, M531	594	165	150-171	122-244	2	26	-

	No.	TL Mean of means	Central 50%	Range	No.	Weight Mean of means	Range
Age II (cont.)							
MA, NY: E9, G25, M87, M146, S289	121+	167	163-170	132-206	39	79	34-148
MN: C35, E3, G110, C270, S384	262	168	137-196	96-262	-	-	-
KS: C226, H332, S207	62	181	127-251	127-264	4	235	-
IA, IL Mississippi R.: U4, U8	108+	183	-	145-251	-	-	-
OH: L14, R40, T143	521+	183	175-185	140-241	-	-	-
WI: M20, F15, C117, T76, E97, H250	826+	188	170-205	145-274	442	110	64-190
OK: B220, C272, H127, H156, H442, M107	524+	189	147-218	109-239	443	34	18-77
MN Mississippi R.: C356, U8	119	190	-	152-251	-	-	-
IA: L29, M195, E48, H200, N99, M372, I10, M323, M530, J104, J110	292+	192	179-196	127-295	87	146	28-431
NC, SC: D47, D68, H97, H228, R217, S343, S395, T103	237+	193	173-213	155-224	-	-	-
IN: J9, H62, R34, G38, R152, W220, B416	1,367+	193	188-198	142-287	23	62	54-82
IL: L129, T16, B86, S166, W105, G80	16+	200	172-246	102-246	1	48	-
CA: H146, K59, M140	25+	215	137-292	107-310	-	-	-
FL: F74, H226	362	225	-	188-254	-	-	-
CO: E51	+	229	-	-	-	-	-
TN: E22, E26, E28, K71, S13, S97, S136	613+	230	180-263	147-333	60	128	85-184
TX: E97	128	236	-	163-305	127	221	121-275
AZ: E148	+	316	-	298-381	+	680	514-912
Combined	6,837+	188	160-206	66-381	1,228+	121	18-912
Age III							
OR: B236, B317, C279, K80	106+	135	-	79-237	-	-	-
MD: E76	97	155	-	-	-	-	-
SD: V91, W192	218	163	-	137-251	-	-	-
MT: B108	30	168	-	-	-	-	-
MN: E3, G110, S384	579	186	160-198	132-360	-	-	-
MA, NY: G25, M87, M146, S289	111	190	168-201	168-221	80	117	85-181
OH: L14	159	201	-	-	-	-	-
France: D28	+	201	-	-	-	-	-
IA: E48, H200, J104, J110, L29, M195, M323, M235, M372, M503, M530, N99, U8	248+	205	190-221	135-305	87	125	57-254
IL: L129, U4, B86, S166, W105, G80	10+	207	167-216	150-312	8	51	34-68
KS: C226, H332	54	209	-	185-269	-	-	-
MI: B75, B31, C213, C215, M531	639	215	203-224	160-279	26	57	-

		TL				Weight	
	No.	Mean of means	Central 50%	Range	No.	Mean of means	Range
Age III (cont.)							
WI: M20, F15, T76, E97, H250	904+	215	198-245	157-284	713	121	82-142
AR, OK: B220, C272, H127, H156, H442, J42, M107, O37	278+	219	180-262	130-358	205	107	14-254
CO: C208, G185, P207	31	221	-	167-257	7	88	60-120
IN: J9, R34, R40, H62, U1, G38, R152, W220, B416	439+	231	221-257	188-295	97	125	69-167
NC, SC: D47, D68, H97, H228, R217, S343, S395, T103	354+	236	214-249	170-340	-	-	-
MN Mississippi R.: C356, C270, U8	92	237	-	170-315	-	-	-
FL: F74, H226	297	251	-	231-267	-	-	-
TN: E26, K71, S13, S97, S136	325+	258	226-303	178-387	217	190	127-269
TX: E97	82	259	-	163-315	82	287	247-337
CA: H146, M140	36	264	-	122-188	-	-	-
AZ: E148	+	408	-	373-429	+	1,178	814-1,213
Combined	5,089+	218	193-244	79-429	1,522+	152	14-1,213
Age IV							
MT: B108	14	173	-	-	-	-	-
MD: E76	75	190	-	-	-	-	-
KS: C226, H332	6	212	-	206-218	-	-	-
SD: S330, V91, W192	101	218	190-282	185-282	-	-	-
MN: C35, E3, G110, S384	545	223	188-277	196-394	-	-	-
OR: B236, B317, C279, K80	61+	230	218-255	201-270	-	-	-
MA, NY: G25, M87, M146, S289	61	232	216-448	198-269	47	222	164-337
IA: E48, H220, J110, J104, L29, M323, M235, M372, M530, N99, U8	169+	232	215-241	160-351	62	142	71-224
OH: L14	48	251	-	-	-	-	-
MI: B31, B75, C213, C215, M531	294	253	229-270	173-287	1	65	-
WI: M20, B47, C117, E97, H250	664+	253	234-277	196-315	664	385	195-649
AR, OK: B220, C272, H156, H127, H442, M107	50+	253	213-287	168-292	16	100	57-299
IN: R40, H62, R34, U1, G38, R152, W220, B416	97+	256	246-259	224-297	61	207	136-281
FL: F74, H226, H228	350	258	239-275	239-302	-	-	-
IL: L129, T16, U4, W105, G80	3+	261	218-316	152-367	1	800	-
TN: K71, S13, S97, S136	210+	261	239-330	198-413	163	235	170-312
CO: C208, E51, G158, P207	32+	266	198-292	185-330	21	110	60-150

	No.	TL Mean of means	Central 50%	Range	No.	Weight Mean of means	Range
Age IV (cont.)							
MN Mississippi R.:							
C270, C356, U8	52	272	-	228-335	-	-	-
NC, SC: S395, S343,							
H97, D47, D68	191+	286	254-290	178-368	-	-	-
TX: E97	36	290	-	254-396	35	351	318-434
CA: M140	7	351	-	330-363	-	-	-
Combined	3,066+	247	216-277	152-413	1,071	228	57-800
Age V							
MD: E76	60	216	-	-	-	-	-
OK: C272, H127	9	218	-	142-272	4	147	32-295
CO: E51, G158, P207	7+	224	-	178-348	7	115	60-180
IL: L129, T16, U4	+	241	-	152-279	-	-	-
SD: S330, W192, V91	91	241	-	203-328	-	-	-
MA, NY: G25, M87,							
M146, S289	47	246	218-272	218-306	32	279	209-468
WI: M20, C117, E97,							
H250	246+	247	240-257	208-305	231	295	281-312
MN: E3, G110, S384	196	257	236-306	201-394	-	-	-
IA: E48, J104, J110,							
H200, L29, M235,							
M323, M372, M530,							
N99	39+	259	230-292	173-334	28	200	82-348
OR: B236	19	277	-	272-295	-	-	-
OH, IN: L14, H62, R34,							
R152	28	277	269-295	236-300	12	241	-
FL: F74, H226, H228	247	277	257-295	257-323	-	-	-
TX: E97	2	279	-	274-287	2	354	-
TN: S13, S97, S136	41	280	264-282	239-335	40	359	227-417
MI: B31, C215, M531	750	282	251-310	251-333	-	-	-
MN Mississippi R.:							
C270, C356, U8	38	301	295-310	254-361	-	-	-
SC, NC: D47, D68, H97,							
R217, S343, T103	144+	318	284-332	284-363	-	-	-
CA: M140	1	368	-	-	-	-	-
Combined	1,965+	267	236-295	142-394	356	256	32-468
Age VI							
OK: C272	1	165	-	-	1	59	-
MD: E76	36	234	-	-	-	-	-
MA: M87, M146, S289	15	250	-	224-297	6	271	136-406
SD: V91, W192	25	250	-	229-315	-	-	-
OR: C279, B236	5+	270	-	206-320	-	-	-
WI: M20, B47, C117,							
E97	90	276	254-290	254-356	87	407	264-907
IL, IN, OH: L14, L129,							
H62, R152, T16, U4	7+	282	264-305	165-345	4	275	-
IA: E48, J104, L29,							
M323, M372, M530,							
N99	18+	285	239-324	183-421	13	246	151-326
MI: B31, C215	115	286	-	272-325	-	-	-
FL: F74, H226, H228	132	288	272-307	-	-	-	-
TN: S13	26	297	-	279-325	26	403	255-561
MN Mississippi R.: U8	15	305	-	279-335	-	-	-
MN: E3, C35, G110,							
S348	56	315	267-324	230-400	-	-	-
CO: E51	+	335	-	-	-	-	-

		TL				Weight	
	No.	Mean of means	Central 50%	Range	No.	Mean of means	Range
Age VI (cont.)							
NC, SC: D47, D68,							
H97, S343	38+	347	345-353	318-376	-	-	-
Combined	579+	289	259-318	165-421	136	319	136-907
Age VII							
MA: M146, S289	5	255	-	241-262	1	170	-
MD: E76	4	269	-	-	-	-	-
MI, MN, SD, WI: B31,							
B47, C35, C117,							
E3, M20, S384, V91	135	305	-	269-460	20	458	346-907
FL: F74, H228	90	308	-	290-328	-	-	-
IA: E48, M320, M372,							
N99	7	318	-	224-399	6	646	148-1,088
TN: S13	9	345	-	335-361	9	601	595-695
CO: E51	+	368	-	-	-	-	-
SC: S343	12	369	-	330-472	1	2,183	-
Combined	262+	313	277-343	224-472	37	645	170-2,183
Age VIII							
IA: L29, N99	2	259	-	216-302	1	156	-
MI, MN, SD, WI: B31,							
C35, C117, E3, M20,							
V91	44	321	300-360	287-425	5	408	-
FL: F74, H228	48	322	-	305-338	-	-	-
SC: S343	3	364	-	351-391	-	-	-
Combined	97	321	300-351	216-425	6	345	156-408
Age IX							
MN, WI, SD: C35, M20,							
V91	10	337	315-375	300-396	2	477	-
FL: F74, H228	5	358	-	-	-	-	-
Age X							
MD: S327	1	358	-	-	1	488	-
Age XI							
MD: S327	1	483	-	-	-	-	-
Age XIII							
MA: S289	1	368	-	-	-	-	-

		Mean calculated TL at each annulus									
	No.	1	2	3	4	5	6	7	8	9	10
Western waters											
B232 MT 3 Forks L.	55	28	61	94	122	145	168	178			
B108 MT ponds	45	28	86	140	160						
G185 CA Salt Spring L.	+	69	130	155	178						
P159 MT streams	24	56	117	165	221	218					
P159 MT lakes	139	84	135	170	198	213	211	206			
G185 CA Mendota Pool	+	-	122	178	196						
G161 OR 28 waters	547	53	135	183	211	231	224				
G77 OR 6 waters	+	56	142	206	264						
B134 CO Loveland L.	18	102	155	216	300						
G185 CA L. Havasu	+	76	180	231	264	287					
G185 CA Clear L.	+	41	221	292	345	351					
E149 AZ Roosevelt L.	+	210	260	310							
K152 AZ Roosevelt L.	77	218	275	348	378						

		Mean TL and increment								
	No.	1	2	3	4	5	6	7	8	9	10

Western waters (cont.)

	No.	1	2	3	4	5	6	7	8	9	10
K80 OR pond		90	160	200	208						
	Incr.	90	70	30	15						
	No.	8	8	3	2						
Mean for western waters		85	156	206	234	241	201	192			

		Mean calculated TL at each annulus									
	No.	1	2	3	4	5	6	7	8	9	10

Northern waters

	No.	1	2	3	4	5	6	7	8	9	10
W192 SD Lewis & Clark L.	+	58	114	150	190	229	246				
G78 MA Indian L.	+	76	127	160	178	188	198	211	221		
After thinning	+	76	140	185	203	216	234	254	262		
M292 MA Quabbin L.	52	79	135	170	196	213	229	241	267	284	297
After smelt added	4	71	127	201	224	246					
K33, B365, M258: MN	2,166	61	122	173	211	241	267	295	312	325	343
S62 MN	+	69	127	175	221	264	300	333			
C35 MN Vermillion L.	23	76	133	188	240	278	311	335	354	364	
E3 MN	1,783	74	136	189	239	285	321	357	257		
S213 MN Clear L.											
1949 yr. cl.	+	79	130	190	221	239					
D89 WI Murphy Flowage	+	91	130	196	251	254	267	282	284	287	287
C270 MN Miller L.	89	51	112	206							
G180 SD Francis Case L.	870	80	167	218	247	274	284	295	288		

		Mean TL and increments									
	No.	1	2	3	4	5	6	7	8	9	10

		1	2	3	4	5	6	7	8	9	10
F105 SD Oahe L.		38	86	122	178						
	Incr.	38	43	36	61						
	No.	52	7	5	4						
S228 SD Gavins Point	1958	64	119	124							
	Incr.	64	56	36							
	No.	144	141	1							
N94, N115 SD Gavins L.		52	101	136	170	203	216				
	Incr.	52	49	37	33	33	30				
	No.	126	126	101	66	20	1				
V91 SD Lewis & Clark L.		68	118	148	185	220	239	267	286	314	
	Incr.	68	54	41	38	32	29	30	21	25	
	No.	1,224	901	493	267	172	80	52	17	4	
S226 SD Ft. Randall L.		79	170	175	127	188					
	Incr.	79	97	58	25	61					
	No.	271	130	3	1	1					
Tailwaters		81	152								
	Incr.	81	91								
	No.	25	16								
S225 SD Ft. Randall L.		89	135	178	213						
Incr. before impoundment		61	71	48							
	No.	19	14	7							
Incr. after impoundment		94	112	56	46						
	No.	140	5	7	7						
Tailwaters		94	160	203							
	Incr.	94	71	48							
	No.	59	5	4							
S227 SD Gavins Point 1956		76	178	183							
	Incr.	76	99	43							
	No.	82	32	1							

			Mean TL and increments									
	No.	1	2	3	4	5	6	7	8	9	10	
Northern waters (cont.)												
F100, 124, 125: SD Oahe L.		85	150	208	203							
	Incr.	85	76	51	28							
	No.	139	39	10	1							
S393 SD Ft. Randall L.		79	155	213	249	282						
	Incr.	79	74	58	38	28						
	No.	127	91	78	35	6						
C270 MN Mississippi R.		71	145	216	262	290						
	Incr.	71	81	74	46	28						
	No.	132	70	32	14	9						
E97 WI Chetek L.		91	160	218	244	257	279					
	Incr.	91	79	43	28	18	23					
	No.	76	47	43	41	31	5					
N133 SD L. Oahe		88	177	231	272	304	326	371	394			
unweighted	Incr.	88	90	53	37	32	26	30	10			
	No.	473										
S228 SD Ft. Randall L.		71	173	246	274	318						
	Incr.	71	102	79	48	51						
	No.	176	148	31	6	3						
S229 SD Gavins Point L.		79	170	254								
	Incr.	79	97	122								
	No.	147	3	1								
Mean for northern waters		74	139	187	217	249	265	295	302	309		

			Mean calculated TL at each annulus								
	No.	1	2	3	4	5	6	7	8	9	10
Central waters											
E30 OH poorest lake	+	36	89	127	168	211	229				
H194 NB Kimball L.	388	41	94	132	157	173	203	259			
H194 NB Box Butte L.	149	58	114	155	178	193	203	218	236		
C284 OH St. Mary's L.	+	66	117	157	185	234	269	312			
E30 OH average	500	56	119	160	198	234	251	292	302		
H265 IA Fairport ponds	57	58	110	165	220						
H194 NB Whitney L.	211	51	130	173	198	213	226	249			
M198 IA Keomah L.	106	97	142	178	206						
M201 IA Thayer L.	67	89	147	178	203	267					
E48 IA Clear L.	229	58	130	183	221	267	295	302			
H194 NB McConoughy L.	2	89	127	190							
L29 IA East L.	31	74	142	190	217						
E30 OH best lake	+	69	132	198	257	330					
M320 IA East Osceola L.	21	114	165	203	241						
R59 OH	+	119	190	234	279	305	330	348	366		
P113 MO	282	79	157	211	224	244					
middle rivers	162	79	157	211	234	259					
lower rivers	120	79	160	211	211	226					
poorest station	+	74	142	190	196	226					
best station	+	81	178	236	241	259					
H194 NB Minatare L.	16	76	175	229	284	323					
P204 IA Red Rock L.	36	141	193	235	262	277					
L29 IA Red Haw L.	132	79	176	240	287	303	287	296	302		
G80 IL Horseshoe L.	6	135	188	255	310						
B86 IL Onized L.	353	86	203	290							
S207 KS Fall River L.	7	132	239								

	No.	Mean TL and increment									
		1	2	3	4	5	6	7	8	9	10

Central waters (cont.)

	No.	1	2	3	4	5	6	7	8	9	10
E84 OH St. Mary's L.		53	112	152	183	229	236	216	254	272	
	Incr.	53	61	51	43	48	48	28	23	18	
	No.	186	136	104	79	48	9	2	1	1	
C193, M235 IA Backbone L.	16	56	104	155	198	254					
year after removal of 19 kg/ha	38	64	142	152	183	208					
year after additional removal of 40 kg/ha	21	56	155	198	216						
incr. before removals		56	51	51	43	20					
	No.	76	36	30	28	1					
incr. after removals		53	102	38	30	38					
	No.	6	40	11	3	11					
N99 IA Clear L.		84	137	155	201	216	224	216	211		
	Incr.	84	56	33	28	18	13	12	8		
	No.	58	53	50	40	33	14	4	1		
H200 IA Ahquabi L.		84	140	170	201	246					
	Incr.	84	56	33	33	43					
	No.	146									
N138 IA Clear L.		77	130	173	196	213	224	229	237		
	Incr.	77	56	41	26	19	15	18	15		
	No.	570	473	340	237	129	37	13	3		
S419 IA Backbone L. 1959-61		97	145	183	208	234	251	241			
	Incr.	97	48	38	33	30	28	38			
	No.	146	146	144	94	67	36	1			
R40 IN Foot's P.		71	145	193	241						
	Incr.	71	74	53	17						
	No.	18	18	8	1						
J9 IN Greenwood L.		114	178	196							
	Incr.	114	64	23							
	No.	1,876	1,156	224							
J110 IA Spirit L.		71	145	203	251	279					
	Incr.	71	74	58	38	25					
	No.	297	264	110	18	3					
I10 IA Missouri R. cutoffs		97	193								
	Incr.	97	47								
	No.	30	11								
Mean for central waters		80	147	188	220	248	248	265	273	272	

	No.	Mean calculated TL at each annulus									
		1	2	3	4	5	6	7	8	9	10
AR, OK waters											
H153 OK	2,406	79	160	208	251	295	343	386			
slowest average	+	38	97	150	180	211	307				
fastest	+	216	302	335	376	399	411				
7 new reservoirs	332	112	208	234	297	272					
10 old reservoirs	16	71	165	224	282	345	411				
8 lakes, 45-202 ha	340	64	135	180	234	287	338	384			
4 new lakes, 2-44 ha	26	152	221	274							
31 old lakes, 2-44 ha	1,194	69	152	198	241	292	315				
17 ponds	173	79	160	203	249	211					
6 streams	225	69	147	218	295	356					
O37 OK Heyburn L.	2	81	127	173							
F64 OK Little R.	48	48	124	180	231	290					
G161 OK 28 waters	547	53	135	183	211	231	224				
J34 OK Ardmore L.	29	61	132	188	226	279	338	386			

	No.	Mean calculated TL at each annulus									
		1	2	3	4	5	6	7	8	9	10
AR, OK waters (cont.)											
L195 OK Arkansas R.	29	71	137	196	208	257	239	257			
H442 AR Catherine L.	17	66	145	201	236						
E61 OK Salt Creek	1	84	152								
J37 OK Rod & Gun Club L.	55	81	165	208	231						
J42 OK Ft. Gibson	63	61	145	211							
H442 AR Hamilton L.	62	79	157	216	236						
P87 OK Little R.	1	38	104	216							
H156 OK Canton L.	40	94	163	221	264	292					
B220 OK Canton L.	+	58	180	226							
J45 OK tributaries	9	69	170								
H442 AR Ouachita L.	75	109	173	229	246						
B220 OK Ft. Supply L.	32	74	175	239							
H220 OK Ft. Gibson	346	74	165	241	290						
J37 OK Chickasaw L.	106	114	216	251	272						
F62 OK Sub-Prison L. renovated	41	84	203	257							
E126 OK Rocket Plant L.	9	130	213	282	328						
J101 OK Spavinaw L.	131	91	208	284	333	376	411				
J101 OK Eucha L.	190	84	221	295	368	406					

	No.	Mean TL and increments									
		1	2	3	4	5	6	7	8	9	10
B220 OK pond		51	89	112	142						
	Incr.	51	41	23	23						
	No.	54	41	33	2						
C272 OK Country Club L.		58	102	119	122	160	163				
	Incr.	58	41	25	3	25	18				
	No.	211	197	152	123	4	1				
C272 OK Sanborn L.		46	124	173	203	254					
	Incr.	46	79	43	38	20					
	No.	650	529	139	15	1					
H303 OK Lawtonka L.		119	163	188							
	Incr.	119	44	33							
	No.	23	23	11							
B183 OK Canton L.		81	173	216	249	272					
	Incr.	81	97	51	66	20					
	No.	190									
Mean for AR, OK waters		76	158	212	246	283	286	343			

	No.	Mean calculated TL at each annulus									
		1	2	3	4	5	6	7	8	9	10
KY, TN waters											
K71 TN White Oak L.	765	91	160	193	302						
E28 TN, Chickamauga L.	132	56	122								
S100 TN Douglas L.	28	119	178	206							
T54 KY lakes	51	76	196	229							
C110 KY Herrington	29	74	208	231	254						
F151, T127 TN, Eastern TVA, 1956-62	136	74	190	239	287	338					
S110 TN Cherokee L.	85	46	183	262							
C169 TN Norris L.	+	64	234	292	323	348					
S97 TN Norris L.	925	81	241	300	323	348					
E26 TN Norris L.	211	127	277	310							
Mean for KY, TN waters		81	199	251	298	345					

	No.	1	2	3	4	5	6	7	8	9	10
			Mean calculated TL at each annulus								
DE, MD, NC, SC waters											
B218 MD Cash L.	+	74	112	130							
L157 NC Waccamaw L.	7	53	122	168							
D68 NC	259	74	122	173	216	254	292				
D47 NC Little R.	49	64	119	180	224	277	325				
S192 MD Potomac R.	+	94	147	198	224						
L157 NC White L.	1	53	147	231							
S110 NC Hiwassee L.	9	74	190	259	292						
H158 DE	+	168	241	269	292	310	323				

		1	2	3	4	5	6	7	8	9	10
			Mean TL and increment								
S395 NC Roanoke Rapids L.	No.	66	142	145	165						
	Incr.	66	76	43	38						
	No.	43	35	9	1						
R217 NC James L.		64	127	173							
	Incr.	64	71	41							
	No.	38	11	4							
R127 NC Rhodhiss L.		76	150	180							
	Incr.	76	69	33							
	No.	8	3	1							
R217 NC Lure L.		48	140								
	Incr.	48	102								
	No.	31	18								
T103 NC Blewett Falls L.		69	145								
	Incr.	69	86								
	No.	11	4								
R217 NC Lookout Shoals L.		46	157								
	Incr.	46	117								
	No.	22	9								
T103 NC Catawba L.		102	160								
	Incr.	102	61								
	No.	14	7								
T103 NC Mt. Island L.		104	170								
	Incr.	104	69								
	No.	26	2								
T103 NC Tillery L.		86	130	185	203	246					
	Incr.	86	43	48	33	43					
	No.	16	15	3	2	2					
R217 NC Hickory L.		71	152	190	231	325					
	Incr.	71	94	58	25	94					
	No.	42	13	6	1	1					
S395 NC Kerr L.		76	150	196							
	Incr.	76	74	38							
	No.	39	36	19							
S343 SC Marion L.		46	122	201	257	290	320	318	340		
	Incr.	46	76	86	64	51	36	28	33		
	No.	239	237	218	140	72	38	12	2		
S343 SC Moultrie L.		58	160	269	315	335	356	381	381		
	Incr.	58	109	107	56	38	48	30	28		
	No.	198	142	101	75	45	9	3	1		
Mean for DE, MD, NC, SC waters		75	148	197	242	291	323	350	360		

	No.	Mean calculated TL at each annulus									
		1	2	3	4	5	6	7	8	9	10
FL, TX waters											
F74 FL	418	48	102	160	206	239	269	297	318	351	
H226 FK	943	112	206	251	292	318	307				
H139 FL George L.	943	117	201	257	292	328					

	No.	Mean TL and increment									
		1	2	3	4	5	6	7	8	9	10
H229 FL Harris L.		48	107	168	216	249	284	310	330	351	
Incr.		48	64	64	53	41	33	30	25	23	
No.	394	327	301	250	155	80	43	17	3		
H229 FL Eustis L.		51	112	173	211	239	262	284	297		
Incr.		51	61	66	48	41	28	23	15		
No.	289	281	259	200	123	70	33	10			
E97 TX Koon Kreek Klub L.		94	178	229	257	272					
Incr.		94	97	53	30	36					
No.	439	243	116	26	7						
Mean for FL, TX waters		78	151	206	246	274	281	297	315	351	
Mean of regional means		78	157	207	243	276	267	290	312	311	

The first year growth of 858 crappies from Larto L., LA (L74) was described as $TL = 1.182\ \text{mm} + 0.009A$, where A = days after April 1st. Growth of age I crappies in South Dakota was 25 mm in 19 days (S227), 23 days (S228), 24 days (S225), and 35 days (S229); and of age II, was 25 mm in 33 days (S227), and 46 days (S228, S276); and of age III, was 25 mm in 88 days (S228).

	No.	Mean weight at each annulus									
		1	2	3	4	5	6	7	8	9	10
H153 OK lowest average	+	-	9	42	78	132	463				
V91 SD Lewis & Clark L.	1,224	4	22	45	90	153	197	277	342	456	
H200 IA Ahquabi L.	146	7	36	66	112	213					
Incr.		7	29	32	48	97					
E48 IA Clear L.	229	3	34	91	167	287	363	417			
M198 IA Keomah L.	106	28	74	99	113						
M201 IA Thayer L.	67	28	57	99	127	227					
H153 OK	2,406	5	52	123	239	406	662	890			
M320 IA East Osceola L.	21	28	65	130	247						
E149 AZ Roosevelt L.	+	154	314	680							
H153 OK fastest	+	142	438	620	907	1,105	1,225				

Scales were reported to be difficult to read on black crappies from Roosevelt L., AZ (G184), and vertebrae proved to be a promising but a time-consuming method of calculation of age and growth. Age and growth was therefore determined from spine sections (E149). Marked crappies grew 22 mm or 12.5 g per month during part of the 1st year, but spines indicated an average of 16 mm per month for the 1st year (E149). Scales were used for age and growth calculation in most studies. The first two readings of scales of black crappies from Clear L., IA, agreed in 67% of the cases (N138) and from Lewis and Clark L., SD, in 76.7% (V91).

Criteria for annuli were listed as: (1) erosion of the posterior part of the last circuli followed by a series of complete circuli; (2) crowding of the circuli in the anterior field; (3) anastomosis of circuli on the lateral fields and a clear narrow gap between circuli (V91). Consistencies in the data attested to the validity of the interpretations.

The pattern of 1st scale formation differed somewhat (B220, C375, W83), but the first scales appeared in the caudal peduncle. The lateral line was scaled rather late (C375), or scalation proceeded along the lateral line (B220, W83). Most 4-week-old crappies, TL 18.8 mm, had scales and scalation was about complete at TL 26 mm (W83). Scales did not appear until 5 weeks, at 16-20 mm, in one pond, but did appear at 3.5 weeks, 21-23 mm, in another (B220). The 1st scales did not appear until the crappies were 28 mm, and scalation was completed at 38-40 mm (C375). The largest crappie without scales was 34 mm (C375). The body-scale intercept for 89 young-of-the-year crappies, 24-47 mm, was 19.9 mm (B220).

In Tennessee (S97), the 1st annulus was formed by the 2nd week of May, but older crappies did not form annuli until late May. In Indiana (J9), annuli were formed from late April to mid-June, with the younger fish usually earlier in the period. In Clear L., IA, annulus formation occurred in mid-June (N138); in SD, in mid-May to the 1st week of June (S225, S227-9, S276); and in MN, the last half of June (S213). Ages 0 and I grew mostly in early summer and had completed their growth by September, but age II crappies made much of their annual growth in September and October in TN (S97). In Oklahoma, growth occurred from April through September but not into October (B220).

Most growth calculations were made on a direct proportion basis, but Fraser-type corrections were used in IA (H220, L29, N99, N138), IN (J9), MA (M292), NC (T103), SD (N144), and OK (B183, B220). A correlation between the correction factor and the mean calculated TL at the 1st annulus is shown in T103:

Correction factor	5	33	53	64
Mean TL at age I	69	86	104	102

Lee's phenomenon was reported by E48 and N138 but not by V91.

The oldest black crappie in captivity was 7.5 years (F6), but the tabulated data indicate ages to age XIII. Few were reported over age VIII. The tendency for slower growth in the northern part of the range was not as evident for black as for white crappies. Growth is more rapid in southern than in northern Wisconsin (S574). Growth was slower in turbid lakes of 45 to 200 ha than in other Oklahoma waters (H153). Growth was slower in turbid than in clear ponds (H332). Generally, years of good growth of black crappies were years of poor growth for white crappies (N138). No correlations could be found between annual growth of black crappies in Clear L., IA, with April to October temperatures nor with water levels (N138). A significant increase in growth was reported with a 1 meter water level increase in Big Portage L., MI (M531). Growth of black crappies was slower in the Arkansas R. than the Oklahoma average (L195). Lowering of water levels in July slowed the growth of crappies in Greenwood L., IN (J9).

Every California water with a stunted crappie population had a paucity of small forage fish (G185). Growth was better where insects composed a greater percentage of the good than where copepods did (C272). Crappies at 211 mm were age IV in a dense population but age III where less crowded, and there was more food per stomach in the less dense population (S464).

Growth was slower in L. Marion than in L. Moultrie probably because of higher population density (S343). Growth improved when population density was reduced in Goddard L., OK (M107), and in Backbone L., IA (S566). Carp removal did not give any clear-cut improvement of crappies in 4 Minnesota lakes (S384). Growth and condition factors were poorest when carp became abundant in Pymatuning L., PA (C355). Addition of smelt in Quabbin L., MA

(M292), perhaps reduced the growth of black crappies to age III but improved the growth thereafter.

Growth was better in a KS lake after elimination of the population and restocking (S340). At fall age I crappies averaged 251 mm. Growth was better when bass reduced the number of crappies (T146).

In new impoundments, growth usually was good (B220, S97, H220, G180). In Lewis and Clark L., SD (V91), growth was not rapid the 1st few years of impoundment but improved 8 to 10 years later. In Norris L., TN (S97), growth was rapid then declined, but it increased some years later when water levels were raised, flooding new areas. In Francis Case L., SD (G180), black crappies were replaced by white crappies as the impoundment became more turbid. Black crappies transferred from a crowded population grew rapidly in a new impoundment, and age II crappies in September averaged 241 mm and 255 g, compared to the same age fish in Rocky Fork L. at 140 mm, from which the fish were taken about June 1st (T143). Growth was very rapid when crappies were introduced into Roosevelt L., AZ (E148), and into Jumbo L., CO (E51).

Growth usually was more rapid in reservoirs that stratified thermally than in windswept unstratified reservoirs or in municipal water-supply reservoirs (M323). In L. Oahe, SD, only 32% of the variation in annual growth could be explained by percent change in surface area, average depth, and water level fluctuation (N144), with the latter having little correlation.

Males were reported slightly larger than females at equivalent ages (C208, H226) and females larger than males (N144), but no sex differences in growth were noted elsewhere (V91, S226, S228).

Crappies mature at age I (S226, S393) or at age II (S228, S276, G185, N144). They spawn from March to July when water temperatures reach 14.4°C to 17.8°C (G185). In Texas, they spawn from late March to early May (S511); in Florida, April to autumn (M242); in Ohio, late March to early June (M541); in Minnesota, May to July (B365); in South Dakota, late June to mid-July (S226, S228, S393). Water temperatures during the spawning period ranged from 4.4°C to 15.6°C (M541), but the most favorable temperatures were 17.8°C to 20.0°C (S574). The 1st hatch occurred when the water was 20°C, 15 cm from the surface (S334). In Wisconsin, larval crappies were taken in the limnetic zone in the 1st half of June to July at temperatures of 18°-20°C (F156); and in Ontario, from mid-May through June at 13°-23°C (A102). There were two peaks of larval abundance in 1960, early June and early July. In Roosevelt L., AZ, there was apparently a 2nd spawn in late October or early November 1968, because in May 1969 there were two size groups of age I crappies, one averaged 220 mm, 200 g and the other 177 mm and 104 g (E148). Apparently, the crappies grew significantly in the winter (E148).

	No.	Size	Mean	Eggs per female Range
U1 IN	5	age III	33,712	26,000-41,562
	5	age IV	41,879	30,139-65,520
C77	1	227 g	16,950	-
	1	454 g	158,696	-
V21 MN	2	28-115 g	19,000	16,000-21,000
	24	120-230 g	30,000	11,000-48,000
	20	235-570 g	45,000	22,000-66,000
	5	575-680 g	137,000	77,000-188,000
E4 MN	+	227 g	-	20,000-60,000

	No.	Size	Mean	Eggs per female Range
E4 MN	1	680 g	140,000	-
E148 AZ Roosevelt L.	+	-	29,300	27,700-30,300
E149 AZ Roosevelt L.				
April 1968	18	240-636 g	29,800	15,000-42,000
E150 AZ Roosevelt L.	1	73 g	6,900	-
	10	159-245 g	24,800	23,000-27,000
	5	259-358 g	27,400	25,000-29,000
	3	391-518 g	30,800	30,000-32,000
	1	712 g	40,000	-
B365 MN	+	-	-	11,000-188,000

Buildup of the ovaries takes place in the fall, and little further development is needed in the spring (M541).

Eggs were 0.93 ± 0.082 mm in diameter, which makes them smaller than the eggs of warmouth, bluegill, and largemouth bass (M529). No increase in egg size was noted with increase in TL of the female (M529).

Bowl-shaped nests are built on gravel, sand, or rarely softer bottoms at depths of 25 cm to 6.1 m (S574). As the temperature approached 18°C, males moved to the shallows, usually the smaller males first, to establish territories (G184). The nests were usually built after attracting females to the area, and although each male established and guarded a territory, the nests were often close together, as many as 30 nests in 9 square meters (G184). Vegetation may be a critical limiting factor as most spawning is at the base of vegetation. Males guard the nest and even drive carp from the vicinity (E148). However, carp were found to invade the nests to feed on *Corbicula* snails (E148, G184), which did not interfere with crappie reproduction.

The median hatch was at 57.5 hours and at 18.3°C, with the first at 48.1 hours and the last at 67.8 hours (M544). From 5% to 20% of the fry showed some deformity (M544).

Dominant year classes appeared in Clear L., MN, in 1949 and 1953, following heavy mortality and a removal program (S213). A big year class occurred the 1st year of impoundment in Lewis and Clark L., SD, and there was successful reproduction each of the next 10 years without evidence of dominance (V91). Strong year classes of crappies do not occur when there is a strong age I bass population because these bass are effective predators (S561). Crappies spawn as early as or earlier than bass, and the fry go to the limnetic zone where they are not subject to predation by age 0 bass.

There was evidence that there were subpopulations of black crappie in Lakes Moultrie and Marion, SC (S343). Standing crops in White R., IN, were 0.8 and 2.7 kg/ha compared to 0.8 and 3.9 kg/ha in thermally affected areas (B417). The standing crops of black crappies in Alanconnie L., PA (C355), were 6.7 to 51.4 kg/ha. There were 167 crappies over 179 mm per hectare in Grove L., MN, (S464) and an annual angler harvest of 22 per hectare. Harvest rates from Quabbin L., MA, ranged from 0.007 to 0.45 kg per ha (M296). When Mills L., MI, was opened after 5 years of no fishing, there were 29 black crappie/ha, and 14% were caught in 3 days with 96 hours of fishing/ha (S595).

Annual mortality rates were 69% and 79% for black crappies age II-IV in Spear (R152) and Oliver Lakes (G38) and 64% for crappies ages II=III and 91% for ages III-IV in Muskellung L., WI (R34).

Black crappies did survive at 34°C (T102). In Louisiana, turbidity was not considered a major limiting factor, but the maximum salinity where black crappies were found was 2 ppt (C368). Under an oxygen atmosphere, 64-102 mm crappies showed no distress with 44 ppm dissolved oxygen (W237).

Black crappies were classed as a species of slightly alkaline eutrophic waters with limits of 900 ppm total alkalinity, 250 ppm carbonates, and 200 ppm potassium and sodium (M552).

Young crappies feed mostly on entomostraca (D183, R301, S357, B413, B220, N94, N93, and T146). Copepods, chironomid larvae, and one larval fish were taken by 2-week-old crappies, cladocerans at 3 weeks, *Chaoborus* at 5 weeks, ostracods at 6 weeks, and oscillatorial algae at 10 weeks; and all items were still taken at 13 weeks (B220).

Insects and forage fish become a more important item of diet as black crappies get older (J9, N99, J53, K74, H229, D163, D161, B220, S357, F91, H250, L127, C272, S464, M422, S343, B402, E148, K129, P205, F148, D183, B315, K153, R301, and G185). Cladocera still provided over 50% of the food volume in adult fish (H250, B317, B368); in crappies, 125-255 mm in summer (S357); and in crappies, over 235 mm in February (B402). The gill rakers are well adapted to plankton feeding. The first gill arch has 25-29 rakers compared to 8-12 on rock bass, bluegills, and largemouth bass (K153). *Chaoborus* larvae were often a significant part of the diet (H250, B401, K129, B220, D163), indicating feeding in midwater at night (J9, K129, K153). Significant numbers of flying insects were eaten in the summer (K137, K129).

In Ft. Supply L., OK, black crappies fed mostly on *Hybognathus* sp. minnows, while white crappies were eating threadfin shad (B220). Threadfin shad was the principal forage fish eaten by black crappies in some other waters (B220, G185, B402, E148, B315, and B413). In Roosevelt L., AZ, 56% of the food of black crappies under 100 mm was largemouth bass fry, 32% was threadfin shad, and 10% was plankton; but in black crappies over 100 mm, threadfin shad constituted 87% of the food (E148). More fish were eaten by crappies in years when young threadfin shad were abundant (B402). Young gizzard shad were eaten by crappies over 230 mm long (B402). Atlantic shad, striped bass, white catfish, and channel catfish were also eaten (G185), as well as cyprinids (K129, F148), brook silversides, and centrarchids (F91). Fish are eaten mostly in the late summer and fall (D163, B413, N99, and B402) or in July (K129).

In Wisconsin, black crappies were reported to stop feeding in late autumn or early winter (P7), and in Ontario, they were reported to start feeding about April 14 when the water was 6.5°C (K137). Crappies had food in their stomachs in January in South Carolina (S343); in February, in Arkansas (B402); and all winter, in Arizona (B315). The lowest volume of food was reported to be in February and March with the low maximum in September (B402). Food consumption was reported to be low in August in a Minnesota lake (S464).

Feeding is often extensive at night (G184, K153). Digestion was 24% completed in 6 hours and 48% in 7 hours (S464).

A literature review suggested that the food habits tend to be similar from Ontario to Texas (K153).

Black crappies preferred water 24°-30.5°C in L. Monona and 26.5°-30.5°C in the laboratory tests (N141).

The sizes of black crappies taken in gillnets of different mesh size were (L216):

Bar measure	No. fish	Mean TL	Range
25	14	173	127-318
38	17	259	203-368
51	19	305	267-368
64	22	343	318-368
76	5	361	356-394

The number of chromosomes is reported as 48 (R286).

Conversion Tables

INCHES (BY TENTHS) TO NEAREST MILLIMETER

	0	0.1	0.2	0.3	0.4	0.5	0.6	0.7	0.8	0.9
0	0	3	5	8	10	13	15	18	20	23
1	25	28	30	33	36	38	41	43	46	48
2	51	53	56	58	61	64	66	69	71	74
3	76	79	81	84	86	89	91	94	97	99
4	102	104	107	109	112	114	117	119	122	124
5	127	130	132	135	137	140	142	145	147	150
6	152	155	157	160	163	165	168	170	173	175
7	178	180	183	185	188	190	193	196	198	201
8	203	206	208	211	213	216	218	221	224	226
9	229	231	234	236	239	241	244	246	249	251
10	254	257	259	262	264	267	269	272	274	277
11	279	282	284	287	290	292	295	297	300	302
12	305	307	310	312	315	318	320	323	325	328
13	330	333	335	338	340	343	345	348	351	353
14	356	358	361	363	366	368	371	373	376	378
15	381	384	386	388	391	394	396	399	401	404
16	406	409	411	414	417	419	421	424	427	429
17	432	434	437	439	442	445	447	450	452	455
18	457	460	462	465	467	470	472	475	478	480
19	483	486	488	490	493	495	498	500	503	505
20	508	511	513	516	518	521	523	526	528	531
21	533	536	538	541	544	546	549	551	554	556
22	559	561	564	566	569	572	574	577	579	582
23	584	587	589	592	594	597	599	602	605	607
24	610	612	615	617	620	622	625	627	630	632
25	635	638	640	643	645	648	650	653	655	658
26	660	663	665	668	671	673	676	678	681	683
27	686	688	691	693	696	698	701	703	706	709
28	711	714	716	719	721	724	726	729	732	734
29	737	739	742	744	747	749	752	754	757	759
30	762	765	767	770	772	775	777	780	782	785
31	787	790	792	795	798	800	803	805	808	810

INCHES (BY TENTHS) TO NEAREST MILLIMETER (continued)

	0	0.1	0.2	0.3	0.4	0.5	0.6	0.7	0.8	0.9
32	813	815	818	820	823	826	828	831	833	836
33	838	841	843	846	848	851	853	856	859	861
34	864	866	869	871	874	876	879	881	884	886
35	889	892	894	897	899	902	904	907	909	912
36	914	917	919	922	925	927	930	932	935	937
37	940	942	945	947	950	953	955	958	960	963
38	965	968	970	973	975	978	980	983	986	988
39	991	993	996	998	1001	1003	1006	1008	1011	1013
40	1016	1019	1021	1024	1026	1029	1031	1034	1036	1039
41	1041	1044	1046	1049	1052	1054	1057	1059	1062	1064
42	1067	1069	1072	1074	1077	1080	1082	1085	1087	1090
43	1092	1095	1097	1100	1102	1105	1107	1110	1113	1115
44	1118	1120	1123	1125	1128	1130	1133	1135	1138	1140
45	1143	1146	1148	1151	1153	1156	1158	1161	1163	1166
46	1168	1171	1173	1176	1179	1181	1184	1186	1189	1191
47	1194	1196	1199	1201	1204	1207	1209	1212	1214	1217
48	1219	1222	1224	1227	1229	1232	1234	1237	1240	1242
49	1245	1247	1250	1252	1255	1257	1260	1262	1265	1267

FEET TO METERS

	0	1	2	3	4	5	6	7	8	9
		0.30	0.61	0.91	1.22	1.52	1.83	2.13	2.44	2.74
10	3.05	3.35	3.66	3.96	4.27	4.57	4.88	5.18	5.49	5.79
20	6.10	6.40	6.71	7.01	7.32	7.62	7.92	8.23	8.53	8.84
30	9.14	9.45	9.75	10.05	10.36	10.66	10.97	11.28	11.58	11.89
40	12.19	12.50	12.80	13.11	13.41	13.72	14.02	14.33	14.63	14.94
50	15.24	15.54	15.85	16.15	16.46	16.76	17.07	17.37	17.68	17.98
60	18.29	18.59	18.90	19.20	19.51	19.81	20.12	20.42	20.73	21.03
70	21.34	21.64	21.95	22.25	22.56	22.86	23.16	23.47	23.77	24.08
80	24.38	24.69	24.99	25.30	25.60	25.91	26.21	26.52	26.82	27.13
90	27.43	27.74	28.04	28.35	28.65	28.96	29.26	29.57	29.87	30.18
100	30.48	30.78	31.09	31.39	31.70	32.00	32.31	32.61	32.92	32.22

MILES TO KILOMETERS

	0	1	2	3	4	5	6	7	8	9
		1.6	3.2	4.8	6.5	8.1	9.7	11.3	12.9	14.5
10	16.1	17.8	19.4	21.0	22.6	24.2	25.8	27.4	29.1	30.7
20	32.3	33.9	35.5	37.1	38.7	40.4	42.0	43.6	45.2	46.8
30	48.4	50.0	51.7	53.3	54.9	56.5	58.1	59.7	61.3	63.0
40	64.6	66.2	67.8	69.4	71.0	72.6	74.3	75.9	77.5	79.1
50	80.7	82.3	83.9	85.6	87.2	88.8	90.4	92.0	93.6	95.2
60	96.9	98.5	100.1	101.7	103.3	104.9	106.5	108.2	109.8	111.4
70	113.0	114.6	116.2	117.8	119.5	121.1	122.7	124.3	125.9	127.5
80	129.1	130.8	132.4	134.0	135.6	137.2	138.8	140.4	142.1	143.7
90	145.3	146.9	148.5	150.1	151.7	153.4	155.0	156.6	158.2	159.8
100	161.4	163.0	164.8	166.3	167.9	169.5	171.1	172.7	174.3	176.0

OUNCES TO GRAMS

Pounds

Ounces	0	1	2	3	4	5	6
0	0	454	907	1361	1814	2268	2722
1	28	482	936	1389	1843	2296	2750
2	57	510	964	1418	1871	2325	2778
3	85	539	992	1446	1899	2353	2807
4	113	567	1021	1474	1928	2381	2835
5	142	595	1049	1503	1956	2410	2863
6	170	624	1077	1531	1985	2438	2892
7	198	652	1106	1559	2013	2466	2920
8	227	680	1134	1588	2041	2495	2948
9	255	709	1162	1616	2070	2523	2977
10	284	737	1191	1644	2098	2552	3005
11	312	765	1219	1673	2126	2580	3033
12	340	794	1247	1701	2155	2608	3062
13	369	822	1276	1729	2183	2637	3090
14	397	851	1304	1758	2211	2665	3119
15	425	879	1332	1786	2240	2693	3147

HUNDREDTHS OF POUND TO NEAREST GRAM

	0	0.01	0.02	0.03	0.04	0.05	0.06	0.07	0.08	0.09
0.0		5	9	14	18	23	27	32	36	41
0.1	45	50	54	59	64	68	73	77	82	86
0.2	91	95	100	104	109	113	118	122	127	132
0.3	136	141	145	150	154	159	163	168	172	177
0.4	181	186	191	195	200	204	209	213	218	222
0.5	227	231	236	240	245	249	254	259	263	268
0.6	272	277	281	286	290	295	299	304	308	313
0.7	318	322	327	331	336	340	345	349	354	358
0.8	363	367	372	376	381	386	390	395	399	404
0.9	408	413	417	422	426	431	435	440	445	449

POUNDS (BY TENTHS) TO NEAREST GRAM

	0.0	0.1	0.2	0.3	0.4	0.5	0.6	0.7	0.8	0.9
0		45	91	136	181	227	272	318	363	408
1	454	499	544	590	635	680	726	771	816	862
2	907	953	998	1043	1089	1134	1179	1225	1270	1315
3	1361	1406	1451	1497	1547	1588	1633	1678	1724	1769
4	1814	1860	1905	1950	1996	2041	2087	2132	2177	2223
5	2268	2313	2359	2404	2449	2495	2540	2585	2631	2676
6	2722	2767	2812	2858	2903	2948	2994	3039	3084	3130
7	3175	3220	3266	3311	3357	3402	3447	3443	3538	3583
8	3629	3674	3719	3765	3810	3856	3901	3946	3992	4037
9	4082	4128	4173	4218	4264	4309	4354	4400	4445	4491
10	4536	4581	4627	4672	4717	4763	4808	4853	4899	4944
11	4989	5035	5080	5125	5171	5216	5262	5307	5352	5398

CONVERSION TABLES

POUNDS (BY TENTHS) TO NEAREST GRAM (continued)

	0.0	0.1	0.2	0.3	0.4	0.5	0.6	0.7	0.8	0.9
12	5443	5488	5534	5579	5625	5670	5715	5761	5806	5851
13	5897	5942	5987	6033	6078	6123	6169	6214	6260	6305
14	6350	6396	6441	6486	6532	6577	6622	6668	6713	6758
15	6804	6849	6895	6940	6985	7031	7076	7121	7167	7212
16	7257	7303	7348	7394	7439	7484	7530	7575	7620	7666
17	7711	7756	7802	7847	7892	7938	7983	8029	8074	8119
18	8165	8210	8255	8301	8346	8391	8437	8482	8527	8573
19	8618	8663	8709	8754	8800	8845	8890	8936	8981	9026
20	9072	9117	9163	9208	9253	9299	9344	9389	9435	9480
21	9525	9571	9616	9661	9707	9752	9798	9843	9888	9934
22	9979	10024	10070	10115	10160	10206	10251	10296	10342	10387
23	10433	10478	10523	10569	10614	10659	10705	10750	10795	10841
24	10886	10932	10977	11022	11068	11113	11158	11204	11249	11294
25	11340	11385	11430	11476	11521	11567	11612	11657	11703	11748
26	11793	11839	11884	11929	11975	12020	12065	12111	12156	12202
27	12247	12292	12338	12383	12428	12474	12519	12564	12610	12655
28	12701	12746	12791	12837	12882	12927	12973	13018	13063	13109
29	13154	13200	13245	13290	13336	13381	13426	13472	13517	13562
30	13608	14061	14515	14968	15422	15876	16329	16783	17236	17690
40	18144	18597	19051	19504	19958	20412	20865	21319	21772	22226
50	22680	23133	23587	24040	24494	24947	25401	25855	26308	26762
60	27215	27669	28123	28576	29030	29483	29937	30391	30844	31298
70	31751	32205	32658	33112	33566	34019	34473	34926	35380	35834
80	36287	36744	37195	37648	38102	38555	30990	39463	39916	40370
90	40823	41277	41731	42184	42638	43091	43545	43998	44452	44906

ACRES TO HECTARES

	0	1	2	3	4	5	6	7	8	9
		0.40	0.81	1.21	1.62	2.02	2.43	2.83	3.24	3.64
10	4.05	4.45	4.86	5.26	5.67	6.07	6.48	6.88	7.28	7.69
20	8.09	8.50	8.90	9.31	9.71	10.01	10.52	10.93	11.33	11.74
30	12.14	12.55	12.95	13.36	13.76	14.16	14.57	14.97	15.38	15.78
40	16.19	16.59	17.00	17.40	17.81	18.21	18.62	19.02	19.43	19.83
50	20.24	20.64	21.04	21.45	21.85	22.26	22.66	23.07	23.47	23.88
60	24.28	24.69	25.10	25.50	25.90	26.31	26.71	27.11	27.52	27.92
70	28.33	28.73	29.14	29.54	29.95	30.35	30.76	31.16	31.57	31.97
80	32.38	32.78	33.19	33.59	33.99	34.40	34.80	35.21	35.61	36.02
90	36.42	36.83	37.23	37.64	38.04	38.45	38.85	39.26	39.66	40.07

POUNDS/ACRE TO KILOGRAMS/HECTARE

1 pound/acre = 1.12085 kilo/hectare

	0	1	2	3	4	5	6	7	8	9
		1.1	2.2	3.4	4.5	5.6	6.7	7.8	9.0	10.1
10	11.2	12.3	13.5	14.6	15.7	16.8	17.9	19.1	20.2	21.3
20	22.4	23.5	24.7	25.8	26.9	28.0	29.1	30.3	31.4	32.5
30	33.6	34.7	35.9	37.0	38.1	39.2	40.4	41.5	42.6	43.7
40	44.8	46.0	47.1	48.2	49.3	50.4	51.6	52.7	53.8	54.9
50	56.0	57.2	58.3	59.4	60.5	61.6	62.8	63.9	65.0	66.1
60	67.3	68.4	69.5	70.6	71.7	72.9	74.0	75.1	76.2	77.3
70	78.5	79.6	80.7	81.8	82.9	84.1	85.2	86.3	87.4	88.5
80	89.7	90.8	91.9	93.0	94.2	95.3	96.4	97.5	98.6	99.8
90	100.9	102.0	103.1	104.2	105.4	106.5	107.6	108.7	109.8	111.0

CONDITION FACTOR C TO K

	0	1	2	3	4	5	6	7	8	9
10	0.28	0.30	0.33	0.36	0.39	0.42	0.44	0.47	0.50	0.53
20	0.55	0.58	0.61	0.64	0.66	0.69	0.72	0.75	0.78	0.80
30	0.83	0.86	0.89	0.91	0.94	0.97	1.00	1.02	1.05	1.08
40	1.11	1.14	1.16	1.19	1.22	1.25	1.27	1.30	1.33	1.36
50	1.39	1.41	1.44	1.47	1.50	1.52	1.55	1.58	1.61	1.63
60	1.66	1.69	1.72	1.75	1.77	1.80	1.83	1.86	1.88	1.91
70	1.94	1.97	1.99	2.02	2.05	2.08	2.11	2.13	2.16	2.19
80	2.22	2.24	2.27	2.30	2.33	2.35	2.38	2.41	2.44	2.47
90	2.49	2.52	2.55	2.58	2.60	2.63	2.66	2.69	2.71	2.74

TEMPERATURE—FAHRENHEIT TO CENTIGRADE

	0	1	2	3	4	5	6	7	8	9
30	-1.1	-0.6	0	0.6	1.1	1.7	2.2	2.8	3.3	3.9
40	4.4	5.0	5.6	6.1	6.7	7.2	7.8	8.3	8.9	9.4
50	10.0	10.6	11.1	11.7	12.2	12.8	13.3	13.9	14.4	15.0
60	15.6	16.1	16.7	17.2	17.8	18.3	18.9	19.4	20.0	20.6
70	21.1	21.7	22.2	22.8	23.3	23.9	24.4	25.0	25.6	26.1
80	26.7	27.2	27.8	28.3	28.9	29.4	30.0	30.6	31.1	31.7
90	32.2	32.8	33.3	33.9	34.4	35.0	35.6	36.1	36.7	37.2
100	37.8	38.3	38.9	39.4	40.0					

Citations

[Abbreviations cited follow those listed in Biosis 1974, List of Serials. BioSciences Information Service, Philadelphia. 218 pp.]

A1   Adams, C. C. and T. L. Hankinson, 1928. The ecology and economics of Oneida Lake fish. Roosevelt Wildl. Ann. 1(3 & 4):241-358.

A8   Aldrich, A. D., 1948. Fish quality vs. quantity? Okla. Game Fish News Feb. 1948:4-5.

A10   ———. 1949. Progress report: improvement of fish-cultural practices in Oklahoma. Prog. Fish-Cult. 11(1):25-30.

A39   Alabama Department of Conservation. 1957. Report for the fiscal year October 1, 1955-September 30, 1956. State Ala. 205 pp.

A40   ———. 1958. Report for fiscal year October 1, 1956-September 30, 1957. State Ala. 200 pp.

A42   Allen, E. R. and W. T. Neil, 1953. A xanthic largemouth bass (*Micropterus*) from Florida. Copeia, 1953 (2):116.

A47   Anderson, R. O., 1958. A seasonal study of food consumption and growth of the bluegill. Abstr. Midw. Wildl. Conf. 20:5-6.

A87   Applegate, R. L., J. W. Mullan, and D. I. Morais, 1967. Food and growth of six centrarchids from shoreline areas of Bull Shoals Reservoir. Proc. SE Assoc. Game Fish Comm. 20:469-482.

A88   Albertson, R. D. and F. Schultz, 1968. Fishes of Little Wall Lake, 1967. Proc. Iowa Acad. Sci. 75:164-169.

A89   Andrews, A. K., C. C. Van Valin, and B. E. Stebbings, 1966. Some effects of heptochlor on bluegills (*Lepomis machrochirus*). Trans. Am. Fish. Soc. 95:297-309.

A90   Anderson, R. O., 1971. Stocking strategies for warmwater fishes in lentic environments. Proc. North Cent. warmwater fish cult. manage. workshop, Iowa Coop. Fish. Unit. 247 pp.

A91   Anderson, R. O., H. S. Mohler, and G. Devine. 1971. Growth and survival of different geographic stocks of smallmouth bass in ponds in Missouri (Abstract). P. 200 in Proc. North Cent. warmwater fish cult. manage. workshop, Iowa State Univ.

A92   Applegate, R. L., 1966. Pyloric caeca counts as a method for separating the advanced fry and fingerlings of largemouth and spotted basses. Trans. Am. Fish. Soc. 95(2):226.

A93   Applegate, R. L. and J. W. Mullan. 1967. Food of young largemouth
       bass, *Micropterus salmoides*, in a new and old reservoir. Trans. Am.
       Fish. Soc. 96(1):74-77.

A94   Anderson, R. O. and M. L. Heman. 1969. Angling as a factor influenc-
       ing catchability of largemouth bass. Trans. Am. Fish. Soc. 98(2):317-
       320.

A95   Al-Rawi, T. R. and D. W. Toetz. 1972. Growth of white crappie and
       gizzard shad in Keystone, Oklahoma. Proc. Okla. Acad. Sci. 52:1-5.

A96   Applegate, R. L. and J. W. Mullan. 1969. Ecology of *Daphnia* in Bull
       Shoals Reservoir. U.S. Fish. Wildl. Ser. Res. Rep. 74:23 pp.

A97   Al-Rawi, T. R., 1971. Investigating the validity of the scale method in
       determining the growth of two species of fish in Oklahoma and its rela-
       tion to temperature and water level. Ph.D. thesis, Okla. State Univ.
       152 pp.

A98   Anderson, R. O., 1974. Influence of mortality rate on production and
       potential sustained harvest of largemouth bass populations. North
       Cent. Div. Am. Fish. Soc. Spec. Publ. 3:18-28.

A99   Aggus, L. R., 1973. Food of angler harvested largemouth, spotted and
       smallmouth bass in Bull Shoals Reservoir. Proc. S.E. Assoc. Game
       Fish Comm. 26:519-529.

A100   Addison, J. H. and S. L. Spencer, 1972. Preliminary evaluation of
       three strains of largemouth bass, *Micropterus salmoides* (Lacepede),
       stocked in ponds in south Alabama. Proc. S.E. Assoc. Game Fish
       Comm. 25:366-374.

A101   Abbott, C. C., 1883. On the habits of certain sunfishes. Am. Nat. 17:
       1254-1257.

A102   Amundrud, J. R., D. J. Faber, and A. Keast, 1974. Seasonal succes-
       sion of free-swimming perciform larvae in Lake Opinicon, Ontario. J.
       Fish. Res. Board Can. 31(10):1661-1665.

A103   Anderson, R. J., 1974. Feeding artificial diets to smallmouth bass.
       Prog. Fish-Cult. 36(3):145-151.

A104   Ackerman, G., 1974. Turkey River investigations. Iowa State Conserv.
       Comm. Fish. Manage. Proj. 73-11-C-21, pp. 83-99.

B2   Bailey, R. M., 1943. Progress report—Fisheries research in Spirit
       and Okoboji Lakes. Q. Rep. Iowa Coop. Wildl. Res. Unit Iowa Fish.
       Res. Unit. July-Sept. 1943:11-13.

B3   ———. 1943. Fisheries research in Clear Lake, Iowa. Q. Rep. Iowa
       Coop. Wildl. Res. Unit Iowa Fish. Res. Unit. July-Sept. 1943:14-17.

B4   ———. 1943. Fisheries research in Clear Lake, Iowa. Q. Rep. Iowa
       Coop. Wildl. Res. Unit Iowa Fish. Res. Unit. Oct.-Dec. 1943:16-17.

B5   Bailey, R. M. and H. M. Harrison, Jr., 1945. The fishes of Clear Lake,
       Iowa. Iowa State Coll. J. Sci. 20(1):57-77.

B6   ———. 1948. Food habits of the southern channel catfish (*Ictalurus
       lacustris punctatus*) in the Des Moines River, Iowa. Trans. Am. Fish.
       Soc. 75:11-138.

B7   Bailey, R. M. and K. F. Lagler, 1938. An analysis of hybridization in
       a population of stunted sunfishes in New York. Pap. Mich. Acad. Sci.
       Arts Lett. 23:577-606.

B11   Bajkov, A., 1930. A study of the whitefish (*Coregonus clupeaformis*) in
       Manitoba and lakes. Contrib. Can. Biol. Fish. NS 5(15):442-455, 1929.

B13   Ball, R. C., 1948. A summary of experiments in Michigan lakes on the
       elimination of fish populations with rotenone, 1934-1942. Trans. Am.
       Fish. Soc. 75:139-146.

B15   Ball, R. C., 1948. Relationship between available fish food, feeding
       habits of fish and total fish production in a Michigan Lake. Mich. State
       Coll. Tech. Bull. 206:1-59.

B16   ——. 1949. Experimental use of fertilizer in the production of fish-
       food organisms and fish. Mich. Agric. Exp. Stn. Tech. Bull. 210:1-28.

B18   Barney, R. L. and B. J. Anson, 1920. Life history and ecology of the
       pigmy sunfish, *Elassome zonatum*. Ecology 1(4):241-256.

B19   Barney, R. J. and B. J. Anson, 1923. Life history and ecology of the
       orange-spotted sunfish (*Lepomis humilis*). Rep. U.S. Fish. Comm.
       1922: App. 15, 1-16 pp.

B25   Beckman, W. C., 1941. Increased growth rate of rock bass, *Amblop-
       lites rupestris* (Rafinesque) following reduction in the density of the
       population. Trans. Am. Fish. Soc. 70:143-148.

B26   ——. 1941. Meet Mr. Bluegill. Mich. Conserv. 10(7):6-7, 11.

B29   ——. 1943. Further studies on the increased growth rate of the rock
       bass, *Ambloplites rupestris* (Rafinesque), following the reduction in
       density of the populations. Trans. Am. Fish. Soc. 72:72-78.

B30   ——. 1948. The length-weight relationship, factors for conversions
       between standard and total lengths, and coefficients of condition for
       seven Michigan fishes. Trans. Am. Fish. Soc. 75:237-256.

B31   ——. 1949. The rate of growth and sex ratio for seven Michigan
       fishes. Trans. Am. Fish. Soc. 76:63-81.

B37   Belding, D. L. and M. P. Clark, 1938. Observations on the salmon
       parr of the Margaree River. Trans. Am. Fish. Soc. 67:184-194.

B39   Bennett, G. W., 1937. The growth of the large-mouthed black bass,
       Huro salmoides (Lacepede), in the waters of Wisconsin. Copeia, 1937
       (2):104-118.

B40   ——. 1938. Growth of the small-mouthed black bass, *Micropterus
       dolomieu* Lacepede, in Wisconsin waters. Copeia, 1938(4):157-170.

B41   ——. 1948. The bass blue-gill combination in a small artificial lake.
       Ill. Nat. Hist. Surv. Bull. 24(3):377-412.

B42   Bennett, G. W., D. H. Thompson, and S. A. Parr, 1940. Lake manage-
       ment reports. 4. A second year of fisheries investigations at Fork
       Lake, 1939. Ill. Nat. Hist. Surv. Biol. Notes. 14:1-24.

B44   Berg, G., 1924. Fish culture on the farm. Publ. Indiana Dep. Conserv.
       40:1-25.

B53   Bishop, S. C., 1936. Fisheries investigations in the Delaware and
       Susquehanna rivers. Suppl. 25th Ann. Rep. N.Y. Conserv. Dep. Biol.
       Surv. 10:122-139.

B54   Bishop, S. C. and R. E. James, 1938. Fisheries investigations in the
       Allegheny and Chemung rivers. Suppl. 27th Ann. Rep. N.Y. Conserv.
       Dep. Biol. Surv. 12:102-112.

B60   Bolen, H. R., 1924. The relation of size to age in some common fresh-
       water fishes. Proc. Indiana Acad. Sci. (1923) 39:307-309.

B61   Bonham, K., 1946. Management of a small fish pond in Texas. J.
       Wildl. Manage. 10(1):1-4.

B65   Bower, S., 1897. The propagation of smallmouth black bass. Trans.
       Am. Fish. Soc. 25:127-136.

B69   Breder, C. M., 1936. Long-lived fishes in the aquarium. Bull. N.Y.
       Zool. Soc. 39:116-117.

B72   Breder, C. M. and A. C. Redmond, 1929. The blue-spotted sunfish. A
       contribution to the life history and habits of *Enneacanthus* with notes on
       other Lepominae. Zoologic (NY) 9(10):379-401.

B73   Brice, J. J., 1898. A manual of fish-culture, based on the methods of
        the United States Commission of Fish and Fisheries with chapters on
        the cultivation of oysters and frogs. Rep. U.S. Comm. Fish. 1897:
        Appendix pp. 1-340.

B75   Brown, C. J. D., 1942. A fisheries Survey of Hopkins Lake, Shiawassee
        County. Mich. Conserv. 11(11):6-7.

B77   Brown, C. J. D. and R. C. Ball, 1943. A fish population study of Third
        Sister Lake. Trans. Am. Fish. Soc. 72:177-185.

B80   Brown, G. W. N., 1916. The construction of a pond cultural station, and
        the propagation and distribution of largemouth black bass in South
        Carolina. Trans. Am. Fish. Soc. 46(1):30-34.

B86   Bennett, G. W., 1945. Overfishing in a small artificial lake, Onized
        Lake near Alton, Illinois. Ill. Nat. Hist. Surv. Bull. 23(3):372-406.

B92   Ball, R. C. and H. D. Tait, 1952. Production of bass and bluegills in
        Michigan ponds. Mich. State Coll. Agric. Exp. Stn. Tech. Bull. 231:1-
        24.

B93   Ball, R. C. and H. A. Tanner, 1951. The biological effects of fertilizer
        on a warm-water lake. Mich. State Coll. Agric. Exp. Stn. Tech. Bull.
        233:1-32.

B94   Barnickol, P. G. and W. C. Starrett, 1951. Commercial and sport
        fishery of the Mississippi River between Caruthersville, Missouri, and
        Dubuque, Iowa. Ill. Nat. Hist. Surv. Bull. 25(5):267-350.

B96   Beckman, W. C., 1950. Changes in growth rates of fishes following re-
        duction in population densities by winterkill. Trans. Am. Fish. Soc. 78
        (1948):82-90.

B97   Bennett, G. W., 1948. Winterkill of fishes in an Illinois Lake. Ill. Nat.
        Hist. Surv. Biol. Notes. 19:1-9.

B98   ———. 1951. Experimental largemouth bass management in Illinois.
        Trans. Am. Fish. Soc. 80:231-239.

B99   Bennett, G. W. and L. Durham, 1951. Cost of Bass fishing at Ridge
        Lake, Coles County, Illinois. Ill. Nat. Hist. Surv. Biol. Notes. 23:1-16.

B101   Bilton, T. H., 1951. Creel census studies in Lakelse Lake, Skeena
        River. Fish Res. Board Can. Prog. Rep. Pac. Coast Stn. 87:39-41.

B102   Binkley, L. E., 1949. The interpretation of fish scale readings in fish-
        ery management. Proc. W. Va. Acad. Sci. 21:45-47.

B103   ———. 1951. Planned Aquaculture. Proc. W. Va. Acad. Sci. 23:42-47.

B105   Bridge, T., 1943. Feeding experiments on Kamloops trout (Salmo
        gairdnerii kamloops Jordan). B.C. Game Comm. Rep. 1942:13 pp.

B108   Brown, C. J. D. and N. A. Thoreson, 1951. Ranch fish ponds in Mon-
        tana; their construction and management. Mont. Agric. Exp. Stn. Bull.
        480:1-30.

B109   Brown, W. H., 1951. Results of stocking largemouth black bass and
        channel catfish in experimental Texas farm ponds. Trans. Am. Fish.
        Soc. 80:210-217.

B110   ———. 1952. Rate of survival of largemouth black bass fingerlings
        stocked in experimental farm ponds. Prog. Fish-Cult. 14(2):79-80.

B111   ———. 1952. Rate of survival of largemouth black bass fry stocked in
        experimental farm ponds. Prog. Fish-Cult. 14(4):177-179.

B114   Burress, R. M., 1949. The growth rates of bluegills and largemouth
        bass in fertilized and unfertilized ponds in central Missouri. M.A.
        thesis, U. Mo. 79 pp. typewritten.

B115   ———. 1950. The largemouth bass. Mo. Conserv. 11(7):16.
B116   ———. 1951. The longear sunfish. Mo. Conserv. 12(5):16.

B122 Beland, R. D., 1953. The occurrence of two additional centrarchids in the lower Colorado River, California. Calif. Fish Game 39(1):149-151.

B131 Ball, R. C. and J. R. Ford, 1953. Production of food-fish and minnows in Michigan ponds. Mich. Agric. Exp. Stn. Q. Bull. 35(3):384-391.

B134 Barnhart, R., 1955. Survey of Lake Loveland Reservoir, Larimer County, Colorado. Colo. Fish. Res. Unit Q. Rep. 2(1 & 2):31-40.

B142 Bennett, G. W., 1954. The effects of a late summer drawdown on the fish population of Ridge Lake, Coles County, Illinois. Trans. N. Am. Wildl. Conf. 19:259-270.

B143 ———. 1954. Largemouth bass in Ridge Lake, Coles County, Illinois. Ill. Nat. Hist. Surv. Bull. 26(2).

B144 Bennett, G. W. and W. F. Childers, 1957. The smallmouth bass, *Micropterus dolomieui,* in warmwater ponds. J. Wild. Manage. 21(4): 414-424.

B150 Birkenholz, D. and A. Fritz, 1956. Fishes of Little Wall Lake, Iowa. Iowa State U. Coop. Fish. Unit. 21 pp. typewritten.

B176 Brown, M. W., 1940. Smallmouth black bass propagation in California. Trans. Am. Fish Soc. 69:119-124.

B181 Buck, D. H., 1956. Effects of turbidity on fish and fishing. Okla. Fish. Res. Lab. Rep. 56.

B182 ———. 1956. Effects of turbidity on fish and fishing. Trans. N. Am. Wildl. Conf. 21:249-261.

B183 Buck, H. and F. Cross, 1952. Early limnological and fish population conditions of Canton Reservoir, Oklahoma, and fishery management recommendations. Okla. A. M. Coll. Res. Found. 110 pp.

B189 Byrd, I. B., 1959. Angling success and seasonal distribution of catch in Alabama's public fishing lakes. Trans. N. Am. Wildl. Conf. 24:225-237.

B192 Brown, E. H., Jr., 1960. Little Miami River headwater-stream investigations. Ohio Dept. Nat. Res. Div. Wildl. 1-143.

B200 Black, E. C., 1953. Upper lethal temperatures of some British Columbia freshwater fishes. J. Fish. Res. Board Can. 10(4):196-210.

B201 Brett, J. R., 1944. Some lethal temperature relations of Algonquin Park fishes. Publ. Ont. Fish. Res. Lab. 63:1-49.

B218 Bradley, M. C., 1948. An analysis of the growth and development of the fish in the fresh water impoundments at the Patuxent Research Refuge. M.S. thesis, Univ. Md. 23 pp.

B219 Brower, A., 1952. Observations on the growth rate of the largemouth bass (*Micropterus salmoides*) in Wintergreen Lake, Kalamazoo County, Michigan. M.S. thesis, Mich. State Coll. 82 pp.

B220 Burris, W. E., 1956. Studies of the age, growth, and food of known-age young-of-year black crappie and of stunted and fast-growing black and white crappies of some Oklahoma lakes. Ph.D. thesis, Okla. A. M. Coll. 88 pp.

B221 Bruns, P. M., 1958. Seasonal changes in growth rates of bluegill (*Lepomis macrochirus*) in Felt Lake. Stanford, California. M.A. thesis, Stanford Univ. 83 pp.

B222 Bernhardt, R. W., 1957. Growth of fish in the waters of the Huntington Wildlife Forest. M.S. thesis, Syracuse Univ. 93 pp.

B231 Bonde, T. and J. E. Maloney, 1960. Food habits of burbot. Trans. Am. Fish. Soc. 89(4):374-376.

B232 Brown, C. J. D. and S. M. Logan, 1960. Age and growth of four species of warm-water game fish from three Montana ponds. Trans. Am. Fish. Soc. 89(4):379-382.

B233   Bond, Carl E., 1948. Fish management problems of Lake of the Woods, Oregon. M.S. thesis, Oreg. State Coll. 109 pp.

B236   Bisbee, L., 1960. Southeast Oregon. Oreg. State Game Comm. Fish. Div. Ann. Rep. 1959:109-131.

B239   Bowman, M. L., 1954. Some aspects of the life history of the black redhorse (*Moxostoma duquesni* LeSueur), with reference to its association with the smallmouth bass (*Micropterus dolomieu* Lacepede) in two south central Missouri streams, the Niangua and the Big Piney. M.S. thesis, Univ. Mo. 39 pp.

B252   Brown, E. H., Jr., 1961. Movements of native and hatchery-reared game fish in a warm-water stream. Trans. Am. Fish. Soc. 90(4):449-456.

B268   Brown, B. E. and W. M. Tatum, 1962. Growth and survival of young spotted bass in Lake Martin, Alabama. Trans. Am. Fish. Soc. 91(3):324-326.

B286   Balon, E. K., 1959. Die Entwicklung des akklimatisierten *Lepomis gibbosus* (Linne 1748) wahrend der embryonalen Periode in den Donauseitenwassern. Z. Fisch. Hilfswiss 8 N.F. (1-3):1-27.

B305   Brittan, M. R., 1958. Sacramento perch, *Archoplites interruptus* Girard. Material for Handb. Biol. Data. 4 pp.

B307   Bennett, G. W. and W. F. Childers, 1966. The lake chubsucker as a forage species. Prog. Fish-Cult. 28(2):89-92.

B315   Beers, G. D. and Wm. J. McConnell, 1966. Some effects of threadfin shad introduction on black crappie diet and condition. J. Ariz. Acad. Sci. 4(2):71-74.

B317   Bisbee, L. E., 1962. Harney-Malheur District. Ore. State Game Comm. Fish. Div. Ann. Rep. 1962:135-163.

B349   Bunting, D. R., II and W. H. Irwin, 1965. The relative resistances of seventeen species of fish to petroleum refinery effluents and a comparison of some possible methods of ranking resistances. Proc. S.E. Assoc. Game Fish Comm. 17:293-307.

B350   Brown, B. E., 1965. Two-year study of a bass, sunfish, channel catfish population exposed to flooding and angling. Proc. S.E. Assoc. Game Fish Comm. 17:367-372.

B351   Behmer, D. J., 1965. Movement and anglers harvest of fishes in the Des Moines River, Boone County, Iowa. Proc. Iowa Acad. Sci. 71:259-263.

B358   Breder, C. M., Jr. and D. E. Rosen, 1966. Modes of reproduction in fishes. Natural History Press. 941 pp.

B365   Burrows, C. and J. Moyle, 1967. Special fishing edition. Conserv. Volunteer. 30(172):64.

B368   Biggins, R. G., 1968. Centrarchid feeding interactions in a small desert impoundment. M.S. thesis, Univ. Ariz. 44 pp.

B370   Breder, C. M., Jr., 1936. The reproductive habits of the North American sunfishes (family Centrarchidae). Zoologica (NY) 21(1):1-48.

B372   Bennett, G. W., H. W. Adkins, and W. F. Childers, 1969. Largemouth bass and other fishes in Ridge Lake, Illinois, 1941-1963. Ill. Nat. Hist. Surv. Bull. 30(1):1-67.

B373   Buck, D. H. and C. F. Thoits, III, 1970. Dynamics of one-species populations of fishes in ponds subjected to cropping and additional stocking. Ill. Nat. Hist. Surv. Bull. 30(2):68-165.

B374   Beyerle, G. B., 1971. A study of two northern pike-bluegill populations. Trans. Am. Fish. Soc. 100(1):69-73.

B375 Beyerle, G. B. and J. E. Williams, 1937. Attempted control of bluegill reproduction in lakes by the application of copper sulfate crystals to spawning nests. Prog. Fish-Cult. 29(3):150-155.

B376 Benda, R. S. and J. R. Gammon, 1967. The fish populations of Big Walnut Creek. Proc. Indiana Acad. Sci. 77:193-205.

B377 Brown, B. E. and J. J. Jossel, Jr., 1970. Condition factors and growth in length of a stunted white crappie population in Boomer Lake, Payne Co., Oklahoma. Proc. Okla. Acad. Sci. 49:156-162.

B378 Bacon, E. J., Jr. and R. V. Kilambi, 1968. Some aspects of the age and growth of the longear sunfish, *Lepomis megalotis,* in Arkansas waters. Proc. Ark. Acad. Sci. 22:44-57.

B380 Benson, N. G., 1959. Fish management on Wood's Reservoir. J. Tenn. Acad. Sci. 34(3):172-189.

B381 Bright, W. M., 1940. Spermatogenesis in sunfish. Trans. Ky. Acad. Sci. 8:37-38.

B382 Bennett, G. W., 1971. Management of lakes and ponds, 2nd ed. Van Nostrand Reinhold Co., N.Y. 375 pp.

B383 Boyer, R. L. and L. E. Vogele, 1969. Longear sunfish. U.S. Bur. Sport Fish. Wildl. Resour. Publ. 77:76-78.

B384 Boyer, R. L., 1969. Aspects of the behavior and biology of the longear sunfish, *Lepomis megalotis* (Rafinesque) in two Arkansas reservoirs. M.S. thesis, Okla. State Univ. 103 pp.

B385 Berra, T. M. and G. E. Gunning, 1970. Repopulation of experimentally decimated sections of streams by longear sunfish, *Lepomis megalotis megalotis* (Rafinesque). Trans. Am. Fish. Soc. 99(4):776-781.

B386 Boyer, R. L. and L. E. Vogele, 1971. Longear sunfish behavior in two Ozark reservoirs. Am. Fish. Soc. Publ. No. 8:13-25.

B388 Brynildson, C., 1972. Year classes make your fishing. Wis. Conserv. Bull. 37(2):20-21.

B389 Birdsong, R. S. and R. W. Yerger, 1967. A natural population of hybrid sunfishes: *Lepomis macrochirus* X *Chaenobryttus gulosus.* Copeia, 1967 (1):62-71.

B390 Beeman, H. W., 1924. Habits and propagation of the small-mouthed black bass. Trans. Am. Fish. Soc. 43:92-107.

B392 Burdick, G. E., M. Lipschuetz, H. F. Dean, and E. F. Harris, 1954. Lethal oxygen concentrations for trout and smallmouth bass. N.Y. Fish Game J. 1(1):84-97.

B393 Brown, D., 1931. The basses of Arkansas and some experiments in their propagation. Trans. Am. Fish. Soc. 61:83-85.

B394 Becker, D.A., R. G. Heard, and P. D. Holmes, 1966. A preimpoundment survey of the helminth and copepod parasites of *Micropterus* spp. of Beaver Reservoir in Northwest Arkansas. Trans. Am. Fish Soc. 95(1):23-34.

B395 Badenhuizen, T. R., 1969. Effect of incubation temperature on mortality of embryos of the largemouth bass, *Micropterus salmoides* (Lacepede). M.S. thesis, Cornell Univ. 88 pp.

B396 Beamish, F. W. H., 1970. Oxygen consumption of largemouth bass, *Micropterus salmoides,* in relation to swimming speed and temperature. Can. J. Zool. 48(6):1221-1228.

B397 Bliss, Q. P., 1971. Survival of stocked largemouth bass in southeast Nebraska. Proc. North Cent. warmwater fish cult. manage. workshop, Iowa State Univ. Pp. 201-203.

B398 Bryant, H. E. and A. Houser, 1971. Population estimates and growth of largemouth bass in Beaver and Bull Shoals Reservoirs. Am. Fish. Soc. Spec. Publ. 8:349-357.

B399 Bailey, R. M., H. E. Winn, and C. L. Smith, 1954. Fishes from the Escambia River, Alabama and Florida, with ecologic and taxonomic notes. Proc. Acad. Nat. Sci., Philadelphia. 106:109-164.

B400 (Quoting work of Buck, H., R. Baur, and C. R. Rose), 1972. Where bass grow best. Ill. Nat. Hist. Surv. Rep. 116:2-3.

B402 Ball, R. L. and R. V. Kilambi, 1970. Food habits of the white and black crappies in Beaver Reservoir. U.S. Fish. Wildl. Ser. Res. Publ. 106:296-297.

B403 Bennett, D. H., 1972. Length-weight relationships and condition factors of fishes from a South Carolina reservoir receiving thermal effluent. Prog. Fish-Cult. 34(2):85-87.

B404 Borges, H. M., 1950. Fish distribution studies, Niangua Arm of the Lake of the Ozarks, Missouri. J. Wildl. Manage. 14(1):16-33.

B406 Bennett, D. H. and J. W. Gibbons, 1972. Food of largemouth bass (*Micropterus salmoides*) from a South Carolina reservoir receiving heated effluent. Trans. Am. Fish. Soc. 101(4):650-654.

B407 Buck, D. H., R. J. Baur, and C. R. Rose, 1974. Interactions of intensive cultures of channel catfish with largemouth bass and bluegills in 1-acre ponds. Ill. Nat. Hist. Surv. Biol. Notes 84:8 pp.

B408 Bennett, G. W., H. W. Adkins, and W. F. Childers, 1973. The effects of supplemental feeding and fall drawdowns on the largemouth bass and bluegills at Ridge Lake, Illinois. Ill. Nat. Hist. Surv. Bull. 31(1):1-28.

B409 Bennett, D. H., 1971. Preliminary examination of body temperatures of largemouth bass (*Micropterus salmoides*) from an artificially heated reservoir. Arch. Hydrobiol. 68(3):376-381.

B410 Ball, R. L., 1974. Largemouth bass survival and exploitation rates in two heavily fished southern Indiana reservoirs. Paper presented at 36th Midwest Fish Wildl. Conf. 11 pp. mimeo.

B411 Bennett, G. W., 1974. Ecology and management of largemouth bass, *Micropterus salmoides*. North Cent. Div. Am. Fish. Soc. Spec. Publ. 3:10-17.

B412 Buck, D. H., R. J. Baur, and C. R. Rose, 1973. An experiment in the mixed culture of channel catfish and largemouth bass. Prog. Fish-Cult. 35(1):19-21.

B413 Ball, R. L. and R. V. Kilambi, 1973. The feeding ecology of the black and white crappies in Beaver Reservoir, Arkansas, and its effect on the relative abundance of the crappie species. Proc. S.E. Assoc. Game Fish Comm. 26:577-590.

B414 Beamish, F. W. H., 1974. Apparent specific dynamic action of largemouth bass, *Micropterus salmoides*. J. Fish. Res. Board Can. 31(11):1763-1769.

B415 Bouck, G. R., 1972. Effects of diurnal hypoxia on electrophoretic protein fractions and other health parameters of rock bass (*Ambloplites rupestris*). Trans. Am. Fish. Soc. 101(3):488-493.

B416 Benda, R. S., 1974. Growth and movement of fish in the vicinity of a thermal discharge. Indiana Acad. Sci. Proc. 83:185-191.

B417 Benda, R. S. and M. A. Proffitt, 1974. Effects of thermal effluents on fish and invertebrates. Pp. 438-477, Thermal Ecology, J. W. Gibbons and R. R. Sharitz (eds.). A.E.C. Symp. Ser. (Conf. 730505).

B418   Beyerle, G. B. and J. E. Williams, 1972. Survival, growth and production by bluegills subjected to population reduction in ponds. Mich. Dep. Nat. Res. R. D. Rep. 273. 28 pp.

B419   Brauhn, J. L., 1972. A suggested method for sexing bluegills. Prog. Fish-Cult. 34(1):17.

B420   Burton, D. T., E. L. Morgan, and J. Cairns, Jr., 1972. Mortality curves of bluegills (*Lepomis macrochirus* Rafinesque) simultaneously exposed to temperature and zinc stress. Trans. Am. Fish. Soc. 101(3): 435-441.

B421   Baumann, P. C. and J. F. Kitchell, 1974. Diel patterns of distribution and feeding of bluegill (*Lepomis macrochirus*) in Lake Wingra, Wisconsin. Trans. Am. Fish. Soc. 103(2):255-260.

B422   Banner, A. and J. A. Van Arman, 1973. Thermal effects on eggs, larvae and juveniles of bluegill sunfish. U.S. Env. Prot. Agency, Ecol. Res. Ser. EPA-R3-73-041:111 pp.

B423   Beitinger, T. L., 1974. Thermoregulatory behavior and diel activity patterns of bluegill, *Lepomis macrochirus*, following thermal shock. U.S. Fish. Bull. 72(3):1087-1093.

B424   Bietinger, T. L., J. J. Magnuson, W. H. Neill, and W. R. Shaffer, 1975. Behavioural thermoregulation and activity patterns in the green sunfish, *Lepomis cyanellus*. Anim. Behav. 23:222-229.

B425   Bennett, D. H. and J. W. Gibbons, 1975. Reproductive cycles of largemouth bass (*Micropterus salmoìdes*) in a cooling reservoir. Trans. Am. Fish. Soc. 104(1):77-82.

B426   Burr, B. M., 1974. A new intergeneric hybrid combination in nature: *Pomoxis annularis* X *Centrarchus macropterus*. Copeia, 1974(1):369-370.

B427   Buchanan, T. M. and K. Strawn, 1969. A field test of the use of scale size at the formation of the first annulus to permanently mass-mark smallmouth bass. Proc. S.E. Assoc. Game Fish Comm. 23:303-311.

B428   Barans, C. A. and R. A. Tubb, 1973. Temperatures selected seasonally by four fishes from western Lake Erie. J. Fish. Res. Board Can. 30(11):1697-1703.

B429   Bennett, G. W. and W. F. Childers, 1972. Thirteen-year yield of smallmouth bass from a gravel pit pond. J. Wild. Manage. 36(4):1249-1253.

B430   Bennett, D. H. and J. W. Gibbons, 1974. Growth and condition of juvenile largemouth bass from a reservoir receiving thermal effluent. Pp. 246-254, Thermal Ecology, A.E.C. Symp. Ser. (Conf. 730505).

C6   Carbine, W. F., 1939. Observations on the spawning habits of centrarchid fishes in Deep Lake, Oakland County, Michigan. Trans. N. Am. Wildl. Conf. 4:275-287.

Ç13   Carbine, W. F. and V. C. Applegate, 1948. The fish population of Deep Lake, Michigan. Trans. Am. Fish. Soc. 75:200-227.

C20   Carlander, K. D., 1943. Length-weight relationship of Minnesota fishes. Minn. Bur. Fish. Res. Invest. Rep. 17: revised: 23 pp. typewritten.

C22   ——. 1944. Notes on the minor species of fish taken in the commercial fisheries at Lake of the Woods 1939 to 1943. Minn. Bur. Fish. Res. Suppl. 2 of Fish. Res. Invest. Rep. 42: 27 pp. typewritten.

C23   ——. 1944. Notes on the coefficient of condition K of Minnesota fishes. Minn. Bur. Fish. Res. Invest. Rep. 41: revised: 40 pp. typewritten.

346 CITATIONS

C30 Carlander, K. D., 1949. Project No. 39. Yellow pike-perch management. Prog. Rep. Iowa Coop. Wildl. Res. Unit. Jan.-Mar.:44-57.

C31 ——. 1950. Growth rate studies of saugers, *Stizostedion canadense
canadense* (Smith) and yellow perch, *Perca flavescens* (Mitchill),
from Lake of the Woods, Minnesota. Trans. Am. Fish. Soc. 79:30-42.

C32 ——. 1950. Some considerations in the use of growth data derived
from scale studies. Trans. Am. Fish. Soc. 79:187-194.

C33 Carlander, K. D. and R. A. Fredin, 1948. Project No. 37: Management
of small ponds for fish production. Prog. Rep. Iowa Coop. Wildl. Res.
Unit. Oct.-Dec.:101-105.

C34 Carlander, K. D. and L. E. Hiner, 1943. Preliminary report on fisheries investigations, Leech Lake, Cass County. Minn. Bur. Fish. Res.
18 pp. typewritten.

C35 ——. 1943. Fisheries investigation and management report for Lake
Vermillion, St. Louis County. Minn. Bur. Fish. Res. Invest. Rep. 54:
1-175.

C36 Carlander, K. D. and J. Parsons, 1949. Project No. 39: Yellow pike-
perch management. Prog. Rep. Iowa Coop. Wildl. Res. Unit. April-
June 1949:49-52.

C37 Carlander, K. D. and L. L. Smith, Jr., 1945. Some factors to consider
in the choice between standard, fork, or total lengths in fishery investigations. Copeia, 1945(1):7-12.

C38 Carlander, K. D. and G. Sprugel, 1948. Project No. 42: Bullhead
management: shallow lake investigations. Prog. Rep. Iowa Coop.
Wildl. Res. Unit. Oct.-Dec.:112-117.

C39 Carr, A. F. Jr., 1939. Notes on the breeding habits of the warmouth
bass. Proc. Fla. Acad. Sci. 4:108-112.

C40 Carr, M. H., 1942. The breeding habits, embryology and larval development of the largemouthed black bass in Florida. Proc. N. Engl. Zool.
Club. 20:43-77.

C41 ——. 1946. Notes on the breeding habits of the eastern stump-
knocker, *Lepomis punctatis punctatis* (Cuvier). Q. J. Fla. Acad. Sci.
9(2):101-106.

C42 Carter, E. N., 1914. An experiment in feeding young largemouth bass.
Trans. Am. Fish. Soc. 44(1):71-72.

C60 Cooper, G. P., 1937. Food habits, rate of growth and cannibalism of
young largemouth bass (*Aplites salmoides*) in state-operated rearing
ponds in Michigan during 1935. Trans. Am. Fish. Soc. 66:242-266.

C65 Cornish, H., 1940. Green sunfish, age, rate of growth and food. Ohio
Conserv. Bull. 4(4):18-19.

C72 Creaser, C. W., 1926. The structure and growth of the scales of fishes
in relation to the interpretation of their life history, with special reference to the sunfish, *Eupomatis gibbosus*. Mus. Zool. Univ. Mich.,
Misc. Publ. No. 17.

C75 Cross, S. X., 1935. The effect of parasitism on growth of perch in the
Trout Lake region. J. Parasitol. 21(4):267-273.

C77 Culler, C. F., 1938. Notes on warm-water fish culture. Prog. Fish-
Cult. 36:19-24.

C78 Curran, H. W., J. Bardach, R. I. Bowman, and H. G. Lawler, 1947. A
biological survey of Lake Opinicon—Progress Report. Queens Univ.
Biol. Stn. (Chaffey's Lock) Ont. 48 pp.

C85 Churchill, W., 1949. Do little bass always grow big? Wis. Conserv.
Bull. 14(10):17-19.

C89   Carlander, K. D. and R. Moorman, 1949. Project No. 37: Management
        of small ponds for fish production. Iowa Coop. Wildl. Res. Unit. Q.
        Rep. July-Sept. 1949:54-83.

C91   Carlander, K. D. and J. Parsons, 1949. Project No. 39: Yellow pike-
        perch management. Iowa Coop. Wildl. Fish. Res. Units. Q. Rep. Oct.-
        Dec. 1949:34-45.

C92   Clark, M., 1950. Bass production in a Kentucky fish hatchery pond.
        Prog. Fish-Cult. 12(1):33-34.

C94   Calhoun, A. J., 1950. California angling catch records from postal
        card surveys: 1936-1948; with an evaluation of postal card nonre-
        sponse. Calif. Fish Game 36(3):177-234.

C99   Carlander, K. D., 1951. An unusually large population of fish in a
        gravel pit lake. Proc. Iowa Acad. Sci. 58:435-440.

C101   Carlander, K. D., T. S. English, and J. G. Erickson, 1950. Project No.
        39: Yellow pike-perch management. Q. Rep. Iowa Coop. Wildl. Fish.
        Res. Unit. 16(1):40-48.

C102   Carlander, K. D., J. Forney, and W. Pearcy, 1951. Project No. 39:
        Clear Lake Investigations. Q. Rep. Iowa Coop. Wildl. Fish. Res. Units.
        17(1):37-43.

C103   Carlander, K. D. and R. C. Hennemuth, 1952. Artificial lakes. Iowa
        Coop. Wildl. Fish. Res. Unit. Q. Rep. 18(1):32-35.

C105   Carlander, K. D. and R. B. Moorman, 1950. Project No. 37: Manage-
        ment of small ponds of fish production. Q. Rep. Iowa Coop. Wildl. Fish.
        Res. Units. 15(4):36-40.

C106   ———. 1951. Project No. 37: Management of small ponds for fish
        production. Q. Rep. Iowa Coop. Wildl. Fish. Res. Unit. 16(3):43-45.

C107   ———. 1952. Project No. 37: Management of farm ponds for fish pro-
        duction. Q. Rep. Iowa Coop. Wildl. Fish. Res. Unit. 17(3):24-27.

C109   Carlander, K. D. and R. R. Whitney, 1952. Clear Lake Investigations.
        Iowa Coop. Wildl. Fish. Res. Unit. Q. Rep. 18(1):27-30.

C110   Carter, B. T., 1951. Studies of fish populations. Ky. Happy Hunting
        Ground. 7(2):15, 17, 20.

C111   Chance, C. J., 1950. Fish catch in Bedford and Tullahoma Lakes,
        Tennessee, with special reference to soil productivity. J. Tenn. Acad.
        Sci. 25(2):157-168.

C112   Cheatum, E. P., 1952. Golden shiner, bluegill and green sunfish pro-
        duction in a small lake. Field Lab (S. Methodist Univ.). 20(3):103-104.

C115   Churchill, W., 1949. The effect of perch removal on a small north-
        eastern Wisconsin lake. Wis. Div. Fish Manage. Invest. Rep. 714.
        7 pp.

C117   ———. 1950. Experimental stocking of predators for population con-
        trol. Wis. Div. Fish Manage. Invest. Rep. 728:1-8. mimeo.

C118   ———. 1951. A review of fish management techniques on stunted pan-
        fish lakes in northeast Wisconsin. 13th Midw. Wildl. Conf. 2 pp. ab-
        stract.

C122   Cleary, R. E., 1951. The age and growth of the smallmouth bass in the
        streams of northeast Iowa. Iowa State Conserv. Comm. Q. Biol. Rep.
        3(2):32-33.

C123   ———. 1951. Monthly success and effort on stream fishing in N.E.
        Iowa. Iowa State Conserv. Comm. Q. Biol. Rep. 3(2):29-31.

C124   ———. 1952. Comparison of 1950 and 1951 angling success and corre-
        lation between angling and netting success on rivers of northeast Iowa.
        Iowa Q. Biol. Rep. 4(1):50-53.

348 CITATIONS

C134   Cross, F. B., 1950. Effects of sewage and of a headwaters impound-
        ment on the fishes of Stillwater Creek in Payne County, Oklahoma.
        Am. Midl. Nat. 43(1):128-145.

C148   Caldwell, D. K., H. T. Odum, T. R. Hellier, Jr., and F. H. Berry,
        1957. Populations of spotted sunfish and Florida largemouth bass in a
        constant temperature spring. Trans. Am. Fish. Soc. 85:120-134.

C154   Carlander, K. D., 1955. The standing crop of fish in lakes. J. Fish.
        Res. Board Can. 12(4):543-570.

C157   Carlander, K. D. and G. Sprugel, 1955. Fishes of Little Wall Lake,
        Iowa, prior to dredging. Proc. Iowa Acad. Sci. 62:555-566.

C159   Carter, B. T., 1953. The status of gigging and snagging in Kentucky.
        Ky. Fish. Bull. 14: 7 pp. mimeo.

C160   Carter, E. R., 1953. Growth rates of the white crappie, *Pomoxis
        annularis*, in Kentucky Lake. Ky. Fish. Bull. 12: 7 pp.

C162   ———. 1955. Growth rates of the white crappie, *Pomoxis annularis*,
        in the Tennessee River. Ky. Fish. Bull. 17: 5 pp.

C168   Chance, C. J., 1955. Unusually high returns from fish tagging experi-
        ments on two TVA reservoirs. J. Wildl. Manage. 19(4):500-501.

C169   ———. 1955. Growth of fishes in the Tennessee Valley. T.V.A. 1 p.
        mimeo.

C175   Charles, J. R., 1957. Final report on population manipulation studies
        on three Kentucky streams. Ky. Fish. Bull. 22: 45 pp.

C176   ———. 1957. Final report on population manipulation studies in three
        Kentucky streams. Proc. S.E. Assoc. Game Fish Comm. 11:155-185.

C182   Clark, C. F., 1956. Sandusky River Report. Ohio Div. Wildl.

C193   Cleary, R. E. and T. Moen, 1957. The effect of mechanical reduction
        on the growth of the crappie in Backbone Lake. Iowa State Conserv.
        Comm. Q. Biol. Rep. 9(4):13-19.

C197   Cochrun, K., 1956. Klamath District. Ore. State Game Comm. Fish.
        Div. Ann. Rep. 1955:160-167.

C198   ———. 1957. Klamath District. Ore. State Game Comm. Fish. Div.
        Ann. Rep. 1956:163-169.

C202   Cole, C. F., J. Trenary, and S. Finkelstein, 1958. Experimental in-
        troduction of threadfin shad. Ark. Game Fish Comm. Ann. Rep. 56
        pp. mimeo.

C208   Cook, E. P., 1955. A lake inventory of Meredith Reservoir, Crowley
        County, Colorado. Colo. Dep. Game Fish Publ. 46 pp.

C213   Cooper, G. P., 1953. Population estimates of fish in Sugar-Loaf Lake,
        Washtenaw County, Michigan, and their exploitation by anglers. Pap.
        Mich. Acad. Sci. Arts. Lett. 1952. 38:163-185.

C214   Cooper, G. P. and W. C. Latta, 1954. Further studies on the fish
        population and exploitation by angling in Sugar-Loaf Lake, Washtenaw
        County, Michigan. Pap. Mich. Acad. Sci. Arts Lett. 39:209-223.

C215   Cooper, G. P. and R. N. Schafer, 1954. Studies on the population of
        legal-size fish in Whitmore Lake, Washtenaw and Livingstone Counties,
        Michigan. Trans. N. Am. Wildl. Conf. 19:239-258.

C225   Counselman, J., 1957. A method for evaluating fresh water sport fish
        utilization. Proc. S.E. Assoc. Game Fish Comm. 10:119-123.

C226   Cross, F. B., J. E. Deacon, and C. M. Ward, 1959. Growth data on
        sport fishes in twelve lakes in Kansas. Trans. Kans. Acad. Sci. 62(2):
        162-164.

C232   Crowe, W. R., 1955. Numerical abundance and use of a spawning run of
        walleyes in the Muskegan River, Michigan. Trans. Am. Fish. Soc. 84:
        125-136.

C247 Carlander, K. D. and R. E. Cleary, 1949. The daily activity patterns of some freshwater fishes. Am. Midl. Nat. 41:447-452.

C251 Cleary, R. E., 1959. Weekend creel census and economic evaluation of northeast Iowa rivers—1959. Iowa State Conserv. Comm. Q. Biol. Rep. 11(3):23-28.

C256 Cooper, W. A., Jr., 1950. Age, growth and food habits of the large-mouthed black bass (*Micropterus salmoides*) and the spotted bass (*Micropterus punctulatus*) in north and east Texas lakes. M.S. thesis, N. Texas State Coll. 90 pp.

C257 Cargo, D. G., 1949. Comparison of the condition factors of four game fish species from different areas of the Pymatuning Lakes. M.S. thesis, Univ. Pittsburgh. 29 pp.

C260 Cole, V. W., 1951. Contributions to the life history of the redear sun-fish (*Lepomis microlophus*) in Michigan waters. M.S. thesis, Mich. State Coll. 54 pp.

C265 Calhoun, A. J., 1942. The biology of the black-spotted trout (*Salmo clarkii henshawi* Gill and Jordan) in two Sierra lakes. Ph.D. thesis, Stanford Univ. 218 pp.

C270 Christenson, L. M., 1957. Some characteristics of the fish populations in backwater areas of the Upper Mississippi River. M.S. thesis, Univ. Minn. 125 pp.

C271 Cross, Frank, 1951. Early limnological and fish population conditions of Canton Reservoir, Oklahoma, with special reference to carp, channel catfish, largemouth bass, green sunfish and bluegill, and fishery man-agement recommendations. Ph.D. thesis, Okla. A. M. Coll. 92 pp.

C272 Crawley, H. D., 1954. Causes of stunting of crappie (*Pomoxis nigro-maculatus* and *Pomoxis annularis*) in Oklahoma lakes. Ph.D. thesis, Okla. A. M. Coll. 94 pp.

C279 Corthell, R. A., 1960. Coos-Coquille District. Ann. Rep. Fish. Div. Ore. State Game Comm. 1960:278-290.

C284 Clark, C. F., 1960. Lake St. Marys and its management. Ohio Dep. Nat. Resour. Div. Wildl. Publ. W-324:107.

C285 Childers, W. F. and G. W. Bennett, 1961. Hybridization between three species of sunfish (*Lepomis*). Ill. Nat. Hist. Surv. Biol. Notes. 46:1-15.

C286 Cooper, E. L., J. A. Boccardy, and J. K. Anderson, 1962. Growth rate of brook trout at different population densities in a small infertile stream. Prog. Fish-Cult. 24(2):74-80.

C302 Cooper, E. L., H. Hidu, and J. K. Anderson, 1963. Growth and pro-duction of large mouth bass in a small pond. Trans. Am. Fish. Soc. 92:391-400.

C305 Clugston, J. P., 1964. Growth of the Florida largemouth bass, *Micro-pterus salmoides floridanus* (LeSueur), and the northern largemouth bass, *M. S. salmoides* (Lacepede), in subtropical Florida. Trans. Am. Fish. Soc. 93(2):146-154.

C327 Cook, S. F., Jr., R. L. Moore, and J. D. Conners, 1966. The status of the native fishes of Clear Lake, Lake County, California. Wasmann J. Biol. 24(1):141-160.

C348 Carlander, K. D., 1966. Relationship of limnological features to growth of fishes in lakes. Verh. Int. Verein. Limnol. 16:1172-1175.

C349 Cross, F. B., 1967. Handbook of fishes of Kansas. Univ. Kans. Mus. Nat. Hist. Misc. Publ. 45:1-357.

C351 Conley, J. M. and A. Witt, Jr., 1966. The origin and development of scales in the flier, *Centrarchus macropterus* (Lacepede). Trans. Am. Fish. Soc. 95(4):433-434.

C352 Carnes, W. C., 1966. Preliminary observations on supplementary feeding of pond fishes. N.C. Wildl. Res. Comm., Raleigh, 8 pp. Presented at S.E. Division, A.F.S. 1966.

C353 Cohen, M. and B. E. Brown, 1969. Scale and body growth of young-of-year centrarchids in two Oklahoma farm ponds. Proc. Okla. Acad. Sci. 48:199-205.

C354 Childers, W. F., 1967. Hybridization of four species of sunfishes (Centrarchidae). Ill. Nat. Hist. Surv. Bull. 29(3):159-214.

C355 Cooper, E. L., C. C. Wagner, and G. E. Krantz, 1971. Bluegills dominate production in mixed populations of fishes. Ecology 52(2):280-290.

C356 Christenson, L. M. and L. L. Smith, 1965. Characteristics of fish populations in Upper Mississippi River backwater areas. U.S. Bur. Sport Fish Wildl. Circ. 212:53 pp.

C357 Crowe, W. R., 1959. The bluegill in Michigan. Mich. Dep. Conserv. Fish Div. Pamphlet. 31:6 pp.

C358 Coble, D. W., 1970. False annulus formation in bluegill scales. Trans. Am. Fish. Soc. 99(2):363-368.

C359 Cope, O. B., J. P. McCraren, and L. L. Ellen, 1969. Effects of dichlobenil on two fishpond environments. Weed Sci. 17(2):158-165.

C360 Cope, O. B., E. M. Wood, and G. H. Wallen, 1970. Some chronic effects of 2, 4-D on the bluegill (*Lepomis macrochirus*). Trans. Am. Fish. Soc. 99(1):1-12.

C361 Clugston, J. P., 1966. Centrarchid spawning in the Florida everglades. Q. J. Fla. Acad. Sci. 29(2):137-144.

C362 Clady, M. D. and G. U. Ulrichson, 1968. Mortality of recently hatched bluegill fry as a result of hydra. Prog. Fish-Cult. 30(1):39-40.

C363 Calabrese, A., 1969. Effect of acids and alkalies on survival of bluegills and largemouth bass. U.S. Bur. Sport Fish. Wildl. Tech. Pap. 42:10 pp.

C364 Carothers, J. L. and R. Allison, 1968. Control of snails by the redear (shellcracker) sunfish. F.A.O. Fish. Rep. 44(5):399-406.

C365 Clark, F. W. and M. H. A. Keenleyside, 1967. Reproductive isolation between the sunfish, *Lepomis gibbosus* and *L. macrochirus*. J. Fish. Res. Board Can. 24(3):495-514.

C366 Childers, W. F. and G. W. Bennett, 1967. Hook-and-line yield of largemouth bass and redear x green sunfish hybrids in a one-acre pond. Prog. Fish-Cult. 29(1):27-35.

C367 Chastain, G. A. and J. R. Snow, 1966. Nylon mats as spawning sites for largemouth bass, *Micropterus salmoides*, Lac. Proc. S.E. Assoc. Game Fish. Comm. 20:390-404.

C368 Carver, D. C., 1967. Distribution and abundance of the centrarchids in the recent delta of the Mississippi River. Proc. S.E. Assoc. Game Fish. Comm. 20:390-404.

C369 Cleary, R. E., 1956. Observations on factors affecting smallmouth bass production in Iowa. J. Wildl. Manage. 20(4):353-359.

C370 Coble, D. W., 1971. Effects of fin clipping and other factors on survival and growth of smallmouth bass. Trans. Am. Fish. Soc. 100(3):460-473.

C371 ——. 1967. Relationship of temperature to total annual growth in adult smallmouth bass. J. Fish. Res. Board Can. 24(1):87-99.

C372   Cairns, J., Jr. and W. T. Waller, 1971. The use of fish movement
        patterns to monitor zinc. Water Pollut. Control Res. Ser. Proj. 18050.
        EDP12/71:55 pp.

C373   Carter, B. T., 1967. Growth of three centrarchids in Lake Cumber-
        lands, Kentucky. Ky. Fish. Bull. 44:30.

C374   Cole, C. F., 1966. Virtual population estimations of largemouth bass in
        Lake Fort Smith, Arkansas, 1957-60. Trans. Am. Fish. Soc. 95(1):52-
        55.

C375   Cooper, J. A., 1971. Scale development as related to growth of juvenile
        black crappie, *Pomoxis nigromaculatus* Lesueur. Trans. Am. Fish.
        Soc. 100(3):570-572.

C376   Childers, W. and H. H. Shoemaker, 1953. Time of feeding of the black
        crappie and the white crappie. Trans. Ill. Acad. Sci. 46:227-230.

C377   Cooper, L. J., 1952. A histological study of the reproductive organs of
        crappies (*Pomoxis nigro-maculatus* and *Pomoxis annularis*). Trans.
        Am. Microsc. Soc. 71(4):393-404.

C379   Chen, T., 1969. Age and growth of the white crappie, *Pomoxis annu-
        laris* Rafinesque in Lake Waco, Texas. M.S. thesis, Baylor Univ. 64
        pp. (not seen).

C380   Carlson, A. R. and J. G. Hale, 1972. Successful spawning of large-
        mouth bass, *Micropterus salmoides* (Lacepede) under laboratory con-
        ditions. Trans. Am. Fish. Soc. 101(3):539-542.

C381   Carlson, A. R., 1973. Induced spawning of largemouth bass, *Micro-
        pterus salmoides* (Lacepede). Trans. Am. Fish. Soc. 102(2):442-444.

C382   Coutant, C. C., H. M. Ducharme, Jr., and J. R. Fisher, 1974. Effects
        of cold shock on vulnerability of juvenile channel catfish (*Ictalurus
        punctatus*) and largemouth bass (*Micropterus salmoides*) to predation.
        J. Fish. Res. Board Can. 31(3):351-354.

C383   Carlson, A. R. and R. E. Siefert, 1974. Effects of reduced oxygen on
        the embryos and larvae of lake trout (*Salvelinus namaycush*) and large-
        mouth bass (*Micropterus salmoides*). J. Fish. Res. Board Can. 31(8):
        1393-1396.

C384   Chew, R. L., 1973. The failure of largemouth bass, *Micropterus sal-
        moides floridanus* (LeSueur), to spawn in eutrophic, overcrowded en-
        vironments. Proc. S.E. Assoc. Game Fish Comm. 26:306-319.

C385   Crawford, J. K. and R. L. Butler, 1973. Cover response of fish in
        bioassay of acid water. U.S. Environ. Prot. Agency Ecol. Res. Ser.
        R3-73-032:1-71.

C386   Cooper, E. L. and C. C. Wagner, 1973. The effects of acid mine
        drainage on fish populations. U.S. Environ. Prot. Agency Ecol. Res.
        Ser. R3-73-032:73-124.

C387   Colgan, P. and D. Ealey, 1973. Role of woody debris in nest site
        selection by pumpkinseed sunfish, *Lepomis gibbosus*. J. Fish. Res.
        Board Can. 30(6):853-856.

C388   Carson, J. B., 1968. The green sunfish. Underwater Nat. 5(1):29.

C389   Coble, D. W., 1972. Vulnerability of fin clipped bluegill to largemouth
        bass predation in tanks. Trans. Am. Fish. Soc. 101(3):503-505.

C390   Cairns, J., Jr., E. L. Morgan, and R. E. Sparks, 1974. The response
        of bluegills (*Lepomis macrochirus* Rafinesque) in a pollution monitor-
        ing system to a diurnal temperature change. Trans. Am. Fish. Soc.
        103(1):138-140.

C391   Cooper, G. P. and G. N. Washburn, 1949. Relation of dissolved oxygen
        to winter mortality of fish in Michigan lakes. Trans. Am. Fish. Soc.
        76:23-33.

C392   Chiszar, D., M. Moody, and J. T. Windell, 1972. Failure of bluegill
        sunfish, *Lepomis macrochirus*, to habituate to handling. J. Fish. Res.
        Board Can. 29(5):376-378.

C393   Cairns, J., Jr. and R. E. Sparks, 1971. The use of bluegill breathing
        to detect zinc. Water Pollut. Control Res. Ser. (U.S. EPA) 18050 EDQ.
        12-71:45 pp.

C394   Chiszar, D. and J. T. Windell, 1973. Predation by bluegill sunfish
        (*Lepomis macrochirus* Rafinesque) upon mealworm larvae (*Tenebrio
        molitor*). Anim. Behav. 21(3):536-543.

C396   Cherry, D. S., K. L. Dickson, and J. Cairns, Jr., 1975. Temperatures
        selected and avoided by fish at various acclimation temperatures. J.
        Fish. Res. Board Can. 34(4):485-491.

C397   Clady, M. D., 1975. Early survival and recruitment of smallmouth
        bass in northern Michigan. J. Wildl. Manage. 39(1):194-200.

C398   Chew, L. E. and J. G. Stanley, 1973. The effects of methyltestosterone
        on sex reversal in bluegill. Prog. Fish. Cult. 35(1):44-47.

C399   Chew, R. L., 1974. Early life history of the Florida largemouth bass.
        Fla. Game Freshwater Fish. Comm. Bull. 7:76 pp.

C400   Childers, W. F. and J. A. Tranquilli, 1973. Hybrid vigor in bass. Ill.
        Nat. Hist. Surv. Rep. 124:3.

C401   Christie, W. J. and H. A. Regier, 1973. Temperature as a major fac-
        tor influencing reproductive success of fish—two examples. Int. Counc.
        Explor. Sea Rep. Proc. 164:208-218.

C402   Clady, M. D., 1974. Food habits of yellow perch, smallmouth bass and
        largemouth bass in two unproductive lakes in northern Michigan. Am.
        Midl. Nat. 91(2):453-459.

D5   Davis, H. S. and A. H. Wiebe, 1930. Experiments in the culture of the
        black bass and other pond fish. Rep. U.S. Fish. Comm. 1930-App. 9.
        Doc. 1085:M-203.

D13   Dill, W. A., 1944. The fishery of the Lower Colorado River. Calif.
        Fish Game. 30(3):109-211.

D19   Doan, K. H., 1938. Lake Erie smallmouths. Ohio Conserv. Bull.
        2(12):18, 39.

D20   ———. 1939. Growth of bass fry. Copeia, 1939. (2):81-87.

D21   ———. 1940. Studies of the smallmouth bass. J. Wildl. Manage. 4(3):
        241-266.

D28   Duclos, G., 1889. American silver perch or calico bass. Bull. U.S.
        Comm. Fish. 7:215-216.

D31   Dyche, L. L., 1914. Notes on the new Kansas fish hatchery and the
        first year's output. Trans. Am. Fish. Soc. 44(1):5-12.

D37   Dequine, J. F., 1950. Fish Management. Fla. Game Freshwater Fish
        Comm. Bienn. Rep. 1949-50:43-93.

D43   Dryer, W., 1951. Bass reproduction in selected Tennessee streams.
        Tenn. Game Fish Comm. 1-23. mimeo.

D47   Dickson, A. W., 1953. Fish management investigations of coastal
        streams. Sept. 16, 1952-Dec. 15, 1952. Q. Prog. Rep. N.C. Fed. Aid
        11(2):57-108.

D56   Darnell, R. M., 1958. Food habits of fishes and larger invertebrates
        of Lake Pontchartrain, Louisiana, an Estuarine Community. Inst. Mar.
        Sci. 5:353-416.

D62   Dendy, J. S., 1954. How large do redeye bass grow? Ala. Conserv.
        26(3):12.

D67 Dickson, A. W., 1953. Fish management investigations of coastal
     streams. N.C. Wildl. Resour. Comm. Q. Prog. Rep. Fed. Aid 3(2):
     A1-A23.

D68 ———. 1955. Fish management investigations of coastal streams.
     N.C. Wildl. Resour. Comm. Proj. Compl. Rep. F-2-R. 178 pp.

D72 Di Costanzo, C. J., 1957. Growth of bluegill, *Lepomis macrochirus*,
     and pumpkinseed, *L. gibbosus*, of Clear Lake, Iowa. Iowa State J. Sci.
     32(1):19-34.

D88 Durham, L., 1955. Ecological factors affecting the growth of small-
     mouth bass and longear sunfish in Jordan Creek. Ill. Acad. Sci. Trans.
     47:25-34.

D89 Dunham, D. K., 1956. How old is that fish? Wis. Conserv. Bull. 21(7):
     11-13.

D93 Doan, K. H., 1942. Some meteorological and limnological factors in the
     abundance of certain fishes in Lake Erie. Ecol. Monogr. 12:293-314.

D103 Dillon, J. F., 1955. Studies on the growth of the fish population of
     Snowden Pond, Patuxent Research Refuge, Laurel, Md. M.S. thesis,
     Univ. Md. 58 pp.

D161 Davis, J. T., 1960. Fish populations and aquatic conditions in polluted
     waters in Louisiana. La. Wildl. Fish Comm. Bull. 1-1960:121 pp.

D163 Dendy, J. S., 1946. Food of several species of fish, Norris Reservoir,
     Tennessee. J. Tenn. Acad. Sci. 21(1):105-127.

D167 Dickson, A. W., 1961. Coastal plain lakes of northeastern North Caro-
     lina. N.C. Wildl. Resour. Comm. Fed. Aid Proj. F5R and F6R Job
     Compl. Rep. 1:59-72.

D168 Dobie, J., 1962. Role of the tiger salamander in natural ponds used in
     Minnesota for rearing suckers. Prog. Fish-Cult. 24(2):85-87.

D172 Dendy, J. S., 1956. Bottom fauna in ponds with largemouth bass only
     and with a combination of largemouth bass plus bluegill. J. Tenn.
     Acad. Sci. 31(3):198-207.

D178 Durham, L., 1957. Green sunfish, bluegill, largemouth bass. Sum-
     maries for Handbook Biol. Data. 20 pp.

D179 Dietz, E. M. C. and K. C. Jurgens, 1963. An evaluation of selective
     shad control of Medina Lake, Texas. Tex. Parks Wildl. Dep. 1 F Rep.
     5:1-32.

D183 Dobie, J., 1966. Food and feeding habits of the walleye, *Stizostedion v
     vitreum*, and associated game and forage fishes in Lake Vermilion,
     Minnesota, with special reference to the tullibee, *Coregonus (Leuci-
     chthys) artedi*. Minn. Fish. Invest. 4:39-71.

D200 Dunst, R. C., 1969. Cox Hollow Lake. The first eight years of im-
     poundment. Wis. Dep. Nat. Resour. Res. Rep. 47:19 pp.

D201 Dillard, J. G., 1971. Evaluation of two stocking methods for Missouri
     farm ponds. Proc. N.C. Warmwater Fish Cult. Manage. Workshop.
     Iowa Coop. Fish Unit, Ames. 247 pp.

D202 Dymond, J. R., 1931. The smallmouth bass and its conservation. Ont.
     Dept. Game Fish, Biol. Fish. Cult. Branch, Bull. 2:10 pp.

D203 Dahlberg, M. L., D. L. Shumway, and P. Doudoroff, 1968. Influence of
     dissolved oxygen and carbon dioxide on swimming performance of
     largemouth bass and coho salmon. J. Fish. Res. Board Can. 25(1):
     49-70.

D204 Di Giovanni, M. V., 1969. First observations on the growth of large-
     mouth bass (*Micropterus salmoides* Lacepede) in some agricultural
     lakes in Umbria. Riv. Ital. Piscicoltura Ittiopatologia. 4(2):46-50.

D205   Dudley, R. G., 1969. Survival of largemouth bass embryos at low dissolved oxygen concentrations. M.S. thesis, Cornell Univ. 61 pp.

D206   Denyes, H. A. and J. M. Joseph, 1956. Relationships between temperature and blood oxygen in the largemouth bass. J. Wildl. Manage. 20 (1):56-64.

D207   Dendy, J. S., 1948. Predicting depth distribution of fish in three TVA storage type reservoirs. Trans. Am. Fish. Soc. 75:65-71.

D208   Davis, H. S., 1949. *Cytophaga columnaris* as a cause of fish epidemics. Trans. Am. Fish. Soc. 77:102-104.

D209   Dubets, H., 1954. Feeding habits of the largemouth bass as revealed by a gastroscope. Prog. Fish Cult. 16(3):134-136.

D210   Davis, J. R., 1972. The spawning behavior, fecundity rates, and food habits of the redbreast sunfish in southeastern North Carolina. Proc. S.E. Assoc. Game Fish Comm. 25:556-560.

D211   Davis, R. E., 1962. Daily rhythm in the reaction of fish to light. Science 137(3528):430-432.

D212   Davis, R. C., 1964. Daily predawn peak of locomotion in fish. Anim. Behav. 12(2 & 3):272-283.

D213   Dudley, R. G. and A. W. Eipper, 1975. Survival of largemouth bass embryos at low dissolved oxygen concentrations. Trans. Am. Fish. Soc. 104(1):122-128.

E2   Eddy, S. and K. D. Carlander, 1939. Growth of Minnesota fishes. Minn. Conservationist 69:8-10.

E3   ———. 1942. Growth rates studies of Minnesota fish. Minn. Dep. Conserv. Fish Res. Invest. Rep. 28:64 pp. mimeo.

E4   Eddy, S. and T. Surber, 1947. Northern Fishes, Rev. ed. Univ. Minn. Press. 276 pp.

E9   Embody, G. C., 1915. The farm fishpond. Cornell Reading Courses Country Life Ser. 3:213-252.

E20   Eschmeyer, R. W., 1939. Analysis of the complete fish population from Howe Lake, Crawford County, Michigan. Pap. Mich. Acad. Sci. Arts Lett. 24(2):117-137.

E22   ———. 1948. Growth of fishes in Norris Lake, Tennessee. Proc. Tenn. Acad. Sci. 15(3):329-341.

E23   ———. 1942. The catch, abundance, and migration of game fishes in Norris Reservoir, Tennessee, 1940. J. Tenn. Acad. Sci. 17(1):90-114.

E24   ———. 1944. Fish migration into the Clinch River below Norris Dam, Tennessee. J. Tenn. Acad. Sci. 19(1):31-41.

E26   Eschmeyer, R. W. and A. M. Jones, 1941. The growth of game fishes in Norris Reservoir during the first five years of impoundment. Trans. N. Am. Wildl. Conf. 6:222-240.

E27   Eschmeyer, R. W. and D. E. Manges, 1945. Fish migrations into the Norris Dam tailwater in 1943. J. Tenn. Acad. Sci. 20(1):92-96.

E28   Eschmeyer, R. W., R. H. Stroud, and A. M. Jones, 1944. Studies of the fish population on the shoal area of a TVA main-stream reservoir. J. Tenn. Acad. Sci. 19(1):70-122.

E30   Evans, I. M., 1948. Rates of growth of game and pan fishes in Ohio, their practical interpretation and some technical problems involved. Mss. Ohio Conserv. Comm.

E32   Evermann, B. W. and H. W. Clark, 1920. Lake Maxinkuckee. Indiana Dep. Conserv. 1:1-660.

E37   Eschmeyer, R. W., 1935. Analysis of the game-fish catch in a Michigan lake. Trans. Am. Fish. Soc. 65:207-223.

E38   Eschmeyer, R. W., 1937. A second season of creel census on Fife
      Lake. Trans. Am. Fish. Soc. 66:324-334.

E48   Erickson, J. G., 1952. Age and growth of the black and white crappies,
      Pomoxis nigro-maculatus (LeSueur) and *P. annularis* Rafinesque, in
      Clear Lake, Iowa. Iowa State J. Sci. 26(3):491-505.

E51   Evans, B., 1951. Jumbo fish. Colo. Conserv. July 1951, 20-22.

E52   Everhart, E. H., 1950. Relation between body length and scale meas-
      urements in the smallmouth bass. J. Wildl. Manage. 14(3):266-276.

E59   Eipper, A. W., 1953. Fish production in New York fish ponds. Paper
      given at 7th ann. meeting Empire State Chap. Soil Conserv. Soc. Am.
      3 pp. mimeo.

E60   Elder, D. E. and W. M. Lewis, 1955. An investigation and comparison
      of the fish populations of two farm ponds. Am. Midl. Nat. 53(2):390-
      395.

E61   Elkin, R. E., 1954. The fish population of two cutoff pools in Salt
      Creek, Osage County, Oklahoma. Proc. Okla. Acad. Sci. 35:25-29.

E73   Elser, H. J. and P. Gale, 1953. Growth of Potomac River smallmouth
      bass. Md. Dep. Res. Educ. Solomons, 3 pp. mimeo.

E76   Elser, H. J. and R. D. Van Deusen, 1954. Deep Creek Lake report,
      1948-1952. Md. Dep. Res. Educ. Res. Study Rep. 3:1-19.

E79   El-Zarka, S. E., 1959. Fluctuations in the population of yellow perch,
      Perca flavescens (Mitchill), in Saginaw Bay Lake Huron. U.S. Fish
      Wildl. Ser. Fish. Bull. 59(151):365-415.

E84   Erickson, J. G. and W. M. Zarbock, 1954. A preliminary evaluation of
      the effects of the removal of rough fish upon crappies in Lake St.
      Marys, Ohio. Midw. Wildl. Conf. 16: 19 pp. mimeo.

E93   Everhart, W. H., 1958. Fishes of Maine, 2nd ed. Maine Dep. Inland
      Fish. Game 1958:5-94.

E94   Eddy, S. and K. D. Carlander, 1940. The effect of environmental fac-
      tors upon the growth rates of Minnesota fishes. Proc. Minn. Acad. Sci.
      8:14-19.

E97   Elliot, J. M., 1948. The age and rate of growth of the black crappie,
      Pomoxis nigro-maculatus (LeSueur), and the white crappie, *Pomoxis
      annularis* Rafinesque, in the Koon Kreek Club Lakes, Texas. M.S.
      thesis, N. Texas State Teachers Coll. 56 pp.

E98   Estes, C. M., 1949. The fecundity of the bluegill (*Lepomis macro-
      chirus*) in certain small east Texas reservoirs. M.S. thesis, N. Texas
      State Coll. 39 pp.

E101 Elser, H. J., 1960. Escape of fish over spillways: Maryland, 1958-
      1960. Proc. Ann. Conf. S.E. Assoc. Game Fish Comm. 14:174-185.

E105 ———. 1961. Record Maryland fish. Md. Conservationist 38(2):15-17.

E112 Eipper, A. W. and H. A. Regier, 1962. Fish management in New York
      farm ponds. Cornell Ext. Bull. 1089:40 pp.

E119 Eddy, S., 1957. How to know the freshwater fishes. Wm. C. Brown Co.
      253 pp.

E121 Ewers, L. A. and M. W. Boesel, 1935. The food of some Buckeye Lake
      fishes. Trans. Am. Fish. Soc. 65:57-70.

E126 Elkin, R. E., 1955. An estimate of the fish population of a 16 acre lake
      based on recovery during draining. Proc. Okla. Acad. Sci. 36:53-59.

E135 Etnier, D. A., 1971. Food of three species of sunfishes (*Lepomis*, Cen-
      trarchidae) and their hybrids in three Minnesota lakes. Trans. Am.
      Fish. Soc. 100(1):124-128.

E136   Elwood, J. W. and G. D. Holton, 1964. An age and growth analysis of pumpkinseed (*Lepomis gibbosus*) from Horseshoe Lake, Montana. Mont. Fish Game Dep. Job Compl. Rep. F-7-R-13-II. 36 pp.

E137   Erickson, J. G., 1967. Social hierarchy, territoriality and stress reactions in sunfish. Physiol. Zool. 40:40-48.

E138   Emig, J. W., 1966. Bluegill sunfish. Pp. 375-392, Inland fisheries management, A. Calhoun ed. Calif. Dep. Fish Game.

E139   ———. 1966. Red-ear sunfish. Inland Fisheries Management, A. Calhoun ed. Calif. Dep. Fish Game. Pp. 392-399.

E141   Eller, L. L., 1969. Pathology in redear sunfish exposed to Hydrothol 191. Trans. Am. Fish. Soc. 98(1):52-59.

E142   Etnier, D. A., 1968. Reproductive success of natural populations of hybrid sunfish in three Minnesota lakes. Trans. Am. Fish. Soc. 97(4):466-471.

E143   Emig, J. W., 1966. Smallmouth bass. Pp. 354-366, Inland fisheries management, A. Calhoun ed. Calif. Dep. Fish Game.

E144   Everhart, W. H., 1949. Body length of the smallmouth bass at scale formation. Copeia, 1949. (2):110-115.

E145   Emig, J. W., 1966. Largemouth bass. Pp. 332-353, Inland Fisheries Management, A. Calhoun ed. Calif. Dep. Fish Game.

E146   Elrod, J. H., 1971. Dynamics of fishes in an Alabama pond subjected to intensive angling. Trans. Am. Fish. Soc. 100(4):757-768.

E147   Ellis, M. M., 1937. Detection and measurement of stream pollution. U.S. Bur. Fish. Bull. 48(22):365-437.

E148   Ercole, J. A., 1970. Investigation of factors limiting population growth of crappie. Ariz. Game Fish Dep. Fish. Res. Ariz. 1969-70:1-12.

E149   ———. 1969. Investigation of factors limiting population growth of crappie. Fish. Res. Ariz. 1968:1-12.

E150   ———. 1968. Investigation of factors limiting population growth of crappies. Fish. Res. Ariz. 1967:1-12.

E151   Eckblad, J. W. and M. H. Shealy, Jr., 1972. Predation on largemouth bass embryos by the pond snail *Viviparus georgianus*. Trans. Am. Fish. Soc. 101(4):734-738.

E152   Espinosa, F. A., Jr. and J. E. Deacon, 1973. The preference of largemouth bass (*Micropterus salmoides* Lacepede) for selected bait species under experimental conditions. Trans. Am. Fish. Soc. 102(2):355-362.

E153   Emery, A. R., 1973. Preliminary comparisons of day and night habits of freshwater fish in Ontario lakes. J. Fish. Res. Board Can. 30(6):761-774.

E154   Ellis, J. E., 1974. The jumping ability and behavior of green sunfish (*Lepomis cyanellus*) at the outflow of a 1.6-ha pond. Trans. Am. Fish. Soc. 103(3):620-623.

F1   Fessler, F. R., 1949. Fish populations in some Iowa farm ponds. Prog. Fish-Cult. 12(1):3-11.

F2   Fischthal, J. H., 1948. Stunted bluegills become prize fish. Wis. Conserv. Bull. 13(11):16-17.

F5   Fish, M. P., 1932. Contributions to the early life histories of sixty-two species of fishes from Lake Erie and its tributary waters. Bull. U.S. Bur. Fish. 47(10):293-398.

F6   Flower, S. S., 1925. Contributions to our knowledge of the duration of life in vertebrate animals. I. Fishes. Proc. Zool. Soc. London. 1925:247-268.

F7 Flower, S. S., 1935. Further notes on duration of life in animals. I. Fishes as determined by otolith and scale readings and direct observation on live individuals. Proc. Zool. Soc. London. 1935(2):265-304.

F11 Ford, T., 1947. King size bream. Ala. Conserv. 19(4):7, 44.

F15 Frey, D. G. and L. Vike, 1941. A creel census on Lakes Waubesa and Kegonsa, Wis. in 1939. Trans. Wis. Acad. Sci. Arts Lett. 34:339-362.

F22 Fry, F. E. J., 1947. South Bay Experiment-Advance report on smallmouth bass (Micropterus dolomieu). Ont. Dep. Lands For. 14 pp. mimeo.

F26 Fuller, J. L. and G. P. Cooper, 1946. A biological survey of the lakes and ponds of Mount Desert Island, and the Union and Lower Penobscot River drainage systems. Maine Dep. Inland Fish. Game. Fish Surv. Rep. 7:1-221.

F31 Fessler, F. R., 1949. A survey of fish populations in small ponds by two methods of analysis. M.S. thesis, Iowa State Coll. 47 pp.

F43 Fisheries Service Bulletin, 1929. Production of bass fry at Fairport. Fish. Serv. Bull. 166:2-3.

F45 Fisheries Service Bulletin, 1929. Large bass. Fish. Serv. Bull. 167:3.

F62 Finnell, J. C., 1954. Comparison of growth-rates of fishes in Stringtown Sub-prison Lake prior to, and three years after draining and restocking. Proc. Okla. Acad. Sci. 35:30-36.

F64 Finnell, J. C., R. M. Jenkins, and G. E. Hall, 1956. The fishery resources of the Little River system, McCurtain County, Oklahoma. Okla. Fish. Res. Lab. Rep. 55:82 pp. mimeo.

F70 Foye, R. E., 1958. Largemouth bass (Micropterus salmoides Lacepede). Pp. 84-86, Fishes of Maine, 2nd ed., W. H. Everhart.

F73 Fraser, J. M., 1955. The smallmouth bass fishery of South Bay, Lake Huron. J. Fish Res. Board Can. 12(1):147-177.

F74 Freeman, B. O. and M. T. Huish, 1953(?) A summary of a fish population control investigation conducted in two Florida lakes. Fla. Game Freshwater Fish Comm. 109 pp. mimeo.

F82 Fry, F. E. J. and K. E. F. Watt, 1957. Yields of year classes of the smallmouth bass hatched in the decade of 1940 in Manitoulin Island waters. Trans. Am. Fish. Soc. 85:135-143.

F87 Finnell, J. C., 1955. Growth of fishes in cutoff lakes and streams of the Little River System, McCurtain County, Oklahoma. Proc. Okla. Acad. Sci. 35:61-66.

F88 Funk, J. L., C. A. Purkett, Jr., P. E. Robinson, G. C. Fleener, and O. Fajen, 1959. The Missouri smallmouth bass. Mo. Conserv. Comm. 17 pp.

F91 Finkelstein, S. L., 1960. A food study of four centrarchidae at Lake Fort Smith, Crawford County, Arkansas. M.S. thesis, Univ. Arkansas. 63 pp.

F96 Fabian, M. W., 1954. An investigation of the bluegill population in Ford Lake, Michigan. M.S. thesis, Mich. State Coll. 69 pp.

F97 Fetterolf, C. M., Jr., 1952. A population study of the fishes of Wintergreen Lake, Kalamazoo County, Michigan: with notes on movement and effect of netting on condition. M.S. thesis, Mich. State Coll. 127 pp.

F100 Fogle, N. E., 1961. Report of fisheries investigations during the third year of impoundment of Oahe Reservoir, South Dakota, 1960. S.D. Dep. Game Fish Parks. D. J. Project F-1-R-10. Jobs 9-12, 57 pp. mimeo.

F103 Forney, J. L., 1961. Growth, movements and survival of smallmouth bass (Micropterus dolomieu) in Oneida Lake, New York. N.Y. Fish Game. 8(2):88-105.

F105 Fogle, N. E., 1961. Report of fisheries investigations during the second year of impoundment of Oahe reservoir, South Dakota, 1959. Dept. Game, Fish Parks, S.D. Dingell-Johnson Project F-1-R-9. Job 12, 13, and 14: 43 pp.

F124 ———. 1963. Report of fisheries investigations during the fourth year of impoundment of Oahe Reservoir, S. Dak. 1961. S.D. Dingell-Johnson Project F-1-R-11, Jobs 10, 11, and 12. 43 pp.

F125 ———. 1963. Report of fisheries investigations during the fourth year of impoundment of Oahe Reservoir, S. Dak. 1962. S.D. Dingell-Johnson Project F-1-R-12, Jobs 10, 11, 12. 43 pp.

F148 Flemer, D. A. and W. S. Woolcott, 1966. Food habits and distribution of the fishes of Tuckahoe Creek, Virginia, with special emphasis on the bluegill, Lepomis m. macrochirus Rafinesque. Chesapeake Sci. 7(2): 75-89.

F151 Fitz, R. B., 1965. Growth of fishes in eastern Tennessee Valley reservoirs. 1 p. T.V.A. Norris, Tenn.

F152 Funk, J. L. and R. S. Campbell, 1953. The population of larger fishes in Black River, Missouri. Univ. Mo. Stud. 26(2):69-82.

F155 Forbes, S. A., 1903. The food of fishes, 2nd ed. Ill. Lab. Nat. Hist. Bull. 1(3):19-70.

F156 Faber, D. J., 1967. Limnetic larval fish in northern Wisconsin lakes. J. Fish. Res. Board Can. 24(5):927-937.

F158 Funk, J. L., 1957. Movement of stream fishes in Missouri. Trans. Am. Fish. Soc. 85:39-57.

F159 Fedoruk, A. N., 1966. Feeding relationship of walleye and smallmouth bass. J. Fish. Res. Board Can. 23(6):941-943.

F160 Fieldhouse, R. D., 1971. Results of stocking largemouth bass in Nassau Lake. N.Y. Fish Game J. 18(1):68-69.

F161 Ferguson, R. G., 1958. The preferred temperature of fish and their midsummer distribution in temperate lakes and streams. J. Fish. Res. Board Can. 15(4):607-624.

F162 Fry, J. P. and W. D. Hanson, 1968. Lake Taneycomo: a cold-water reservoir in Missouri. Trans. Am. Fish. Soc. 97(2):138-145.

F163 Farabee, G. B., 1974. Effects of a 12-inch length limit on largemouth bass and bluegill populations in two northeast Missouri lakes. Symposium on overharvest of largemouth bass in small impoundments. N.C. Div. Am. Fish. Soc. Spec. Publ. 3:95-99.

F164 Forney, J. L., 1974. Interactions between yellow perch abundance, walleye predation and survival of alternate prey in Oneida Lake, New York. Trans. Am. Fish. Soc. 103(1):15-24.

F165 Fontana, F., B. Chiarelli, and A. Rossi, 1970. Some new data on the number of chromosomes of teleost fish obtained by means of tissue culture in vitro. Experentia 26(9):1021.

F166 Funk, J. L. and G. C. Fleener, 1974. The fishery of a Missouri Ozark stream, Big Piney River, and the effects of stocking fingerling smallmouth bass. Trans. Am. Fish. Soc. 103(4):757-771.

F167 Fajen, O. F., 1962. Homing ability of smallmouth bass. Trans. Am. Fish. Soc. 91(4):346-349.

F168 Forney, J. L., 1972. Biology and management of smallmouth bass in Oneida Lake, New York. N.Y. Fish Game J. 19(2):132-154.

F169 Fleener, G. C., J. L. Funk, and P. E. Robinson, 1974. The fishery of Big Piney River and the effects of stocking fingerling smallmouth bass. Mo. Dep. Conserv. Aquatic Ser. 9:32 pp. (Same as F166).

G19 Greeley, J. R., 1934. Fishes of the Raquette watershed. Pp. 53-108, A Biological Survey of the Raquette watershed. Suppl. 23d Ann. Rep. N.Y. Conserv. Dep. Biol. Surv. 8.

G20 ———. 1935. Fishes of the watershed. A Biol. Surv. Mohawk-Hudson watershed. Suppl. 24th Ann. Rep. N.Y. Conserv. Dep. Biol. Surv. 9:63-101.

G21 ———. 1936. Fishes of the area with annotated list, pp. 45-88, A biological survey of the Delaware and Susquehanna watersheds. Suppl. 25th Ann. Rep. N.Y. Conserv. Dep. Biol. Surv. 10.

G22 ———. 1937. Fishes of the area with annotated list, pp. 45-104, A biological survey of the Lower Hudson watershed. Suppl. 26th Ann. Rep. N.Y. Conserv. Dep. Biol. Surv. 11.

G23 ———. 1938. Fishes of the area with annotated list, pp. 48-73, A biological survey of the Allegheny and Chemung watersheds. Suppl. 27th Ann. Rep. N.Y. Conserv. Dep. Biol. Surv. 12.

G24 ———. 1939. Fresh-water fishes of Long Island and Staten Island with annotated list, pp. 29-44, A biological survey of the fresh waters of Long Island. Suppl. 28th Ann. Rep. N.Y. Conserv. Dep. Biol. Surv. 13.

G25 ———. 1940. Fishes of the watershed with annotated list, pp. 42-81, A biological survey of the Lake Ontario watershed. Suppl. 29th Ann. Rep. N.Y. Conserv. Dep. Biol. Surv. 16.

G27 Greeley, J. R. and S. C. Bishop, 1933. Fishes of the Upper Hudson watershed, pp. 64-101, A Biological survey of the Upper Hudson watershed. Suppl. 22 Ann. Rep. N.Y. State Conserv. Dep.

G32 Grimm, W. W. and R. V. Bangham, 1931. Growth of Buckeye Lake fishes in 1930, six common species compared. Ohio Div. Conserv. Bull. 71:1 p. mimeo.

G36 Gabrielson, I. N. and F. LaMonte, 1950. The Fisherman's Encyclopedia. 698 pp.

G38 Gerking, S. D., 1950. Populations and exploitation of fishes in a marl lake. Invest. Ind. Lakes. 3(1):389-434.

G39 ———. 1950. A carp removal experiment at Oliver Lake, Indiana. Invest. Ind. Lakes. 3(10):373-388.

G42 Greenbank, J., 1950. The length-weight relationship of some upper Mississippi River fishes. Upper Miss. River Conserv. Comm. 12 pp. ms.

G48 Gerking, S. D., 1953. Vital statistics of the fish population of Gordy Lake, Indiana. Trans. Am. Fish. Soc. 82:48-67.

G49 ———. 1954. The food turnover of a bluegill population. Ecology 35 (4):490-497.

G74 Greer, J. K. and F. B. Cross, 1956. Fishes of El Dorado City Lake, Butler County, Kansas. Trans. Kan. Acad. Sci. 59(3):358-363.

G75 Grenfell, R. A., 1956. Warm-water game fish. Oreg. State Game Comm. Fish Div. Ann. Rep. 1955:221-234.

G76 ———. 1957. Warm-water fish. Oreg. State Game Comm. Fish. Div. Ann. Rep. 1956:241-255.

G77 ———. 1959. Warm-water game fish. Oreg. State Game Comm. Fish. Div. Ann. Rep. 1958:262-270.

G78 Grice, F., 1958. Effect of removal of panfish and trashfish by fyke nets upon fish populations of some Massachusetts ponds. Trans. Am. Fish. Soc. 87:108-115.

G80 Gunning, G. E., 1954. The fishes of Horseshoe Lake, Illinois. M.S. thesis, Southern Illinois Univ., Carbondale.

G82   Gunning, G. E. and W. M. Lewis, 1956. Age and growth of two important bait species in a coldwater stream in southern Illinois. Am. Midl. Nat. 55(1):118-120.

G83   Grice, F., 1959. Elasticity of growth of yellow perch, chain pickerel, and largemouth bass in some reclaimed Massachusetts waters. Trans. Am. Fish. Soc. 88(4):332-335.

G85   Gerking, S. D., 1956. Mortality rates of fishes. 17 pp. table to Handb. Biol. Data.

G94   Galbraith, M. G., Jr., 1952. An age and growth study of rock bass, *Amploplites r. rupestris* (Rafinesque), from Lower Loch Alpine, Michigan. M.S. thesis, Univ. Michigan. 17 pp.

G97   Gross, R. W., 1959. A study of the alewife, *Alosa pseudoharenqus* (Wilson), in some New Jersey lakes, with special reference to Lake Hopatcong. M.S. thesis, Rutgers. 52 pp.

G99   Grenfell, R. A., 1960. Warm-water game fish. Oreg. State Game Comm. Fish. Div. Ann. Rep., 1959:277-283.

G110   Groebner, J. F., 1960. Appraisal of the sport fishery catch in a bass-panfish lake of southern Minnesota. Lake Francis, LeSueur County, 1952-1957. Minn. Dep. Conserv. Invest. Rep. 225:17 pp.

G112   Grudniewski, C., 1961. The development of some morphological features during the larvae stage of Wdzydze Lake trout (*Salmo trutta morpha lacustris L.*). Rocz. Nauk. Roln. 93:595-626 (Engl. summ.).

G114   Galligan, J. P., 1962. Depth distribution of lake trout and associated species in Cayuga Lake, New York. N.Y. Fish Game J. 9(1):44-68.

G127   Gerking, S. D., 1964. Timing and magnitude of the production of a bluegill sunfish population and its food supply. Verh. Int. Verein. Limnol. 15:496-503.

G133   Gill, T., 1906. Parental care among fresh-water fishes. Ann. Rep. Smithson. Inst. 1905:403-531.

G136   Grenfell, R. A., 1965. Warm-water game fish. Ann. Rep. Oreg. State Game Comm. Fish Div. 1964:277-283.

G137   Ganssle, D., 1966. Fishes and decapods of San Pablo and Suisun Bays. Calif. Fish Bull. 133:64-94.

G152   Gammon, J. R. and A. D. Hasler, 1965. Predation by introduced muskellunge on perch and bass, 1: years 1-5. Trans. Wis. Acad. Sci. Arts Lett. 54:249-272.

G156   Gould, W. R. III and W. H. Irwin, 1965. The suitabilities and relative resistances of twelve species of fish as bioassay animals for oil-refinery effluents. Proc. S.E. Assoc. Game Fish. Comm. 16:333-348.

G158   Gregory, R. W., D. T. Weber, T. G. Powell, 1970. Warm-water investigations. Colo. Fish. Res. Rev. 1969(6):1-11.

G160   Grenfell, R. A., 1961. Warm-water game fish. Oreg. State Game Comm. Fish. Div. Ann. Rep. 1961:322-327.

G161   ———. 1962. Warm-water game fish. Oreg. State Game Comm. Fish. Div. Ann. Rep. 1962:369-379.

G162   Grover, J. H., 1968. Hemosiderin in bluegill spleen. Trans. Am. Fish. Soc. 97(1):48-50.

G163   ———. 1966. Splenic variations in the bluegill, *Lepomis macrochirus*, from Iowa Farm ponds. M.S. thesis, Iowa State Univ., Ames. 33 pp.

G164   Graves, E. and B. Haines, 1969. Fishery surveys of Navajo Reservoir and tailwaters. N.M. Dep. Game Fish. Job compl. Rep. A-6 (a) (b). 103 pp.

G165   Gilderhus, P. A., 1967. Effects of diquat on bluegills and their food organisms. Prog. Fish. Cult. 29(2):67-74.

G166   Gerking, S. D., 1966. Annual growth cycle, growth potential, and growth compensation in the bluegill sunfish in northern Indiana lakes. J. Fish. Res. Board Can. 23(12):1923-1956.

G167   Gray, W. and R. P. Ward, 1968. Age and rate of growth of bluegill in selected farm ponds in Obion County, Tennessee. J. Tenn. Acad. Sci. 43(1):5-6.

G168   Gerking, S. D., 1966. Length of the growing seasons of the bluegill sunfish in northern Indiana. Verh. Internat. Verein Limnol. 16:1056-1064.

G169   Gilderhus, P. A., 1966. Some effects of sublethal concentrations of sodium arsenite on bluegills and the aquatic environment. Trans. Am. Fish. Soc. 95(3):289-296.

G170   Gerking, S. D., 1952. The protein metabolism of sunfishes of different ages. Physiol. Zool. 25(4):358-372.

G171   ――――. 1955. Influence of rate of feeding on body composition and protein metabolism of bluegill sunfish. Physiol. Zool. 28(4):267-282.

G172   ――――. 1955. Endogenous nitrogen excretion of bluegill sunfish. Physiol. Zool. 28(4):283-289.

G173   ――――. 1950. Stability of a stream fish population. J. Wildl. Manage. 14:193-202.

G174   ――――. 1953. Evidence for the concepts of home range and territory in stream fishes. Ecology 34:347-365.

G175   Gunning, G. E., 1959. The sensory basis for homing in the longear sunfish, *Lepomis megalotis megalotis* (Rafinesque). Invest. Ind. Lakes Streams 5:103-130.

G176   Gunning, G. E. and C. R. Shoop, 1963. Occupancy of home range by longear sunfish *Lepomis megalotis megalotis* (Rafinesque) and bluegill *Lepomis macrochirus macrochirus* (Rafinesque). Anim. Behav. 11:325-330.

G177   Gasaway, C. R., 1968. Comparison of bass-bluegill and bass-redear sunfish stocking in Oklahoma farm ponds. Proc. Okla. Acad. Sci. 47:397-406.

G178   Goodson, L. F., Jr., 1966. Redeye bass. Pp. 371-373, Inland Fisheries Management, A. Calhoun ed. Calif. Dep. Fish Game.

G179   Gerking, S. D., 1959. The restricted movement of fish populations. Biol. Rev. 34:221-242.

G180   Gasaway, C. R., 1970. Changes in the fish population in Lake Francis Case in South Dakota in the first 16 years of impoundment. U.S. Bur. Sport Fish. Wild. Tech. Pap. 56:30 pp.

G181   ――――. 1971. Estimating the costs of sustained stocking of northern Great Plains reservoirs. Proc. N. Cent. warm-water fish cult. manage. workshop, I.S.U. pp. 65-83.

G182   Grinstead, B. G., 1969. The vertical distribution of the white crappie in the Buncombe Creek Arm of Lake Texoma. Okla. Fish. Res. Lab. Bull. 3:37 pp.

G183   Gasaway, C. R., 1965. Growth and abundance of young-of-the-year fishes in 1964 and 1965, determined by other trawling. Fish. Res. Okla. July-Dec. 1965:27-32.

G184   Ginnelly, G. C., 1971. Investigation of factors limiting population growth of crappie. Fish. Res. Ariz. 1970-71:1-15.

G185 Goodson, L. F., Jr., 1966. Crappie. Pp. 312-332, Inland Fisheries Management, A. Calhoun ed. Calif. Dep. Fish Game.

G186 Goodyear, C. P. and C. E. Boyd, 1972. Elemental composition of largemouth bass (*Micropterus salmoides*). Trans. Am. Fish. Soc. 101(3):545-547.

G188 Gibbons, J. W., J. T. Hook, and D. L. Forney, 1972. Winter responses of largemouth bass to heated effluent from a nuclear reactor. Prog. Fish-Cult. 34(2):88-90.

G189 Grinstead, B. G. and G. Wright, 1973. Estimation of black bass, *Micropterus* spp., population in Eufaula Reservoir, Oklahoma with discussion of techniques. Okla. Acad. Sci. Proc. 53:48-52.

G190 Graham, L. K., 1974. Effects of four harvest rates on pond fish populations. Symposium on overharvest of largemouth bass in small impoundments. N.C. Div. Am. Fish. Soc. Spec. Publ. 3:29-38.

G191 Gerald, J. W., 1971. Sound production during courtship in six species of sunfish (Centrarchidae). Evolution 25(1):75-87.

G192 Greenfield, D. W. and G. C. Deckert, 1973. Introgressive hybridization between *Gila orcutti* and *Hesperoleucas symmetricus* (Pisces:Cyprinidae) in the Cuyama River Basin, California: II. Ecological aspects. Copeia, 1973(3):417-427.

G193 Gerking, S. D., 1962. Production and food utilization in a population of bluegill sunfish. Ecol. Monogr. 32:31-78.

G194 Gallepp, G. W. and J. J. Magnuson, 1972. Effects of negative buoyancy on the behavior of the bluegill, *Lepomis macrochirus* Rafinesque. Trans. Amer. Fish. Soc. 101(3):507-512.

G195 Gerking, S. D., 1972. Revised food consumption estimate of a bluegill sunfish population in Wyland Lake, USA. J. Fish. Biol. 4(2):301-308.

G196 ———. 1971. Influence of rate of feeding and body weight on protein metabolism of bluegill sunfish. Physiol. Zool. 44(1):9-19.

G197 Gilbert, R. J., 1973. Systematics of *Micropterus p. punctalatus* and *M. p. henshalli* and life history of *M. p. henshalli*. Auburn Univ. Ph.D. thesis, 146 pp. (only abstract seen).

G198 Greene, D. S. and C. E. Murphy, 1971. Food and feeding habits of the white crappie (*Pomoxis annularis* Rafinesque) in Benbrook Lake, Tarrant County, Texas. J. Tex. Acad. Sci. 25(1):35-51.

H8 Hankinson, T. L., 1919. Notes of life histories of Illinois fish. Trans. Ill. State. Acad. Sci. 12:132-150.

H10 Hansen, D. F., 1942. The angler's catch at Lake Chautauqua near Havana, Illinois, with comparative data on hoopnet samples. Trans. Ill. State Acad. Sci. 35(2):197-104.

H15 Harkness, W. J. K., 1945. Rate of growth of game fish. 6 pp. mimeo.

H18 Harrison, A. C., 1934. Black bass in the Cape Province. Report on the early history of the acclimatization of the American largemouth black bass (*Micropterus salmoides*) in the Cape Province. Union S. Afr. Fish Mar. Biol. Surv. Rep. 11:5-92.

H19 ———. 1938. Maryland smallmouth bass spawn when sixteen months old in South Africa. Prog. Fish-Cult. 42:17-21.

H20 ———. 1939. Bluegills from Maryland make good in South Africa. Prog. Fish-Cult. 45:23-28.

H21 ———. 1944. Acclimatization of fish. Cape Piscatorial Soc. Ann. Rep. 13:1-6.

H34 Hasler, A. D. and W. G. Einsele, 1948. Fertilization for increasing productivity of natural inland waters. Trans. N. Am. Wildl. Conf. 13:527-555.

H40   Hayford, C. O., 1932. Economic bass culture in trout wintering ponds at the Hackettstown State Fish Hatchery. Trans. Am. Fish. Soc. 62: 167-168.

H45   Hazzard, A. S. and R. W. Eschmeyer, 1938. Analysis of the fish catch for one year in the Waterloo Project area. Pap. Mich. Acad. Sci. Arts Lett. 23:633-644.

H49   Heacox, C. E., 1948. What about bass management? N.Y. State Conserv. 3(1):19.

H52   Henshall, J. A., 1904. Book of the black bass. Stewart Kidd Co. 452 pp.

H54   Herman, E. F., D. E. Holtman, and D. J. O'Donnell, 1947. A double-mouth largemouth bass. Wis. Conserv. Bull. 12(10):27-28.

H58   Hey, D., 1947. The culture of freshwater fish in South Africa. Inland Fish. Dep., Cape Good Hope. Stellenbosch. 124 pp.

H61   Hildebrand, S. F. and W. C. Schroeder, 1928. Fishes of Chesapeake Bay. Bull. U.S. Bur. Fish. 43:1-366.

H62   Hile, R., 1931. Rate of growth of fishes of Indiana. Invest. Indiana Lakes 2:9-55.

H65   ———. 1941. Age and growth of the rock bass, *Ambloplites rupestris* (Rafinesque), in Nebish Lake, Wisconsin. Trans. Wis. Acad. Sci. Arts Lett. 33:189-337.

H66   ———. 1942. Growth of the rock bass, *Ambloplites rupestris* (Rafinesque), in five lakes of northeastern Wisconsin. Trans. Am. Fish. Soc. 71:131-143.

H67   ———. 1943. Mathematical relationship between the length and the age of the rock bass, *Ambloplites rupestris* (Rafinesque). Pap. Mich. Acad. Sci. Arts Lett. 28:331-341.

H68   ———. 1948. Standardization of methods of expressing lengths and weights of fish. Trans. Am. Fish. Soc. 75:157-164.

H72   Hiner, L. E., 1943. A creel census on Minnesota lakes, 1938-1942. Minn. Bur. Fish Res. Invest. Rep. 44:20 pp.

H76   Hoffman, D. H. and E. W. Surber, 1948. Effects of an aerial application of wettable DDT on fish and fish-food organisms in Back Creek, West Virginia. Trans. Am. Fish. Soc. 75:48-58.

H77   Hogan, J., 1936. Are young carp of any value as a forage fish? Prog. Fish-Cult. 20:22-23.

H86   Hubbs, C. L., 1921. An ecological study of the life-history of the fresh water atherine fish, *Labidesthes sicculus*. Ecology 2(4):262-276.

H87   ———. 1922. Variation in the number of vertebrae and other meristic characters of fishes correlated with the temperature of water during development. Am. Nat. 56:360-372.

H90   Hubbs, C. L. and G. P. Cooper, 1935. Age and growth of the long-eared and the green sunfishes in Michigan. Pap. Mich. Acad. Sci. Arts Lett. 20:669-696.

H93   Hubbs, C. L. and L. C. Hubbs, 1931. Increased growth in hybrid sunfishes. Pap. Mich. Acad. Sci. Arts Lett. 13:291-301.

H94   ———. 1933. The increased growth, predominant maleness and apparent infertility of hybrid sunfishes. Pap. Mich. Acad. Sci. Arts Lett. 17:613-641.

H96   Huber, W. B. and L. E. Binkley, 1935. Reproduction and growth of the white crappie (*Pomoxis annularis*) in Meander Lake. Ohio Bur. Sci. Res. Bull. 91.

H97   Hueske, E. W., 1948. Fish resources of the Bay lakes. Wildl. in N.C. 12(5):4-6, 17-19.

H126 Harper, D. C., 1938. Crappie and calico bass culture in Texas. Prog. Fish-Cult. 38:12-14.

H127 Hall, G. E., 1951. Preimpoundment fish populations of the Wister Reservoir area in the Poteau River Basin, Oklahoma. Trans. N. Am. Wildl. Conf. 16:266-283.

H128 Hansen, D. F., 1951. Biology of the white crappie in Illinois. Ill. Nat. Hist. Surv. Bull. 25(4):209-265.

H129 Hansen, D. F. and G. W. Bennett, 1949. Sport fishing in fertilized and unfertilized ponds. Midw. Wildl. Conf. 4 pp. mimeo.

H137 Hooper, F. F., 1951. Limnological features of a Minnesota seepage lake. Am. Midl. Nat. 46(2):462-481.

H141 Hart, J. S., 1952. Geographic variations of some physiological and morphological characters in certain freshwater fish. U. Toronto Biol. Ser. 60. Publ. Ont. Fish Res. Lab. 72:1-79.

H146 Hagy, R. H., 1956. A preliminary report on the white crappie, *Pomoxis annularis* (Rafinesque) and black crappie, *Pomoxis nigromaculatus* (LeSueur), in Northern California. Calif. Inland Fish. Admin. Rep. 56-63.

H150 Hall, G. E. and R. M. Jenkins, 1953. Continued fisheries investigation of Tenkiller Reservoir, Oklahoma, during its first year of impoundment, 1953. Okla. Fish. Res. Lab. Rep. 33:1-54.

H153 Hall, G. E., R. M. Jenkins, and J. C. Finnell, 1954. The influence of environmental conditions upon the growth of white crappie and black crappie in Oklahoma waters. Okla. Fish. Res. Lab. Rep. 40:1-56.

H156 Hancock, H. M., 1955. Age and growth of some of the principal fishes in Canton Reservoir, Oklahoma, 1951, with particular emphasis on the white crappie. Okla. Fish and Game Counc. Proj. Rep. Part 2. 110 pp. mimeo.

H158 Harmic, J. L., 1952. Fresh water fisheries survey. Del. Bd. Game and Fish Comm. Fish. Publ. 1:154 pp.

H162 Harrison, A. C., 1953. The acclimatization of smallmouth bass. Piscator 27:89-96.

H163 ———. 1954. The story of the Paarde Vlei, Part II, Introduction of non-indigenous fish. Piscator 8(29):29-32.

H164 ———. 1954. The story of Paarde Vlei, Part III, The change in water condition. Piscator 30:56-64.

H165 ———. 1954. The junction pool. Piscator 31:80-92.

H177 Harrison, H. M., 1959. Progress report of fish populations, Humboldt study area. Iowa Conserv. Comm. Q. Biol. Rep. 11(2):24-29.

H194 Heaton, J. R. and O. E. Orr, 1956. Preliminary surveys of public power and irrigation reservoirs in Nebraska with special reference to the fish populations. Midw. Wildl. Conf. 18th. 23 pp.

H195 Hedges, S. B. and R. C. Ball, 1953. Production and harvest of bait fishes in Michigan. Mich. Dept. Conserv. Misc. Publ. 6:1-30.

H198 Henderson, D. and R. F. Foster, 1957. Studies of smallmouth black bass (*Micropterus dolomieu*) in the Columbia River near Richland, Washington. Trans. Am. Fish. Soc. 86:112-127.

H200 Hennemuth, R. C., 1955. Growth of crappies, bluegill, and warmouth in Lake Ahquabi. Iowa State Coll. J. Sci. 30(1):119-137.

H202 Herke, W. H., 1959. Comparison of the length-weight relationship of several species of fish from two different, but connected habitats. Proc. S.E. Assn. Game Fish. Comm. 13:299-313.

H205 Hewson, L. C., 1955. Age, maturity, spawning and food of burbot (*Lota lota*) in Lake Winnipeg. J. Fish. Res. Board Can. 12(6):930-940.

H218   Hooper, P. L., 1954. Elk River sport fishing conditions. W. Va. Conserv. Comm. 16 pp. mimeo.

H219   Hourston, A. S., 1955. A study of variations in the Maskinonge from three regions in Canada. R. Ont. Mus. Zool. Paleontol. Contrib. 40:13.

H220   Houser, A., 1958. A summary of fisheries investigation of Fort Gibson Reservoir, Oklahoma. Okla. Dep. Wildl. Conserv. Fed. Aid Div. Job Compl. Rep. Proj. F6-R1.

H226   Huish, M. T., 1954. Life history of the black crappie of Lake George, Fla. Trans. Am. Fish. Soc. 83:176-193.

H228   ———. 1957. Life history of the black crappie of Lakes Eustis and Harris, Florida. Proc. S.E. Assn. Game Fish Comm. 11:302-312.

H229   ———. 1957. Food habits of three Centrarchidae in Lake George, Fla. Proc. S.E. Assn. Game Fish. Comm. 11:293-302.

H243   Hoopes, D. T., 1960. Utilization of mayflies and caddis flies by some Mississippi River fishes. Trans. Am. Fish. Soc. 89(1):32-34.

H244   Hunt, B. P., 1960. Digestion rate and food consumption of Florida gar, warmouth, and largemouth bass. Trans. Am. Fish. Soc. 89(2):206-211.

H250   Helm, W. T., 1958. Notes on the ecology of panfish in Lake Wingra with special reference to the yellow bass. Ph.D. thesis, Univ. Wis.

H255   Hathaway, E. S., 1928. Quantitative study of the changes produced by acclimatization in the tolerance of high temperatures by fishes and amphibians. Bull. U.S., Bur. Fish. 43:169-192.

H262   Howland, J. W., 1929. Growth changes in the rock bass. M.S. thesis, Ohio State Univ. 22 pp.

H264   Hoffman, J. M., 1955. Age and growth of the green sunfish, *Lepomis cyanellus* Rafinesque, in the Niangua Arm of the Lake of the Ozarks. M.S. thesis, Univ. Missouri.

H265   Harrison, C., 1923. A study of age determination in fishes of the family Centrarchidae. M.S. thesis, Univ. Ill. 28 pp.

H270   Hooper, A. D. and J. H. Crance, 1960. Use of rotenone in restoring balance to overcrowded fish populations in Alabama lakes. Trans. Am. Fish. Soc. 89(4):351-357.

H272   Hunn, J. B., 1957. Methods for the control of stunted panfish populations. M.S. thesis, Mich. State Coll. 57 pp.

H273   Howland, J. W., 1931. Studies on the life history of the spotted or Kentucky bass, *Micropterus pseudaplites* Hubbs. Ph.D. thesis, Ohio State Univ. 112 pp.

H278   Hansen, D. F., G. W. Bennett, R. J. Webb, and J. M. Lewis, 1960. Hook-and-line catch in fertilized and unfertilized ponds. Ill. Nat. Hist. Surv. Bull. 27(5):345-390.

H280   Havey, K. A., 1950. The freshwater fisheries of Long Pond and Echo Lake, Mount Desert Island, Maine. M.S. thesis, Univ. Maine. 83 pp.

H288   Hathaway, E. S., 1927. The relation of temperature to the quantity of food consumed by fishes. Ecology 8(4):428-434.

H303   Houser, A., 1960. A fishery survey by population estimation techniques in Lake Lawtonka. Okla. Fish. Res. Lab. Rep. 76:18 pp.

H304   Harrison, H. M., C. O'Farrell, and T. Moen, 1961. Progress report, renovation of the Winnebago River, Iowa. Iowa Conserv. Comm. Q. Biol. Rep. 13(1):32-37.

H332   Hastings, C. E. and F. B. Cross, 1962. Farm ponds in Douglas County, Kansas. Univ. Kans. Mus. Nat. Hist. Misc. Publ. 29:1-21.

H341   Hunsaker, D. II and R. W. Crawford, 1964. Preferential spawning behavior of the largemouth bass, *Micropterus salmoides*. Copeia, 1964 (1):240-241.

H342 Harlan, J. R. and E. B. Speaker, 1956. Iowa fish and fishing, 3rd ed. State of Iowa. 377 pp.

H382 Hubbs, C. L. and K. F. Lagler, 1964. Fishes of the Great Lakes region, with a new preface. Univ. Mich. Press. 213 pp.

H389 Hansen, D. F., 1957. Ecological life history table of the white crappie as observed in Illinois. Material for Handb. Biol. Data. 1 p.

H392 Hauser, A., 1963. Loss in weight of sunfish following aquatic vegetative control using the herbicide Silvex. Proc. Okla. Acad. Sci. 43: 232-237.

H398 Helms, D. R., 1966. 1965 annual survey of the Coralville Reservoir fish population. Iowa Conserv. Comm. Biol. Rep. 18(2):27-32.

H419 Houser, A. and M. G. Bross, 1963. Average growth rates and length-weight relationships for fifteen species of fish in Oklahoma waters. Okla. Fish. Res. Lab. Rep. 85:75 pp.

H424 Houser, A. and B. Grinstead, 1961. The effect of black bullhead catfish and bluegill removals on the fish population of a small lake. Proc. S.E. Assoc. Game Fish. Comm. 15:193-200.

H432 Harrington, R. W., Jr., 1956. An experiment on the effects of contrasting daily photoperiods on gametogenesis and reproduction in the centrarchid fish, *Enneacanthus obesus* (Girard) J. Exp. Zool. 131(3): 203-224.

H433 Hasler, A. D. and W. J. Wisby, 1958. The return of displaced largemouth bass and green sunfish to a "home" area. Ecology 39:289-293.

H434 Heimstra, N. W., D. K. Damkot, and N. G. Benson, 1969. Some effects of silt turbidity of behavior of juvenile largemouth bass and green sunfish. U.S. Bur. Sport Fish Wildl. Tech. Pap. 20:9 pp.

H435 Hunn, J. B., R. A. Schoettger, and E. W. Whealdon, 1968. Observations on the handling and maintenance of bioassay fish. Prog. Fish-Cult. 30(3):164-167.

H436 Hunter, J. R., 1963. The reproductive behavior of the green sunfish, *Lepomis cyanellus*. Zoologica (NY) 48(1):13-24.

H437 Hubbell, P. M., 1966. Pumpkinseed sunfish. Pp. 402-404, Inland Fisheries Management, A. Calhoun, ed. Calif. Dep. Fish Game.

H438 ———. 1966. Warmouth. Pp. 405-407, Inland Fisheries Management, A. Calhoun, ed. Calif. Dep. Fish Game.

H439 Hall, D. J., W. E. Cooper, and E. E. Werner, 1970. An experimental approach to the production dynamics and structure of freshwater animal communities. Limnol. Oceanogr. 15(6):839-928.

H440 Hansen, D. F., 1966. Stocking and sport fishing at Lake Glendale (Illinois). Ill. Nat. Hist. Surv. Bull. 29(2):105-158.

H441 Holcik, J., 1970. Standing crop, abundance, production and some ecological aspects of fish populations in some inland waters of Cuba. Acta soc. zool. Bohemoslov. 34(3):184-201.

H442 Hulsey, A. H. and J. H. Stevenson, 1958. Comparison of growth rates of game fish in Lake Catherine, Lake Hamilton, and Lake Ouachita, Arkansas. Proc. Ark. Acad. Sci. 12:17-31.

H443 Heckman, J. R., 1969. Embryological comparison of *Lepomis macrochirus* X *macrochirus* and *Lepomis macrochirus* X *gibbosus*. Trans. Am. Fish. Soc. 98(4):669-675.

H444 Huck, L. L. and G. E. Gunning, 1967. Behavior of the longear sunfish, *Lepomis megalotis* (Rafinesque). Tulane Stud. Zool. 14(3):121-131.

H445 Hester, F. E., 1970. Phylogenetic relationships of sunfishes as demonstrated by hybridization. Trans. Am. Fish. Soc. 99(1):100-104.

H446   Hubbs, C. L. and L. C. Hubbs, 1932. Experimental verification of natural hybridization between distinct genera of sunfishes. Pap. Mich. Acad. Sci. Arts Lett. 15:427-437.

H447   Hewkin, J. A., 1961. John Day District. Oreg. State Game Comm. Fish. Div. Ann. Rep. 1961:208-225.

H448   Haines, T. A. and R. L. Butler, 1969. Responses of yearling smallmouth bass (*Micropterus dolomieui*) to artificial shelter in a stream aquarium. J. Fish. Res. Board Can. 26(1):21-31.

H449   Hunter, G. W. III and W. S. Hunter, 1938. Studies on host reactions to larval parasites. I. The effect on weight. J. Parasit. 24(6):477-481.

H450   Hubbs, C. L. and R. M. Bailey, 1938. The smallmouthed bass. Cranbrook Inst. Sci. Bull. 10:89 pp.

H451   Harrison, H. M., 1954. Smallmouth bass studies, Des Moines River. Iowa Conserv. Comm. Q. Biol. Rep. 5:6-11.

H452   Howland, J. W., 1931. Studies on the Kentucky black bass (*Micropterus pseudaplites* Hubbs). Trans. Am. Fish. Soc. 61:89-94.

H453   ———. 1932. Experiments in the propagation of spotted black bass. Trans. Am. Fish. Soc. 62:185-188.

H454   Hanson, W. D., 1962. Dynamics of the largemouth bass population in Bull Shoals Reservoir, Missouri. Proc. S.E. Assoc. Game Fish Comm. 16:398-404.

H455   Huston, J. E., 1965. Investigation of two Clark Fork River hydroelectrical impoundments. Proc. Mont. Acad. Sci. 25:20-40.

H456   Heman, M. L., R. S. Campbell, and L. C. Redmond, 1969. Manipulation of fish populations through reservoir drawdown. Trans. Am. Fish. Soc. 98(2):293-304.

H457   Hornbeck, R. G., W. White, and F. P. Meyer, 1966. Control of *Apus* and fairy shrimp in hatchery rearing ponds. Proc. S.E. Assoc. Game Fish Comm. 19:401-403.

H458   Hensen, D. F., 1937. The date of annual ring formation in the scales of the white crappie. Trans. Am. Fish. Soc. 66:227-236.

H459   ———. 1943. On nesting of the white crappie, *Pomoxis annularis*. Copeia, 1943(4):259-260.

H460   Hellier, T. R., Jr., 1967. The fishes of the Sante Fe River system. Bull. Fla. State Mus. Biol. Ser. 2(1):46 pp.

H461   Hallam, J. C., 1959. Habitat and associated fauna of four species of fish in Ontario streams. J. Fish. Res. Board Can. 16(2):147-173.

H462   Hunter, J. R. and W. J. Wisby, 1961. Utilization of the nests of green sunfish (*Lepomis cyanellus*) by the redfin shiner (*Notropis umbratilis cyanocephalus*). Copeia, 1961(1):113-115.

H463   Hunter, J. R. and A. D. Hasler, 1965. Spawning association of the redfin shiner, *Notropis umbratilis*, and the green sunfish, *Lepomis cyanellus*. Copeia, 1965(3):265-281.

H464   Hocutt, C. H., 1973. Swimming performance of three warm-water fishes exposed to a rapid temperature change. Chesapeake Sci. 14(1):11-16.

H465   Hashagen, K. A., Jr., 1973. Population structure changes and yields of fishes during the initial eight years of impoundment of a warm-water reservoir. Calif. Fish Game 59(4):221-244.

H466   Heidinger, R. C. and W. M. Lewis, 1972. Potentials of the redear sunfish X green sunfish hybrid in pond management. Prog. Fish-Cult. 34(2):107-109.

H467   Hickman, G. D. and J. C. Congdon, 1974. Effects of length limits on the fish populations of five north Missouri lakes. Symposium on over-harvest of largemouth bass in small impoundments. N.C. Am. Fish. Soc. Spec. Publ. 3:84-94.

H468   Hoey, J. W. and L. C. Redmond, 1974. Evaluation of opening Binder Lake with a length limit for bass. Symposium on overharvest of large-mouth bass in small impoundments. N.C. Am. Fish. Soc. Spec. Publ. 3:100-105.

H469   Holbrook, J. A. II and D. Johnson, 1973. Management implications of bass fishing tournaments. Proc. S.E. Assoc. Game Fish Comm. 26: 320-324.

H470   Hiranvat, S., 1973. Preimpoundment age and growth of the redbreast sunfish, *Lepomis auritus*, in the proposed West Point Reservoir, Alabama and Georgia. (Abstract) Ga. Acad. Sci.

H471   Hubbs, C., 1971. Survival of intergroup Percid hybrids. Jap. J. Ichthyol. 18(2):65-75.

H472   Hokanson, K. E. and L. L. Smith, 1971. Some factors influencing toxicity of linear alkylote sulfonate (LSA) to the bluegill. Trans. Am. Fish. Soc. 199(1):1-12.

H473   Heidinger, R. C., 1971. Use of ultraviolet light to increase the avail-ability of aerial insects to caged bluegill sunfish. Prog. Fish-Cult. 38(4):187-192.

H474   Hickman, G. D. and M. R. Dewey, 1973. Notes on the upper lethal temperature of the duskystripe shiner, *Notropis pilsbryi*, and the bluegill, *Lepomis macrochirus*. Trans. Am. Fish. Soc. 102(4): 838-840.

H475   Haines, T. A., 1973. Effects of nutrient enrichment and a rough fish population (carp) on a game fish population (smallmouth bass). Trans. Am. Fish. Soc. 102(2):346-354.

H476   ———. 1973. An evaluation of RNA-DNA ratio as a measure of long-term growth in fish populations. J. Fish. Res. Board Can. 30(2):195-199.

H477   Horning, W. B. II and R. E. Pearson, 1973. Growth temperature re-quirements and lower lethal temperatures for juvenile smallmouth bass (*Micropterus dolomieu*). J. Fish. Res. Board Can. 30(8):1226-1230.

H478   Hile, R., 1970. Body-scale relation and calculation of growth in fishes. Trans. Am. Fish. Soc. 99(3):468-474.

H479   Heidinger, R. C., 1974. An indexed bibliography of the largemouth bass, *Micropterus salmoides* (Lacepede). Bass Research Foundation. P. O. Box 3385, Montgomery, AL. 36109. 84 pp. $9.95.

I7   Irving, R. B. and P. Cuplin, 1956. The effect of hydroelectric develop-ments on the fishery resources of Snake River. Idaho Fish Game Dep. Proj. F-8-R.

I10   Iowa Conservation Commission, Biology Section 1961. Calculated total lengths at each annulus for various species of fish taken from DeSoto Bend, September 25-29, 1961. 13 pp. mimeo.

I16   Isaac, G. W. and C. E. Bond, 1963. Standing crops of fish in Oregon farm ponds. Trans. Am. Fish. Soc. 92(1):25-29.

I17   Ingram, W. M. and E. P. Odum, 1941. Nests and behavior of *Lepomis gibbosus* (Linnaeus) in Lincoln Pond, Rensselaerville, New York. Am. Midl. Nat. 26(1):182-193.

I18   Imler, R. R., 1972. Sacramento perch introductions. Colo. Fish. Res. Rev. 7:33-34.

J9   Johnson, W. L., 1945. Age and growth of the black and white crappies of Greenwood Lake, Indiana. Invest. Indiana Lakes streams. 2(15): 297-324.

J10   Jones, A. M., 1941. The length of the growing season of largemouth and smallmouth black bass in Norris Reservoir, Tennessee. Trans. Am. Fish. Soc. 70:183-187.

J13   Jordan, D. S. and B. W. Evermann, 1908. American food and game fishes. Doubleday, Page, and Co. 571 pp.

J15   Juday, C. and C. L. Schloemer, 1938. Growth of game fish in Wisconsin waters—fifth report. Notes Limnolog. Lab. Wisc. Geol. Nat. Hist. Surv. 26 pp. mimeo.

J16   Juday, D., D. L. Schloemer, and C. Livingston, 1938. Effect of fertilizers on plankton production and on fish growth in a Wisconsin lake. Prog. Fish-Cult. 40:24-27.

J25   Jenkins, R. M., E. M. Leonard, and G. E. Hall, 1952. An investigation of the fisheries resources of the Illinois River and pre-impoundment study of Tenkiller Reservoir, Oklahoma. Okla. Fish. Res. Lab. Rep. 26:136 pp. mimeo.

J26   Jenkins, R. M., 1951. A fish population study of Claremore City Lake. Proc. Okla. Acad. Sci. 30:84-93.

J27   Jackson, S. W., Jr., 1954. Rotenone survey of Black Hollow on Lower Spavinaw Lake, November, 1953. Proc. Okla. Acad. Sci. 35:10-14.

J28   ———. 1957. Comparison of the age and growth of four fishes from lower and upper Spavinaw Lakes, Oklahoma. Proc. S.E. Assoc. Game Fish Comm. 11:232-249.

J31   Jenkins, R. M., 1953. Growth histories of the principal fishes in Grand Lake (O' the Cherokees), Oklahoma, through thirteen years of impoundment. Okla. Fish. Res. Lab. Rep. 34:1-87.

J33   ———. 1954. An estimate of the fish population in a forty-five-year-old Oklahoma pond. Proc. Okla. Acad. Sci. 35:69-76.

J34   ———. 1955. A summary of fish population studies conducted during 1954 at Ardmore City Lake, Stringtown sub-prison lake, Fairfax City Lake, and Pawhuska City Lake. Okla. Fish. Res. Lab. Rep. 48:31 pp.

J35   ———. 1955. An eleven year growth history of white crappie in Grand Lake, Oklahoma. Proc. Okla. Acad. Sci. 34:40-47.

J37   ———. 1957. The effect of gizzard shad on the fish population of a small Oklahoma lake. Trans. Am. Fish. Soc. 85:58-74.

J38   ———. 1957. A preliminary study of the standing crop of fish in Oklahoma waters. Okla. Fish. Res. Lab. Spec. Rep. mimeo.

J40   ———. 1958. The standing crop of fish in Oklahoma ponds. Proc. Okla. Acad. Sci. 38:157-172.

J41   Jenkins, R. M. et al., 1953. A pre-impoundment survey of Fort Gibson Reservoir, Oklahoma (Summer 1952). Okla. Fish. Res. Lab. Rep. 29: 1-53 pp. mimeo.

J42   Jenkins, R. M. et al., 1953. A report on the growth of fishes in Fort Gibson Reservoir collected in July and October, 1953, the first year of complete impoundment. Okla. Fish. Res. Lab. Rep. 32:10 pp. mimeo.

J44   Jenkins, R., R. Elkin, and J. Finnell, 1955. Growth rates of six sunfishes in Oklahoma. Okla. Fish. Res. Lab. Rep. 49:1-73 mimeo.

J45   Jenkins, R. M. and J. C. Finnell, 1957. The fishery resources of the Verdigris River in Oklahoma. Okla. Fish. Res. Lab. Rep. 59:1-46.

J46   Jenkins, R. M. and G. E. Hall, 1953. Growth of largemouth bass in Oklahoma. Okla. Fish. Res. Lab. Rep. 30:44 pp. mimeo.

J64   Johnson, M. C., 1954. Preliminary experiment on fish culture in brackish water ponds. Prog. Fish-Cult. 16(3):131-133.

J77   Jones, R. S. and W. F. Hettler, 1959. Bat feeding by green sunfish. Tex. J. Sci. 11(1):48.

J83   Johnson, M. G. and W. H. Charlton, 1960. Some effects of temperature on the metabolism and activity of the largemouth bass, *Micropterus salmoides* Lacepede. Prog. Fish-Cult. 22(4):155-163.

J92   John, K. R., 1957. Comparative rates of survival of normal and deformed chub, *Gila atraria* Girard, in Two Ocean Lake, Teton County, Wyoming. Proc. Pa. Acad. Sci. 31:77-82. (Ages should be one year older than reported—John 1959.)

J94   Jordan, D. S., B. W. Evermann, and H. W. Clark, 1930. Check list of the fishes and fishlike vertebrates of North and Middle America. Append. X Rep. U.S. Comm. Fish. 1928: (reprint. 1955 U.S. Govt. Print. Off.). 1-670.

J95   Jenkins, R. M., 1955. Expansion of the crappie population in Ardmore City Lake following a drastic reduction in numbers. Proc. Okla. Acad. Sci. 36:70-76.

J96   ——. 1959. Some results of the partial fish population removal techniques in lake management. Proc. Okla. Acad. Sci. 36:164-173.

J101 Jackson, S. W., Jr., 1966. Summary of fishery management activities on Lakes Eucha and Spavinaw, Oklahoma 1951-1964. Proc. S.E. Assoc. Game Fish. Comm. 19:315-343.

J104 Jennings, T., 1968. Successful renovation of a small natural Iowa lake. Proc. Iowa Acad. Sci. 74:92-98.

J105 Johnson, C. E., 1971. Factors affecting fish spawning. Wis. Conserv. Bull. 36(4):16-17.

J106 James, M. C., 1930. Spawning reactions of smallmouthed bass. Trans. Am. Fish. Soc. 60:62-63.

J107 Johnson, M. G. and H. R. McCrimmon, 1967. Survival, growth, and reproduction of largemouth bass in southern Ontario ponds. Prog. Fish-Cult. 29(4):216-221.

J108 Jester, D. B., T. M. Moody, C. Sanchez, Jr., and D. E. Jennings, 1969. A study of game fish reproduction and rough fish problems in Elephant Butte Lake. N.M. Job Compl. Rep. Fed. Aid. proj. F-22R-9-Job F-1. 73 pp.

J109 Jester, D. B., 1971. Effects of commercial fishing, species introductions, and drawdown control on fish populations in Elephant Butte Reservoir, New Mexico. Am. Fish. Soc. Spec. Publ. 8:265-285.

J110 Jennings, T., 1969. Age and growth of black crappie in Spirit Lake, Iowa. Iowa Q. Biol. Rep. 21(4):60-64.

J111 Jossel, J., Jr., 1969. Age, growth and condition factors of stunted white crappie. M.S. thesis, Okla. State Univ. (not seen)

J112 Johnson, D. L. and R. O. Anderson, 1974. Evaluation of a 12-inch length limit on largemouth bass in Philips Lake, 1966-1973. N.C. Div. Am. Fish. Soc. Spec. Publ. 3:106-113.

J113 Jurgens, K. C. and W. H. Brown, 1954. Chilling the eggs of the largemouth bass. Prog. Fish-Cult. 16(4):172-175.

J114 Jude, D. J., 1973. Sublethal effects of ammonia and cadmium on growth of green sunfish. Ph.D. thesis, Mich. State Univ. 193 pp.

J115 James, M. F., 1946. Histology of gonadal changes in the bluegill, *Lepomis macrochirus* Rafinesque, and the largemouth bass, *Huro salmoides* (Lacepede). J. Morphol. 79:63-92.

J116   Johnson, J. E. and B. L. Swanson, 1974. Length and weight changes of preserved black crappie and yellow perch. Prog. Fish-Cult. 36(4): 201-206.

K23   Krumholz, L. A., 1948. Variations in size and composition of fish populations in recently stocked ponds. Ecology 29(4):401-414.

K24   ———. 1949. Rates of survival and growth of bluegill yolk fry stocked at different intensities in hatchery ponds. Trans. Am. Fish. Soc. 76: 190-203.

K33   Kuehn, J. H., 1949. Statewide average total length in inches at each year. Minn. Fish. Res. Lab. Suppl. Invest. Rep. 51 (2nd rev.).

K41   Krumholz, L. A., 1950. New fish stocking policies for Indiana ponds. Trans. N.A. Wildl. Conf. 15:251-269.

K42   ———. 1950. Further observations on the use of hybrid sunfish in stocking small ponds. Trans. Am. Fish. Soc. 79:112-124.

K43   ———. 1950. Some practical considerations in the use of rotenone in fisheries research. J. Wildl. Manage. 14(4):413-424.

K44   Kuehn, J. H., 1949. A study of a population of longnose dace (*Rhinichthys c. cataractae*). Proc. Minn. Acad. Sci. 17:81-87.

K47   Kmiotek, S. and C. L. Cline, 1952. Growth of Southern Wisconsin largemouth bass with creel census results from lakes with liberalized regulations. Wis. Cons. Dept. fish. Invest. Rep. 670:11 pp. mimeo.

K58   Kimsey, J. B., 1954. The life history of the tui chub, *Siphateles bicolor* (Girard), from Eagle Lake, California. Calif. Fish Game 40(4):395-410.

K59   ———. 1955. Results of population sampling at Millerton Lake, Fresno-Madera counties, July 1955. Calif. Inland Fish. Admin. Rep. 55-20.

K60   ———. 1957. The status of the redeye bass in California. Calif. Fish Game. 43(1):99-100.

K61   Kimsey, J. B., et al., 1956. A survey of the fish population of Pardee Reservoir, Amador/Calaveras Counties. Calif. Inland Fish. Admin. Rep. 56-18.

K62   Kimsey, J. B. and R. R. Bell, 1956. Notes on the status of the pumpkinseed sunfish, *Lepomis gibbosus*, in the lower Susan River, Lassen County, California. Calif. Inland Fish. Admin. Rep. 56-1:1-19.

K63   ———. 1956. Observations on the ecology of the largemouth black bass and tui chub in Big Sage Reservoir, Modoc County. Calif. Inland Fish. Admin. Rep. 55-15:1-17.

K64   Kimsey, J. B., R. H. Hagy, and G. W. McCammon, 1957. Progress report on the Mississippi threadfin shad, *Dorosoma petenensis atchafaylae*, in the Colorado River for 1956. Calif. Inland Fish. Admin. Rep. 57-23:48 pp.

K65   King, J. E., 1954. Three years of partial fish population removal at Lake Hiwassee, Oklahoma. Proc. Okla. Acad. Sci. 35:21-24.

K66   ———. 1955. Growth rates of fishes of Lake Hiwassee, Oklahoma, after two years of attempted population control. Proc. Okla. Acad. Sci. 34:53-56.

K71   Krumholz, L. A., 1956. Observations on the fish population of a lake contaminated by radioactive wastes. Am. Mus. Nat. Hist. Bull. 110 (4):281-367.

K72   Kruse, T. E., 1959. Grayling of Grebe Lake, Yellowstone National Park, Wyoming. U.S. Fish Bull. 59(149):307-351.

K73   Kutkuhn, J. H., 1955. Food and feeding habits of some fishes in a dredged Iowa Lake. Proc. Iowa Acad. Sci. 62:576-588.

K74   Kutkuhn, J. H., 1958. Utilization of gizzard shad by game fishes. Proc. Iowa Acad. Sci. 54:571-579.

K78   Kramer, R. H. and L. L. Smith, Jr., 1960. First-year of the large-mouth bass, *Micropterus salmoides* (Lacepede), and some related ecological factors. Trans. Am. Fish. Soc. 89(2):222-233.

K80   Klavano, W. C., 1958. Age and growth of fish from Oregon farm ponds. M.S. thesis, Ore. State Coll. 41 pp.

K81   Karvelis, E. G., 1952. Growth characteristics of a bluegill population in a Michigan trout lake. M.S. thesis, Mich. State Coll. 72 pp.

K82   Kruse, T. E., 1952. Age and growth of the pumpkinseed, *Lepomis gibbosus* (Linnaeus), in Lower Loch Alpine Pond, Washtenaw County, Michigan. M.S. thesis, Univ. Mich. 17 pp.

K94   Kelley, J. W., 1962. Sexual maturity and fecundity of the largemouth bass, *Micropterus salmoides* (Lacepede), in Maine. Trans. Am. Fish. Soc. 91(1):23-28.

K95   Kramer, R. H. and L. L. Smith, Jr., 1962. Formation of year classes in largemouth bass. Trans. Am. Fish. Soc. 91(1):29-41.

K121   Kimsey, J. R., 1954. Introduction of the redeye black bass and the threadfin shad into California. Calif. Fish Game. 40(2):203-204.

K125   Keast, A. and D. Webb, 1966. Mouth and body form relative to feeding ecology in the fish fauna of a small lake, Lake Opinicon, Ontario. J. Fish Res. Board Can. 23(12):1845-1874.

K129   Keast, A., 1965. Resource subdivision amongst cohabiting fish species in a bay, Lake Opinicon, Ontario. Univ. Mich. Great Lakes Res. Div. Publ. 13:106-132.

K132   Knapp, F. T., 1950. Survey of systems used in measuring lengths and weights of fishes and of systems proposed as standard in the Southwest. Prog. Fish-Cult. 12(4):207-208.

K135   Keast, A. and L. Welsh, 1968. Daily feeding periodicities, food uptake rates, and dietary changes with hour of day in some lake fishes. J. Fish Res. Board Can. 25(6):1133-1144.

K137   Keast, A., 1968. Feeding of some Great Lakes fishes at low temperatures. J. Fish. Res. Board Can. 25(6):1199-1218.

K139   Kudrna, J. J., 1967. Movement and homing of sunfishes in Clear Lake. Proc. Iowa Acad. Sci. 72:263-271.

K140   Keenleyside, M. H. A., 1967. Behaviour of male sunfish (genus *Lepomis*) towards females of three species. Evolution 21:688-695.

K141   Kitchell, J. F. and J. T. Windell, 1968. Rate of gastric digestion in pumpkinseed sunfish, *Lepomis gibbosus*. Trans. Am. Fish. Soc. 97 (4):489-492.

K142   Kennedy, H. D., L. L. Eller, and D. F. Walsh, 1970. Chronic effects of methoxychlor on bluegills and aquatic investebrates. U.S. Sport Fish. Wildl. Tech. Pap. 53:1-18.

K143   Krull, W. H., 1934. *Cercaria bessiae* Cort and Brooks, 1928, an injurious parasite of fish. Copeia, 1934(2):69-73.

K144   Kirkwood, J. B., 1955. The age and growth of bluegill, *Lepomis macrochirus* (Rafinesque), from farm fish ponds in Kentucky. Trans. Ky. Acad. Sci. 17(1):57-65.

K145   Kitchell, J. F. and J. T. Windell, 1970. Nutritional value of algae to bluegill sunfish, *Lepomis macrochirus*. Copeia, 1970(1):86-190.

K146   Kolehmainen, S. E. and D. J. Nelson, 1969. The balances of 137 Co, stable cesium, and the feeding rates of bluegill (*Lepomis macrochirus* Raf.) in White Oak Lake. Oak Ridge Nat. Lab. Rep. 4445. 109 pp. Also Ph.D. thesis, Univ. Tenn.

K147 Keating, J. F., Jr., 1970. Growth rates and food habits of smallmouth bass in the Snake, Clearwater, and Salmon Rivers, Idaho, 1965-1967. M.S. thesis, Univ. Idaho. 40 pp.
K148 Kirkland, L., 1965. Results of a tagging study on the spotted bass, *Micropterus punctulatus*. Proc. S.E. Assoc. Game Fish Comm. 17: 242-255.
K149 Kimsey, J. B., 1956. Largemouth bass tagging. Calif. Fish Game 42(4):337-346.
K150 Kelley, J. W., 1968. Effects of incubation temperature on survival of largemouth bass eggs. Prog. Fish-Cult. 30(3):159-163.
K151 Kilby, J. D., 1955. The fishes of two gulf coastal march areas of Florida. Tulane Stud. Zool. 2(8):175-247.
K152 Kemmerer, A. J., 1967. Investigation of factors limiting population growth of crappie. Fish. Res. Ariz. 1966:1-20.
K153 Keast, A., 1968. Feeding biology of the black crappie, *Pomoxis nigromaculatus*. J. Fish. Res. Board Can. 25(2):285-297.
K154 Krummrich, J. T. and R. C. Heidinger, 1973. Vulnerability of channel catfish to largemouth bass predation. Prog. Fish-Cult. 35(3):173-175.
K155 Keith, W. E. and S. K. Barkley, 1971. Predation of stocked rainbow trout by chain pickerel and largemouth bass in Lake Ouachita, Arkansas. Proc. S.E. Assoc. Game Fish Comm. 24:401-407.
K156 Kaya, C. M., 1973. Effects of temperature on responses of the gonads of green sunfish (*Lepomis cyanellus*) to treatment with carp pituitaries and testosterone proprionate. J. Fish. Res. Board Can. 30(7):905-912.
K157 Kaya, C. M. and A. D. Hasler, 1972. Photoperiod and temperature effects on the gonads of green sunfish, *Lepomis cyanellus* (Rafinesque), during the quiescent, winter phase of its annual sexual cycle. Trans. Am. Fish. Soc. 101(2):270-275.
K158 Kaya, C. M., 1973. Effects of temperature and photoperiod on seasonal regression of gonads of green sunfish, *Lepomis cyanellus*. Copeia, 1973(2):369-373.
K159 Kapoor, N. N., 1971. Locomotory patterns of fish (*Lepomis gibbosus*) under different levels of illumination. Anim. Behav. 19:744-749.
K160 Kramer, B., W. Molenda, and K. Fiedler, 1969. Behavioral effect of the antiandrogen cyproterone acetate (Schering) in *Tilapia mossambica* and *Lepomis gibbosus*. (Abstract). Gen. Comp. Endrocrinol. 13(3):515.
K161 Kitchell, J. F., J. F. Koonce, R. V. O'Neill, H. H. Shugart, Jr., J. J. Magnuson, and R. S. Booth, 1974. Model of fish biomass dynamics. Trans. Am. Fish. Soc. 103(4):786-798.
K162 Kolehmainen, S. E., 1974. Daily feeding rates of bluegill (*Lepomis macrochirus*) determined by a refined radioisotope method. J. Fish. Res. Board Can. 31(1):67-74.
K163 Keenleyside, M. H. A., 1972. Intraspecific intrusions into nests of spawning longear sunfish (Pisces:Centrarchidae). Copeia, 1972(2): 272-278.
L8 Langlois, T. H., 1932. Problems of pondfish culture. Trans. Am. Fish. Soc. 62:156-166.
L9 ———. 1934. Standard methods of computing bass production. Trans. Am. Fish. Soc. 64:163-166.
L11 ———. 1936. A study of smallmouth bass (*Micropterus dolomieu*) in rearing ponds in Ohio. Ohio Biol. Surv. Bull. 6(33):191-225.
L12 ———. 1936. Length-weight relationships in Ohio State fish ponds. Copeia, 1936(2):120.

L13   Langlois, T. H., 1939. Ohio fish management progress report. Ohio Conserv. Bull. 3(1):16-19.

L14   ———. 1945. Ohio's fish program. Ohio Div. Conserv. Nat. Resour. 40 pp.

L18   Leach, G. C., 1923. Artificial propagation of brook trout and rainbow trout, with notes on three other species. Rep. U.S. Comm. Fish 1923, App. VI, pp. 1-74.

L22   Leary, J. L., 1908. Description of San Marcos Station with some of the methods of propagation in use at that station. Trans. Am. Fish. Soc. 37:75-81.

L29   Lewis, W. M., 1950. Fisheries investigations on two artificial lakes in southern Iowa. II. Fish populations. Iowa State J. Sci. 24(3):287-324.

L32   Lewis, W. M. and T. S. English, 1949. The warmouth, *Chaenobryttus coronarius* (Bartram), in Red Haw Hill Reservoir, Iowa. Iowa State J. Sci. 23(4):317-322.

L40   Lydell, D., 1903. The habits and culture of the black bass. Bull. U.S. Comm. Fish. 22:39-44.

L48   Lachner, E. A., 1950. Food, growth and habits of fingerling northern smallmouth bass, *Micropterus dolomieu dolomieu* Lacepede, in trout waters of western New York. J. Wildl. Manage. 14(1):50-56.

L54   Lachner, E. A., G. L. Trembley, and P. S. Handwerk, 1947. Some biological problems concerning the management of the northern smallmouth bass in Pennsylvania. Penn. Angler June:1, 16-18.

L56   Lagler, K. F., A. S. Hazzard, W. E. Hazen, and W. A. Tompkins, 1950. Outboard motors in relation to fish behavior, fish production and angling success. Trans. N.A. Wildl. Conf. 15:280-303.

L63   Leonard, E. M., 1950. Ten years of management and fishing on an Oklahoma farm pond. Midw. Wildl. Conf. 12th:4 pp. mimeo.

L68   Lane, C. E., Jr., 1952. Growth rate of the bluegill in Clearwater Lake, Missouri. Midw. Wildl. Conf. 14th:2 pp. mimeo. abstr.

L72   Lagler, L. F. and G. C. De Roth, 1953. Populations and yield to anglers in a fishery for largemouth bass, *Micropterus salmoides* (Lacepede). Pap. Mich. Acad. Sci. Arts Lett. 38:235-253.

L73   Lagler, K. F. and T. E. Kruse, 1953. Food conversion in black basses of the genus *Micropterus*. J. Wildl. Manag. 17(2):217.

L74   Lambou, V. W., 1958. Growth rate of young-of-the-year largemouth bass, black crappie, and white crappie in some Louisiana lakes. La. Acad. Sci. Apr. 1958. 9 pp.

L76   ———. 1959. Fish population of backwater lakes in Louisiana. Trans. Am. Fish. Soc. 88(1):7-16.

L78   Lane, C. E., Jr., 1954. Age and growth of bluegill in a new impoundment. J. Wildl. Manage. 18(3):358-365.

L79   Langlois, T. H., 1935. The production of small-mouth bass under controlled conditions. Prog. Fish-Cult. 7:1-7.

L84   Larimore, R. W., 1954. Dispersal, growth, and influence of smallmouth bass stocked in a warm-water stream. J. Wildl. Manage. 18(2):207-216.

L86   ———. 1957. Ecological life history of the warmouth (Centrarchidae). Ill. Nat. Hist. Survey Bull. 27(1):1-83.

L95   Latta, W. C., 1959. Significance of trap-net selectivity in estimating fish population statistics. Pap. Mich. Acad. Sci. Arts. Lett. 64:123-138.

L97   Lawrence, J. M., 1957. Estimated sizes of various forage fishes largemouth bass can swallow. Proc. S.E. Assoc. Game Fish. Comm. 11:220-226.

L105   Lennon, R. E. and P. S. Parker, 1959. The reclamation of Indian and Abrams Creeks Great Smoky Mountains National Park. U.S. Fish. Wildl. Serv. Spec. Sci. Rep. Fish.:306-1-22.

L106   Leonard, E. M. and R. M. Jenkins, 1954. Growth of the basses of the Illinois River, Oklahoma. Proc. Okla. Acad. Sci. 33:21-29.

L108   Lewis, W. M., 1957. The fish population of a spring-fed stream system in southern Illinois. Trans. Ill. State Acad. Sci. 50:23-29.

L110   Lewis, W. M. and D. Elder, 1953. The fish population of the headwaters of a spotted bass stream in southren Illinois. Trans. Am. Fish. Soc. 82:193-202.

L113   Lichens, A. B., 1958. Central Oregon-Columbia District. Ore. State Game Comm. Fish. Div. Ann. Rep. 1957:121-132.

L117   Locke, F., 1958. The largemouth bass. Ore. State Game Comm. Bull. 13(12):3-4, 7.

L121   Louder, D. E., and W. M. Lewis, 1957. Study of a stocking of the red-ear and bluegill in southern Illinois. Prog. Fish-Cult. 19(3):140-141.

L122   Louisiana Conservationist, 1954. The biggest bass in Louisiana. La. Conserv. 7(1):6.

L123   Lowry, E. M., 1953. The growth of the smallmouth bass (*Micropterus dolomieu* Lacepede) in certain Ozark streams of Missouri. Ph.D. thesis, Univ. Mo. 316 pp. typewritten.

L126   Lynch, T. M., P. A. Buscemi, and D. G. Lemons, 1953. Limnological and fishery conditions of Two Buttes Reservoir, Colorado, 1950 and 1951. Colo. Game and Fish Dep. Rep. 92 pp.

L127   Lux, F. E. and L. L. Smith, Jr., 1960. Some factors influencing seasonal changes in angler catch in a Minnesota lake. Trans. Am. Fish. Soc. 89(1):67-79.

L129   Lopinot, A., 1958. How fast do Illinois fish grow? Outdoors Ill. 5(4): 8-10.

L131   Lambou, V. W. and H. Stern, Jr., 1959. Preliminary report on the effects of the removal of rough fishes on the Clear Lake Sport Fishery. Proc. S.E. Assn. Game Fish Comm. 12:36-56.

L132   Lux, F. E., 1960. Notes on first-year growth of several species of Minnesota fish. Prog. Fish-Cult. 22(2):81-82.

L135   Lagler, K. F., 1956. Freshwater Fishery Biology, 2d ed. Wm. C. Brown Co. 421 pp.

L137   Lambou, V. W., 1952. Food and habitat of gar fish in the tide water of southeastern Louisiana. M.S. thesis, La. State Univ. 54 pp.

L140   ——. 1961. Utilization of macrocrustaceans for food by freshwater fishes in Louisiana and its effects on the determination of predator-prey relations. Prog. Fish-Cult. 23(1):18-25.

L142   Latta, W. C., 1957. The ecology of the smallmouth bass, *Micropterus d. dolomieu* Lacepede, at Waugoshance Point, Lake Michigan. Ph.D. thesis, Univ. Mich. 114 pp.

L151   Lopinot, A., 1961. The red-ear sunfish. Ill. Wildl. 17(1):3-4.

L154   Lewis, W. M., G. E. Gunning, E. Lyles, and W. L. Bridges, 1961. Food choice of largemouth bass as a function of availability and vulnerability of food items. Trans. Am. Fish. Soc., 90(3):227-280.

L157   Louder, D. E., 1961. Coastal plain lakes of southeastern North Carolina. N. Ca. Wildl. Res. Comm. Fed. Aid Job Compl. Rep. Proj. F5R and F6R. No. 1:9-55.

L173 Lambou, V. W., 1963. Distribution of sex ratios of bluegill and redear sunfish. Trans. Am. Fish. Soc. 92:435-436.

L174 Lemke, A. E. and D. Mount, 1963. Some effects of alkyl benzene sulfonate on the bluegill, *Lepomis macrochirus*. Trans. Am. Fish. Soc. 92:372-378.

L178 Latta, W. C., 1963. The life history of the smallmouth bass, *Micropterus d. dolomieui*, at Waugoshance Point, Lake Michigan. Mich. Inst. Fish. Res. Bull. 5:1-56.

L183 Lindquist, A. W., C. C. Deonier, and J. E. Haney, 1943. The relationship of fish to the Clear Lake gnat in Clear Lake, California. Calif. Fish Game 29(4):196-202.

L187 Lamby, K., 1941. Zur Fishchereibiologie des Myvatn, Nord-Island. Publ. Circ. Cons. Explor. Mer. 53:7-174.

L189 Lawrence, J. M., 1957. Life history and ecology of centrarchid fishes. Data for Handb. Biol. Data. 9 pp.

L190 Lowry, E. M., 1958. The life history and ecology of the smallmouth bass. Data for Handb. Biol. Data 29 pp.

L195 Linton, L., 1961. A study of fishes of the Arkansas and Cimarron Rivers in the area of the proposed Keystone Reservoir. Okla. Fish Res. Lab. Rep. 81:30 pp.

L214 La Faunce, D. A., 1965. Long-term retention of tags by some freshwater fish. Calif. Fish Game 51(1):52-3.

L216 Lambou, V. W., 1961. Efficiency and selectivity of flag gill nets fished in Lake Bistineau, Louisiana. Proc. S.E. Assoc. Game Fish Comm. 15:319-59.

L217 Lewis, W. M., M. Anthony, and D. R. Helms, 1965. Selection of animal forage to be used in the culture of channel catfish. Proc. S.E. Assoc. Game Fish Comm. 17:364-7.

L222 Linton, K. J. and R. C. Ball, 1965. A study of the fish populations in a warm-water stream. Mich. Agric. Exp. Stn. Q. Bull. 48(2):255-285.

L223 Linton, K. J., 1964. Dynamics of fish populations in a warm-water stream. M.S. thesis, Mich. State Univ. 71 pp.

L226 Lawrence, J. M., 1957. Mouth size of largemouth bass in relation to size of forage fishes. Ph.D. thesis, Iowa State Univ. 126 pp.

L227 LaFaunce, D. A., J. B. Kimsey, and H. K. Chadwick, 1964. The fishery at Sutherland Reservoir, San Diego County, California. Cal. Fish Game 50(4):271-291.

L228 Lopinot, A. C., 1967. Pond fish and fishing in Illinois. Ill. Dep. Conserv. Fish. Bull. 5:62 pp.

L229 Laarman, P. W., 1971. Seasonal variation in protein content of bluegill pituitary glands. Trans. Am. Fish. Soc. 100(2):360-362.

L230 Lewis, W. M. and J. Nickum, 1964. The effect of *Posthodiplostomum minimum* upon the body weight of the bluegill. Prog. Fish-Cult. 26 (3):121-3.

L231 Lewis, W. M. and R. Heidinger, 1971. Supplemental feeding of hybrid sunfish populations. Trans. Am. Fish. Soc 100(4):619-623.

L232 Lagler, K. F. and C. Steinmetz, Jr., 1957. Characteristics and fertility of experimentally produced sunfish hybrids, *Lepomis gibbosus* X *L. macrochirus*. Copeia, 1957(4):290-292

L233 Larimore, R. W. and M. J. Duever, 1968. Effects of temperature acclimation on swimming ability of smallmouth bass fry. Trans. Am. Fish. Soc. 97(2):175-84.

CITATIONS 377

L234 Langlois, T. H., 1931. The problem of efficient management of hatcheries used for the production of pond fishes. Trans. Am. Fish. Soc. 61:106-115.

L236 Lewis, W. M. and D. R. Helms, 1964. Vulnerability of forage organisms to largemouth bass. Trans. Am. Fish. Soc. 93(3):315-8.

L237 Lawrence, J. M., 1961. Estimated lengths of various forage fishes spotted bass can swallow. Proc. S.E. Assoc. Game Fish Comm. 15: 235-236.

L238 Lewis, W. M., R. Heidinger, and M. Konekoff, 1969. Artificial feeding of yearling and adult largemouth bass. Prog. Fish. Cult. 31(1):44-6.

L239 Lewis, W. M. and S. Flickinger, 1967. Home range tendency of the largemouth bass (Micropterus salmoides). Ecology 48(6):1020-3.

L240 Lewis, W. M., 1967. Predation as a factor in fish populations. Pp. 386-90, Reservoir Fishery Resources Symposium, Am. Fish. Soc. Southern Div.

L241 Laurence, G. C., 1969. The energy expenditure of largemouth bass larvae, Micropterus salmoides, during yolk absorption. Trans. Am. Fish. Soc. 98(3):398-405.

L242 ———. 1971. Digestion rate of larval largemouth bass. N.Y. Fish Game J. 18(1):52-6.

L243 Litt, B. D., 1952. Histological changes in the ovary of white crappie, Pomoxis annularis Rafinesque, correlated with age and season. M.S. thesis, Univ. Mo. 87 p. not seen, quoted by W238.

L244 Leary, J. L., 1910. Propagation of crappie and catfish. Trans. Am. Fish. Soc. 39:143-148.

L245 Lewis, W. M., R. Heidinger, W. Kirk, W. Chapman, and D. Johnson, 1974. Food intake of the largemouth bass. Trans. Am. Fish. Soc. 103(2):277-80.

L246 Laurence, G. C., 1972. Comparative swimming abilities of fed and started larval largemouth bass, (Micropterus salmoides). J. Fish. Biol. 4(1):73-8.

L247 LaBastille, A., 1974. Ecology and management of the Atitlan grebe, Lake Atitlan, Guatemala. Wildl. Monogr. 37:66pp.

L248 Lingenfeltzer, D. P. and R. C. Summerfelt, 1973. Angler harvest in heated fishing docks on an Oklahoma Reservoir. Proc. S.E. Assoc. Game Fish Comm. 26:611-621.

L249 Lotrich, V. A., 1973. Growth, production and community composition of fishes inhabiting a first-, second-, and third-order stream of eastern Kentucky. Ecol. Monogr. 43(3):377-397.

L250 Lewis, W. M. and R. C. Heidinger, 1973. Fish stocking combinations for farm ponds. Southern Ill. Univ. Fish. Bull. 4:17 pp.

L251 LeTendre, G. C., C. P. Schneider, and N. F. Ehlinger, 1972. Net damage and subsequent mortality from furunculosis in smallmouth bass. N.Y. Fish Game J. 19(1):73-82.

M1 McCabe, B. C., 1942. Section 3. Fishes. Pp. 30-68 in Fish Sur. Rep. 1942, Mass. Dept. Conserv.

M11 Macfarlane, P. R. C., 1928. Salmon (Salmo salar) of the River Moisie (Eastern Canada) 1926-27. Proc. Roy. Soc. Edinb. 48(2):134-139.

M14 McGavock, A. M., 1932. Discussion (on bass rearing on Daphnia). Trans. Am. Fish. Soc. 62:168-170.

M17 MacKay, H. H., 1930. The present status of fish culture in the Province of Ontario. Trans. Am. Fish. Soc. 60:33-44.

M19   Mackenthun, K. M., 1947. Age and growth of southern Wisconsin blue-
      gills, *Lepomis macrochirus.* Wis. Conserv. Bull. 12(5):20-22.

M20   ——. 1948. Age-length and length-weight relationship of southern
      area lake fishes. Wis. Conserv. Dep. Fish. Biol. Invest. Rep. 586
      (rev.) 1-7 pp.

M31   Meehean, O. L., 1935. The life history of the bluegill sunfish (*Helio-
      perca incisor*) (Cuvier and Valenciennes) as determined from the
      scales. Proc. La. Acad. Sci. 2:139-145.

M32   ——. 1939. A method for the production of largemouth bass on na-
      tural foods in fertilized ponds. Prog. Fish-Cult. 47:1-19.

M33   ——. 1943. Gain in weight per day as a measure of production in
      fish rearing ponds. Trans. Am. Fish. Soc. 72:220-230.

M75   Moffett, J. W., 1942. A fishery survey of the Colorado River below
      Boulder Dam. Calif. Fish Game 28(2):76-86.

M75   Morris, D. M. and A. Hale, 1942. Production of largemouth bass in
      ponds, in relation to their chemical characteristics and plankton crops.
      Invest. Ind. Lakes Streams 2:145-159.

M81   Mottley, C. M. and T. K. Chamberlain, 1948. Management of game
      fish of Conchas Reservoir. Prog. Fist-Cult. 10(4):177-186.

M84   Moyle, J. B., J. H. Kuehn, and C. R. Burrows, 1950. Fish population
      and catch data from Minnesota lakes. Trans. Am. Fish. Soc. 78:163-
      75.

M85   Murphy, G. I., 1948. A contribution to the life history of the Sacra-
      mento perch (*Archoplites interruptus*) in Clear Lake, Lake County,
      California. Calif. Fish Game 34(3):93-100.

M87   McCabe, B. C., 1946. Fisheries Report for Lakes of Central Massa-
      chusetts, 1944-1945. Mass. Dep. Conserv. 254 pp.

M92   Miller, L. F. and P. Bryan, 1947. The harvesting of crappie and white
      bass in Wheeler Reservoir, Alabama. J. Tenn. Acad. Sci. 22(1):62-69.

M94   Meehean, O., 1935. Some factors controlling largemouth bass produc-
      tion. Prog. Fish-Cult. 16:1-6.

M96   Meehean, O. L., 1942. Fish populations in five Florida lakes. Trans.
      Am. Fish. Soc. 71:184-194.

M97   McCutchin, T., 1949. Balancing an unbalanced lake. Wis. Conserv.
      Bull. 14(11):3-5.

M99   McClane, A. M., ed., 1951. The Wise Fishermen's Encyclopedia. Wm.
      H. Wise and Co. 1336 pp.

M107 Martin, M., 1952. The Goddard Lake Story. Okla. Game Fish News
      8(6):3-4.

M111 Miller, L. F., 1951. Fish harvesting on two mainstream reservoirs.
      Trans. Am. Fish. Soc. 80:2-10.

M115 Missouri Conservationist, 1939. Scale study reveals rate of growth
      of Lake of Ozarks fish. Mo. Conserv. 2(1):6.

M117 Missouri River Basin Studies, 1950. Reservoir fishery investigations
      for summer, 1949, Fort Pick Reservoir, Montana. U.S. Fish. Wildl.
      Serv. Billings, Mont. 26 pp. mimeo.

M130 Moen, T. E., 1950. Notes on the growth of Lost Island Lake bullheads.
      Biol. Sem. Iowa State Conserv. Comm. Oct. 1950, 11 pp. mimeo.

M131 ——. 1950. Progress Fisheries Report. Biol. Sem. Iowa State Con-
      serv. Comm. Jan. 1950. Pp. 28-32.

M134 Morgan, G. D., 1951. The life history of the bluegill sunfish, *Lepomis
      macrochirus,* of Buckeye Lake (Ohio). Dennison Univ. Bull., J. Sci.
      Lab. 42(4):21-59.

CITATIONS   379

M136   Moyle, J. B., 1952. Age and growth of fishes. Conservation Volunteer 15(86):14-17.

M140   Murphy, G. I., 1951. The fishery of Clear Lake, Lake County, California. Calif. Fish Game 37(4):439-484.

M146   McCabe, B. C., 1953. Fisheries report for lakes and ponds of north Central Massachusetts. (1950). Mass. Div. Fish Game. 122 pp.

M148   McCammon, G. W., 1957. Further observations on the food of fingerling largemouth bass (*Micropterus salmoides*). Calif. Fish Game, Inland Fish. Adm. Rep. 57-7:5-14.

M172   Mansueti, R. and H. J. Elser, 1953. Ecology, age and growth of the mud sunfish, *Acantharchus pomotis*, in Maryland. Copeia, 1953 (2): 117-118.

M174   Marcy, D. E., 1954. The food and growth of the white crappie, *Pomoxis annularis* in Pymatuning Lake, Pennsylvania and Ohio. Copeia, 1954 (3):236-239.

M178   Martin, R. G., 1957. Influence of fishing pressure on bass fishing success. Proc. S.E. Assoc. Game Fish Comm. 11:76-82.

M182   Maryland Dept. of Research and Education, 1955. Annual report, 1954. Md. Educ. Ser. 39:32 pp.

M187   May, O. D., Jr., (n.d., 1956). An evaluation of two years commercial trapping in the Oconee, Ocmulgee and Altmaha Rivers of Georgia. Ga. Fish Game Comm. D-J. Proj. F-4-R-2, 16 pp.

M191   Mayhew, J. 1956. The bluegill, *Lepomis macrochirus* (Rafinesque) in West Okoboji Lake, Lake, Iowa. Proc. Iowa Acad. Sci. 63:705-713.

M195   ———. 1957. Population studies—fish population of a southern Iowa artificial lake. Iowa Conserv. Comm. Q. Biol. Rep. 9(1):1-5.

M197   ———. 1958. The fish population of a southern Iowa artificial lake. Proc. Iowa Acad. Sci. 65:565-570.

M198   ———. 1958. The eradication of the fish population from Lake Keomah. Iowa Conserv. Comm. Q. Biol. Rep. 10(3):18-23.

M200   ———. 1959. A preliminary report on the increased growth of bluegills in a southern Iowa artificial lake following reduction in population density. Iowa Conserv. Comm. Q. Biol. Rep. 11(1):37-40.

M201   Mayhew, J. and D. Stufflebeam, 1958. The eradication of a fish population in a southern Iowa artificial lake. Iowa Conserv. Comm. Q. Biol. Rep. 10(2):1-4.

M215   Minckley, W. L., 1959. Fishes of the Big Blue River Basin, Kansas. U. Kansas Publ., Mus. Nat. Hist. 11(7):401-442.

M218   Missouri River Basin, 1953. Fishery Investigation, Harry Strunk Lake, Nebraska, 1952. U.S. Fish. Wildl. Serv. River Basins Invest. 23 pp. mimeo.

M219   Missouri River Basin, 1953. Fisherman expenditure study, West Gallatin River, Montana, 1949-1950. U.S. Fish. Wildl. Serv. River Basin Invest. 11 pp. mimeo.

M235   Moen, T. and R. E. Cleary, 1957. Preliminary notes on the crappies of Backbone Lake. Iowa Conserv. Comm. Q. Biol. Rep. 9(1):18-20.

M242   Moody, H. L., 1957. A fishery study of Lake Panasoffkee, Florida. Q. J. Fla. Acad. Sci. 20(1):21-88.

M243   Moore, Emmeline, 1922. The primary sources of food of certain food and game, and bait fishes, of Lake George. Pp. 52-63, biological survey of Lake George, New York. N.Y. State Conserv. Comm.

M245   Moorman, R. B., 1953. Fish populations in some Iowa farm ponds in relation to past history and management. Ph.D. thesis, Iowa State Univ. 150 pp. typewritten.

M246 Moorman, R. B., 1957. Reproduction and growth of fishes in Marion County, Iowa, farm ponds. Iowa State Coll. J. Sci. 32(1):71-88.

M247 ———. 1957. Some factors related to success of fish population in Iowa farm ponds. Trans. Am. Fish. Soc. 86:361-370.

M248 Morgan, G. D., 1954. The life history of the white crappie (*Pomoxis annularis*) of Buckeye Lake, Ohio. J. Sci. Lab. Denison Univ. 43, (6,7,8):113-144.

M258 Moyle, J. B. and C. R. Burrows, 1954. Manual of instructions for lake survey. Minn. Bur. Fish; Fish Res. Unit. Spec. Publ. 1:70 pp.

M265 Mraz, D. and E. L. Cooper, 1957. Natural reproduction and survival of carp in small ponds. J. Wildl. Manage. 21(1):66-69.

M267 Mraz, D. and C. W. Threinen, 1957. Angler's harvest, growth rate and population estimate of the largemouth bass of Brown's Lake, Wisconsin. Trans. Am. Fish. Soc. 85:241-256.

M292 McCaig, R. S. and J. W. Mullan, 1960. Growth of eight species of fishes in Quabbin Reservoir Massachusetts, in relation to age of reservoir and introduction of smelt. Trans. Am. Fish. Soc. 89(1):27-31.

M296 McCaig, R. S., J. W. Mullan, and C. O. Dodge, 1960. Five-year report on the development of the fishery of a 25,000-acre domestic water supply reservoir in Massachusetts. Prog. Fish-Cult. 22(1):15-23.

M302 Moen, T., 1960. Length-weight tables for fishes from northwest Iowa lakes. typewritten mss.

M304 Moody, H. L., 1960. Recaptures of adult largemouth bass from the St. Johns River, Florida. Trans. Am. Fish. Soc. 89(3):295-300.

M306 Manning, J. H., 1951. A study of the populations and growth phenomena of centrarchid fishes in a soft-water impoundment of Maryland. M.S. thesis, Univ. Md. 85 pp.

M308 Meehan, O. L., 1932. The structure of scales in relation to the life histories of certain fish from the sloughs on the Upper Mississippi Wildlife and Fish Refuge. M.S. thesis, Univ. Minn. 86 pp.

M310 Marcy, D. E., 1953. The food and growth of some fishes of Pymatuning Lake, Pennsylvania. Ph.D. thesis, Univ. Pittsburgh, 100 pp.

M311 Montgomery, A. B., 1956. Age and growth of the bluegill, *Lepomis m. macrochirus* Rafinesque, in the Niangua Arm of the Lake of the Ozarks. M.S. thesis, Univ. Mo. 76 pp.

M317 McReynolds, H. E., 1960. Muscatatuck River Studies, Final Report. Ind. Dep. Conserv. 111 pp. mimeo.

M320 Mayhew, J., 1960. The eradication of a fish population in a small artificial lake. Iowa State Conserv. Comm. Q. Biol. Rep. 12(2):4-8.

M323 ———. 1960. A comparison of the growth of four species of fish in three different types of Iowa artificial lakes. Iowa State Conserv. Comm. Q. Biol. Rep. 12(1):58-72.

M336 Maupin, J. K., J. R. Wells, Jr., and C. Leist, 1954. A preliminary survey of food habits of the fish and physico-chemical conditions of three strip-mine lakes. Trans. Kansas Acad. Sci. 57(2):164-171.

M340 McDonald, D. B. and P. A. Dotson, 1960. Fishery investigations of the Glen Canyon and Flaming Gorge impoundment areas. Utah State Dep. Fish. Game Info. Bull. 60-63. 70 pp.

M342 Miller, J. and K. Buss, 1961. The age and growth of the pumpkinseed in Pennsylvania. Penn. Angler Feb. 1961.

M343 ———. 1960. The age and growth of the bluegill in Pennsylvania. Penn. Angler Dec. 1960.

M344 ———. 1960. The age and growth of the largemouth bass in Pennsylvania. Penn. Angler Oct. 1960.

M351   Moss, D. D. and D. C. Scott, 1961. Dissolved-oxygen requirements of
       three species of fish. Trans. Am. Fish. Soc. 90(4):377-393.

M355   Moody, H. L., 1961. Exploited fish populations of the St. Johns River,
       Florida. Q. J. Fla. Acad. Sci. 24(1):1-18.

M359   Maloney, J. E., D. R. Schupp, and W. J. Scidmore, 1962. Largemouth
       bass population and harvest, Gladstone Lake, Crow Wing County, Minne-
       sota. Trans. Am. Fish. Soc. 91(1):42-52.

M368   Messer, J., 1961. Tennessee River drainage reservoirs. N.C. Wildl.
       Res. Comm. Fed. Aid Proj. F5R Job. Compl. Rep. No. 1:233-297.

M372   Mayhew, J., 1960. The fish population in a southern Iowa water supply
       reservoir. Iowa Conserv. Comm., Q. Biol. Rep. 12(3):42-46.

M374   MacPhee, C., 1961. An experimental study of competition for food in
       fish. Ecology 42(4):666-681.

M376   Mayhew, J., 1962. Thermal stratification and its effects on fish and
       fishing in Red Haw Lake, Iowa. Iowa State Conserv. Comm. Biol. Sec.
       23 pp. mimeo.

M399   McConnell, W. J., 1963. Primary productivity and fish harvest in a
       small desert impoundment. Trans. Am. Fish. Soc. 92(1):1-12.

M404   Minckley, W. L., 1963. The ecology of a spring stream, Doe Run,
       Meade County, Kentucky. Wildl. Monog. 11:1-124.

M418   Minckley, W. L. and F. B. Cross, 1959. Distribution, habitat and
       abundance of the Topeka shiner, Notropis topeka (Gilbert), in Kansas.
       Am. Midl. Nat. 61(1):210-217.

M420   Moore, G. A., 1957. Fishes, pp. 31-210, Blair, et al., Vertebrates of
       the United States. McGraw-Hill Book Co.

M422   McConnell, W. J. and J. H. Gerdes, 1964. Threadfin shad, Dorosoma
       petenense, as food of yearling centrarchids. Calif. Fish Game. 50(3):
       170-175.

M427   Mraz, D., 1957. Largemouth bass. Data for Handb. Biol. Data 4 pp.
M428   Morgan, G. D., 1958. Bluegill, Lepomis macrochirus macrochirus
       Rafinesque. Data for Handb. Biol. Data 11 pp.

M448   Mayhew, J., 1965. Pre-impoundment studies of the Chariton River in
       the vicinity of Rathbun dam and reservoir. Iowa Conserv. Comm. Biol.
       Rep. 17(4):4-10.

M499   Muench, B., 1963. Length-weight relationship of eighteen species of
       fish in northeastern Illinois. Typed Mss. NE Area Fish. Hdqtrs.
       Marengo, Ill.

M503   Mayhew, J., 1964. Coralville Reservoir fisheries investigations, 1963.
       Part II: Limnology and fish populations. Iowa Conserv. Comm. Q. Biol.
       Rep. 16(1):25-31.

M517   Martin, R. G. and R. S. Campbell, 1953. The small fishes of Black
       River and Clearwater Lake, Missouri. Univ. Mo. Studies 26(2):45-66.

M520   McCarraher, D. B. and R. W. Gregory, 1970. Adaptability and current
       status of introductions of Sacramento perch, Archoplites interruptus,
       in North America. Trans. Am. Fish. Soc. 99(4):700-707.

M521   McCarraher, D. B., 1963. Sacramento perch, Archoplites interruptus,
       studies in Nebraska for 1961-2 with notes on introduction potential for
       North American waters. Nebr. Game, Forest Parks Comm. 29 pp.
       mimeo.

M522   Mathews, S. B., 1965. Reproductive behavior of the Sacramento perch,
       Archoplites interruptus. Copeia, 1965(2):224-228.

M523   Mullan, J. W. and R. L. Applegate, 1970. Food habits of five cen-
       trarchids during filling of Beaver Reservoir, 1965-66. U.S. Bur. Sport
       Fish Wildl. Tech. Paper 50:16 pp.

M524 Mullan, J. W. and R. L. Applegate, 1968. Centrarchid food habits in a new and old reservoir during and following bass spawning. Proc. S.E. Assoc. Game Fish Comm. 21:332-342.

M525 Meyer, F. A., 1970. Development of some larval centrarchids. Prog. Fish-Cult. 32(3):130-136.

M526 McKechnie, R. J. and R. C. Tharratt, 1966. Green sunfish. Pp. 399-401, Inland Fisheries Management, A. Calhoun, ed. Calif. Dep. Fish Game.

M527 Merriner, J. V., 1969. Constant-bath malachite green solution for incubating sunfish eggs. Prog. Fish-Cult. 31(4):223-225.

M528 Miller, H. C., 1963. The behavior of the pumpkinseed sunfish, *Lepomis gibbosus* (L.), with notes on the behavior of other species of Lepomis and the pigmy sunfish, *Elassoma evergladei*. Behaviour 22:88-151.

M529 Merriner, J. V., 1971. Egg size as a factor in intergeneric hybrid success of centrarchid. Trans. Am. Fish. Soc. 100(1):29-32.

M530 Mayhew, J., 1967. Comparative growth of four species of fish in three different types of Iowa artificial lakes. Proc. Iowa Acad. Sci. 72: 224-229.

M531 Merna, J. W., 1964. The effect of raising the water level on the productivity of a marl lake. Pap. Mich. Acad. Sci. Arts, Lett. 44:217-227.

M532 Mayhew, J., 1963. Further studies of the effects of thermal stratification on the growth of bluegill. Iowa Conserv. Comm. Q. Biol. Rep. 15 (3):28-31.

M533 McCraren, J. P., O. B. Cope, and L. L. Eller, 1969. Some chronic effects of duiron on bluegills. Weed Sci. 17(4):497-504.

M534 Miller, R. J., 1967. Nest building and breeding activities of some Oklahoma fishes. Southwest Nat. 12(4):463-468.

M535 Morgan, G. D., 1960. A study of the effects of fertilizers on vegetation growth, plankton population and numbers, and pounds of bass harvested in eight one-acre ponds. J. Sci. Lab. Denison Univ. 45(2):3-17.

M536 ———. 1958. A study of six different pond stocking ratios of largemouth bass, *Micropterus salmoides* (Lacepede), and bluegill, *Lepomis macrochirus* Rafinesque; and the relation of the chemical, physical, and biological data to pond balance and productivity. J. Sci. Lab. Denison Univ. 44:151-202.

M537 Montgomery, M., 1967. Problem fish. Ore. State Game Comm. Bull. 22(5):3, 6-7.

M538 McComish, T. S., 1968. Sexual differentiation of bluegills by the urogenital opening. Prog. Fish-Cult. 30(1):28.

M539 Muncy, R. J., 1966. Aging and growth of largemouth bass, bluegill and redear sunfish from Louisiana ponds of known stocking history. Proc. S.E. Assoc. Game Fish Comm. 19:343-349.

M540 Moffett, J. W. and B. P. Hunt, 1943. Winter feeding habits of bluegills (*Lepomis macrochirus*) and yellow perch (*Perca flavescens*) in Cedar Lake, Washtenaw County, Michigan. Trans. Am. Fish. Soc. 73:231-242.

M541 Morgan, G. D., 1951. A comparative study of the spawning periods of the bluegill, *Lepomis macrochirus*, the black crappie, *Pomoxis nigromaculatus* (Le Seuer), and the white crappie, *Pomoxis annularis* (Rafinesque), of Buckeye Lake, Ohio. J. Sci. Lab. Dennison Univ. 42:112-118.

M542 Moore, W. G., 1941. Studies on the feeding habits of fishes. Ecology 22(1):91-96.

M543 Moore, W. G., 1942. Field studies on the oxygen requirements of certain fresh-water fishes. Ecology 23(3):319-329.

M544 Merriner, J. V., 1971. Development of intergeneric centrarchid hybrid embryos. Trans. Am. Fish. Soc. 100(4):611-8.

M545 Meldrim, J. W. and J. J. Gift, 1971. Temperature preference, avoidance and shock experiments with estuarine fishes. Ichthyological Assoc. Bull. 7. 75 pp.

M546 McCarraher, D. B., M. L. Madsen, and R. E. Thomas, 1971. Ecology and fishery management of McConaughy Reservoir, Nebraska. Am. Fish. Soc. Spec. Publ. 8:299-311.

M547 Marshall, T. L. and R. P. Johnson, 1971. History and results of fish introductions in Saskatchewan 1900-1969. Sask. Dep. Nat. Resource Fish. Rep. 8. 29 pp.

M548 Meehan, W. E., 1911. Observations on the small-mouth black bass in Pennsylvania during the spawning season of 1910. Trans. Am. Fish. Soc. 40:129-132.

M549 Munther, G. L., 1970. Movement and distribution of smallmouth bass in the Middle Snake River. Trans. Am. Fish. Soc. 99(1):44-53.

M550 McKechnie, R. J., 1966. Spotted bass. Pp. 366-70, Inland Fisheries Management, A. Calhoun, ed. Calif. Dep. of Fish Game.

M551 Mraz, D., S. Kmiotek, and L. Frankenberger, 1961. The largemouth bass: its life history, ecology and management. Wis. Conserv. Dep. Publ. 232. 13 pp.

M552 McCarraher, D. B., 1971. Survival of some freshwater fishes in the alkaline eutrophic waters of Nebraska. J. Fish. Res. Bd. Canada 28(11):1811-14.

M553 McDowall, R. M., 1968. The proposed introduction of the largemouth black bass, *Micropterus salmoides* (Lacepede) into New Zealand. N. Z. J. Marine Freshwater Res. 2(2):149-61.

M554 Miller, K. D. and R. H. Kramer, 1971. Spawning and early life history of largemouth bass (*Micropterus salmoides*) in Lake Powell. Am. Fish. Soc. Spec. Publ. 8:73-83.

M555 Mraz, D., 1964. Evaluation of liberalized regulations on largemouth bass, Brown's Lake, Wisconsin. Wis. Conserv. Dep., Tech. Bull. 31: 24 pp.

M556 Mitzner, L., 1972. Some vital statistics of the crappie population in Coralville Reservoir with an evaluation of management. Iowa Fish. Res. Tech. Ser. 72-1:35 pp.

M557 Markus, H. C., 1932. The extent to which temperature changes influence food consumption in largemouth bass. Trans. Am. Fish. Soc. 62:202-10.

M558 McCammon, G. W., D. LaFaunce, and C. M. Seeley, 1964. Observations on the food of fingerling largemouth bass in Clear Lake, Lake County, California. Calif. Fish Game 50(3):158-169.

M559 McLane, W. M., 1949. Notes on the food of the largemouth black bass, *Micropterus salmoides floridanus* (LeSueur), in a Florida lake. Q. J. Fla. Acad. Sci. 12:195-201.

M560 Mathur, D. and T. W. Robbins, 1971. Food habits and feeding chronology of young white crappie, *Pomoxis annularis* Rafinesque, in Conowingo Reservoir. Trans. Am. Fish. Soc. 100(2):307-311.

M561 Mayhew, J., 1974. 0-age fish production at Lake Rathbun. Iowa Conserv. Comm. Fish. Sec. Fed. Aid Proj. F-88-R-1 Study 701-3. 1 July 1973 - 30 June 1974. 83 pp. mimeo.

M562   Mathur, D., 1972. Seasonal food habits of adult white crappie, *Pomoxis annularis* Rafinesque, in Conowingo Reservoir. Am. Midl. Nat. 87(1): 236-241.

M563   Mitchell, G. C., 1945. Food habit analysis of the two species of Texas crappie. M.S. thesis, N. Tex. State Univ. 44 pp. (not seen).

M564   McCann, J. A. and K. D. Carlander, 1970. Mark and recovery estimates of fish populations in Clear Lake, Iowa, 1958 and 1959. Iowa State J. Sci. 44(3):369-403.

M565   Meyer, F. P., K. E. Sneed, and P. T. Eschmeyer, 1973. Second report to the fish farmers. U.S. Bur. Sport Fish Wildl. Resource Publ. 113. 123 pp.

M566   Moyle, P. B. and R. D. Nickols, 1973. Ecology of some native and introduced fishes of the Sierra Nevada foothills in central California. Copeia, 1973(3):478-90.

M567   Mitzner, L., 1974. Life history and dynamics of largemouth bass in man-made lakes. Iowa Conserv. Comm. Fed. Aid Proj. F-88-R-1, Study No. 503-1:26 pp.

M568   Morgan, W. S. G. and P. C. Kuhn, 1974. A method to monitor the effects of toxicants upon breathing rate of largemouth bass (*Micropterus salmoides* Lacepede). Water Res. 8:67-77.

M569   Ming, A., 1974. Regulation of largemouth bass harvest with a quota. Symposium on overharvest of largemouth bass in small impoundments. N.C. Div. Am. Fish. Soc. Spec. Publ. 3:39-53.

M570   May, B. E., 1973. Evaluation of large scale release programs with special reference to bass fishing tournaments. Proc. S.E. Assoc. Game Fish Comm. 26:325-335.

M571   Moyle, P. B., S. B. Mathews, and N. Bonderson, 1974. Feeding habits of the Sacramento perch, *Archoplites interruptus*. Trans. Am. Fish. Soc. 103(2):399-402.

M572   Moody, H. L., 1975. Tournament catch of largemouth bass from St. Johns River, Florida. Proc. S.E. Assoc. Game Fish Comm. 28:73-82.

M573   McNeeley, D. L. and W. D. Pearson, 1974. Distribution and condition of fishes in a small reservoir receiving heated waters. Trans. Am. Fish. Soc. 103(3):518-530.

M574   McComish, T. S., R. O. Anderson, and F. G. Goff, 1974. Estimation of bluegill (*Lepomis macrochirus*) proximate composition with regression models. J. Fish. Res. Bd. Canada 31(7):1250-1254.

M575   Mitzner, L., 1974. Effects of cage reared and released channel catfish on established fish and benthic fish food populations. Iowa Conserv. Comm. Fish Sec. Fed. Aid Proj. F-88-R-1. Study 501-2, 29 pp.

M576   Martin, N. V. and F. E. J. Fry, 1973. Lake Opeongo. The ecology of the fish community and of man's effects on it. Great Lakes Fish. Comm. Tech. Rep. 24:34 pp.

N25   Needham, J. G., 1920. Clean waters for New York State. Cornell Rural School Leaflet 13:153-182.

N37   Nelson, M. N. and A. D. Hasler, 1942. The growth, food distribution and relative abundance of the fishes of Lake Geneva, Wisconsin, in 1941. Trans. Wis. Acad. Sci. Arts Lett. 34:137-148.

N50   New Jersey Fisheries Survey, 1950. Lakes and ponds. N.J. Dep. Conserv. Rep. 1:1-189.

N52   Noland, W. E., 1951. The hydrography, fish, and turtle population of Lake Wingra. Trans. Wis. Acad. Sci. Arts Lett. 40(2):5-58.

N93   Nelson, W. R., 1961. Report of fisheries investigations during the eighth year of impoundment of Fort Randall Reservoir, South Dakota, 1960. S.D. Dep. Game, Fish, Parks. 30 pp. mimeo.

N94   ——. 1961. Report of fisheries investigations during the sixth year of impoundment of Gavins Point Reservoir, South Dakota, 1960. S.D. Dep. Game, Fish, Parks, D. J. F-1-R-10 Jobs 2,3,4.

N95   Newell, A. E., 1960. Biological survey of the lakes and ponds in Coos, Grafton and Carroll Counties. N.H. Fish Game Dep., Surv. Rep. 8a: 297 pp.

N99   Neal, R. A., 1962. White and black crappies in Clear Lake, summer, 1960. Proc. Iowa Acad. Sci. 68:247-253.

N100   Nicholson, H. P., et al., 1962. Insecticide contamination in a farm pond. Trans. Am. Fish. Soc. 91(2):213-222.

N115   Nelson, W. R., 1962. Report of fisheries investigations during the seventh year of impoundment of Gavins Point Reservoir, South Dakota, 1961. S.D. D.J. Proj. F-1-R-11 Jobs 1,2,3, and 7. 40 pp.

N117   Nurnberger, P. K., 1930. The plant and animal food of the fishes of Big Sandy Lake. Trans. Am. Fish. Soc. 60:253-259.

N133   Nikol'skii, G. V., 1961. Special ichthyology. Transl. from Russian, Nat. Sci. Found., Smithsonian Inst. 538 pp.

N135   Nelson, D. J., J. W. Gooch, N. A. Griffith, and S. A. Rucker, 1971. White Oak Lake studies. Oak Ridge Nat. Lab. Rep. 4634:104-6.

N136   Nelson, W. R., R. E. Siefert, and D. V. Swedberg, 1967. Studies of the early life history of reservoir fishes. Reservoir Fishery Resources Symposium, Am. Fish. Soc. pp. 374-385.

N138   Neal, R. A., 1963. Black and white crappies in Clear Lake, 1950-1961. Iowa State J. Sci. 37(4):425-45.

N139   Nelson, J. T., R. W. Bowker, and J. D. Robinson, 1974. Rearing pellet-fed largemouth bass in a raceway. Prog. Fish-Cult. 36(2):108-110.

N140   Nelson, D. J., 1974. Part II. Aquatic studies program. Pp. 24-33, Oak Ridge National Lab. Environmental Sciences Division, Ann. Prog. Rep. Sept. 30, 1973.

N141   Neill, W. H. and J. J. Magnuson, 1974. Distributional ecology and behavioral thermoregulation of fishes in relation to heated effluent from a power plant at Lake Monona, Wisconsin. Trans. Am. Fish. Soc. 103 (4):663-710.

N142   Neale, G., 1931. Sacramento perch. Calif. Fish, Game 17(4):409-411.

N143   Norris, J. S., D. O. Norris, and J. T. Windell, 1973. Effect of simulated meal size on gastric acid and pepsin secretory rates in bluegill (*Lepomis macrochirus*). J. Fish. Res. Bd. Canada 30(2):201-204.

N144   Nelson, W. R., 1974. Age, growth and maturity of thirteen species of fish from Lake Oahe during the early years of impoundment, 1963-1968. U.S. Fish, Wildl. Serv. Tech. Pap. 77:1-29.

N145   Neves, R. J., 1975. Factors affecting fry production of smallmouth bass (*Micropterus dolomieu*) in South Branch Lake, Maine. Trans. Am. Fish. Soc. 104(1):83-87.

O7   Ohio Bureau of Scientific Research, 1934. Length-weight relationship of several Ohio food and game fishes. Bull. 70. 2 pp. mimeo.

O8   Ohio Bureau of Scientific Research, Age and rate of growth of several game fishes in the inland fishing district of Ohio. Bull. 9. 1 p. mimeo.

O9   Ohio Conservation Bulletin, 1939. Sandusky Bay fishing survey, 1938. Ohio Conserv. Bull. 3(4):4-5.

O24   Oregon State Game Commission, 1952. Annual report, Ore. Fishery Division. 1951. 238 pp.

O33   ———. 1953. Annual report, Ore. Fishery Division. 1952. 308 pp.

O34   ———. 1954. Annual report, Ore. Fishery Division. 239 pp.

O36   Oakley, A. L., 1956. Farm fish pond production as influenced by climatic conditions. M.S. thesis, Ore. State Coll. 89 pp.

O37   Orr, O. E., 1958. The populations of fishes and limnological conditions of Heyburn Reservoir with reference to productivity. Ph.D. thesis, Okla. State Univ. 68 pp.

O49   O'Hara, J., 1968. The influence of weight and temperature on the metabolic rate of sunfish. Ecology 49(1):159-161.

O50   O'Rear, R. S., 1970. A growth study of redbreast, *Lepomis auritus* (Gunther), and bluegill, *Lepomis macrochirus* (Rafinesque), populations in a thermally influenced lake. Proc. S.E. Assoc. Game Fish Comm. 23:545-553.

O51   Oseid, D. and L. L. Smith, Jr., 1972. Swimming endurance and resistance to copper and malathion of bluegills treated by long term exposure to sublethal levels of hydrogen sulfide. Trans. Am. Fish. Soc. 101(4):620-625.

O52   Olmsted, L. L. and D. G. Cloutman, 1974. Reproduction after a fish kill in Mud Creek, Washington County, Arkansas, following pesticide pollution. Trans. Am. Fish. Soc. 103(1):79-87.

P7   Pearse, A. S., 1919. Habits of the black crappie in inland lakes of Wisconsin. Rep. U.S. Comm. Fish. 1918. Append. 3:1-16.

P8   ———. 1925. The chemical composition of certain freshwater fishes. Ecology 6(1):7-16.

P12   Potter, G. S., 1925. Scales of the bluegill (*Lepomis pallidus* Mitchill). Trans. Am. Micro. Soc. 44(1):31-37.

P17   Prevost, G., 1944. First Report of the Biological Bureau, Province of Quebec (1943). Pp. 44-69.

P18   Price, J. W., 1931. Growth and gill development of the small-mouthed black bass. Franz Theodore Stone Lab. Contrib. No. 4:1-46.

P27   Padfield, J. H., Jr., 1951. Age and growth differentiation between the sexes of the largemouth black bass, *Micropterus salmoides* (Lacepede). J. Tenn. Acad. Sci. 26(1):42-54.

P28   Parsons, J. W., 1952. Growth rates and notes on the redeyed bass, *Micropterus coosae*, Hubbs and Bailey. Proc. S.E. Assoc. Game Fish Comm. 6th, 15 pp. mimeo.

P30   Patriarche, M., 1951. Reservoir research required. Mo. Conserv. 12(12):1-3, 14-15.

P34   Pearce, E. W., 1951. Big largemouth bass from Mbuto Dam, Transkeian Territories. Piscator 5(11):8.

P39   Piscator, 1950. Secretarial notes. Piscator 4(13):14-15; (15):67-68; (16):98-99.

P41   ———. 1951. Secretarial notes. Piscator 5(17):17-19.

P42   ———. 1951. Big fish records and notable fish. Piscator 5(17):3-5; (19):72-74.

P53   ———. 1952. Big fish records and notable fish. Piscator 6(21):7.

P54   ———. 1952. Letters to the editor, and Liesbeek Notes, and Lower Eerste River Notes. Piscator 6(22):37, 38, 42, 55; 6(21):4-7; 6(23):68-70.

P66   Parsons, J. W., 1954. The fish species composition and chemical and physical features of a seventy-two acre reservoir in Tennessee. J. Tenn. Acad. Sci. 29(1):55-65.

P67   ———. 1954. Growth and habits of the redeye bass. Trans. Am. Fish Soc. 83:202-211.

P70   Parsons, J. W. and E. Crittenden, 1959. Growth of the redeye bass in Chipola River, Florida. Trans. Am. Fish. Soc. 88(3):191-192.

P71   Patriarche, M. H. and R. S. Campbell, 1958. The development of the fish population in a new flood-control reservoir in Missouri, 1948 to 1954. Trans. Am. Fish. Soc. 87:240-258.

P72   Pasko, D. G., 1957. Carry Falls Reservoir Investigation. N.Y. Fish, Game J. 4(1):1-30.

P73   Patriarche, M. H., 1953. The fishery in Lake Wappapello, a flood-control reservoir on the St. Francis River, Missouri. Trans. Am. Fish. Soc. 82:242-254.

P74   Patriarche, M. H. and E. M. Lowry, 1953. Age and rate of growth of five species of fish in Black River, Missouri. Univ. Mo. Studies 26(2):26-109.

P87   Phillips, R. W., 1956. Lower Willamette. Ore. State Game Comm. Fish Div. Ann. Rep. 1955:62-72.

P91   Pintler, H. E., 1957. A summary of the 1955 Clear Lake Fishery Lake County, California. Calif. Dep. Fish, Game, Inland Fish. Admin. Rep. 57-27:1-14.

P92   Piscator, 1952. (Various notes). Piscator 6(24):98-100, 103.

P93   ———. Secretarial notes. Piscator 7(25):30-32.

P95   ———. 1953. Various notes. Piscator 28:99, 100, 103, 107, 125.

P96   ———. 1954. Various notes. Piscator 29:4, 8, 32.

P98   ———. 1954. Various notes. Piscator 30:36-55; 31:71-77.

P99   ———. 1955-61. Various notes. Piscator 33-50.

P112 Purkett, C. A., Jr., 1958. Growth of the fishes in the Salt River, Missouri. Trans. Am. Fish. Soc. 87:116-131.

P113 ———. 1958. Growth rates of Missouri stream fishes. Mo. D.J. Ser. 1:46 pp.

P122 Parker, R. A., 1958. Some effects of thinning on a population of fishes. Ecology 39(2):304-317.

P124 Proffitt, M. A., 1950. The comparative morphometry and growth of scales in the bluegill, Lepomis m. macrochirus Rafinesque, with special reference to related body growth. Ph.D. thesis, Univ. Mich. 96 pp.

P159 Montana Fish and Game Dept. Fisheries Division, John C. Peters, ed. 1964. (Summary of calculated growth data on Montana fishes 1948-61). Job Compl. Rep. F-23-R-6 Job I and II Mont. Fish. Game Dep. 76 pp. mimeo.

P165 Pearse, A. S., 1918. The food of the shore fishes of certain Wisconsin lakes. Bull. U.S. Bur. Fish 35:249-292.

P189 Pierce, P. C., J. E. Frey, and H. M. Yawn, 1965. An evaluation of fishery management techniques utilizing winter drawdowns. Proc. S.E. Assoc. Game Fish Comm. 17:347-363.

P191 Perlmutter, A., E. E. Schmidt, and E. Leff, 1967. Distribution and abundance of fish along the shores of the lower Hudson River during the summer of 1965. N.Y. Fish, Game J. 14(1):47-75.

P192 Papadopol, M. and Gh. Ignat, 1967. Contributii la studiul biologiei reproducerii si cresterii bibanului-soare (Lepomis gibbosus (L.)), din Dunarea Inferioara (Zona Inundabila). Bull. Inst. Cercetari Pisc. 26(4):55-68.

P193   Patriarche, M. H., 1968. Production and theoretical equilibrium yields for the bluegill (*Lepomis macrochirus*) in two Michigan lakes. Trans. Am. Fish. Soc. 97(3):242-251.

P194   Prather, E. E., 1967. A note on the accuracy of the scale method in determining the ages of largemouth bass and bluegill from Alabama waters. Proc. S.E. Assoc. Game Fish Comm. 20:483-486.

P195   Pflieger, W. L., 1966. Reproduction of the small mouth bass (*Micropterus dolomieui*) in small Ozark Stream. Am. Midl. Nat. 76(2):410-419.

P196   Parsons, J. W. and N. G. Benson, 1960. Fertilization of Obed River, Tennessee. J. Tenn. Acad. Sci. 35(1):63-76.

P197   Peek, F., 1966. Age and growth of the smallmouth bass, *Micropterus dolomieu* Lacepede, in Arkansas. Proc. S.E. Assoc. Game Fish. Comm. 19:422-431.

P198   Pardue, G. B. and F. E. Hester, 1967. Variation in the growth rate of known-age largemouth bass (*Micropterus salmoides* Lacepede) under experimental conditions. Proc. S.E. Assoc. Game Fish Comm. 20:300-310.

P199   Pageau, G., 1967. Comportement, alimentation et croissance de l'achigan a petite bouche (*Micropterus dolomieui* Lacepede) dans la Plaine de Montreal et dans les Laurentides. Ph.D. thesis, Univ. Montreal. (Abstract only seen.)

P200   Parker, W. D., 1971. Preliminary studies on sexing adult largemouth bass by means of an external characteristic. Prog. Fish-Cult. 33(1):54-55.

P201   Parker, F. R., Jr., 1973. Reduced metabolic rates in fishes as a result of induced schooling. Trans. Am. Fish. Soc. 102(1):125-131.

P202   Petit, G. D., 1973. Effects of dissolved oxygen on survival and behavior of selected fishes of western Lake Erie. Ohio Biol. Surv. Bull. 4 (4):1-76.

P203   Paragamian, V. L., 1974. Evaluation of the 14-inch size limit on largemouth bass at Big Creek Reservoir. Ia. Conserv. Comm. Fed. Aid F-88-R-1, Study No. 601-1:28 pp.

P204   ———. 1974. Vital statistics of fish populations in Red Rock Reservoir and Rathbun Reservoir following initial impoundment. Iowa Conserv. Comm. Fed. Aid F-88-R-1, Study No. 702-703, Job 1:1-46.

P205   Parks, C. E., 1949. The summer food of some game fishes of Wisiona Lake. Invest. Ind. Lakes, Streams 3(4):235-245.

P206   Pessah, E. and P. M. Powles, 1974. Effect of constant temperature on growth rates of pumpkinseed sunfish (*Lepomis gibbosus*). J. Fish. Res. Bd. Canada 31:1678-1682.

P207   Powell, T. G., 1972. Northern pike introductions. Colo. Fish. Res. Rev. 7:25-58.

P208   Proffitt, M. A. and R. S. Benda, 1971. Growth and movement of fishes, and distribution of invertebrates, related to a heated discharge into the White River at Petersburg, Indiana. Indiana Univ. Water Resources Rep. Inves. 5:94 pp.

P209   Pardue, G. B., 1973. Production response of bluegill sunfish, *Lepomis macrochirus* Rafinesque, to added attachment surface for fish-food organisms. Trans. Am. Fish. Soc. 102(3):622-666.

P210   Pierce, R. J. and T. E. Wissing, 1974. Energy cost of food utilization in the bluegill (*Lepomis macrochirus*). Trans. Am. Fish. Soc. 103(1):38-45.

P211   Petrosky, B. R. and J. J. Magnuson, 1973. Behavioral responses of northern pike, yellow perch and bluegill to oxygen concentrations under simulated winterkill conditions. Copeia, 1973(1):124-133.

P212   Petty, L. L. and J. J. Magnuson, 1974. Lymphocystis in age 0 bluegills (*Lepomis macrochirus*) relative to heated effluents in Lake Monona, Wisconsin. J. Fish. Res. Bd. Canada 31(7):1189-1193.

P213   Paragamian, V. L. and D. W. Coble, 1975. Vital statistics of smallmouth bass in two Wisconsin rivers, and other waters. J. Wildl. Manage. 39(1):201-210.

P214   Powell, T. G., 1973. Effect of northern pike introductions on an overabundant crappie population. Colo. Div. Wildl. Spec. Rep. 31:6 pp.

Q5   Qualls, C., 1965. Comparison of day and night catches of fishes by seining from Canton Reservoir, June, 1965. Okla. Fish. Res. Lab. Semiann. Rep. Jan.-June 1965:111-119.

R15   Rawson, D. S., 1932. The pike of Waskesiu Lake, Saskatchewan. Trans. Am. Fish. Soc. 62:323-330.

R16   ———. 1938. Natural rearing enclosures for smallmouth black bass. Trans. Am. Fish, Soc. 67:96-104.

R22   Reid, H., 1930. A study of *Eupomotis gibbosus* (L.) as occurring in the Chamcook Lakes, N. B. Contri. Canad. Biol. Fish. NS 5(16):459-466.

R24   Reighard, J., 1915. An ecological reconnaissance of the fishes of Douglas Lake, Cheboygan County, Michigan, in midsummer. Bull. U.S. Comm. Fish 33:215-249.

R25   Reighard, J., 1929. A biological examination of Loon Lake, Gogebic County, Michigan, with suggestions for increasing its yield of smallmouth bass (*Micropterus dolomieu*). Pap. Mich. Acad. Sci. Arts Lett. 10:589-612.

R26   Richardson, R. E., 1913. Observations on the breeding of the European carp in the vicinity of Havana, Illinois. Bull. Ill. State Lab. Nat. Hist. 9(7):387-404.

R27   ———. 1913. Observations on the breeding habits of fishes at Havana, Illinois, 1910 and 1911. Bull. Ill. State Lab. Nat. Hist. 9:405-416.

R31   Ricker, W. E., 1942. Fish populations of two artificial lakes. Invest. Ind. Lakes, Streams 2(13):255-265.

R32   ———. 1942. Creel census, population estimates and rate of exploitation of game fish in Shoe Lake, Indiana. Invest. Ind. Lakes, Streams 2:215-253.

R33   ———. 1942. The rate of growth of bluegill sunfish in lakes of northern Indiana. Invest. Ind. Lakes, Streams 2:161-214.

R34   ———. 1945. Abundance, exploitation and mortality of the fishes in two lakes. Invest. Ind. Lakes, Streams 2(17):345-448.

R35   ———. 1945. Natural mortality among Indiana bluegill sunfish. Ecology 26(2):111-121.

R36   ———. 1945. Fish catches in three Indiana lakes. Invest. Ind. Lakes, Streams 2(16):325-344.

R37   ———. 1947. Tri-Lakes test proves fallacy of over fishing. Outdoor Ind. 14(8):4-5.

R38   ———. 1948. Hybrid sunfish for stocking small ponds. Trans. Am. Fish. Soc. 75:84-96.

R39   ———. 1949. Effects of removal of fins upon the growth and survival of spiny-rayed fishes. J. Wildl. Manage. 13(1):29-39.

R40   Ricker, W. E. and K. F. Lagler, 1942. The growth of spiny-rayed fishes in Foots Pond. Invest. Ind. Lakes, Streams 2:85-97.
R41   Ricker, W. E. and D. Merriman, 1945. On the methods of measuring fish. Copeia, 1945(4):184-191.
R42   Roach, L. S., 1941. Growth and food of black bass. Ohio Conserv. Bull. 5(7):20-21.
R44   ———. 1947. The bluegills. Ohio Conserv. Bull. 11(8):13.
R47   ———. 1947. In fishing circles. Ohio Conserv. Bull. 11(11):12-13.
R48   ———. 1947. *Huro salmoides*. Ohio Conserv. Bull. 11(7):13.
R49   ———. 1947. White crappie. Ohio Conserv. Bull. 11(9):13.
R53   ———. 1948. White bass. Ohio Conserv. Bull. 12(8):13.
R54   ———. 1948. Green sunfish. Ohio Conserv. Bull. 12(10):13.
R55   ———. 1948. Golden mullet. Ohio Conserv. Bull. 12(2):13.
R57   ———. 1948. Yellow perch. Ohio Conserv. Bull. 12(9):13.
R58   ———. 1948. In fishing circles. Ohio Conserv. Bull. 12(6):12-13.
R59   ———. 1948. Black crappie. Ohio Conserv. Bull. 12(7):13.
R71   Roszman, F. D., 1939. Home life of Ohio finny favorites. Ohio Conserv. Bull. 3(3):10-11.
R74   Royce, W. F., 1942. Standard length versus total length. Trans. Am. Fish. Soc. 71:270-274.
R77   Roach, L. S., 1946. Analysis and significance of fish populations and harvest in the nine liberalized fishing lakes, 1946. Ohio Div. Conserv. Nat. Res. Leaflet 219, 3 pp.
R78   ———. 1949. Lake management reports. 4. Meander Lake. Ohio Div. Conserv. 37 pp. mimeo.
R79   Roach, L. S. and I. M. Evans, 1947. Growth of game and panfish in Ohio. 1. Bluegills. Ohio Div. Conserv. 26 pp. mimeo.
R80   ———. 1948. Growth of game and panfish in Ohio. 3. Largemouth bass. Ohio Div. Conserv. 18 pp. mimeo.
R81   ———. 1948. Growth of game and panfish in Ohio. 2. Crappies. Ohio Div. Conserv. 29 pp. mimeo.
R82   Roach, L. S. and J. Z. Pelton, 1947. Lake management reports. 1. Lake Alma. Ohio Div. Conserv. 27 pp. mimeo.
R83   ———. 1948. Lake Management Reports 2. Lake Vesuvius. Ohio Div. Conserv. 30 pp. mimeo.
R95   Raver, D. D. and J. H. Cornell, 1951. Comparative growth rates of largemouth bass (*Micropterus salmoides* Lacepede) in North Carolina waters. S.E. Assoc. Game Fish Comm. 5:7 pp. mimeo.
R101   Roach, L., 1950. Pumpkinseed sunfish. Ohio Conserv. Bull. 14(1):13.
R115   Rose, E. T. and T. Moen, 1951. Results of increased fish harvest in Lost Island Lake. Trans. Am. Fish. Soc. 80:50-55.
R116   Roseberry, D. A., 1950. Game fisheries investigation of Clayton Lake, a main stream impoundment of New River, Pulaski County, Virginia, with emphasis on *Micropterus punctulatus* Rafinesque. Ph.D. thesis, Va. Poly. Inst. 268+xxxviii pp.
R117   ———. 1951. Fishery management of Clayton Lake, an impoundment on the New River in Virginia. Trans. Am. Fish. Soc. 80:194-209.
R118   ———. 1952. Back Bay fish and fishing. Va. Wildl. 13(5):4-7.
R119   Roseberry, D. A. and R. R. Bowers, 1952. Under the cover of Lake Drummond. Va. Wildl. 13(8):21-23.
R124   Runnstrom, S., 1950. Director's report for the year 1949. Inst. Freshwater Res., Drottningholm, Rep. 31:5-18.
R125   Ruhr, C. E., 1952. Fish population of a mining pit lake, Marion County, Iowa. Iowa State Coll. J. Sci. 27(3):55-77.

R127 Rawson, D. S., 1952. Mean depth and the fish production of large lakes. Ecology 33(4):513-521.

R147 Regier, H. A., 1959. An evaluation of the scale method for age and growth determination of bluegills in New York farm ponds. M.S. thesis, Cornell Univ. 140 pp.

R152 Ricker, W. E., 1955. Fish and fishing in Spear Lake, Indiana. Invest. Ind. Lakes, Streams 4:117-162.

R166 Rose, E. T., 1954. Iowa lakes creel census. Iowa State Conserv. Comm. Q. Biol. Rep. 6(1):36-40.

R178 Roseberry, D. A., 1954. The largemouth bass, *Micropterus salmoides* (Lacepede) of Back Bay, a Virginia estuary. Midwest Wildl. Conf. 16, 16 pp. mimeo.

R196 Ricker, W. E., 1949. Mortality rates in some little-exploited populations of freshwater fishes. Trans. Am. Fish. Soc. 66:114-128.

R200 Roszman, F. D., 1935. The age, rate of growth and food of five species of game fish in Lake Erie. M.S. thesis, Ohio State Univ. 31 pp.

R201 Robinson, P. E., 1949. Age and rate of growth of the largemouth black bass, *Micropterus salmoides* (Lacepede), of various impoundments in Oklahoma. M.S. thesis, Okla. A. M. College. 23 pp.

R210 Ryer, R. III, 1938. Contributions to the life history of *Notropis cornutus cornutus* (Mitchill). M.S. thesis, Cornell, Univ. 41 pp.

R212 de Ryke, Willis, 1923. Foods of fishes and the relation to fish culture. Proc. Iowa Acad. Sci. 30:163-166.

R217 Richardson, F. and H. M. Ratledge, 1961. Upper Catawba River Reservoirs and Lake Lure. N. C. Wildl. Res. Comm. Fed. Aid Proj. F5R and F6R Job Compl. Rep. 1:161-231.

R218 Robinson, D. J., 1960. Roadside impoundment studies in Fremont and Mills County, Iowa. Iowa Conserv. Comm., Q. Biol. Rep. 12(3):38-41.

R219 Regier, H. A., 1962. Some aspects of the ecology and management of warm-water fish in New York farm ponds. Ph.D. thesis, Cornell Univ. 420 pp.

R231 ———. 1962. Validation of the scale method for estimating age and growth of bluegills. Trans. Am. Fish. Soc. 91(4):362-374.

R237 ———. 1963. Ecology and management of largemouth bass and bluegills in farm ponds in New York. N.Y. Fish, Game J. 19(1):1-89.

R239 ———. 1963. Ecology and management of largemouth bass and golden shiners in farm ponds in New York. N.Y. Fish, Game J. 10(2):139-169.

R280 Rawstrom, R. R., 1967. Harvest, mortality, and movement of selected warm-water fishes in Folsom Lake, California. Calif. Fish., Game 53(1):40-8.

R282 Rounsefell, G. A. and W. H. Everhart, 1953. Fishery Science: Its Methods and Applications. John Wiley & Sons. 444 pp.

R285 Raney, E. C., 1965. Some pan fishes of New York—rock bass, crappies and other sun fishes. N.Y. State Conserv. Dep. Info. Leaflet D-47:10-16.

R286 Roberts, F. L., 1967. Chromosome cytology of the Osteichthyes. Prog. Fish-Cult. 29(2):75-83.

R288 Ranthum, R. G., 1969. Distribution and food habits of several species of fish in Pool 19, Mississippi River. M.S. thesis, Iowa State Univ. 207 pp.

R289 Roland, J. V., 1970. Some effects of the introduction of hard water into Corvin Cove Reservoir, Virginia. M.S. thesis, Va. Polytech. Inst. 103 pp. (Abstract only seen.)

R290 Reed, R. J., 1971. Underwater observations of the population density and behavior of pumpkinseed, *Lepomis gibbosus* (Linnaeus) in Cranberry Pond, Massachusetts. Trans. Am. Fish. Soc. 100(2):350-353.

R291 Ricker, W. E., 1949. Utilization of food by bluegills. Invest. Ind. Lakes, Streams 3(8):313-318.

R292 Range, J. D., 1971. The possible effect of the threadfin shad, *Dorosoma petense* (Gunther), on the growth of five species of game fish in Dale Hollow Reservoir. M.S. thesis, Tenn. Tech. Univ. 52 pp.

R293 Reynolds, J. B., 1965. Life history of smallmouth bass, *Micropterus dolomieui* Lacepede, in the Des Moines River, Boone County, Iowa. Iowa State J. Sci. 39(4):417-436.

R294 ———. 1963. Life history of smallmouth bass, *Micropterus dolomieui* Lacepede, in the Des Moines River, Boone County, Iowa. M.S. thesis, Iowa State Univ. 85 pp.

R295 Rawson, D. S., 1945. The experimental introduction of smallmouth black bass into lakes of the Prince Albert National Park, Saskatchewan. Trans. Am. Fish. Soc. 73:19-31.

R297 Reighard, J. E., 1905. The breeding habits, development and propagation of the black bass. Mich. State Bd. Fish. Comm. Bien. Rep. 16 (Append. pp. 1-73).

R298 Ryan, P. W., J. W. Avault, Jr., and R. O. Smitherman, 1970. Food habits and spawning of the spotted bass in Tchefuncte River, southeastern Louisiana. Prog. Fish-Cult. 32(3):162-167.

R299 Rogers, W. A., 1967. Food habits of young largemouth bass (*Micropterus salmoides*) in hatchery ponds. Proc. S.E. Assoc. Game Fish. Comm. 21:543-553.

R300 Robinson, D. W., 1961. Utilization of spawning box by bass. Prog. Fish-Cult. 23(3):119.

R301 Reid, G. K., Jr., 1950. Food of the black crappie *Pomoxis nigromaculatus* (LeSueur), in Orange Lake, Florida. Trans. Am. Fish. Soc. 79:145-54.

R302 Riggs, C. D. and K. E. Sneed, 1959. The effects of controlled spawning and genetic selection on the fish culture of the future. Trans. Am. Fish. Soc. 88(1):53-57.

R303 Rawstron, R. R. and R. A. Reavis, 1974. First year harvest rates of largemouth bass at Folsom Lake and Lake Berryessa, California. Calif. Fish, Game 60(1):52-53.

R304 Rawstron, R. R. and K. A. Hashagen, Jr., 1972. Mortality and survival rates of tagged largemouth bass (*Micropterus salmoides*) at Merle Collins Reservoir. Calif. Fish, Game 58(3):221-230.

R305 Redmond, L. C., 1974. Prevention of overharvest of largemouth bass in Missouri impoundments. Symposium on overharvest of largemouth bass in small impoundments. N.C. Div. Am. Fish. Soc. Spec. Publ. 3:54-68.

R306 Rasmussen, J. L. and S. M. Michaelson, 1974. Attempts to prevent largemouth bass overharvest in three northwest Missouri lakes. Symposium on overharvest of largemouth bass in small impoundments. N.C. Div. Am. Fish. Soc. Spec. Publ. 3:69-83.

R307 Ramsey, J. S. and R. O. Smitherman, 1972. Development of color pattern in pond-reared young of five *Micropterus* species of southeastern U.S. Proc. S.E. Assoc. Game Fish Comm. 25:348-356.

R308 Range, J. D., 1973. Growth of five species of game fishes before and after introduction of threadfin shad into Dale Hollow Reservoir. Proc. S.E. Assoc. Game Fish Comm. 26:510-518.

R309   Rickett, J. D., 1974. Trophic relationships involving crayfish of the genus *Orconectes* in experimental ponds. Prog. Fish-Cult. 36(4):207-211.

R310   Rice, L. A., 1941. The food of six Reelfoot Lake fishes in 1940. Tenn. Acad. Sci. J. 16(1):22-26.

R311   Ramsey, J. S., 1973. The *Micropterus coosae* complex in southeastern U.S. (Osteichthyes, Centrarchidae). A.S.B. Bull. 20(2):76.

R312   Robbins, W. H. and H. R. MacCrimmon, 1974. The blackbass in America and overseas. Biomanag. and Research Enterp., Box 2300 Saulte Ste Marie, Ont. P6A5Pa. 196 pp.

R313   Rutledge, W. P. and J. C. Barron, 1972. The effects of the removal of stunted white crappie on the remaining crappie population of Meridian State Park Lake, Bosque, Texas. Texas Parks, Wildl. Dep. Techn. Ser. 12:41 pp.

R314   Ricker, W. E., 1973. Linear regressions in fishery research. J. Fish. Res. Bd. Canada 30(3):409-434.

R315   ———. 1975. Computation and interpretation of biological statistics of fish populations. Fish. Res. Bd. Canada Bull. 191:382 pp.

S11   Schoffman, R. J., 1938. Age and growth of the bluegills and largemouth black bass in Reelfoot Lake. J. Tenn. Acad. Sci. 13:81-103.

S12   ———. 1939. Age and growth of the red-eared sunfish in Reelfort Lake. J. Tenn. Acad. Sci. 14(3):71-71.

S13   ———. 1940. Age and growth of the black and white crappie, the warmouth bass, and the yellow bass in Reelfort Lake. Rep. Reelfoot Lake Biol. Stn. 4:22-42.

S18   ———. The size distribution of the bluegill and the largemouth black bass in Reelfoot Lake, Tennessee. J. Tenn. Acad. Sci. 20(1):98-102.

S19   ———. 1948. Age, growth, and size distribution of bluegills in Reelfoot Lake for 1937 and 1947. J. Tenn. Acad. Sci. 22(1):12-19.

S26   Seaman, E. A., 1949. Fish facts. W. Va. Conserv. 12(2):17.

S29   Sharp, J., 1898. The large-mouthed black bass in Utah. Bull. U.S. Fish Comm. 17:363-368.

S53   Smith, E. V. and H. S. Swingle, 1941. Winter and summer growth of bluegills in fertilized ponds. Trans. Am. Fish. Soc. 70:335-338.

S54   ———. 1943. Results of further experiments on the stocking of fish ponds. Trans. N. Am. Wildl. Conf. 8:168-179.

S57   Smith, H. M., 1907. The fishes of North Carolina. N.C. Geol., Econ. Surv. 2:453 pp.

S62   Smith, L. L., Jr. and N. L. Moe (compilers), 1944. Minnesota fish facts. Minn. Dep. Conserv. Bull. 7:1-31.

S63   Smith, L. L., Jr. and J. B. Moyle, 1945. Factors influencing production of yellow pikeperch, *Stizostedion vitreum vitreum*, in Minnesota rearing ponds. Trans. Am. Fish. Soc. 73:243-261.

S66   Smith, M. W., 1942. The smallmouth black bass in the Maritime Provinces. Fish. Res. Bd. Can. Prog. Rep. Atlantic 32:3-4.

S86   Speaker, E. B., 1936. Growth of bass and pike-perch in ponds. Prog. Fish-Cult. 24:27.

S97   Stroud, R. H., 1948. Growth of the basses and black crappie in Norris Reservoir, Tenn. J. Tenn. Acad. Sci. 23(1):31-99.

S99   ———. 1949. Growth of Norris Reservoir walleye during the first twelve years of impoundment. J. Wildl. Manage. 13(2):157-177.

S100   ———. 1949. Rate of growth and condition of game and panfish in Cherokee and Douglas Reservoirs, Tennessee, and Hiwassee Reservoir, North Carolina. J. Tenn. Acad. Sci. 24(1):60-74.

S106   Surber, E. W., 1939. A comparison of four eastern smallmouth bass streams. Trans. Am. Fish. Soc. 68:322-335.

S108   ———. 1941. Productivity of three smallmouth bass streams. Trans. N. Am. Wildl. Conf. 6:179-189.

S109   ———. 1941. A quantitative study of the food of the smallmouth black bass, *Micropterus dolomieu,* in three eastern streams. Trans. Am. Fish Soc. 70:311-324.

S110   ———. 1948. Increasing production of bluegill sunfish for farm pond stocking. Prog. Fish-Cult. 10(4):199-203.

S117   Swingle, H. S., 1949. Experiments with combinations of largemouth black bass, bluegills, and minnow in ponds. Trans. Am. Fish. Soc. 76: 46-62.

S118   Swingle, H. S. and E. V. Smith, 1939. Increasing fish production in ponds. Trans. N. Am. Wildl. Conf. 4:332-338.

S120   ———. 1942. The management of ponds with stunted fish populations. Trans. Am. Fish. Soc. 71:102-105.

S121   ———. 1943. Effect of management practices on the catch in a 12 acre pond during a 10 year period. Trans. N. Am. Wildl. Conf. 8:141-155.

S124   Surber, E. W., 1949. Results of varying the ratio of largemouth black bass and bluegills in the stocking of experimental farm ponds. Trans. Am. Fish. Soc. 77:141-151.

S130   Scott, D. C., 1949. A study of a stream population of rock bass, *Ambloplites rupestris.* Invest. Ind. Lakes, Streams 3(3):169-234.

S131   Speaker, E. B., 1948. A fish population study of an artificial lake. Proc. Iowa Acad. Sci. 55:437-444.

S133   Surber, E. W. and E. A. Seaman, 1949. The catches of fish in two smallmouth bass streams in West Virginia. W. Va. Conserv. Comm. Tech. Bull. 1:1-37.

S136   Schoffman, R. J., 1952. Growth of the bluegills and crappies in Reel-foot Lake, Tennessee. J. Tenn. Acad. Sci. 27(1):15-26.

S138   Scott, W. B., D. N. Ormand, and G. H. Lawler, 1957. Experimental rearing of yellow pikeperch fry in natural waters. Can. Fish-Cult. 10:38-43.

S150   Sivells, H. C., 1949. Food studies of black crappie fry (*Pomoxis nigromaculatus*). Tex. J. Sci. 1(3):38-40.

S151   Smith, C. G., W. M. Lewis, and H. M. Kaplan, 1952. A comparative morphologic and physiologic study of fish blood. Prog. Fish-Cult. 14 (4):169-172.

S157   Smith S., 1952. Fish management in Sardis Reservoir. Miss. Game, Fish 16(11):5, 10.

S166   Starrett, W. C. and P. L. McNeil, Jr., 1952. Sport fishing at Lake Chautauqua, near Havana, Illinois in 1950-1951. Ill. Nat. Hist. Surv. Biol. Notes 30:1-31.

S169   Stone, U. B., 1949. 1,000 Islands bass. N.Y. State Conserv. 3(6):18-19.

S171   Surber, E. W. and M. H. Everhart, 1950. Biological effects of nitro-sine used for control of weeds in hatchery ponds. Prog. Fish-Cult. 12 (3):135-140.

S176   Swingle, H. S., 1951. Experiments with various rates of stocking blue-gills, *Lepomis macrochirus* Rafinesque, and largemouth bass, *Micropterus salmoides* (Lacepede) in ponds. Trans. Am. Fish. Soc. 80:218-230.

S177   Schultz, Vincent, 1952. A limnological study of an Ohio farm pond. Ohio J. Sci. 52(5):267-285.

S182   Stone, U. B., D. G. Pasko, and R. M. Roecker, 1951. A study of small-mouth bass, Lake Ontario-St. Lawrence River. N.Y. Div. Fish, Game Res. Ser. 2:1-25.

S183   Stokely, P. S., 1952. The vertebral axis of two species of Centrarchid
        fishes. Copeia, 1952(4):255-261.

S184   Saila, S. B., 1950. A survey of farm fishpond possibilities in New York.
        Conserv. Dep. Cornell Univ. 5 pp. mimeo.

S187   ———. 1958. Size limits in largemouth black bass management.
        Trans. Am. Fish. Soc. 87:229-239.

S188   Saila, S. B. and D. Horton, 1957. Fisheries investigations and manage-
        ment in Rhode Island lakes and ponds. R.I. Div. Fish, Game. Fish
        Publ. 3:1-134.

S192   Sanderson, A. E., 1958. Smallmouth bass management in the Potomac
        River Basin. Trans. N. Am. Wildl. Conf. 23:248-262.

S199   Schneidermeyer, F. and W. M. Lewis, 1956. Utilization of gizzard
        shad by largemouth bass. Prog. Fish-Cult. 18(3):137-138.

S205   Schoffman, R. J., 1959. Age and rate of growth of the bluegills in Reel-
        foot Lake, Tennessee for 1950 and 1958. J. Tenn. Acad. Sci. 34(1):73-
        77.

S207   Schoonover, R. and W. H. Thompson, 1954. A post-impoundment study
        of the fisheries resources of Fall River Reservoir, Kansas. Trans.
        Kans. Acad. Sci. 57(2):172-179.

S212   Schwartz, F. J. and J. Norvell, 1958. Food, growth and sexual dimor-
        phism of the redside dace *Clinostomus elongatus* (Kirtland) in Lines-
        ville Creek, Crawford County, Pennsylvania. Ohio J. Sci. 58(5):311-
        316.

S213   Scidmore, W. J., 1955. Notes on the fish population structure of a
        typical rough fish-crappie lake of southern Minnesota. Minn. Dep.
        Conserv., Invest. Rep. 162:1-11.

S217   Seaman, E. A., and staff, 1954. Fishery observations on Wheeling
        Creek. W. Va. Div. Fish. Manage. 19+ pp. mimeo.

S225   Shields, J. T., 1955. Report of fisheries investigations during the
        second year of impoundment of Fort Randall Reservoir, South Dakota,
        1954. S.D. Dep. Game, Fish, Parks. 100 pp. mimeo.

S226   ———. 1956. Report of fisheries investigation during the third year
        of impoundment of Fort Randall Reservoir, South Dakota, 1955. S.D.
        Dep. Game, Fish, Parks. D.J. Proj. F-R-5 91 pp.

S227   ———. 1957. Report of fisheries investigations during the second
        year of impoundment of Gavins Point Reservoir, South Dakota, 1956.
        S.D. Dep. Game, Fish, Parks. D.J. Proj. F-1-R-6 34 pp. mimeo.

S228   ———. 1957. Report of fisheries investigations during the fourth year
        of impoundment of Fort Randall Reservoir, South Dakota, 1956. S.D.
        Dep. Game, Fish, Parks. D.J. Proj. F-1-R-6:1-60.

S229   ———. 1958. Report of fisheries investigations during the third year
        of impoundment of Gavins Point Reservoir, South Dakota, 1957. S.D.
        Dep. Game, Fish, Parks. D.J. Proj. F-1-R-7:1-48.

S230   ———. 1958. Report of fisheries investigations during the fifth year
        of impoundment of Fort Randall Reservoir, South Dakota, 1957. S.D.
        Dep. Game, Fish, Parks. D.J. Proj. F-1-R-7.

S241   Slack, K. V., 1955. A study of the factors affecting stream productiv-
        ity by the comparative method. Invest. Ind. Lakes, Streams 4:3-47.

S224   Smith, E. V. and H. S. Swingle, 1943. Percentages of survival of blue-
        gills (*Lapomis macrochirus*) and largemouth black bass (*Huro sal-
        moides*) when planted in new ponds. Trans. Am. Fish. Soc. 72:63-67.

S248   Smith, L. L., Jr., D. R. Franklin, and R. H. Kramer, 1958. Determi-
        nation of factors influencing year class strength in northern pike and
        largemouth bass. Minn. D.J. Proj. F-1-2-R Jobs 11 and 111, compl.
        rep. 202 + A 116 pp.

S258   Smith, R. F., 1957. Lakes and ponds. N.J. Fish. Surv. Rep. 3:198 pp.

S266   Smith, W. A., Jr., J. B. Kirkwood, and J. F. Hall, 1955. A survey of the success of various stocking rates and ratios of bass and bluegill in Kentucky farm ponds. Ky. Fish. Bull. 16:42 pp. mimeo.

S276   Sprague, J. W., 1959. Report of fisheries investigations during the fourth year of impoundment of Gavins Point Reservoir, South Dakota, 1958. S.D. Dep. Game, Fish, Parks. 42 pp. mimeo.

S277   Sprugel, G., Jr., 1954. Growth of bluegills in a new lake with particular reference to false annuli. Trans. Am. Fish. Soc. 83:58-75.

S278   ———. 1955. The growth of green sunfish (Lepomis cyanellus) in Little Wall Lake, Iowa. Iowa State Coll. J. Sci. 29(4):707-719.

S286   Stone, U. B., D. G. Pasko, and R. M. Roecker, 1954. A study of Lake Ontario-St. Lawrence River smallmouth bass. N.Y. Fish, Game J. 1(1):1-26.

S289   Stroud, R. H., 1955. Fisheries report for some central, eastern and western Massachusetts lakes, ponds, and reservoirs, 1951-1952. Mass. Div. Fish, Game. 447 pp.

S298   Surber, E. W. and G. E. Klak, 1939. Experiments with forage minnows in bass ponds. Prog. Fish-Cult. 47:31-37.

S299   Suttkus, R. D., 1955. Age and growth of a small stream population of "stunted" smallmouth black bass, Micropterus dolomieu dolomieu (Lacepede). N.Y. Fish, Game J 2(1):83-94.

S307   Swingle, H. S., 1954. Experiments on commercial fish production in ponds. Proc. S.E. Assoc. Game Fish Comm. 69-74.

S308   ———. 1954. Fish populations in Alabama rivers and impoundments. Trans. Am. Fish. Soc. 83:47-57.

S323   Snow, H., 1960. Bluegill at Murphy Flowage. Wis. Conserv. Bull. 25 (3):11-14.

S325   Swingle, H. S., 1960. Comparative evaluation of two tilapias as pond-fishes in Alabama. Trans. Am. Fish. Soc 89(2):142-148.

S327   Schwartz, F. J., 1960. The crappies. Chesapeake Biol. Lab. Educ. Ser. 42:4 pp.

S329   Schoffman, R. J., 1960. Age and rate of growth of the white crappie in Reelfoot Lake, Tennessee, for 1950 and 1959. J. Tenn. Acad. Sci. 35 (1):3-8.

S330   Sprague, J. W., 1959. Report of fisheries investigations during the sixth year of impoundment of Fort Randall Reservoir, South Dakota, 1958. S.D. Dep. Game, Fish, Parks. D.J. Proj. F-1-R-8. 32 pp. mimeo.

S333   Strawn, K., 1958. Optimum and extreme temperatures for growth and survival: various fishes. Data for Handb. Biol. Data. 1 p.

S334   Swingle, H. S., 1952. Pounds of fish per acre in Central Alabama power reservoirs. Pounds of fish per acre in Alabama rivers. Temperatures of surface water of ponds at Auburn, Alabama when the first young fish hatch in the spring. Data for Handb. Biol. Data.

S340   Schoonover, R., 1958. The rehabilitation of warm-water lakes for fishery improvement. Central Mts. and Plains Sec. Wildl. Soc. Proc. 3d Ann. Conf. 12 pp. mimeo.

S342   Smith, W. A., Jr., 1959. Shad management in reservoirs. Proc. S.E. Assoc. Game Fish Comm. 12:143-147.

S343   Stevens, R. E., 1959. The black and white crappies of the Santee-Cooper Reservoir. Proc. S.E. Assoc. Game Fish Comm. 12:158-168.

S345   Swingle, H. S., 1959. Experiments on growing fingerling channel cat-
        fish to marketable size in ponds. Proc. S.E. Assoc. Game Fish Comm.
        12:63-72.
S348   Schloemer, C. L., 1939. The age and rate of growth of the bluegill,
        Helioperca macrochira (Rafinesque). Ph.D. thesis, Univ. Wis. 113 pp.
S350   Sigler, W. F., 1960. The largemouth bass in northern Utah. Ms. 9 pp.
S354   Saunders, J. W., 1960. The effect of impoundment on the population
        and movement of Atlantic salmon on Ellerslie Brook, Prince Edward
        Island. J. Fish Res. Bd. Can. 17(4):453-473.
S357   Scidmore, W. J. and D. E. Woods, 1960. Some observations on com-
        petition between several species of fish for summer foods in four
        southern Minnesota Lakes in 1955, 1956, and 1957. Minn. Fish, Game
        Invest. Fish. Ser. 2:13-24.
S358   Scidmore, W. J., 1960. Evaluation of panfish removal as a means of
        improving growth rates and average size of stunted fish. Minn. Fish,
        Game Invest. Fish. Ser. 2:42-51.
S363   Sayre, R. C., 1960. Northeastern Oregon. Ore. State Game Comm.
        Fish. Div. Ann. Rep. 1959:86-108.
S366   Salyer, J. T., 1958. Factors associated with the decline of the large-
        mouth bass, *Micropterus salmoides* (Lacepede), in San Vicente Reser-
        voir, San Diego County, California. M.A. thesis, San Diego State Coll.
        103 pp.
S372   Swenson, E. A., Jr., 1954. Analysis of the fish populations of two farm
        ponds in central New York. M.S. thesis, State Univ. N.Y., College
        Forestry. 100 pp.
S374   Stockinger, N. F. and H. A. Hays, 1960. Plankton, benthos and fish in
        three strip-mine lakes with varying pH values. Trans. Kans. Acad.
        Sci. 63(1):1-11.
S376   Snow, J. R., 1960. An exploratory attempt to rear largemouth black
        bass fingerlings in a controlled environment. Proc. S.E. Assoc. Game
        Fish Comm. 14:253-257.
S381   Schwartz, F. J., 1961. Food, age, growth, and morphology of the black-
        banded sunfish, *Enneacanthus c. chaetodon,* in Smithville Pond, Mary-
        land. Chesapeake Sci. 2(1-2):82-88.
S383   Strawn, K., 1961. Growth of largemouth bass fry at various tempera-
        tures. Trans. Am. Fish. Soc. 90(3):334-335.
S384   Scidmore, W. J. and D. E. Woods, 1961. Changes in the fish popula-
        tions of four southern Minnesota lakes subjected to rough fish removal.
        Minn. Fish. Game Invest. Fish Ser. 3:1-19.
S389   Schoffman, R. J., 1962. Age and rate of growth of the largemouth black
        bass in Reelfoot Lake, Tennessee, for 1952 and 1961. J. Tenn. Acad.
        Sci. 37(1):1-4.
S390   Swingle, H. S., 1956. A repressive factor controlling reproduction in
        fishes. Pac. Sci. Congr. Proc., 8th. IIIA (1953):865-871.
S391   Sandoz, O., 1960. A pre-impoundment study of Arbuckle Reservoir,
        Rock Creek, Murray County, Oklahoma. Okla. Fish. Res. Lab.,
        Norman, Okla., Rep. 77:28 pp.
S393   Sprague, J. W., 1961. Report of fisheries investigations during the
        seventh year of impoundment of Fort Randall Reservoir, South Dakota,
        1959. S.D. Dep. Game, Fish, Parks. D.J. Proj. F-1-R-9 Jobs 5, 6, 7,
        & 8:49 pp.
S395   Smith, W. B., 1961. Roanoke River reservoirs. N.C. Wildl. Res.
        Comm. Fed. Aid Proj. F5R and F6R Job Compl. Rep. 1:75-95.

S398   Schwartz, F. J. and H. J. Elser, 1962. Additions to Maryland list: new record fish. Md. Conserv. 39(2):26-27.

S414   Snow, H. E., 1962. A comparison of the fish population in Murphy Flowage, Wisconsin, before and after a panfish removal program. Wis. Conserv. Dep. 5 pp. mimeo.

S419   Schoumacher, R., 1962. Further notes on the effects of mechanical reduction on the growth of crappies in Backbone Lake, Iowa. Iowa State Conserv. Comm. Q. Biol. Rep. 14(1):16-25.

S438   Smith, L. L., Jr., D. R. Franklin, and R. H. Kramer, 1958. Determination of factors influencing year class strength in northern pike and largemouth bass. Minn. D.J. F-12R 1958(11 and 111):Compl. Rep. 200+ A116 pp.

S450   Sarbahi, D. S., 1951. Studies of the digestive tracts and digestive enzymes of the goldfish, *Carassius auratus* (Linnaeus) and the largemouth black bass, *Micropterus salmoides* (Lacepede). Biol. Bull. 100 (3):244-257.

S462   Smith, B. G., 1908. The spawning habits of *Chrosomus erythrogaster* Rafinesque. Biol. Bull. 15:9118.

S463   Sigler, W. F. and R. R. Miller, 1963. Fishes of Utah. Utah State Dep. Fish, Game. 203 pp.

S464   Seaburg, K. G. and J. B. Moyle, 1964. Feeding habits, digestive rates, and growth of some Minnesota warmwater fishes. Trans. Am. Fish. Soc. 93(3):269-285.

S467   Schloemer, C. L., 1957. Tabular outline of the life history and ecology of the common bluegill, *Lepomis macrochirus* Rafinesque. Data for Handb. Biol. Data. 27 pp.

S472   Swingle, W. E., 1965. Length-weight relationships of Alabama fishes. Auburn Univ. Agric. Exp. Stn. Zool. Ent. Ser. Fish. 3:87 pp.

S481   Stocek, Rudolph and H. R. McCrimmon, 1965. The co-existence of rainbow trout (*Salmo gairdneri* Richardson) and largemouth bass (*Micropterus salmoides* Lacepede) in a small Ontario lake. Can. Fish-Cult. 35:37-58.

S504   Stevenson, J. H., 1964. Fish farming experimental station. U.S. Fish, Wildl. Serv. Circ. 178:79-100.

S526   Smith, W. B., 1969. A preliminary report on the biology of the Roanoke bass, *Ambloplites cavifrons* Cope, in North Carolina. N.C. Wildl. Res. Comm. 11 pp.

S528   Shannon, E. H., 1967. Geographical distribution and habitat requirements of the redbreast sunfish, *Lepomis auritus*, in North Carolina. Proc. S.E. Assoc. Game Fish. Comm. 20:319-323.

S529   Summerfelt, R. C., 1967. Fishes of the Smoky Hill River, Kansas. Trans. Kans. Acad. Sci. 70(1):102-139.

S530   Smith, R. J. F., 1969. Control of prespawning behaviour of sunfish (*Lepomis gibbosus* and *L. megalotis*) I. Gonadal androgen. Anim. Behav. 17(h):279-285.

S531   ——. 1970. Control of prespawning behaviour of sunfish (*Lepomis gibbosus* and *Lepomis megalotis*). 11. Environmental factors. Anim. Behav. 18(3):575-587.

S532   Shoemaker, H. H., 1952. Fish home areas of Lake Myosotis, New York. Copeia, 1952(2):83-87.

S534   Snow, H., A. Ensign, and J. Klingbiel, 1960. The bluegill. Its life history, ecology and management. Wis. Conserv. Dep. Publ. 230. 14 pp.

S535   Shireman, J. V., 1968. Age and growth of bluegills, *Lepomis macro-chirus* Rafinesque, from selected central Iowa farm ponds. Proc. Iowa Acad. Sci. 75:170-178.

S536   Schmittou, H. R., 1968. Some effects of supplemental feeding and controlled fishing in largemouth bass bluegill populations. Proc. S.E. Assoc. Game Fish. Comm. 22:311-320.

S537   Schoffman, R. J., 1966. Age and rate of growth of bluegills in Reelfoot Lake, Tennessee, for 1958 and 1965. J. Tenn. Acad. Sci. 41:32-34.

S538   ———. 1964. Summary of the age and rate of growth of game fish in Reelfoot Lake, Tennessee, from 1937 through 1961. J. Tenn. Acad. Sci. 39(1):11-15.

S539   Smitherman, R. O., 1968. Effect of the strigeid trematode, *Posthodiplostomum minimum*, upon the growth and mortality of bluegill, *Lepomis macrochirus*. F.A.O. Fish. Rep. 44 5:380-388.

S540   Schoffman, R. J., 1965. Age and rate of growth of the white crappie in Reelfoot Lake, Tennessee, for 1959 and 1964. J. Tenn. Acad. Sci. 40:1:6-8.

S541   Stevenson, F., W. T. Momot, and F. J. Svoboda III, 1969. Nesting success of the Bluegill, *Lepomis macrochirus*, Rafinesque, in a small Ohio pond. Ohio J. Sci. 69(6):347-355.

S542   Swingle, H. S. and E. V. Smith, 1943. Factors affecting the reproduction of bluegill bream and largemouth black bass in ponds. Ala. Polytech. Inst. Agr. Exp. Stn. Circ. 87:8 pp.

S543   Schmittou, H. R., 1967. Sex ratios of bluegills in four populations. Trans. Am. Fish. Soc. 96(4):420-421.

S544   Savitz, J., 1971. Effects of starvation on body protein utilization of bluegill sunfish (*Lepomis macrochirus* Rafinesque) with a calculation of calorie requirements. Trans. Am. Fish. Soc. 100(1):18-21.

S545   Spall, R. D., 1970. Possible cases of cleaning symbiosis among freshwater fishes. Trans. Am. Fish. Soc. 99(3):599-600.

S548   Selbig, W., 1970. Chemical rehabilitation of chronic winterkill lakes. Pp. 27-30 in "A Symposium on management of midwestern winterkill lakes." N.C. Div. Am. Fish. Soc. Spec. Publ.

S550   Swingle, H. S., 1949. Some recent developments in pond management. Trans. N. Am. Wildl. Conf. 14:295-312.

S552   Sullivan, C. R., 1956. Population manipulation studies on West Virginia smallmouth bass streams. W. Va. Cons. Comm. Final Rep. Proj. F-1-R-(1-5)53 pp. (Not seen, quoted from B192.)

S553   Surber, E. W., 1943. Observations on the natural and artificial propagation of the smallmouth black bass, *Micropterus dolomieu*. Trans. Am. Fish. Soc. 72:233-245.

S554   ———. 1935. Production of bass fry. Prog. Fish-Cult. 8:1-7.

S555   Smith, P. W. and L. M. Page, 1969. The food of spotted bass in streams of the Wabash River drainage. Trans. Am. Fish. Soc. 98(4):647-651.

S556   Snow, H. E., 1966. The inside story. Wis. Conserv. Bull. 31(4):6-7.

S557   Snow, J. R., 1968. Production of six-to-eight-inch largemouth bass for special purposes. Prog. Fish-Cult. 30(3):144-152.

S558   ———. 1968. The Oregon moist pellet as a diet for largemouth bass. Prog. Fish-Cult. 30(4):235.

S559   Snow, J. R. and J. I. Maxwell, 1970. Oregon moist pellet as a production ration for largemouth bass. Prog. Fish-Cult. 32(2):101-102.

S560 Stewart, N. E., D. L. Shumway, and P. Doudoroff, 1967. Influence of oxygen concentration on the growth of juvenile largemouth bass. J. Fish. Res. Bd. Can. 24(3):475-494.

S561 Swingle, H. S. and W. E. Swingle, 1967. Problems in dynamics of fish populations in reservoirs. Pp. 229-43, Reservoir Fishery Resources Symposium, Am. Fish. Soc. Southern Div.

S562 Snow, J. R., 1961. Forage fish preference and growth rate of largemouth black bass fingerlings under experimental conditions. Proc. S.E. Assoc. Game Fish Comm. 15:303-313.

S563 ——. 1965. Results of further experiments on rearing largemouth bass fingerlings under controlled conditions. Proc. S.E. Assoc. Game Fish. Comm. 17:191-203.

S564 Snow, H. E., 1971. Harvest and feeding habits of largemouth bass in Murphy Flowage, Wisconsin. Wis. Dep. Nat. Resources, Tech. Bull. 50, 25 pp.

S565 Swingle, H. S., 1956. Determination of balance in farm fish ponds. Trans. N. Am. Wildl. Conf. 21:289-322.

S566 Schacht, R., 1968. Backbone Lake renovation project. Iowa Conserv. Comm. Q. Biol. Rep. 20(2):16-18.

S567 Siefert, R. E., 1968. Reproductive behavior, incubation and mortality of eggs, and postlarval food selection in the white crappie. Trans. Am. Fish. Soc. 97(3):252-259.

S568 ——. 1969. Characteristics for separation of white and black crappie larvae. Trans. Am. Fish. Soc. 98(2):326-328.

S569 ——. 1969. Biology of the white crappie in Lewis and Clark Lake. Bur. Sport Fish., Wildl. Tech. Pap. 22:16 pp.

S570 Summerfelt, R. C., 1971. Factors influencing the horizontal distribution of several fishes in an Oklahoma reservoir. Am. Fish. Soc. Spec. Publ. 8:425-439.

S571 Schloemer, C. L., 1947. Reproductive cycles of five species of Texas centrarchids. Science 106:85-86.

S572 Starrett, W. C. and A. W. Fritz, 1957. The crappie story in Illinois. Outdoors Ill. 4(2):11-14.

S573 Scott, W. B. and E. J. Crossman, 1973. Freshwater fishes of Canada. Fish. Res. Bd. Can., Bull. 184 966 pp.

S574 Schneberger, E., 1972. The black crappie. Its life history, ecology and management. Wis. Dep. Nat. Resources. Publ. 243-72. 16 pp.

S575 Schultze, R. F., 1974. Age and growth of largemouth bass in California from ponds. Calif. Fish Game 60(2):94-96.

S575B Schultze, R. F. and C. D. Vanicek, 1975. Age and growth of largemouth bass in California farm ponds. Farm Pond Harvest 9(2):27-29. Copied from Calif. Fish Game.

S576 Smitherman, R. O. and J. S. Ramsey, 1972. Observations on the spawning and growth of four species of basses (*Micropterus*) in ponds. Proc. S.E. Assoc. Game Fish Comm. 25:357-365.

S577 Snow, J. R., 1971. Fecundity of largemouth bass, *Micropterus salmoides* Lacepede, receiving artificial food. Proc. S.E. Assoc. Game Fish Comm. 24:550-559.

S578 ——. 1973. Controlled culture of largemouth bass fry. Proc. S.E. Assoc. Game Fish Comm. 26:392-328.

S579 Smith, W. B., 1972. The biology of the Roanoke bass, *Ambloplites cavifrons* Cope, in North Carolina. Proc. S.E. Assoc. Game Fish Comm. 25:561-570.

S580   Sivak, J. B., 1973. Interrelation of feeding behavior and accommodative lens movements in some species of North American freshwater fishes. J. Fish Res. Bd. Can. 30(8):1141-1146.

S581   Siewert, H. F., 1973. Thermal effects on biological production in nutrient rich ponds. Univ. Wis. Water Resources Center Tech. Compl. Rep. A-020 and A-032. 23 pp.

S582   Smith, P. W., 1968. An assessment of changes in the fish fauna of two Illinois rivers and its bearing on their future. Trans. Ill. State Acad. Sci. 61(1):31-45.

S583   Steele, R. G., and M. H. A. Keenleyside, 1971. Mate selection in two species of sunfish (*Lepomis gibbosus* and *L. megalotis peltastes*). Can. J. Zool. 49(12):1541-1548.

S584   Serns, S. L., 1972. Age, growth and condition of bluegill sunfish. *Lepomis macrochirus* Rafinesque, in four heated reservoirs in Texas. M.S. thesis, Texas A M Univ. 167 pp.

S585   Savitz, J., 1969. Effects of temperature and body weight on the endogenous nitrogen excretion in the bluegill sunfish. J. Fish. Res. Bd. Can. 2(7):1813-1821.

S586   Siefert, R. E., 1972. First food of larval yellow perch, white sucker, bluegill, emerald shiner and rainbow-smelt. Trans. Am. Fish. Soc. 101(2):219-225.

S587   Sparks, R. E., W. T. Walker, and J. Cairns, Jr., 1972. Effect of shelders on the resistance of dominant and submissive bluegills (*Lepomis macrochirus*) to a lethal concentration of zinc. J. Fish. Res. Bd. Can. 29(9):1356-1358.

S588   Schneider, J. C., 1973. Response of the bluegill population and fishery of Mill Lake to exploitation rate and minimum size limit a simulation model. Mich. Dep. Nat. Res., Fish. Res. Rep. 1804:18 pp.

S589   ——. 1973. Response of the bluegill population and fishery of Mill Lake to increased growth; a simulation model. Mich. Dep. Nat. Res., Fish Res. Rep. 1805:17 pp.

S590   Snow, H. E., 1974. Effects of stocking northern pike in Murphy Flowage. Wis. Dep. Nat. Res. Techn. Bull. 79:19 pp.

S591   Smitherman, R. O. and F. E. Hester, 1962. Artificial propagation of sunfishes with meristic comparisons of three species of *Lepomis* and five of their hybrids. Trans. Am. Fish. Soc. 91(4):333-341.

S592   Schneider, C. P., 1971. SCUBA observations of spawning smallmouth bass. N.Y. Fish, Game J 18(2):112-116.

S593   Schneberger, E., 1972. Smallmouth bass. Life history, ecology and management. Wis. Dep. Nat. Res. Publ. 242. 16 pp.

S594   Siefert, R. E., A. R. Carlson, and L. J. Herman, 1974. Effects of reduced oxygen concentrations on the early life stages of mountain whitefish, smallmouth bass, and white bass. Prog. Fish-Cult. 36(4):186-190.

S595   Schneider, J. C., 1973. Angling on Mill Lake, Michigan, after a five-year closed season. Mich. Acad. 5(3):349-355.

S596   Seidensticker, E. P., 1975. Mortality of largemouth bass for two tournaments utilizing a "Don't kill your catch" program. S.E. Assoc. Game, Fish Comm. 28:83-86.

S597   ——. 1975. Texas bass clubs. S.E. Assoc. Game, Fish Comm. 28:96-102.

S598   Sriprasert, R., 1974. Length-weight relationships of walleye, *Stizostedion vitreum*, from Clear Lake, Iowa, 1959-1973. M.S. thesis, Iowa State Univ. 30 pp.

T6   Tate, W. H., 1949. Growth and food habits studies of smallmouth black bass in some Iowa streams. Iowa State J. Sci. 23(4):343-354.

T7   TVA-Forest Log, 1970. Potential egg production of walleyed pike and black bass. Prog. Fish-Cult. 50:46-47.

T8   Tester, A. L., 1930. Spawning habits of the small-mouthed black bass in Ontario waters. Trans. Am. Fish. Soc. 60:53-61.

T9   ———. 1932. Rate of growth of the small-mouthed black bass, (*Micropterus dolomieu*) in some Ontario waters. Univ. Toronto Biol. Ser. 36, Publ. Ont. Fish. Res. Lab. 47:206-221.

T13   Thompson, D. H., 1939. Growth of the largemouth black bass, *Huro salmoides*, in Lake Naivasha, Kenya. Nature (London) 143(3622):561-562.

T16   Thompson, D. H. and G. W. Bennett, 1938. Lake Management Reports. 1. Horseshoe Lake near Cairo, Illinois. Ill. Nat. Hist. Surv. Biol. Notes 8:1-6.

T18   ———. 1939. Lake Management reports. 3. Lincoln Lakes near Lincoln, Illinois. Ill. Nat. Hist. Surv. Biol. Notes 11:1-24.

T19   ———. 1949. Lake Management Reports. 2. Fork Lake near Mount Zion, Illinois. Ill. Nat. Hist. Surv. Biol. Notes 9:1-14.

T20   Thompson, D. H. and D. F. Hansen, 1937. Research on the largemouth black bass in Illinois. Ill. Conserv. 2(2):9-10.

T21   Thompson, W. H., 1948. Secrets in a fish scale. Okla. Game, Fish News Nov. 1948:12-13.

T22   Thorpe, L. M., 1938. Pond fish management program in Connecticut. Trans. N. Am. Wildl. Conf. 3:469-477.

T23   ———. 1942. Application of fishery survey data to heavily fished lakes. Trans. N. Am. Wildl. Conf. 7:436-442.

T27   Titcomb, J. W., 1920. Some fish-cultural notes. Trans. Am. Fish. Soc. 50:200-211.

T31   Trautman, M. B., 1941. Fluctuations in lengths and numbers of certain species of fishes over a five-year period in Whitmore Lake, Mich. Trans. Am. Fish. Soc. 70:193-208.

T35   Turner, C. L. and W. C. Kraatz, 1920. Food of young large-mouth black bass in some Ohio waters. Trans. Am. Fish. Soc. 50:372-380.

T40   Taylor, G. G., 1950. Commission decrees Cacopon Lake for rainbows. W. Va. Conserv. 14(9):19-20, 35-36.

T44   Thompson, W., 1950. Investigation of the fisheries resources of Grand Lake. Okla. Game Fish Dep. Fish Manage. Rep. 18:1-46.

T45   ———. 1950. Present status of fishery management in Oklahoma. Prog. Fish-Cult. 12(4):193-195.

T48   Thompson, W. H., H. C. Ward, and J. F. McArthur, 1951. The age and growth of white crappie, *Pomoxis annularis* (Rafinesque), from four small Oklahoma lakes. Proc. Okla. Acad. Sci. 30:93-101.

T54   Thompkins, W. A. and B. T. Carter, 1951. The growth rates of some Kentucky fishes. Ky. Div. Game, Fish. Fish Bull. 6:1-9.

T64   Tebo, L. B., Jr., 1957. Preliminary experiments on the use of spaghetti tags. Proc. S.E. Assoc. Game Fish. Comm. 10:77-80.

T66   Tennessee Conservationist, 1955. Worlds Record Smallmouth. Tenn. Conserv. 21(8):11.

T67   ———. 1956. Moore's catch has interesting past. Tenn. Conserv. 22(5):18.

T73   Threinen, C. W., 1956. The success of a seine in the sampling of a largemouth bass population. Prog. Fish-Cult. 18(2):81-87.

T76   Threinen, C. W. and W. T. Helm, 1952. Composition of the fish popu-
        lation and carrying capacity of Spauldings Pond, Rock County as deter-
        mined by rotenone treatment. Wis. Invest. Rep. 656:19 pp.

T77   Threinen, C. W. and D. Mraz. 1954. We won't "fish out" bass. Wis.
        Conserv. Bull. 19(12):24-27.

T79   Tiemeier, O. W. and J. B. Elder, 1957. Limnology of Flint Hills farm
        ponds for 1956 and preliminary report on growth studies of fishes.
        Trans. Kan. Acad. Sci. 60(4):379-392.

T89   Tate, W., 1959. Survival of largemouth bass advanced fry in some
        eastern Iowa waters. Iowa Conserv. Comm. Q. Biol. Rep. 11(3):21-22.

T90   Tarrant, R. M., Jr., 1960. Choice between two sizes of forage fish by
        largemouth bass under aquarium conditions. Prog. Fish-Cult. 22(2):
        83-84.

T92   Trenary, J. D., 1958. Growth of three Centrarchidae in Lake Fort
        Smith, Arkansas. M.S. thesis, Univ. Ark. 53 pp.

T94   Turner, W. R., 1960. Standing crops of fishes in Kentucky farm ponds.
        Trans. Am. Fish. Soc. 89(4):333-337.

T99   Tennessee Valley Authority, 1961. Fish and Game Annual Report.
        Norris, Tenn. 18 pp.

T102   Trembley, F. J., 1960. Research project on effects of condenser dis-
        charge water on aquatic life—Progress Report, 1956-1959. Inst. Re-
        search, Lehigh Univ. Prog. Rep. 1956-1959.

T103   Tatum, B., 1961. Yadkin and Lower Catawba River reservoirs. N.C.
        Wildl. Res. Comm. Fed. Aid Proj. F5R and F6R. Job Compl. Rep.
        1:99-158.

T104   Tebo, L. B., Jr., 1961. Discussion: Inventory of fish populations in
        lentic waters. N.C. Wildl. Res. Comm. Fed. Aid Proj. F5R and F6R.
        Job Compl. Rep. 1:298-316.

T105   Tiemeier, O. W., 1962. Supplemental feeding of fingerling channel cat-
        fish. Prog. Fish-Cult. 24(2):88-90.

T113   Trautman, M. B., 1957. The fishes of Ohio. Ohio State Univ. Press.
        683 pp.

T127   Tennessee Valley Authority, 1964. Average growth rates for East
        Tennessee Valley fishes, 1964. 6 pp. mimeo.

T129   Tyus, H. M., 1970. Spawning of rock bass in North Carolina during
        1968. Prog. Fish-Cult. 32(1):25.

T130   Tharratt, R. C. and R. J. McKechnie, 1966. Sacramento perch. Pp.
        373-375, Inland Fisheries Management, A. Calhoun, ed. Calif. Dep.
        Fish, Game.

T131   Taber, C., 1965. Spectacle development in the pygmy sunfish, *Elas-
        soma zonatum*, with observations on spawning habits. Proc. Okla.
        Acad. Sci. 46:73-81.

T132   Tharratt, R. C., 1966. The age and growth of centrarchid fishes in
        Folsom Lake, California. Calif. Fish and Game 52(1):4-16.

T133   Tiemeier, O. W., 1966. Kansas farm ponds. Kansas State Univ. Agri.
        Esp. Stn. Bull. 488:1-64.

T134   Tafanelli, R. J. and J. C. Bass, 1968. Feeding response of *Lepomis
        cyanellus* to blister beetles (Meloidae). Southwest Nat. 13:1:51-55.

T135   Turner, W. B., 1955. Food habits of the bluegill, *Lepomis macrochirus
        macrochirus* (Rafinesque), in 18 Kentucky farm ponds during April and
        May. Trans. Ky. Acad. Sci. 16(4):98-101.

T136   Toetz, D. W., 1966. The change from endogenous to exogenous sources
        of energy in bluegill sunfish larvae. Invest. Ind. Lakes, Streams 7:115-
        146.

T137   Trama, F. B., 1954. The pH tolerance of the common bluegill (*Lepomis macrochirus* Rafinesque). Not. Nat. (Phil.) 256:1-13.

T138   Turner, G. E. and H. R. MacCrimmon, 1970. Reproduction and growth of smallmouth bass, *Micropterus dolomieui*, in a Precambrian Lake. J. Fish. Res. Bd. Can. 27(2):395-400.

T139   Thompson, J. D., 1965. Age and growth of largemouth bass in Clear Lake, Iowa. Proc. Iowa Acad. Sci. 71:252-258.

T140   Taub, S. H., 1972. Exploitation of crayfish by largemouth bass in a small Ohio pond. Prog. Fish-Cult. 34(1):55-58.

T141   Tebo, L. B., Jr. and E. G. McCoy, 1964. Effect of sea-water concentration on the reproduction of largemouth bass and bluegills. Prog. Fish-Cult. 26(3):99-106.

T142   Tagatz, M. E. and E. P. H. Wilkens, 1973. Seasonal occurrence of young gulf menhaden and other fishes in a northwestern Florida estuary. NOAA (Natl. Ocean Atmos. Adm.) Tech. Rep. NMFS (Natl. Mar. Fish. Serv.) SSRF (Spec. Sci. Rep. Fish.) 672:14 pp.

T143   Tennant, D., 1957. The big crappie transfer. Ohio Conserv. Bull. 21 (3):16-17.

T144   Thompson, J. D. and K. D. Carlander, 1970. An estimate of the largemouth bass population in Clear Lake, Iowa, 1964. Iowa State J. Sci. 44(3):411-412.

T145   Taube, S. H., C. F. Clark, D. A. Mayhew, and J. B. Lisiecki, 1974. Suitability of Ohio interstate highway borrow pit ponds for sport fishing. Ohio Agric. Res., Dev. Cent., Res. Bull. 1064:21 pp.

T146   Tucker, W. H., 1973. Food habits, growth and length-weight relationships of young-of-the-year black crappie and largemouth bass in ponds. Proc. S.E. Assoc. Game Fish Comm. 26:565-577.

T147   Tyus, H. M., 1973. Artificial intergeneric hybridization of *Ambloplites rupestris* (Centrarchidae). Copeia, 1973(3):428-430.

U1   Ulrey, L., C. Risk, and W. Scott, 1938. The number of eggs produced by some of our common freshwater fishes. Invest. Ind. Lakes, Streams 1(6):73-78.

U4   Upper Mississippi River Conservation Committee, 1946. Second progress report of the technical committee for fisheries. 27 pp. mimeo.

U8   ———. 1959. Supplemental report, Fish Technical Sub-Committee, Proceedings of the thirteenth annual meeting. 147 pp. mimeo.

U12   Ulrickson, G. U., D. J. Nelson, and N. A. Griffith, 1971. The effect of temperature on elimination rates of 137 Cs in bluegill (*Lepomis macrochirus*). Oak Ridge Nat. Lab. Rep. 4634:106-107.

V2   Van Cleave, J. R. B., 1899. The age of smallmouth black bass. Rep. Ill. State Fish Comm. Oct. 1896-Sept. 30, 1898:25-27.

V5   Van Oosten, J., 1929. Life history of the lake herring (*Leucichthys artedi* LeSueur) of Lake Huron as revealed by its scales, with a critique of the scale method. Bull. U.S. Bur. Fish 44(Doc. 1053):265-428.

V20   Van Oosten, J. and H. J. Deason, 1939. Age, growth, and condition of the wall-eyed pike, yellow perch, and goldeye of Lower Red Lake, Beltrami and Clearwater Counties, Minnesota. Midwest Wildl. Conf. 1939. mimeo.

V21   Vessel, Matt F. and Samuel Eddy, 1941. A preliminary study of the egg production of certain Minnesota fishes. Minn. Bur. Fish. Res. Invest. Rep. 26:26 pp.

V23   Viosca, P., Jr., 1931. The bullhead, *Ameiurus melas catulus*, as a dominant in small ponds. Copeia, 1931(1):17-19.

V25   Viosca, P., Jr., 1936. A new rock bass from Louisiana and Mississippi. Copeia, 1936 (1):37-45.

V26   ——. 1943. Phenomenal growth rates of black bass in Louisiana waters (Abstract). Proc. La. Acad. Sci. 7:75.

V27   ——. 1943. Phenomenal growth rates of largemouth black bass in Louisiana waters. Trans. Am. Fish. Soc. 72:68-71.

V42   ——. 1939. Where to fish in Louisiana for the southern rock bass. La. Conserv. Rev., Summer Issue 1939.

V44   ——. 1952. Eleventh report to International Paper Company, August 5, 1952. 21 pp. typewritten. (Most data also given in a paper at Am. Fish Soc. meetings, Dallas, Tex. Sept. 1952).

V48   Van Meter, H., 1954. Sport fishing survey of the Shenandoah River. W. Va. Conserv. Comm., Div. Fish Manage. 16 pp.

V87   Van Valin, C. C., A. K. Andrews, and L. L. Eller, 1968. Some effects of mirex on two warm-water fishes. Trans. Am. Fish. Soc. 97(2):185-196.

V88   Viosca, P., Jr., 1931. The southern small-mouth black bass, *Micropterus pseudaplites* Hubbs. Trans. Am. Fish. Soc. 61:95-98.

V89   Vogele, L. E., 1969. Underwater observations. U.S. Bur. Sport Fish Wildl. Resour. Publ. 77:75-6.

V90   Von Geldern, C. E., Jr., 1971. Abundance and distribution of fingerling largemouth bass, *Micropterus salmoides*, as determined by electrofishing at Lake Nacimiento, California. Calif. Fish. Game 57(4):228-245.

V91   Vanderpuye, C. J. and K. D. Carlander, 1971. Age, growth and condition of black crappie, *Pomoxis nigromaculatus* (LeSueur) in Lewis and Clark Lake, South Dakota, 1954 to 1967. Iowa State J. Sci. 45(4):541-555.

V92   Vooren, C. M., 1972. Ecological aspects of the introduction of fish species into natural habitats in Europe, with special reference to the Netherlands. A literature review. J. Fish. Biol. 4(4):565-583.

V93   VanDenAvyle, M. J., 1973. Estimates of production of age I and II largemouth bass in a 3-hectare reservoir. M.S. thesis, Iowa State Univ. 102 pp.

V94   Vogele, L. E., 1975. Reproduction of spotted bass, *Micropterus punctulatus*, in Bull Shoals Reservoir, Arkansas. U.S. Fish. Wildl. Serv. Tech. Pap. 84:1-21.

V95   VanDenAvyle, M. J. and K. D. Carlander, 1974. Seasonal variation in length-weight relationships of largemouth bass in a 3-hectare reservoir. Mss. Iowa State Univ. 17 pp.

W1   Wales, J. H., 1946. Castle Lake trout investigations. First Phase: Interrelationships of four species. Calif. Fish, Game 32(3):109-143.

W7   Ward, H. C., 1949. A study of fish populations, with special reference to the white bass, *Lepibema chrysops* (Rafinesque), in Lake Duncan, Oklahoma. M.S. thesis, Univ. Okla. 44 pp. typewritten.

W11   Washington State Game Commission, 1947. Fisheries Biological Progress Report, April 1946 to April 1947. 42 pp.

W13   Webster, D. A., 1942. The life histories of some Connecticut fishes. Pp. 122-227, A fishery survey of important Connecticut Lakes. Conn. Geol. Nat. Hist. Surv. Bull. 63:1-339.

W25   Westerman, F. A. and J. V. Oosten, 1939. Report to the Michigan state Senate on the Fisheries of Potagannissing Bay, Michigan. Mich. Dep. Conserv. 82 pp.

W27   Weyer, A. E., 1940. The Lake of the Ozarks--A problem in fishery management. Prog. Fish-Cult. 51:1-10.

W32   Wickliff, E. L., 1920. Food of young smallmouth black bass in Lake Erie. Trans. Am. Fish. Soc. 50:364-371.

W39   Williamson, L. O., 1940. Length-weight relationship of fish. Wis. Conserv. Bull. 5(9):37-39.

W45   Wright, S., 1929. A preliminary report on the growth of the rock bass, *Ambloplites rupestris* (Rafinesque), in two lakes of northern Wisconsin. Trans. Wis. Acad. Sci. Arts, Lett. 24:581-595.

W48   White, M. O., 1948. Report on fish population of Gilbert Lake, Ohio Div. Conserv. 9 pp. mimeo.

W53   Ward, H. C., 1951. A study of fish populations, with special reference to the white bass, *Lepibema chrysops* (Rafinesque) in Lake Duncan, Oklahoma. Proc. Okla. Acad. Sci. 30:69-84.

W61   West, J. W., 1950. A fish population study of an artificial impoundment in southern Illinois. M.S. thesis, Southern Ill. Univ. 43 pp.

W65   West Virginia Conservation, 1951. Photograph. W. Va. Conserv. 15(9): 40, 15(6):27.

W68   Wilson, C., Jr., 1951. Age and growth of the white crappie (*Pomoxis annularis*) Rafinesque in Lake Texoma, Oklahoma, 1949. Proc. Okla. Acad. Sci. 31:28-38.

W71   Wohlschlag, D. E. and C. A. Woodhull, 1953. The fish population of Salt Springs Valley Reservoir, Calaveras County, California. Calif. Fish., Game 39(1):5-14.

W73   Woods, J. P., 1916. Biennial report of the Missouri State Fish Commission 1-5.

W83   Ward, H. C. and E. M. Leonard, 1954. Order of appearance of scales in the black crappie, *Pomoxis nigromaculatus*. Proc. Okla. Acad. Sci. 33:138-140.

W89   Watson, J. E., 1955. The Maine smallmouth. Maine Fish. Res. Bull. 3:31 pp.

W92   Webster, D. A., 1954. Smallmouth bass, *Micropterus dolomieui*, in Cayuga Lake. Part 1. Life history and environment. N.Y. Agric. Exp. Stn. Ithaca Mem. 327:1-39.

W105   Whitacre, M. A., 1952. The fishes of Crab Orchard Lake, Illinois. Midwest Wildl. Conf. 14:41 pp. mimeo.

W111   Williams, W. E., 1959. Food conversion and growth rates for largemouth and smallmouth bass in laboratory aquaria. Trans. Am. Fish. Soc. 88(2):125-127.

W119   Witt, A., Jr., 1957. Seasonal variation in the incidence of lymphocystis in the white crappie from the Niangua arm of the Lake of Ozarks, Missouri. Trans. Am. Fish. Soc. 85:271-279.

W123   Wohlschlag, D. E. and R. O. Juliano, 1959. Seasonal changes in bluegill metabolism. Limnol. Oceanogr. 4(2):195-210.

W126   Wurtz-Arlet, J., 1952. Le Black-bass en France. Esquisse monographie. Ann. Stn. Centr. Hydrobiol. Appl. 4:203-286.

W131   Wright, Y. E., 1951. Age and growth of the green sunfish, *Lepomis cyanellus* Rafinesque, in northern Utah. M.S. thesis, Utah State Agric. Coll:22 pp.

W132   Ward, C., 1960. A survey of the fishes of Nolichucky River. Tenn. Game, Fish Comm. 30 pp. mimeo.

W137   Wardle, W. D., 1953. An ecological study of a farm fish pond. M.S. thesis, Univ. Utah. 80 pp.

W138 Witt, A., Jr., 1952. Age and growth of the white crappie, *Pomoxis annularis* Rafinesque, in Missouri. Ph.D. thesis, Univ. Mo. 213 pp. +xviii pp.

W142 Walker, G. W., 1951. A fish population study of an artificial lake in southern Illinois. M.S. thesis, Southern Ill. Univ. 33 pp.

W144 Westman, J. R., 1941. A consideration of population life-history studies in their relation to the problem of fish management research, with special reference to the smallmouthed bass, *Micropterus dolomieu* Lacepede, the lake trout, *Cristivomer namaycush* (Walbaum), and the mudminnow *Umbra limi* (Kirtland), Ph.D. thesis, Cornell Univ. 182 pp.

W192 Walburg, C. H., 1964. Fish population studies, Lewis and Clark Lake, Missouri River, 1956 to 1962. U.S. Fish. Wildl. Serv. Spec. Sci. Rep. Fish. 482:1-27.

W213 Wyatt, H. N. and H. A. Zeller, 1965. Fish population dynamics following a selective shad kill. Proc. S.E. Assoc. Game Fish Comm. 16: 411-418.

W214 Ward, C. M. and W. M. Irwin, 1961. The relative resistance of thirteen species of fishes to petroleum refinery effluent. Proc. S.E. Assoc. Game Fish Comm. 15:255-276.

W220 Weaver, D. and H. E. McReynolds, 1964. Age and growth study of four experimental lakes. Ind. Dep. Conserv. Job Compl. Rep. F-4-R-II, A. 22 pp.

W221 Wenke, T. L., 1965. Some ecological relationships of mayflies, caddisflies, and fish in the Mississippi River near Keokuk, Iowa. Ph.D. thesis, Iowa State Univ. 181 pp.

W222 Wirth, T. L., R. C. Dunst, P. D. Uttormark, and W. Hilsenhoff, 1970. Manipulation of reservoir waters for improved quality and fish population response. Wis. Dep. Nat. Resources. Res. Rep. 62:23 pp.

W223 Werner, R. G., 1969. Ecology of limnetic bluegill (*Lepomis macrochirus*) fry in Crane Lake, Indiana. Am. Midl. Nat. 80(1):164-181.

W224 ———. 1967. Intralacustrine movements of bluegill fry in Crane Lake, Indiana. Trans. Am. Fish. Soc. 96(4):416-420.

W225 Windell, J. T., 1966. Rate of digestion in the bluegill sunfish Invest. Ind. Lakes, Streams. 7(6):185-214.

W226 Whitmore, C. M., C. E. Warren, and P. Doudoroff, 1960. Avoidance reactions of salmonid and centrarchid fishes to low oxygen concentrations. Trans. Am. Fish. Soc. 89(1):17-26.

W228 West, J. L., 1970. The gonads and reproduction of three intergeneric sunfish (Family Centrarchidae) hybrids. Evolution 24(2):378-394.

W229 West, J. L. and F. E. Hester, 1966. Intergeneric hybridization of Centrarchids. Trans. Am. Fish. Soc. 95(3):280-288.

W230 Wegener, W. L., 1966. A tag comparison study of largemouth bass in their natural environment. Proc. S.E. Assoc. Game Fish Comm. 19:258-264.

W231 Westman, J. R. and C. B. Westman, 1949. Population phenomena in certain game fishes of Lake Simco, Ontario, and some effects upon angling returns. Part 1, the smallmouth bass, *Micropterus dolomieu* Lacepede. Can. Jour. Res. 27(D):7-29.

W232 Webster, D. A., 1948. Relation of temperature to survival and incubation of the eggs of smallmouth bass (*Micropterus dolomieu*). Trans. Am. Fish. Soc. 75:43-47.

W233 Watt, K. E. F., 1956. The choice and solution of mathematical models for predicting and maximizing the yield of a fishery. J. Fish. Res. Bd. Can. 13(5):613-645.

W234 Wickliff, E. L., 1933. Returns from fish tagged in Ohio. Trans. Am. Fish. Soc. 63:326-331.

W235 Wagner, W. C., 1972. Utilization of alewives by inshore piscivorous fishes in Lake Michigan. Trans. Am. Fish. Soc. 101(1):55-63.

W236 Wright, L. D., 1970. Forage size preference of the largemouth bass. Prog. Fish-Cult. 32(1):39-42.

W237 Wiebe, A. H., 1931. Notes in the exposure of several species of fish to sudden changes in the hydrogen-ion concentration of the water and to an atmosphere of pure oxygen. Trans. Am. Fish. Soc. 61:216-224.

W238 Witt, A., Jr. and R. S. Campbell, 1959. Refinements of equipment and procedures in electro-fishing. Trans. Am. Fish. Soc. 88(1):33-35.

W239 Whiteside, B. G., 1964. Biology of the white crappie, *Pomoxis annularis*, in Lake Texoma, Oklahoma. M.S. thesis, Okla. State Univ. 34 pp.

W240 Welker, B., 1963. Age and growth of Decatur Lake white crappie, 1962. Ia. Conserv. Comm. Q. Biol. Rep. 15(1):7-11.

W241 ———. 1967. Missouri River ox-bow lake fishery. Part 3, Crappie. Iowa Conserv. Comm. Q. Biol. Rep. 19(2):37-39.

W242 Walburg, C. H., G. L. Kaiser, and P. L. Hudson, 1971. Lewis and Clark Lake tailwater biota and some relations of the tailwater and reservoir fish populations. Amer. Fish. Soc. Spec. Publ. 8:449-467.

W243 Wojtalik, T. Z. and C. W. Voigtlander, 1971. The elements of a monitoring program to assess isotope accumulation and thermal effects of a nuclear power plant in a TVA reservoir. Am. Fish. Soc. Spec. Publ. 8:469-480.

W244 Whiteside, B. G., 1973. Age and growth of white crappie in Lake Texoma, Oklahoma, from 1942 to 1962. Texas J. Sci. 24(3):311-318.

W245 Wheeler, A. and P. S. Maitland, 1973. The scarcer freshwater fishes of the British Isles. I. Introduced species. J. Fish. Biol. 5(1):49-68.

W246 Whitaker, J. O., Jr. and R. A. Schlueter, 1973. Effects of heated discharge on fish and invertebrates of White River at Petersburg, Ind. Indiana Univ. Water Resourc. Res. Cent. Invest. Rep. 6:123 pp.

W247 Whiteside, B. G. and N. E. Carter, 1973. Standing crop of fishes as an estimate of fish production in small bodies of water. Proc. S.E. Assoc. Game Fish Comm. 26:414-417.

W248 Werner, R. G., 1972. Bluespotted sunfish, *Enneacanthus gloriosus*, in the Lake Ontario drainage, New York. Copeia, 1972(3):878-879.

W249 Werner, E. E., 1974. The fish size, prey size, handling time relation in several sunfishes and some implications. J. Fish. Res. Bd. Can. 31(9):1531-1536.

W250 White, G. E., 1971. The Texas golden green: A color mutation of the green sunfish. Prog. Fish-Cult. 33(3):155.

W251 Wilbur, R. L. and F. Langford, 1974. Use of human chorionic gonadotropin (HCG) to promote gametic production in male and female largemouth bass. Fla. Game, Fresh Water Fish Comm. Pap. 16. Estis Fisheries Res. Lab. 17 pp. offset.

W252 Wegener, W. and V. Williams, 1975. Fish population responses to improved lake habitat utilizing an extreme drawdown. Proc. S.E. Assoc. Game Fish Comm. 28:144-160.

W253 Werner, E. E. and D. J. Hall, 1974. Optimal foraging and the size selection of prey by bluegill sunfish (*Lepomis macrochirus*). Ecology 55(5):1042-1052.

W254 Wilbur, R. L., 1969. The redear sunfish in Florida. Fla. Game, Fresh Water Fish Comm. Fish. Bull. 5:64 pp.

W255   Whitt, G. S., W. F. Childers, J. Tranquilli, and M. Champion, 1973.
        Extensive heterozygostity at three enzyme loci in hybrid sunfish popu-
        lations. Biochem. Genet. 8(1):55-72.

W256   Witt, A., Jr. and R. C. Marzolf, 1954. Spawning and behavior of the
        longear sunfish, *Lepomis magalotis megalotis*. Copeia, 1954(3):188-
        190.

W257   Wallace, C. R., 1973. Effects of temperature on developing meristic
        structures of smallmouth bass, *Micropterus dolomieui* Lacepede.
        Trans. Am. Fish. Soc 102(1):142-144.

W258   White, W. J. and R. J. Beamish, 1972. A simple fish tag suitable for
        long-term marking experiments. J. Fish. Res. Bd. Can. 29(3):339-
        341.

W259   Watt, K. E. F., 1959. Studies on population productivity. II. Factors
        governing productivity in a population of smallmouth bass. Ecolog.
        Monogr. 29:367-392.

Y13   Youngs, W. D., 1958. Effect of the mandible ring tag on growth and
        condition of fish. N.Y. Fish, Game J. 5(2):184-205.

Z7   Zweiacker, P. L., 1972. Population dynamics of largemouth bass in
        an 808-hectare Oklahoma reservoir. Ph.D. thesis, Okla. State Univ.
        126 pp.

Z8   Zweiacker, P. L. and B. E. Brown, 1971. Production of a minimal
        largemouth bass population in a 3000-acre turbid Oklahoma reservoir.
        Am. Fish Soc. Spec. Publ. 8:481-493.

Z9   Zweiacker, P. L., R. C. Summerfelt, and J. N. Johnson, 1973. Large-
        mouth bass growth in relationship to annual variation in mean pool
        elevation in Lake Carl Blackwell, Oklahoma. Proc. S.E. Assoc. Game
        Fish Comm. 26:530-540.

Author Index

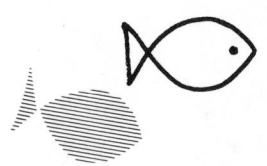

Subject Index